Topics in Safety, Risk, Reliability and Quality

Volume 40

More information about this series at https://link.springer.com/bookseries/6653

Charles B. Keating · Polinpapilinho F. Katina ·
Charles W. Chesterman Jr. · James C. Pyne
Editors

Complex System Governance

Theory and Practice

Springer

Editors
Charles B. Keating
Department of Engineering Management and Systems Engineering
Old Dominion University
Norfolk, VA, USA

Polinpapilinho F. Katina
Department of Informatics and Engineering Systems
University of South Carolina Upstate
Spartanburg, SC, USA

Charles W. Chesterman Jr.
Engineering Management and Systems Engineering
Old Dominion University
Norfolk, VA, USA

James C. Pyne
Engineering Management and Systems Engineering
Old Dominion University
Norfolk, VA, USA

ISSN 1566-0443 ISSN 2215-0285 (electronic)
Topics in Safety, Risk, Reliability and Quality
ISBN 978-3-030-93851-2 ISBN 978-3-030-93852-9 (eBook)
https://doi.org/10.1007/978-3-030-93852-9

This Springer imprint is published by the registered company Springer Nature Switzerland AG
The registered company address is: Gewerbestrasse 11, 6330 Cham, Switzerland

To my parents Bernie and Charlene and my wife Jean

—Charles B. Keating

To my nephews, Raoul and Reuben, and niece, Raïssa Tchomba

—Polinpapilinho F. Katina

To others in the pursuit of knowledge and understanding

—Charles W. (CW) Chesterman Jr.,

To my many colleagues who continue to contribute this domain

—James C. Pyne

Series Foreword

We are very fortunate, and unfortunate, to be witnessing the possibilities and challenges that our complex systems continue to present. Fortunate in that technology and integration of complex systems are creating opportunities to advance society in substantial ways. New advances such as artificial intelligence, augmented and virtual reality, big data, blockchain technology, cloud computing, Industry 4.0, and Internet of things are creating conditions that promise to enhance the quality of life for people across the world. Despite these advances, we are also unfortunate, in that problems stemming from increasingly complex systems seem to be outpacing our ability to 'keep up.' Despite our best efforts, we continue to be confounded by complex systems, unable to 'tame' them. Our complex systems seem to be producing as many problems as they solve. We would be naïve to think that we have mastered our complex systems.

As we attempt to become more effective with complex systems, we need to answer three questions. First, how did we get here? There is not a particularly satisfying answer to this question. What we do know is that complexity is accelerating and what has worked in the past no longer seems capable of addressing the new reality of complex systems. The second question is, why have we not been able to 'get out' of our predicament, despite being the most advanced society ever experienced in human history? Despite being advanced in technology and knowledge, we continue to be confounded by complex systems. We certainly have not managed to keep up with the demands to deal with increasing complexity effectively. The third question is, how can we escape the stranglehold that complex systems appear to be placing on every aspect of society? Irrespective of advances across all sectors serving society, we are left without a reasonable diagnosis and prognosis for escaping the complexity conundrum. While Complex System Governance is not suggested as a universal answer to address this question, it is poised to make considerable contributions toward an answer.

Complex System Governance (CSG) has been introduced as an emerging field, probably not currently known to a wide audience. However, in the spirit of the scientific revolution, perhaps it should be known. CSG is not known for the quantity of scholars and practitioners who have a working knowledge of the field. Instead,

CSG has been quietly unfolding from the first efforts dating to 2014. What has emerged is a new and novel field that has been deliberately and cautiously maturing. The focus of development has been grounded in the underlying conceptual and theoretical foundations as the first point of emphasis. This has set the stage to move beyond conceptual foundations to begin developing tools, methods, and techniques for the application of CSG in operational settings.

It has been interesting to observe CSG maturing over the last several years. What started as a modest undertaking has emerged as a significant body of knowledge with contributions ranging from the theoretical underpinnings to applications. However, the seeds for CSG significantly predate the current state or even the formal instantiation in 2014. In fact, early work in System of Systems Engineering (SoSE) formed an important foundation for CSG. The evolution of CSG was in part in response to the limited acknowledgment of SoSE to consider both the 'hard' and 'soft' components of complex systems. Systems Theory provided a strong and proven set of language that explains the behavior and performance of systems. Additionally, CSG is geared to address the entire spectrum of dimensions of a complex system, spanning the entire range of socio-technical-economic-political dimensions. The inclusion of Management Cybernetics allows CSG to draw upon a proven approach to deal with the structural configuration of functions and communications channels. Finally, the incorporation of governance provides a significant emphasis on direction, oversight, and accountability for complex systems. The unique integration of Systems Theory, Management Cybernetics, and System Governance has resulted in a new and novel approach to better deal with complex systems.

This book represents a necessary push forward in the amalgamation of theory and practice for CSG. The field is still very young and the CSG story continues to evolve. However, this is an important consolidation of the state of knowledge and the field's directions. There is much left to be accomplished as CSG continues to develop. However, although significant challenges remain, this work represents a necessary and critical step forward in the development of the new field.

Zürich, Switzerland
November 2021

Adrian V. Gheorghe
Professor and Batten Endowed Chair
on System of Systems Engineering

Preface

Our present-day systems move faster, are more interconnected, and enable possibilities that have only been visionary desires in the recent past. Yet, here we are. We are frustrated with systems that inevitably fail to meet our expectations and seem to generate as many problems as they address. Out of this frustration, the field of Complex System Governance (CSG) is being pursued as an attempt to enhance our capacity to engage complex situations better. The emphasis of this work is such that the desired potential and promises of complex systems can be better realized. The emerging CSG field is about finding a new and novel path forward in dealing with increasingly complex systems and their problems. It is almost cliché to suggest that we are experiencing difficulties in addressing complex systems as we seem incapable of matching the acceleration of information, interconnectedness, and technology driving our current state of affairs. For all the 'goodness' that complex systems have brought, they have also left a wake of problems that appear to be intractable given our current paradigms and methods of attack (Rainey and Jamshidi, 2018; Warfield, 1976). We have all had experiences of being continually disappointed with our inability to keep pace with rapidly changing systems and with all of our prowess and sophistication, there is still something missing. How could we be so advanced in technology, knowledge, and societal advances yet be continually held hostage to underperforming systems? How can systems and their artifacts be built with such noble intentions but degrade to a state where they seem to do more harm than the 'good' for which they were intended? And what can we possibly do to help alleviate the negative consequences of underperforming systems and thus reduce the suffering they inevitably produce? These are rather lofty questions, nevertheless, they are important questions if we are to build, operate, and maintain complex systems that better serve to advance societal prospects.

In many cases, we talk about poorly performing systems as if they have human qualities and 'are out to get us.' However, with a deeper introspection, one recognizes that the 'systems' are a human construct of convenience, developed to make the infinite more finite. This perspective permits us to better grapple with something we can 'get our hands around,' with a stark realization that our systems have all been designed by us, built by us, executed by us, and maintained by us—humans. Our systems are not guided by a conscionable sense of justice and value. This leads to a powerful realization that, on the contrary, our systems simply do what they do when well-intentioned designs meet the operational setting. If there is disappointment and blame, it should not be directed at the system, as the system is entirely innocent of what it generates. Instead, we should direct criticism and judgment toward our role in designing, executing, and evolving the system to achieve results that we deem desirable. CSG was conceived in part as a remedy to the aforementioned misconception.

The focus on 'governance' for CSG is taken as the active steering of a system through the artful and integrated design, execution, and evolution of the system [6]. Thus, the primary motivation for this book is to share the current state of CSG, and the new approach to governing systems that appear to be ungovernable given our current circumstances. This is the reason for this book—to recognize the human role to provide improved system performance by advancing CSG as an alternative for the design, execution, and evolution of complex systems.

The CSG field, although still in the earliest stages of development, has not been haphazardly developed. Instead, as will be presented in the various chapters, there has been a slow continuous progression of the field to the described current state. This slow progression has permitted CSG to first focus on establishing the theoretical/conceptual grounding. These underpinnings have brought together three previously disparate fields: General Systems Theory [1, 10], Management Cybernetics [2, 12], and Governance [3, 4]. While each of the fields has achieved some notable stature, their intersection has never been exploited. In fact, much of the seminal work, particularly in (general) Systems Theory and Management Cybernetics, has been relegated to an existence off the main stage of developing complex systems. The intersection of these fields was primarily the inspiration for the emergence of the CSG field. Additionally, the time was ripe for this book to capture and share the current stage of CSG development and to project the field into the future. For this book, and the topic of CSG, there was a trade-off that had to be made. Although the field has made significant strides, there is much left to be achieved. The trade-off of waiting for additional field development versus getting the 'message' out to a wider audience across multiple sectors was deliberated. Ultimately, we elected to err on the side of getting the message out rather than waiting for further development. We were sufficiently confident that that time was right to capture the present state of the field, its contributions, and directions for ongoing development. We hope to open new dialogs and applications of CSG to accelerate the advancement of the field and its body of knowledge.

The genesis of CSG can be traced to three pivotal elements that profoundly impacted the development. First, work at Old Dominion University's National Centers for System of Systems Engineering started in 2003. Early work at the center identified several issues that the mainstream development of System of Systems Engineering (SoSE) was not addressing. Among these was the absence of taking a holistic view of SoS that included the entire spectrum of socio-technical-economic-political considerations. Furthermore, the domination of approaches based on 'technology first, technology only' generation of solutions (e.g., technical interoperability) was viewed by the center as limiting. Also, SoSE lacked any grounding in a theoretical/conceptual body of knowledge. Absent this grounding, long-term sustainability of the field would be doubtful. As a result, CSG was born as an evolution of SoSE to remedy the shortcomings identified in the trajectory of SoSE.

The second pivotal element stemmed from the foundational work of the Complex System Governance Learning Community at Old Dominion University. This learning community was composed of faculty, doctoral students, and post-doctoral researchers with interests in CSG. Through the work of this community, the CSG field was significantly enhanced. Several significant accomplishments of the learning community include the production of two special issues of the International Journal of System of Systems Engineering [2015, vol. 6, no.1/2; 2016, vol. 7, no. 1/2/3] [7, 8] dedicated exclusively to CSG; discussions that challenged the foundations of CSG and added rigor to the formulation; and providing a sounding board for doctoral dissertations and an emerging research agenda based on explorations of CSG. The early gains in CSG are largely attributable to the learning community and the CSG field owes much gratitude to this group of scholars who continue to push the boundaries of knowledge for CSG.

The third pivotal element in CSG development was the applications of aspects of CSG in operational settings. This was essential to the tempering of theoretical formulations through the lenses of deployment and application. These deployments in operational settings proved invaluable to working out inconsistencies and accelerating knowledge from applications. In sum, given the state of the CSG field, we had an intuitive sense that the time was right to develop this book. The book stands as a culmination of research, scholarly development, and application experiences. CSG is incomplete and continues to evolve. However, propagation of the field to a wider audience and range of applications/practitioners is an important step in that evolution.

Forging a new field is difficult and CSG is no exception. As described in this book, the current state of CSG is our attempt to create a waypoint. This waypoint stands at the intersection of current development and establishing the future trajectory of the CSG field. Although the journey to this point has been difficult, it has certainly been worth the toil. The opportunity to consolidate the state of our CSG knowledge and define future developmental challenges for the field is an important stepping stone for the field. Thus, this book offers a temporary respite between what has been achieved to advance the field and what is on the horizon for CSG. The book is a challenge to continue the development and propagation of CSG to take its place as an important systems-based methodology for generations to come [5]. This book, born of a passion for improving complex systems, is a significant step in the continuing evolution of the field.

Norfolk, Virginia, USA — Charles B. Keating
Spartanburg, South Carolina, USA — Polinpapilinho F. Katina
Norfolk, Virginia, USA — Charles W. (CW) Chesterman Jr.
Norfolk, Virginia, USA — James C. Pyne

References

1. Adams KM, Hester PT, Bradley JM, Meyers TJ, Keating CB (2014) Systems theory as the foundation for understanding systems. Syst Eng 17(1):112–123 https://doi.org/10.1002/sys.21255
2. Beer S (1981) The brain of the firm: The managerial cybernetics of organization. Wiley
3. Biermann F, Betsill MM, Gupta J, Kanie N, Lebel L, Liverman D, Schroeder H, Siebenhüner B (2009) *Earth system governance: People, places and the planet. Science and implementation plan of the earth system governance project* (Earth System Governance Report 1, IHDP Report 20). IHDP: The Earth System Governance Project. http://www.earthsystemgovernance.org/sites/default/files/publications/files/Earth-System-Governance_Science-Plan.pdf
4. Calida BY (2013) *System governance analysis of complex systems* [Ph.D., Old Dominion University]
5. Jackson MC (2019) Critical systems thinking and the management of complexity (1 edition). Wiley
6. Katina PF (2021) Governance in CPS. In S. Jajodia, P. Samarati, & M. Yung (Eds.), Encyclopedia of Cryptography, Security and Privacy (pp. 1–3). Springer. https://doi.org/10.1007/978-3-642-27739-9_1725-1
7. Keating CB, Ireland V (Eds) (2016) Complex systems governance: Issues and applications. Int J Syst Syst Eng 6(1/2/3):1–21 https://www.inderscience.com/info/dl.php?filename=2016/ijsse-4667.pdf
8. Keating CB, Katina PF (Eds) (2015) Editorial: Foundational perspectives for the emerging complex system governance field. Int J Syst Syst Eng 6(1/2):1–14 https://www.inderscience.com/info/dl.php?filename=2015/ijsse-4265.pdf
9. Rainey LB, Jamshidi M (Eds) (2018). Engineering emergence: A modeling and simulation approach (1 edition). CRC Press
10. von Bertalanffy L (1968) General system theory: Foundations, developments, applications. George Braziller
11. Warfield JN (1976) Societal systems: Planning, policy and complexity. Wiley-Interscience
12. Wiener N (1948) Cybernetics: Or control and communication in the animal and the machine. MIT Press

Acknowledgments

One aspect of academic citation apparatus is acknowledgment of the relevance of the works of others to the topic of discussion. This we have done. However, this apparatus fails to capture the influences of many that were involved in current efforts. With this in mind, the editors wish to acknowledge different people and organizations involved in discussions of this research.

Over the years, several faculty members helped by reviewing changes and providing feedback. Also instrumental in development and propagation of the foundations for CSG was the Complex System Governance Learning Community at Old Dominion University. Several of the chapters in this work are authored by members of that community. Many industrial practitioners also assisted us by commenting on draft chapters. We also wish to acknowledge several students in classes in which we tested teaching/research materials. Finally, we want to recognize stimulating comments and criticisms of Prof. Adrian V. Gheorghe—National Centers for System of Systems Engineering, Old Dominion University, USA and Emeritus Prof. Vernon Ireland—The University of Adelaide, South Australia, Australia.

Indeed, the editors offer apologies for those whom we have forgotten.

Norfolk, Virginia, USA — Charles B. Keating
Spartanburg, South Carolina, USA — Polinpapilinho F. Katina
Norfolk, Virginia, USA — Charles W. (CW) Chesterman Jr.
Norfolk, Virginia, USA — James C. Pyne

Contents

About the Editors

Charles B. Keating is Professor in the Engineering Management and Systems Engineering Department at Old Dominion University (Norfolk, Virginia). As faculty member since 1994, he also serves as Director for the National Centers for System of Systems Engineering (NCSOSE) housed at Old Dominion University. Dr. Keating focuses on teaching and research in the areas of Systems Engineering, System of System of Systems Engineering, Management Cybernetics, and Complex System Governance. He is Fellow and Past President of the American Society for Engineering Management and in 2015 was awarded the society's most prestigious award (the Sarchet Award) for his pioneering efforts in the field.

Dr. Keating holds B.S. in Engineering from the US Military Academy (West Point, New York), M.A. in Management from Central Michigan University (Mount Pleasant, Michigan) and a Ph.D. in Engineering Management from Old Dominion University (Norfolk, Virginia). He is a member of the *American Society for Engineering Management*, the *International Council on Systems Engineering*, and the *International Society for System Sciences.*

Dr. Keating has authored more than 200 peer-reviewed papers, generated over $20 M in research funding, and graduated over 25 Ph.D.s. His research has spanned a variety of organizations, including defense, security, aerospace, healthcare, R&D, and automotive. Prior to joining the faculty at ODU, he served in leadership and technical engineering management positions for over 12 years in both the US Army and private industry.

Polinpapilinho F. Katina currently serves as Assistant Professor in the Department of Informatics and Engineering Systems at the University of South Carolina Upstate (Spartanburg, South Carolina). He previously served as a Postdoctoral Researcher for the National Centers for System of Systems Engineering (Norfolk, Virginia, USA).

Dr. Katina holds B.S. in Engineering Technology, M.Eng. in Systems Engineering and a Ph.D. in Engineering Management and Systems Engineering, all from Old Dominion University (Norfolk, Virginia). His scholarship and training include, among others, *Embry-Riddle Aeronautical University* (Daytona Beach, Florida), *Environmental Systems Research Institute* (Redlands, California), *Politecnico di Milano* (Milan, Italy), *Syracuse University* (Syracuse, New York), *University of Edinburgh* (Edinburgh, UK), and *UNSW University College* (Canberra, Australia).

Dr. Katina's profile includes more than 150 scholarly outputs in peer-reviewed journal articles, conference proceedings, book chapters, and books. His areas of research/teaching interests include, Complex System Governance, critical infrastructure systems, decision making and analysis (under uncertainty), emerging technologies, energy systems (smart grids), engineering management, infranomics, manufacturing systems, system of systems, systems engineering, systems pathology, and systems thinking. He was named in the top *1% for the 2018 Publons Global Peer Review Awards.*

Charles W. Chesterman Jr. is a senior technical program manager at CACI Inc. Prior to joining CACI, Dr. Chesterman served in the US Navy in several capacities, including Surface Warfare Officer and an Engineering Duty Officer.

Dr. Chesterman holds a B.S. in Mechanical Engineering from US Naval Academy (Annapolis, Maryland), M.S. Mechanical Engineering from US Naval Postgraduate School (Monterey, California), and a Ph.D. in Engineering Management and Systems Engineering from Old Dominion University (Norfolk, Virginia).

His areas of research/teaching interests include Complex System Governance, communications in complex situations, and product life cycle management. He continues to support the Navy as a contractor working in the area of maritime maintenance, specifically on the design, development, and deployment of product life cycle management (PLM) applications.

James C. Pyne is Adjunct Assistant Professor of Engineering Management and System Engineering and Research Affiliate at Old Dominion University (Norfolk, Virginia). He was previously employed at Hampton Roads Sanitation District (Virginia Beach, Virginia), first as Plant Manager followed by Chief of Design, then Chief of Small Communities.

Dr. Pyne holds B.S. in Civil Engineering from Virginia Tech (Blacksburg, Virginia), M.E.A. in Engineering Administration from George Washington University (Washington, DC), and Ph.D. in Engineering Management and Systems Engineering from Old Dominion University (Norfolk, Virginia). He is also licensed as Professional Engineer in the Commonwealth of Virginia.

Dr. Pyne has more than 36 years of experience in wastewater management at the technical, managerial, and policy levels. His expertise includes socio-technical systems as well as complex systems analysis. In his role as the HRSD chief of small communities, he has used this knowledge to assist numerous and various stakeholders deal with complex issues.

Introduction

Charles B. Keating, Polinpapilinho F. Katina, Charles W. Chesterman Jr., and James C. Pyne

Abstract Complex System Governance (CSG) is an emerging field that remains in its infancy. Significant progress has been made to advance the field in both theory and practice. Although there is much left to accomplish in maturing this promising field, the time is right to provide the current state of theory and practice to accelerate advancement and share knowledge. Society continues to be dependent on increasingly complex and interconnected systems. This evokes an urgent need to improve the theory and practice to 'tame' modern complex systems and their problems more effectively. This chapter introduces the genesis, present state, and future prospects for the CSG field. Four primary themes are developed. First, the complex system problem domain is examined. The emphasis is on the conditions of this domain that precipitated the development of CSG as a viable response. Second, a brief introduction to CSG as a focused response to the domain challenges is discussed. This discussion establishes the foundations of CSG and its response to address pitfalls in addressing complexity. Third, the current state of the CSG field is examined. The existing literature and works of CSG are explored for the emerging set of themes that delineate the field. Fourth, acknowledging work that has been completed, challenges that must be addressed if the field is to remain viable are explored. The chapter closes with the prospects and promises that CSG theory and practice holds for effectively dealing with increasingly complex and interrelated systems.

Keywords Problem domain · Complexity · Complex System Governance

---

C. B. Keating (✉) · C. W. Chesterman Jr. · J. C. Pyne
Department of Engineering Management and Systems Engineering, Old Dominion University, Norfolk, VA, USA
e-mail: ckeating@odu.edu

C. W. Chesterman Jr.
e-mail: cwjstellami@aol.com

J. C. Pyne
e-mail: jpyne@odu.edu

P. F. Katina
Department of Informatics and Engineering Systems, University of South Carolina Upstate, Spartanburg, SC, USA
e-mail: pkatina@uscupstate.edu

C. B. Keating et al. (eds.), *Complex System Governance*, Topics in Safety, Risk, Reliability and Quality 40, https://doi.org/10.1007/978-3-030-93852-9_1

1 Introduction

Complex System Governance (CSG) is an emerging field that has been described as the *design, execution, and evolution of the metasystem functions necessary to provide control, communication, coordination, and integration of a complex system* [82]. However, prior to delving into this field, it is important to understand the nature of the CSG genesis. CSG was derived from two primary concerns. First, the overreliance on technology as a default response to increasingly complex systems and problems was starting to diminish in returns. Technology has been, and always will be, important to addressing complex systems and their problems. However, *'technology first, technology only'* solutions to complex system problems are overly narrow. Second, complex system problems require an entire range of considerations. The entire spectrum of human/social, organizational/managerial, and political/policy influences must be considered for holistic solution development. Anything less is a miscalculation of the nature of complex systems and their problem domain.

Five enduring themes can characterize the problem domain within which a complex system must operate. While this set is not presented as 'complete' or 'absolute,' nevertheless, it captures the state of the problem domain facing present and future practitioners charged with the design, execution, and development of modern complex systems. The five themes include complexity, ambiguity, uncertainty, context, and holism. The confluence of these themes has been presented in other works [44, 65, 70, 74, 79, 80, 83, 87– 91]. However, they are presented here as an important grounding for the emergence of CSG.

1. *Complexity as a defining characteristic.* It is somewhat naïve to believe that there can be an agreed upon definition of the term 'complexity.' This pursuit was given up as futile after 71 different definitions were discovered and the effort stopped without an end in sight [89]. However, for our purposes, we can provide a perspective of complexity that informs the CSG field perspective taken for complexity related phenomena. This perspective rests on four primary themes. First, a large set of interconnected elements for which the set is not 'finite,' 'definite,' or 'static.' This entails that a complex system can never be completely known. Second, complex systems are dynamic and change over time. Knowledge of complex systems can be fallible, incomplete, and evolves as new knowledge and understanding dynamically develop over time. This does not suggest poor or sloppy engineering. On the contrary, this simply attests to the nature of complex systems, irrespective of efforts to define and understand them. Third, complex systems are richly interconnected, with the structure elaborating as the system operates. Thus, elaboration of rich structural relationships suggests that repeatability, reliability, and confidence in stable cause-effect relationships becomes increasingly doubtful. Fourth, complex systems are subject to emergence. Emergence suggests that behavior, performance, and consequences for a complex system come about through the operation of the system, cannot be known in advance, and are not predictable.

2. *Ambiguity in definition and understanding.* Ambiguity, or a lack of clarity, exists in understanding the structure, behavior, and performance dimensions of a complex system as well as the context within which it is embedded. The formulation of a complex system is subject to multiple perspectives that provide a variety of interpretations. Each of the interpretations is both correct and incorrect depending upon the vantage point and worldview that informs the perspective. Irrespective of the noble objective of 'consensus and agreement' in the definition and perspective of the system, it will always be subject to varying degrees of ambiguity. This ambiguity will also shift over time, as new knowledge of the system and its context continue to evolve. The variability in the understanding of a complex system also extends to the definition of such critical aspects as system identity, boundary conditions, definition of system elements, and the definition of the context within which the system is embedded. For complex systems, ambiguity can never be fully resolved.
3. *Uncertainty as a dominant attribute.* At a fundamental level, uncertainty suggests that the cause-effect relationships cannot be known for truly complex systems. There is doubt in being able to fully understand the variables, relationships, and consequences that exist for a complex system. The precise behavior/performance relationships are difficult, if not impossible to fully understand, analyze, or predict. Additionally, the system will evolve over time, in unpredictable ways, that will further render cause-effect relationships to be indeterminate. In fact, not only are deterministic (e.g., algebra, calculus) methods ineffective, so too are probabilistic (e.g., risk, statistical inference)-based approaches. Thus, traditional reductionist-based approaches become inadequate for complex systems. Instead, more holistic approaches, coupled with a corresponding mindset, are necessary to deal with uncertainty.
4. *Context as an enabling/disabling factor.* Context is taken as the set of circumstances, factors, conditions, trends, or patterns that impact, and likewise are impacted by, a complex system. It is important that context can both constrain and enable structure, behavior, and performance of complex systems. Context impacts thinking, decision, actions, and interpretations taken in response to system challenges. Context also includes the multiple stakeholders and their worldviews related to the system of interest. These worldviews can be convergent or divergent in nature. However, what is certain is that they will have an impact on the design, execution, and development of a system. Ultimately, context cannot be disregarded with respect to the direct and indirect impact on system performance.
5. *Holism considerations across a spectrum of dimensions.* Holism suggests that complex systems are subject to considerations that span multiple dimensions. While technology is important and tends to dominate conversations of complex systems, it is limited in providing the robust range necessary to address complex systems and their problems. Instead, looking beyond technology, complex systems are subject to the holistic influences of human, social, organizational, managerial, political, policy, and information dimensions. To view a complex system narrowly from a 'technology first, technology only' perspective is

shortsighted. The robust range of holistic considerations is necessary to more appropriately understand structure, behavior, and performance generated from complex systems. The interplay between the different holistic dimensions of a system is in keeping with a systems worldview, suggesting that their interaction produces what the individual components cannot.

These five themes capture the landscape within which CSG must provide for design, execution, and development of systems capable of thriving rather than simply surviving. This is especially the case since some of our systems 'have become hopelessly interconnected and overcomplicated, such that in many cases even those who build and maintain them on a daily basis can't fully understand them any longer' (Arbesman [7], p. 2). Table 1 presents a more definitive set of challenges which CSG must meet if a successful response to the problem domain is to be achieved and sustained.

Given the current problem domain and what it is producing for complex systems, we must admit that our responses are falling short. Instead of continuing to indulge responses that fall short of the mark, CSG is suggested as a new and novel approach. This approach offers the potential to minimally 'shift the dominant dialog' and at best generate the foundations for a 'sea change' in dealing with complex systems and their problems.

To capture the current state of affairs and position CSG in response, we suggest an 'issue triad' and a CSG 'response triad' (Fig. 1). The issue triad consists of *Complexity, Entropy, and Control*. The issue with *complexity* is that we seem incapable of addressing the increasing interconnectedness and interdependence of myriad entities. Additionally, complexity suggests that there is a dynamic (temporal) nature to shifts in a system and its context. Complexity also entails the dominance of emergence, suggesting that unpredictable events, behavior, or performance is characteristic of complex systems. *Entropy* captures the concept that a system, absent input of resources, will continue to maximum disorder to achieve the lowest energy state. Complex systems require the continuous import of resources to negate the natural proclivity to move to disorder. As systems become more interconnected and boundaries expand, more energy, matter, and information must be allocated to address the inherent disorder. *Control* involves the introduction of constraints that provide a regulatory capacity to govern a system. This ensures that the system maintains sufficient equilibrium to produce satisficing behavior, structure, and performance to remain viable. Our ineffective appreciation and mastery of system control leaves systems overregulated (stealing autonomy and wasting resources) or underregulated (failing to constrain systems adequately) which permits ineffectiveness in properly constraining a system to produce desired results.

In reply to the issues, CSG provides for a response triad rooted in *design, execution, and development*.

- *Design* involves the purposeful instantiation of the structural configuration of mechanisms for a system. This design configuration determines the degree to which a system can address ongoing and emergent issues being generated both internally and externally to a system. In essence, system design determines

Table 1 Complex system difficulties CSG must address

Difficulty	Description
Conflicting Perspectives	Differing perspectives are to be expected. However, sources of differences may lie well beneath the surface manifestations of differences. These may extend to differences in the basic worldviews held by individuals and entities in conflict. Understanding and potential resolution of conflicting perspectives will not be borne out by superficial surface treatment. Instead, conflicts must be addressed at the 'deep rooted' sources producing the conflicts. This does not suggest selecting a 'right' perspective for resolution of the conflict. Instead, the objective is to understand differences in supporting logic and assumptions as the sources of conflict
Insufficient Information	Given the continuing explosion of information, this difficulty seems unremarkable. However, care must be taken not to confuse availability of large quantities of information with the sufficiency of that information. In fact, extensive quantities confound the information sufficiency dilemma. If there is not accessibility to the right information to facilitate appropriate decision/action, then sheer quantity of information has little to offer
Unstable Resources	Dealing with complex systems and their problems is difficult enough under the best resource conditions. Stable resources support planning for an appropriate response based on scarce resource allocation. However, when resources are unstable and can shift radically, any pretense of detailed planning is rendered innocuous. On the contrary, such planning may be detrimental and a waste of resources given unstable resources
Extreme Uncertainty	Truly complex systems have a high level of uncertainty, where the understanding of cause-effect relationships is questionable. This renders more traditional approaches to decision and analysis to be of limited utility. Non-ergodicity (no clearly defined states or discernible transitions between system states) and non-monotonicity (inherent difficulties in understanding and provoking continuous transition to identified goals) are characteristic [99] of complex systems. Thus, application of traditional approaches, that assume the ability to reduce systems to well understood causal relationships, is a miscalculation. Complex systems represent a type of system where uncertainty is a rule rather than an exception. Assumptions to the contrary should be questioned

(continued)

robustness (the range and degree of perturbations which the design is capable of managing) and resilience (the degree and timing for which a system can return to a satisficing configuration following a perturbation).

- *Execution* involves the 'performance' of the system design. Execution is never optimal, has inherent variability as humans are involved, and is subject to shifts over time. Design inadequacies can be compensated for by shifts in execution

Table 1 (continued)

Difficulty	Description
Unclear Entry Point	Complex system problems are poorly structured and generally difficult to define with specificity. In fact, what 'appears' as a complex system problem may simply be a surface manifestation of a deeper underlying problem or set of intertangled problems. Therefore, even if the decision 'to begin' is made, 'where to' begin becomes problematic. The lack of clear entry point calls into question perspectives rooted in 'defining' the problem as the first and independent step in analysis
Solution Urgency	Modern complex system problems demand urgency in developing a response. While there are certainly circumstances that permit a 'leisurely' pace to completion, it is more likely the case that the solution to the situation is demanding a resolution as soon as possible. This is problematic for complex systems in the case where the depth of the problem requires a tradeoff in time allocated for exploration, understanding, and resolution
Misinformation–Defensiveness	It is a misnomer to think that all information serves noble intentions. On the contrary, information can have a veracity that in the worst case can be misinformation and in the best case might be disingenuous. Additionally, there can be a defensiveness in dealing with information where there is a perceived threat to 'status quo' positions. This defensiveness may rely on interpretations or selective construction/dissemination of information
Ambiguous Boundaries	Boundaries are a basic attribute of systems. They serve to separate a system from all that is external. Matter, information, or energy can cross the boundary as inputs and, once transformed, become outputs of value to be consumed by the environment. A lack of clarity in definition of boundary conditions, and the criteria for inclusion/exclusion for what lies outside the system, is problematic for complex systems
Unintended Consequences	Complex systems are designed with desired outputs and outcomes in mind. However, the realities of complex systems in operation result in behaviors, structure and performance that emerge. This emergence cannot be known or predicted in advance and can often result in unexpected and undesirable behavior. It is systemically naïve to think that systems will not produce consequences that may be far from those desired, intended, or anticipated

(continued)

(e.g., working additional hours to compensate for scheduling issues that the system design was incapable of managing).

- *Development* is concerned with modifications of the system design or execution to ensure that a system remains viable. Modifications are necessary to compensate for environmental/context shifts which can render the design no longer capable of effectively responding. Modifications can be targeted to: (1) *system execution*,

Table 1 (continued)

Difficulty	Description
Emergent Situations	As complex systems operate, they will inevitably produce emergent situations. The precise nature of situations cannot be known or predicted in advance. Instead, they come about as a system operates and must be managed 'in the moment' of their appearance. For complex systems emergence will occur. However, the precise timing, nature, and impacts cannot be known in advance. This does not absolve the responsibility for emergence. Instead, it heightens the emphasis on preparation of the system design to be sufficiently robust and resilient to weather emergence
Shifting Demands	Frequently, the demands placed on complex systems will be dynamic in that they will change over time. This is not necessarily indicative of poor management. On the contrary, it may be the result of complexities which cannot be fully comprehended at the time a systems endeavor was initiated. Shifting demands are a fact of life for complex systems and should be anticipated
Instabilities	There are multiple potential sources of instabilities for a complex system. Instabilities may come from within the system itself (e.g., conflicting procedures), the environment (e.g., stakeholder inconsistencies), or the particular context within which the system is embedded (e.g., weak leadership). Irrespective of source, instabilities are disruptive to maintenance of system performance. Care must be taken to design systems such that instabilities are not capable of incapacitating the system before the results can be mitigated
Divergent Stakeholders	Variability of stakeholders is inevitable in complex systems. However, significant divergence between stakeholders should be accounted for. The assumption that divergence can be overcome might be somewhat naïve for complex systems. As long as unresolvable divergence exists, it will be a source of concern for continuing viability of a complex system
Politically Charged Decisions	All complex systems are subject to 'politics' as the pursuit of strategies to gain or maintain power (influence) over the system. To ignore the fact that politics is in play for the decisions concerning complex systems is naïve at best. Politically charged decisions can and will be a source of concern for complex systems. Designs that fail to account for the existence of politics in complex systems are shortsighted at best and debilitating at worst

where 'first-order' modifications are made in response errors based on detection of 'system performance issues' and correction by adjustments to execution while leaving the system design intact, or (2) *system design*, where 'second-order' modifications are made in response to errors based on detection of 'system design performance issues' and corrections by enacting adjustments to system design.

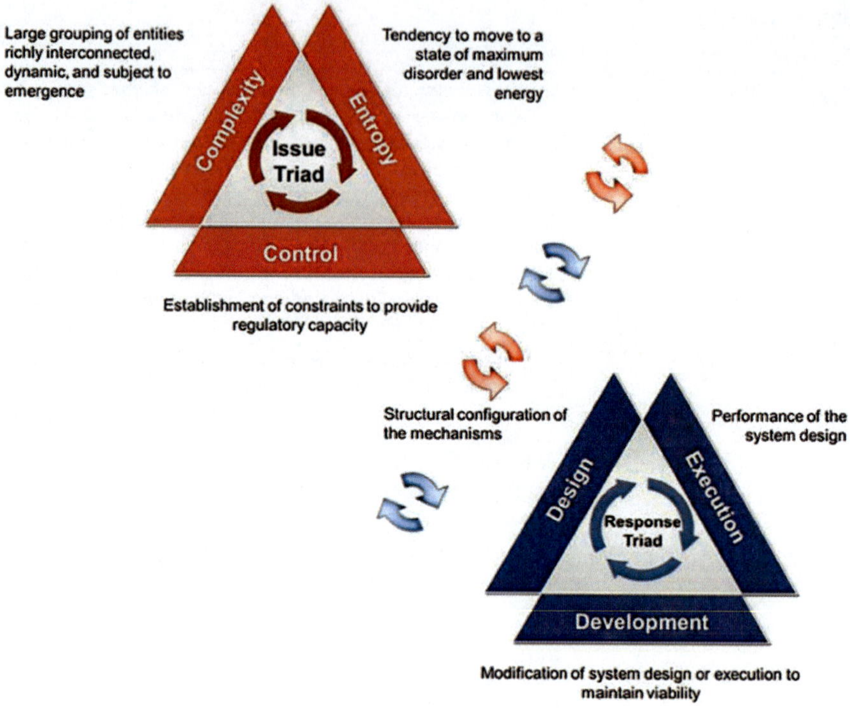

Fig. 1 Issue and response triads for CSG

The interplay of design, execution, and development is the CSG response to the challenges posed by the issues triad.

Thus far, we have set the problem domain targeted by CSG, the need for CSG, and the high-level response of CSG. In the remainder of this chapter, we present an overview of CSG, articulate the present state of the field, and discuss challenges for the future development of CSG.

2 The Emerging CSG Field—A Response to the Future

In this section, four primary themes are developed. First, CSG is put forward as an emerging system-based approach to improve the design, execution, and development of complex systems. This overview briefly introduces CSG as a focused response to the problem domain challenges identified above. This discussion also establishes the foundations of CSG and its response to address pitfalls in addressing complexity. Second, the current state of the CSG field is examined. The existing literature and works of CSG are explored for the emerging set of themes that delineate the field. Third, acknowledging work that has been completed, challenges that must be addressed if the field is to remain viable are explored. The chapter closes with the implications and promise that CSG theory and practice holds for advancing prospects for more effective design, execution, and development of complex systems.

2.1 The Three Fields Upon Which CSG is Based

CSG lies at the intersection of three primary fields, including Systems Theory, Management Cybernetics, and System Governance (Fig. 2). Systems Theory is a primary grounding field for CSG and is focused on effective integration and coordination of disparate elements into a coherent whole. Systems Theory demands that the whole is subject to and must conform to the axioms and corresponding propositions of Systems Theory. The axioms and corresponding propositions define, explain, and govern behavior, both negative and positive, of systems. Management Cybernetics brings an emphasis on communication and control essential to continuing system existence (viability). Viability is necessary for a system as it deals with the inevitable internal flux and environmental turbulence endemic to modern complex systems. Management Cybernetics enables CSG to appreciate and respond to the constant change in the context and environment for a governed system or system of

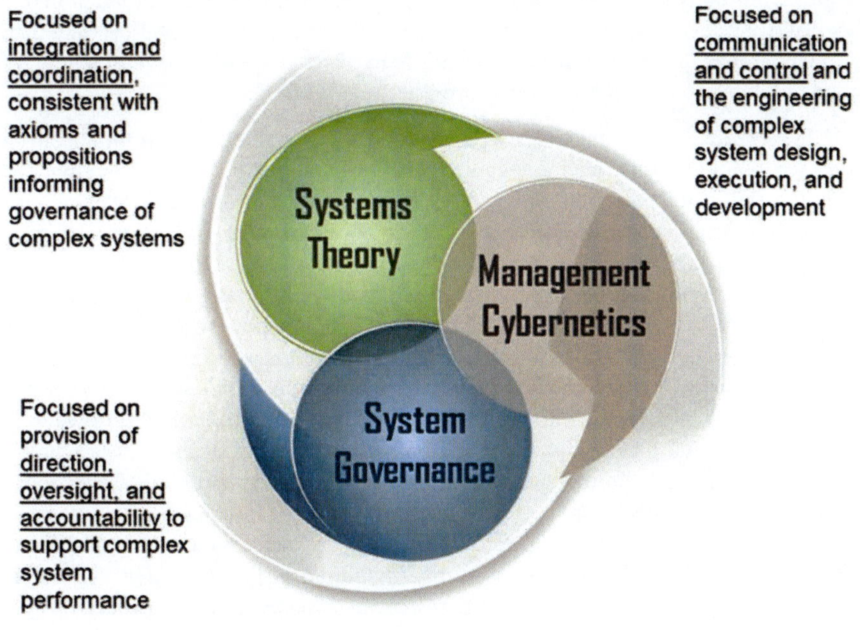

Fig. 2 Three fields contributing to CSG development

systems. Thus, while governance necessarily takes a long view to provide long range stability, it must also monitor and acknowledge the potential impact of near term fluctuations on system viability. Finally, System Governance provides an emphasis on direction, oversight, and accountability for the execution and development of a system. Individually, each of the three fields underpinning CSG has made substantial contributions to the state of human affairs. However, they have not been brought together in meaningful ways that take advantage of their intersection (Fig. 2). CSG evolves from the intersection of these fields to produce a novel alternative for complex system development.

Based on the three underlying theoretical/conceptual foundations, CSG has been constructed as the *design, execution, and evolution of the metasystem functions necessary to provide control, communication, coordination, and integration of a complex system* [82]. Detailed explanations of CSG are covered within this book and in previous works [70, 76, 77, 79, 80, 82]. However, for introductory purposes, in this section, we provide an overview of three essential aspects of CSG, including: (1) the fundamental essence of CSG, (2) the CSG paradigm, and (3) primary contributions of CSG for complex systems.

2.2 *The Fundamental Essence of CSG*

Descriptions and discussion of the nature and specifics of CSG are elaborated in this book and have been explored in a number of prior works [76, 79, 80, 82]. Rather than provide a rehash of the details of the CSG field, instead we offer several summary points of emphasis and distinction to present the CSG field:

- CSG is *holistic* and considers design, execution, and evolution across the spectrum of technology, human, social, organizational, managerial, policy, information, and political dimensions of modern complex systems. Consistent with this perspective is a concentration on ensuring that the governance of a system is targeted correctly to the purpose, problem, or need that the system is intended to address.
- At a most basic level, CSG is concerned with performance of functions that provide *control* (installation of the minimal specification necessary to achieve desirable system performance while providing the greatest degree of autonomy to system constituents), *communications* (exchange of information such that consistent decision, action, and interpretation are supported), *coordination* (provision of sufficient standardization among system constituents such that unnecessary fluctuations are avoided), and *integration* (function of the system as a 'unity' to produce capabilities, behavior, and performance beyond that of individual constituents).
- CSG is built around performance of the set of nine interrelated metasystem functions common to all complex systems, including: *policy and identity, system context, strategic system monitoring, system development, learning and transformation, environmental scanning, system operations, operational performance, and information and communications*. The metasystem functions are achieved

through implementing mechanisms (artifacts that enable performance of functions) that operate within the parameters of Systems Theory axioms and propositions. Inconsistency in application of Systems Theory or violations of its tenets (axioms/propositions) represent CSG pathologies that act to inhibit or degrade system performance [45–50, 53, 76].

- Communications provide for the flow and interpretation of information for the metasystem functions in CSG. CSG communication occurs through support 'channels,' which describe what the communication vehicles must satisfy for continued system viability. Communication support channels for CSG include: *command* (non-negotiable directives), *algedonic* (system threat warning that bypasses all other channels and functions), *dialog* (examination of purpose and essence of the system), *learning* (identification of system adjustments to correct detected variabilities), *environmental scanning* (provides intelligence of external conditions), *resource bargain/accountability* (resource distribution and output expectations), *operations* (providing directions for system operations), *audit* (provides monitoring of routine as well as emergent anomalies in system performance), *coordination* (provides for harmonizing elements within the system to avoid unnecessary fluctuations), and *informing* (providing for routine information in the system).
- CSG functions are performed through mechanisms (artifacts that permit the achievement of metasystem functions and communications). The completeness and performance of the set of mechanisms (e.g., strategic development procedure) determine system performance.
- Effective execution of CSG permits a system to maintain performance amid internal flux and external (environmental) turbulence. CSG assumes inherent instabilities, complexity, and emergence. This requires a sufficiently robust and resilient design to compensate such that performance is maintained and system viability (continued existence) is assured.
- Design for CSG permits the 'active matching' of the infinite variety (a measure of complexity) that is generated by both internal (system flux) and external (environment turbulence) to the system. Left without compensating design (e.g., CSG), this variety has the capacity to destabilize the system and inhibit a system's ability to meet performance expectations.
- All complex systems that continue to exist perform the nine metasystem governance functions mentioned above. The degree of performance of a complex system is determined by the efficacy of the governance functions.
- CSG development is not a 'one time' or 'sporadic' event. CSG development is a continuous process that purposefully advances the maturity of CSG and effectiveness of its execution. CSG development operates at the individual, entity, and system levels.
- CSG development is not an 'all or nothing' proposition. There are benefits that can accrue from different levels of CSG engagement, ranging from the enhancement of individual practitioner effectiveness through structured engagement of organizational system performance.

This concise overview provides a rudimentary backdrop that captures the essence of the CSG field. We now shift to exploring the CSG paradigm.

2.3 The CSG Paradigm

Although the underlying theory, concepts, and execution of CSG are challenging, the essence of CSG is not difficult to grasp. The essence of CSG might be captured in Fig. 3 and described as

> *Subject to fundamental system laws, all systems perform essential governance functions. System performance is determined by effectiveness in the achievement of governance functions consistent with system laws. System performance can be enhanced through purposeful design, execution, and development of governance functions in accordance with system laws.*

The CSG paradigm might be summed up in six essential points:

1. *All systems are subject to the Systems Theory propositions (laws, concepts, principles) of systems*. Just as there are laws governing the nature of matter and energy (e.g., physics law of gravity), so too are our systems subject to laws. These system laws are always there, non-negotiable, unbiased, and explain

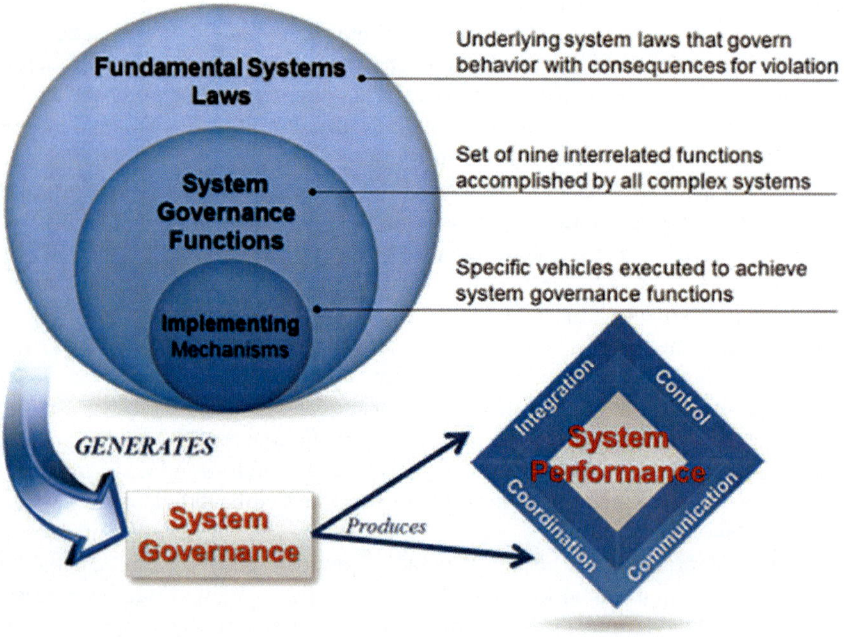

Fig. 3 The CSG paradigm

system performance. Practitioners must ask, 'do we understand systems laws and their impact on our system(s) design and performance?'

2. *All systems perform essential governance functions that determine system performance.* Nine system governance functions are performed by all complex systems, regardless of sector, size, or purpose. These functions define 'what' must be achieved for governance of a system. Every system invokes a set of unique implementing mechanisms (means of achieving governance functions) that determine 'how' governance functions are accomplished. Mechanisms can be formal-informal, tacit-explicit, routine-sporadic, or limited-comprehensive in nature.
3. *CSG produces system performance which is a function of communication, control, integration, and coordination. Control* is implementation of the minimal set of constraints necessary to achieve desirable system performance. This constraint set provides the greatest degree of autonomy to system constituents. *Communications* involves the exchange of information such that consistent decision, action, and interpretation are supported. *Coordination* provides sufficient standardization among system constituents such that unnecessary fluctuations are avoided. *Integration* assures that the system acts as a 'unity' to produce capabilities, behavior, and performance beyond that of individual constituents.
4. *Performance of governance functions, subject to systems laws (propositions) produces system performance necessary to maintain viability.* Since all viable complex systems perform governance functions, the degree to which the functions are effectively performed will determine the performance of the system and continuing viability. Governance provides a heightened state of viability (existence) where continuity of a system is assured.
5. *Violations of systems laws in performance of governance functions carry consequences.* Irrespective of noble intentions, ignorance, or willful disregard, violation of system laws carries real consequences for system performance. In the best case, violations degrade performance. In the worst case, violation can escalate to cause catastrophic consequences or even eventual system collapse.
6. *System performance can be enhanced through purposeful development of governance functions.* When system performance fails to meet expectations, the investigation of deficiencies in governance functions can offer novel insights into the deeper sources of failure. Performance issues can be traced to governance function issues as well as violations of underlying system laws. Thus, system development can proceed in a more informed and purposeful manner.

These six points provide an exploration of the CSG paradigm. The CSG paradigm is the underlying statement that delineates all that follows as the details of CSG are unfolded in the remainder of this book.

2.4 Contributions of CSG for Complex Systems

CSG is a systems-based approach to enable practitioners to better deal with complex systems and their problems. In essence, CSG seeks to improve the design, execution, and development of complex systems. CSG is certainly not alone as a systems-based approach to deal with complex systems. There is not a universally accepted theory, methodology, method, or set of standards to assure success in dealing with the current state of complex systems. CSG does not resolve this dilemma and we expect this is not only the present case but will continue into the future. In fact, Jackson [40] identifies multiple systems-based approaches to deal with complex systems problems—including emphasis on dealing with complex systems across technical, process, structure, organizational, and coercion emphases. Each of these approaches can claim success. However, none of the approaches provides a universally accepted approach. Similarly, Keating [86] has identified 15 different systems-based approaches to deal with complex systems/problems. The different approaches demonstrate the difficult nature of selecting an appropriate methodology(ies) for dealing with complex systems/problems. Nevertheless, complex systems will not wait for a universal approach to emerge. As opposed to 'doing nothing' waiting for a universal approach, something must be done if we are to improve prospects of engaging complex systems and their problems. Although CSG is not a universal approach to dealing with complex systems, it does share some systems DNA as well as provide differentiation from other systems approaches.

CSG is distinct from other systems-based methodologies in several important ways. First, CSG is based on an explicit grounding in Systems Theory. Although many other 'systems-based' approaches purport to be based in underlying 'systems' perspective, they do not elaborate their explicit grounding in Systems Theory. This is not a criticism of other systems-based approaches, but rather a delineation of the degree to which CSG does not leave grounding in Systems Theory to innuendo, assumptions, or speculation. Second, CSG 'qualifies' the level of systems thinking for participants prior to engagement. This qualification stipulates the state of systems thinking capacity held by participants in the design, execution, and development of metasystem functions. For CSG, this qualification establishes a level of expectations as to the capability to engage higher orders of system development. The absence of a compatible level of systems thinking capacity will severely limit the probability of success in CSG development endeavors. Third, prior to engagement for CSG development, both the context and initial state of the metasystem are established. This establishes the CSG baseline, against which progress can be determined. Also, the baseline sets constraints on the nature and type of development activities that can be engaged with a high feasibility of success. Constraints may exist in the context, the individuals performing CSG, or the current state of CSG. In effect, CSG places emphasis on 'the front end' prior to engaging the development. This explicit emphasis departs from more traditional systems-based approaches that do not qualify the level of systems prowess necessary to effectively engage the approach.

Vignette

CSG: Correcting System Drift

CSG is about correcting system drift. System drift denotes systems that, irrespective of the best intentions, have either never been purposefully designed or whose execution continually fails to meet desired performance expectations. In short, these 'drifting' systems fail to deliver minimal value expected, much less produce high performance. We do not need to look far to see examples of drifting systems. In fact, it would be a rare day that we would not be impacted by systems in drift. Consider the following examples: (1) launching of a new Enterprise Resource Planning initiative that collapses due to emergent incompatibilities with existing systems, (2) a costly crisis from discovery of noncompliance to a regulatory requirement that has been in existence for several years but never identified, or (3) introduction of a new purchasing policy that achieves intended reductions in supplier costs but increases overall costs due to schedule delays. Unfortunately, the impacts of system drift are not limited to increased monetary costs. These drifting systems have considerable associated human cost, borne by those destined to suffer as victims of drifting systems. CSG supports thinking, decision, and action to proactively and purposefully address system drift and its consequences.

CSG is holistic in the enhancement of system design, execution and development. In this sense, CSG can provide value across five 'holistic' levels. It is important that these different levels of development are interconnected, each influencing, and being influenced by, the others. These levels and the CSG contributions are identified in Table 2.

The value accrued by CSG stems from: (1) scanning the organization [system] for the capacity to engage in a level of systems thinking compatible with the complexity demands of the system environment, (2) exploration of the design and execution of essential governance functions, (3) identification and prioritization of system performance constraints tracked to problematic governance functions and violations of systems laws, and (4) establishment of feasible developmental strategies across multiple levels essential to holistically improve performance through enhancement of CSG.

3 Distinguishing CSG from Other Systems-Based Approaches

The CSG field is still in the earliest stages of development. While there remains significant development for the CSG field to be viable, there have been substantive advances made. CSG is emerging to take its place among other systems-based approaches. These approaches have been in existence and proven through application in many different circumstances [29, 38–40, 79, 80, 86]. However, CSG has

Table 2 CSG contributions across multiple levels

Level of impact	Description
Practitioner	Enhanced capacity of individual practitioners to engage in the level of systems thinking necessary to more effectively deal with the issues related to design, execution, and evolution of complex systems and their problems
Enterprise	Provide competency development (knowledge, skills, abilities) for targeted entities (units, staff teams, departments) across the enterprise to better engage complex systems and problems
Support Infrastructure	Examination and development of support infrastructure (processes, technologies, systems) for compatibility with system governance design, execution, and development
Context	Identification and development consideration for unique circumstances, factors, and conditions that influence (constrain/enable) achievement of system governance functions and system performance (e.g., stakeholders, regulatory requirements, staff, leadership style, etc.)
System of Interest	Providing identification of impediments to system performance. These impediments are rooted in specific deficiencies in design, execution, and development of governance functions and corresponding system laws

several distinctions that separate it from existing systems-based approaches to development (Fig. 4). This is not a criticism of other systems-based approaches but rather

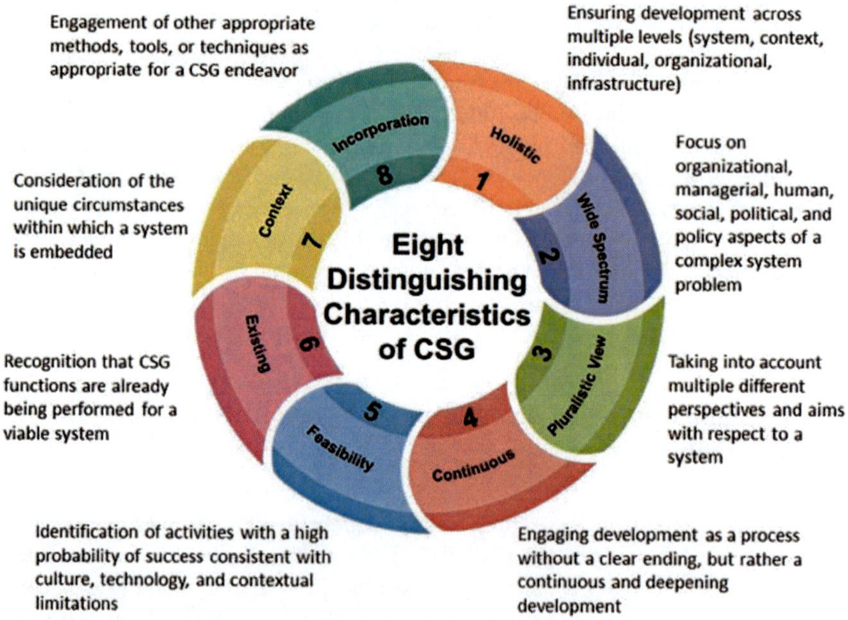

Fig. 4 Distinguishing Characteristics of CSG

serves to distinguish the uniqueness of CSG as a systems-based approach to system development. First, CSG purposefully operates at all levels of system development, ranging from individual to an entire system/enterprise. Thus, CSG is true to the tenets of being 'holistic,' operating across the entire spectrum of complex systems (individual practitioner, context, support infrastructure, entity, system, and enterprise). This emphasis of CSG extends development beyond a specific problem type or system. Instead, the situational aspects that influence, and are influenced by, the system are the subject of CSG development. 'Everything' (all levels) is not simultaneously developed in CSG. However, feasible development is undertaken with an appreciation of the whole. This permits priorities for feasible (culturally, resource, technological, capability) development to be undertaken in CSG applications.

Second, CSG does not narrowly bound systems to address only limited issues (e.g., technology). On the contrary, CSG also appreciates the wider spectrum of system issues, including non-technical aspects (political, human, social, organizational, managerial, policy, and political) of complex system problems. This does not suggest that other systems-based approaches do not take these elements into consideration (tacitly or explicitly). However, this serves to emphasize the explicit consideration CSG makes with respect to these elements. Third, unlike many systems-based approaches, CSG does not assume that a unitary view of system development will be held by those who participate in the system development effort. There are countless different and potentially conflicting motivations, perspectives, and aims that can exist and be pursued (explicitly or tacitly) by individuals and entities engaging in system development (pluralistic view). Instead of ignoring or downplaying the existence of these varying perspectives, CSG embraces and seeks to make explicit the differences. CSG is clearly focused on exploring the logic, assumptions, and implications the differences hold for system development. Fourth, some systems-based efforts are directed to a 'singular' application focused on a particular problem. In contrast, CSG is focused on the continuous development and evolution of the governing (meta)system functions. Therefore, there is no end state to CSG development. Instead, there is a continuing engagement to evolve CSG to achieve more desirable levels of system performance. Fifth, CSG recognizes that some development activities, although desirable and offering significant potential for improvement, are beyond feasible achievement. CSG cautions that development activities pursued must be compatible with the current state of governance capabilities. Otherwise, the potential for failure, irrespective of how enticing an activity might appear, escalates with diminishing feasibility. This suggests targeting the highest priority development areas identified, but only to the extent that their achievement is feasible. Feasibility is a function of practitioner/system competencies, compatibility/congruence of supporting infrastructure, context, and CSG state of the system of interest. Thus, while two systems might identify the same developmental issue, their range of feasible activities that can be undertaken in response will differ based on the state of feasibility unique to each system.

Sixth, CSG metasystem functions are already being performed if a system is viable (currently existing). However, the state of viability might be degraded based on underperforming CSG functions. Even though the language of CSG might not be used to describe the metasystem functions, they are nevertheless being performed. Additionally, the mechanisms used to perform the functions may have been established by processes of 'self-organization' (established without constraint) or 'accretion' (added in a piecemeal or ad hoc manner), resulting in heightened possibility of degradation of overall system performance. Seventh, the context–system interaction is fundamental to CSG development. CSG is intently focused on understanding the system, the system context, and the system–context interrelationship. The system–context interrelationship is essential to understand the enabling and constraining forces that the interrelationship invokes. Finally, CSG does not preclude the incorporation of other methods, tools, or techniques into a development effort. On the contrary, CSG encourages and invites incorporation of a variety of existing, developing, and future vehicles to help achieve CSG developmental goals. In this sense, CSG is entrenched in the concept of '*bricolage,*' or improvising by using what is at hand to create new and novel artifacts. Thus, CSG is not presented as an intractable, prescriptive, linear, or stepwise approach that must be followed without deviation. CSG is not limited to assuring success only if a prescribed sequence of activities are precisely followed, without variance, as directed. Instead, CSG offers a guide that must be tailored to the uniqueness of a specific system and its context for application rather than blindly following a prescribed formula. This also suggests that the results from engagement of CSG cannot be known or predicted in advance of deployment. On the contrary, it would be disingenuous to claim that CSG can, without fail, produce desirable performance improvements. This is consistent with any systems-based endeavor.

3.1 The Current State of the CSG Field

The current state of the CSG field should be expected. The field lies at the convergence of three existing fields, it has only been recently formulated, and the depth of published materials for the field is limited. The earliest foundations for CSG are found in the need to diverge from the mainstream of System of Systems Engineering (SoSE). The need to separate from SoSE was based on three primary divergent perspectives. First, the overemphasis of technology that was dominating the SoSE field. This resulted in the sacrifice of a more holistic view of complex systems appreciating the range of both 'hard' (technology) and 'soft' (human, social, organizational, managerial, policy, political) dimensions of integrated complex systems. Second, the evolving SoSE field was void of grounding in the theoretical underpinnings provided by Systems Theory. This lack of theoretical basis for SoSE was resulting in a field evolving without the science/theory anchoring necessary for long term development and viability. The result was overemphasis on building methods, tools, techniques, and technologies absent of any grounding in a stable theoretical/science basis. Third, the overemphasis of technology, coupled with the military sector drive for technological interoperability, continued to dominate the development of the SoSE field. The 'military/defense' posturing of the SoSE field surrendered the multidisciplinary potential for generalization of the field into other sectors that could benefit from more diverse applications.

CSG draws upon the intersection of the three fields: *Systems Theory* (providing a theoretical grounding), *Management Cybernetics* (providing a Systems Theory-based application model), and *System Governance* (providing a multidisciplinary basis for target system development). As Fig. 5 suggests, CSG exists as both a synthesis of existing fields as well as a novel elaboration that exists beyond the integrated fields. CSG remains: (1) new—never previously suggested or developed, (2) limited—the body of knowledge has not been extensively developed, and (3) localized—development and propagation has been restricted to a narrow set of scholars/practitioners. This is not intended to be a criticism of the emerging field, but rather a recognition that there is much to be done to further develop, gain acceptance and legitimacy, and propagate the CSG field.

Casting the current state of the CSG body of knowledge suggests several dominant themes. First, CSG has been developing across the spectrum of theoretical/conceptual, methodological, models, methods, tools, and application areas. In this sense, the CSG field has made advances across each of those divisions. Second, the bulk of development has been achieved at the level of theoretical/conceptual development. This is expected as the expressed intent of development for the CSG field was first to engage in the underlying foundations prior to moving to more practice and application orientations. This does not suggest that different aspects of CSG have not been developed or applied in practice. In fact, several methods (e.g., M-Path Method) and tools (e. g. Systems Thinking Capacity ST-Cap instrument) have been developed and utilized in applied settings. Third, the CSG Reference Model and the

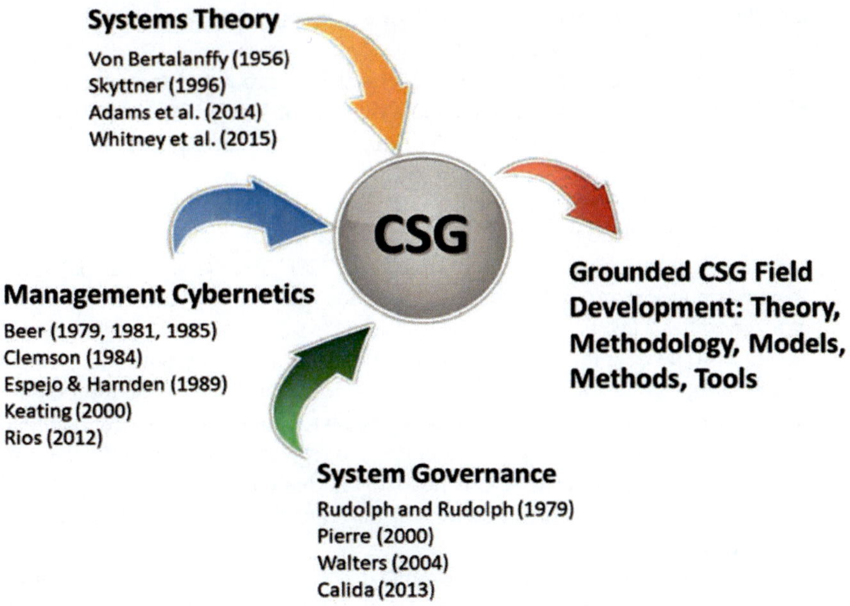

Fig. 5 CSG grounding fields and representative works

CSG Development Methodology have largely driven the development of methods, processes, and applications stemming from CSG activities. This is indicative of a shift in emphasis from the theoretical/conceptual underpinnings to the development of implementing artifacts. Fourth, the 'applications' of CSG have been limited to either conceptual development for sector specific (e.g., acquisition) investigations, exploratory examination of utility for potential deployment areas (e.g., cyberphysical systems), or limited deployment of introductory tools/training (e.g., CSG preliminary demonstration explorations for operational settings). Thus, the current state of CSG has been dominated by demonstration of utility with limited applications as opposed to cases of comprehensive engagement of CSG. However, purposeful CSG field development has been targeted to develop each of the different dimensions, including theoretical/conceptual, methodological, models, methods, tools, and application levels of engagement.

CSG has been rapidly adding to its body of knowledge. The initial inception of CSG can be traced to some early works in a systemic approach to R & D governance [67]. To capture the breadth of CSG, Fig. 5 shows a clustering of literature in the emerging CSG field. This literature is organized along six dimensions. These dimensions are consistent with earlier works suggesting the essential foundations for a developing field [70, 79, 80, 86]. The dimensions include:

1. *Conceptual/Theoretical Foundations*—These foundations draw upon Systems Theory, Management Cybernetics, and System Governance. The essence is found in the underlying CSG paradigm consistent with these informing fields and congruent with the systemic worldview. From these foundations everything else flows, consistent with values and beliefs, systems laws (propositions), and the defining systems paradigm.
2. *Methodological*—Generalizable and theoretically grounded frameworks that provide high-level guidance for conducting CSG. The methodological level identifies 'what' must be performed but stops short of prescriptively identifying precisely 'how' it must be done.
3. *Model*—The representations that articulate the fitting together of the different aspects of CSG. Models are grounded in the underlying theoretical framework for CSG and are supportive of execution of CSG derivative methodologies. However, models are more precise and defined than methodologies but remain at a generalized level.
4. *Methods*—The detailed processes that can be followed to execute some aspect of CSG. They are prescriptive in that they define precisely how execution should occur. Although, in CSG there is certainly room to modify methods for intended utility. Nevertheless, methods are specific and expected to be capable of application in a repeatable fashion. It is noteworthy that methods are also grounded in the underlying conceptual/theoretical base and can support deployment of methodologies and models.
5. *Tools*—These are the artifacts that enable a narrow aspect of CSG deployment to be supported. Tools are prescriptive in nature and have been developed for a specific intended utility for application of CSG. However, tools

are grounded in the underlying conceptual/theoretical base and should not be applied independent of the informing systems worldview upon which they are based.

6. *Applications*—Ultimately, CSG must be deployable if it is to influence the state of system affairs. Applications are the purposeful attempt to develop different aspects of design or execution of CSG. Applications must be consistent with the underlying CSG paradigm and the systems worldview upon which it is based. The range of applications may extend from limited to comprehensive engagements.

Figure 6 demonstrates the breadth and depth of the emerging CSG field. However, it is noteworthy to point out three important facets of this development to date. First, the field has been pushing across the entire spectrum of development, ranging from conceptual/theoretical grounding through application. This is an important feature for continued field development and propagation—to maintain a balance. Second, the field is dominated by a limited set of authors. While this is expected in the early instantiation of a field, continued development suggests the need for an increasingly robust set of authors. This set should include scholars, researchers, and practitioners who will test and push the boundaries of knowledge accumulating for CSG. Third, while application areas have been explored, the primary exploration has been from conceptual application and critiques suggestive of the understanding that CSG

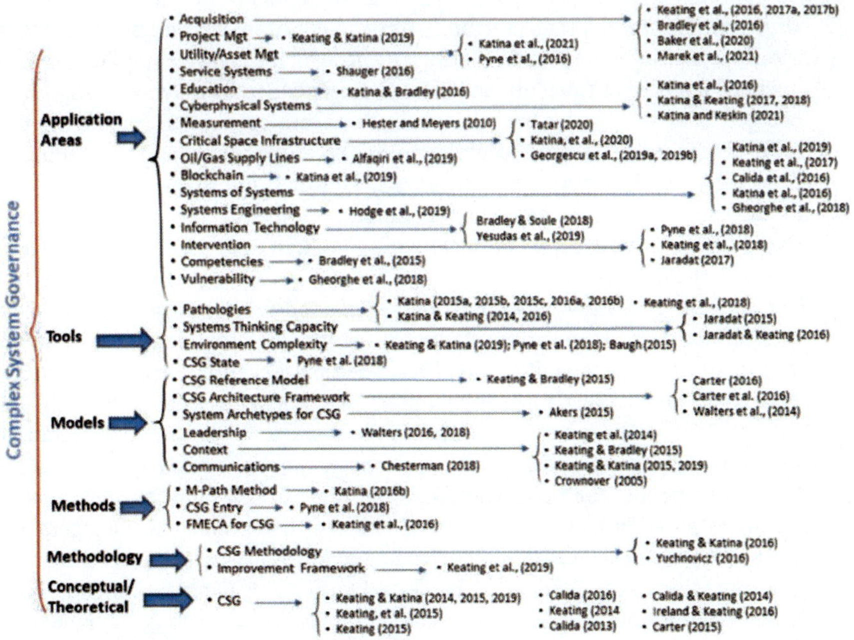

Fig. 6 Current state of the CSG field

can offer from the variety of sectors suggested in the writings. However, there is a compelling need to move CSG closer to deployment in operational settings. Thus, practitioners are a necessary component to push the CSG body of knowledge through deployment of tools, methods, and methodologies to gain an understanding of the practical utility for CSG. Additionally, enhanced focus on operational deployment will help to push research in new and novel directions.

The current state of CSG field development might be summed up with five critical points:

1. *The newness of CSG stems from integration of established fields*—It would be interesting to posture the CSG field as 'totally new and novel' and finding its basis in itself only. However, the reality of CSG is that it lies at the intersection of established fields and seeks to leverage the interface of these fields in new and exciting ways. As such, 'newness' of CSG draws on the strengths, as well as limitations, inherent in the fields upon which it draws (Systems Theory, Management Cybernetics, and System Governance). However, while the informing fields provide grounding, CSG is evolving to provide a uniqueness that exists beyond what any of the supporting fields can offer.
2. *CSG field development has initially focused on building a sustainable foundation*—The bulk of knowledge produced early in the establishment of CSG was focused on establishment of essential foundations (theoretical, conceptual). The developers of CSG have deliberately pursued this course as opposed to engaging development of immediately deployable tools, methods, and techniques (e.g., software). This focus on sustainability through conceptual grounding as the first priority has resulted in a solid platform from which to engage application and practice. The strong foundations for CSG are intended to provide a grounding that will better withstand the tests of time. Absent conceptual grounding, the 'half-life' of the CSG field would certainly be questionable and the ability to avoid early obsolescence doubtful as well. This does not suggest that CSG has not developed methods, tools, and techniques. However, the further development of CSG is now set to resonate from the strong foundations which have been established.
3. *The value in CSG application has not yet allayed perceived practitioner and institutional risks*—The current state of value for CSG does not yet exist at the level of practice. Although the field has been purposefully directed to setting sustainable foundations, these foundations have not yet translated to perceived practice value. This is not a criticism of the work that has been completed in CSG, but rather a recognition that applications will be necessary to accelerate the development of the field. Therefore, the value to risk ratio for practitioners has not been sufficiently recognized to warrant comprehensive engagement. Until such time as the value is perceived to significantly exceed perceived risk, widespread practice, demand, or perceived utility will likely be limited. Only through propagation to practice and application will CSG field development be accelerated.
4. *CSG applicability extends to multiple and diverse problem domains and sectors*—CSG is not easily cast in a particular sector (e.g., transportation,

energy, security, defense, education, healthcare, utilities, commerce) or problem domain. Instead, CSG is 'transdisciplinary,' capable of applicability and crossing multiple sectors and problems. Therefore, the responsibility, accountability, and drive for CSG development exist beyond the more traditional realms of scholarship or practice. The consequence is both enabling and constraining for the CSG field. Enabling in that the field is not shackled or held hostage to preconceived notions of development or applicability to narrow domains. Constraining in that the field also brings the limitations inherent to the fields upon which it draws. In response, the CSG field must not only develop at the intersection of three different fields, but also extend beyond the fields. This requires the balancing of congruities as well as incongruities between the fields as well as their traditional primary audiences.

5. ***CSG has not overcome significant barriers to engagement and deployment in operational settings.*** The decision to engage CSG is not superficial or reversible. Beyond allocation of scarce resources, CSG requires significant investment of time, energy, and access. Unfortunately, to overcome the inherent hurdles for engaging CSG is a difficult proposition for individuals and entities on the 'promise' and 'potential' for increased system performance. Any systemic intervention is difficult, involves uneasiness, and requires risk. As of yet, CSG has not been effective in bridging the perceived risk–benefit–cost gaps. While the concept of engaging CSG is alluring, actually committing the resources, time, and effort to fully employ CSG in the operational setting has been elusive. Despite attempts to make CSG more approachable by practitioners, to date the barriers to comprehensive engagement have proven too difficult to overcome. Even limited CSG engagement by entities, despite understanding the potential contributions, has proven to be a difficult endeavor.

As the CSG field is getting started, we note that across 55 primary publications, there are 700+ citations from a recent Google Scholar review. While this is not astounding, it does point to the field being on an ascendent development path. Based on the current state of the CSG field, we now turn our focus to articulate the challenges that must be confronted to move the field forward.

4 Challenges Moving CSG Forward

CSG field development has not gone without challenges. As with any emerging field, CSG has been exposed to various developmental issues. Some issues have been anticipated, such as accepting the slow pace for effective development of the theoretical and conceptual underpinnings of the field. Some issues have not been anticipated, such as the difficulty in gaining the necessary traction to deploy CSG in operational settings. In looking at the challenges moving forward, we have identified several issues that have emerged in field development. These impediments to field development are not insurmountable. However, the field development strategy has had to be reformulated based on initial deployment discoveries. Among the primary challenges being faced are:

1. *Compatibility of worldviews for constituent fields*—This difficulty follows from the notion that worldview is the system of values and beliefs that allow us to process events, observations, and new experiences such that 'appropriate' meaning can be accorded. 'Growing up' with the predominant worldview that is informed from a primary field colors the way in which we see the world. The corresponding language and interpretative schemas that allow coherence of interpretation lie at the essence of a field. Thus, while there may be overlaps between fields (e.g., Systems Theory and Management Cybernetics), there is also a set of underlying distinctions that may produce compatibilities as well as incompatibilities based on the differences in the underlying worldviews. The discovery and resolution of these field compatibility issues may enable or constrain the utility found at their intersection. In addition, compatibility of worldviews also extends to individual practitioners engaging in CSG. Acceptance of differences and active exploration as to the source and implications for those worldview differences is a continuing challenge for the CSG field.
2. *Traditional perspectives stemming from constituent fields*—The practitioners of the 'governance' field have a different application orientation than both 'systems'-based fields informing CSG. In painting with very broad strokes, governance is more directed at higher level (strategic) understanding and dealing with complexities of organizations (e.g., enterprise governance) and the issues of policy (e.g., corporate governance). In many respects, 'governance' is taken as synonymous with 'government.' In contrast (we believe inappropriately), traditional systems-based applications are more seen as being focused at the detailed, technical, and operational levels of organizations (e.g., systems engineering). Thus, this assumed division, however we may suggest its inappropriateness, may explain the reluctance to engage at CSG, which exists at the intersection of these fields.
3. *Successes of the constituent fields have not necessitated crosspollination between fields*—Each of the fields being intersected by CSG have been successfully developed and applied to address complex issues. Therefore, there has not been the overwhelming drive for exploring their intersection out of perceived necessity or urgency. Perhaps, the anticipated gains have not been seen as substantially beneficial to warrant the investment of resources or intellectual capital necessary to explore the potential for integration. However, we should note the traditional close coupling between Systems Theory and Management Cybernetics, both having the strong 'systems' heritage that is central to their essence.
4. *Hesitance to engage yet 'another' new field*—We must appreciate the reluctance to engage yet another 'new field' with the promised potential to significantly alter the landscape of dealing with modern complex systems and their problems. The landscape of practitioners is littered with promises of new approaches that

will provide resolution to complex system performance issues. Continual disappointment stemming from failure to achieve promised results is certainly cause for skepticism as a new field comes along with similar promised potential. The list of 'new' fields and approaches that have fallen short of promises and expectations to address complex system problems is substantial and litters the literature. This is not intended to disparage any of the existing attempts. On the contrary, this suggests a caution for the development of the CSG field. The caution to be heeded is to avoid the 'faddish' development and corresponding 'over promising' that is sure to fall short of expectations and lead to a premature demise of the burgeoning field.

5. *Extension into and appreciation of related work in different fields*—CSG field development had to start somewhere, and that was found at the intersection of the Systems Theory, Management Cybernetics, and Governance fields. However, further development of the field cannot be arbitrarily closed to injection of knowledge from other fields as the propagation of the CSG field widens in both scope and reach. Likewise, similar concepts and thinking, albeit with different language, should be engaged in the evolving CSG field development. The exclusion of new thinking, approaches, and exemplars would be shortsighted for a field attempting to make a substantial impact.
6. *Difficulties in moving from theory to deployable artifacts to practical implementation*–Although CSG is routinely being performed by viable (continuing to exist) systems, it is not routinely or purposefully thought about or acted upon. There are components in CSG that require deep and introspective examination of the system of interest. This introspection requires investing time, energy, and scarce resources into protracted self-study. Unfortunately, this does not fit an inclination for 'easy, fast, inexpensive, low risk, limited engagement' endeavors. On the contrary, full engagement of CSG requires significant study/learning, commitment, and perseverance for the long view. CSG must find a way to address this imbalance between commitment, expectations, and investment. To date, there is a desire to have limited engagement in the tools and methods levels rather than the required immersion into deep exploration required by CSG.
7. *The desire for maintenance of the 'status quo' is a formidable adversary for CSG*– All viable (existing) systems perform CSG functions. The differences are found in the unique set of mechanisms used to perform the functions and the unique context within which those functions are enacted. However, for existing systems, the status quo has permitted continued viability. Irrespective of the existence of diminished performance states, degraded capacity, and variabilities in performance, the status quo is likely sufficient to provide continued viability. This can create a sufficient level of comfort to deny the need to 'disturb the status quo for system design, execution, and development activities' required for CSG. This desire, both stated and unstated, to maintain the status quo comfort level can be quite compelling. In many cases, CSG rightly suggests a challenge to the existing 'status quo' of design, execution, and development for existing systems may be in order. However, the potential 'disruption' in examining and publicly testing the status quo can create significant resistance. The comfort level of

maintaining the 'status quo,' versus results of an unknown exploration, can be a formidable obstacle to deployment of CSG. In this sense, CSG can be perceived as a threat, and subject to various objections to deployment. CSG has yet to find a viable path past these objections, which frequently manifest themself in superficial surface level objections.

The challenges for moving CSG forward are certainly not insurmountable. These challenges lie beyond the drive to create the coherence of the theoretical/conceptual/application underpinnings which have been the primary focus of the early development of CSG. However, to enhance the legitimacy of the CSG field, overcoming the challenges must be approached with the same vigor that has brought the field to its current state. This book represents a significant step in the direction of launching the next phase in the development of the CSG field. We now look to capture the specific contributions this book makes for CSG.

Vignette

We are Systems People—We get it!

Intervening in complex systems is always a risky endeavor. In this case, an organization had 'systems' in their title and proudly announced that they were in fact already versed in 'all things systems.' They did not need a 'refresher, much less an introduction to systems, prior to engaging the system development effort.' Unfortunately, this pronouncement was taken at face value and a systemically sophisticated development effort was planned and implemented. The plan had the tacit concurrence of the participants and sponsoring organization. Within three months of launching the effort, it became apparent that the organization did not have the requisite systems knowledge to effectively engage at the level of systemic grounding demanded by the approach and previously claimed by participants. The result was an engagement that quickly outpaced the systems thinking capacity held by the participants. Rather than continue down a path of ultimate disappointment, the effort was 'retrenched' to bring the participants up to sufficient systems knowledge necessary to complete the effort. While there were some successes in the endeavor, it achieved far less than what was possible had the initial pronouncement of sufficient systems capacity been questioned. This painful lesson ingrained in the CSG approach the explicit establishment of the state of systems thinking capacity as a first and priority element of engagement.

5 Contributions of This Book

CSG has advanced to the state that a consolidation of the works delineating the body of knowledge is in order. As a field, CSG remains in the embryonic stages of development. Although the field has been maturing rapidly, it is far from being in the mainstream as an accepted systems-based methodology. The roots of the CSG field actually reach back to the foundations of Systems Theory [97, 102, 106] Management Cybernetics [9–11], and System of Systems Engineering [91].

The first formal inception of CSG is found in the [80] article that introduced the CSG construct and opened the exploration of the field. This followed the suggestion of Keating [69] that CSG was a needed evolution to push System of Systems Engineering in new and novel directions, ultimately sowing the seeds for growth of a new and distinct field. Since the initial inception of CSG, there has been significant progression of the field (see Fig. 6). This book offers a consolidation of CSG that captures the current state of the field, including theoretical foundations, the central aspects of CSG, the practice implications, and the future challenges facing the field. In a sense, this book represents a waypoint in the journey of the CSG field. The past history that has gotten the field to this point, the present status of the field, and the future implications are captured in this work.

This book is focused on exploring three primary areas to better understand and utilize current developments in the CSG field (Fig. 7). First, the *conceptual foundations* for CSG, grounded in "System Theory", "Management Cybernetics", and "System Governance" are developed. Following this introduction, chapters "Complexity", "Complex System Governance" and "Complex System Governance Reference Model" each provide an essential aspect of the foundations upon which the CSG field has been built. This grounding provides the theoretical and conceptual foundations essential to propagation and sustainability of CSG as a field in both theory and practice. Second, a set of critical topics that capture the *fundamentals* essential to the CSG field are examined. These chapters

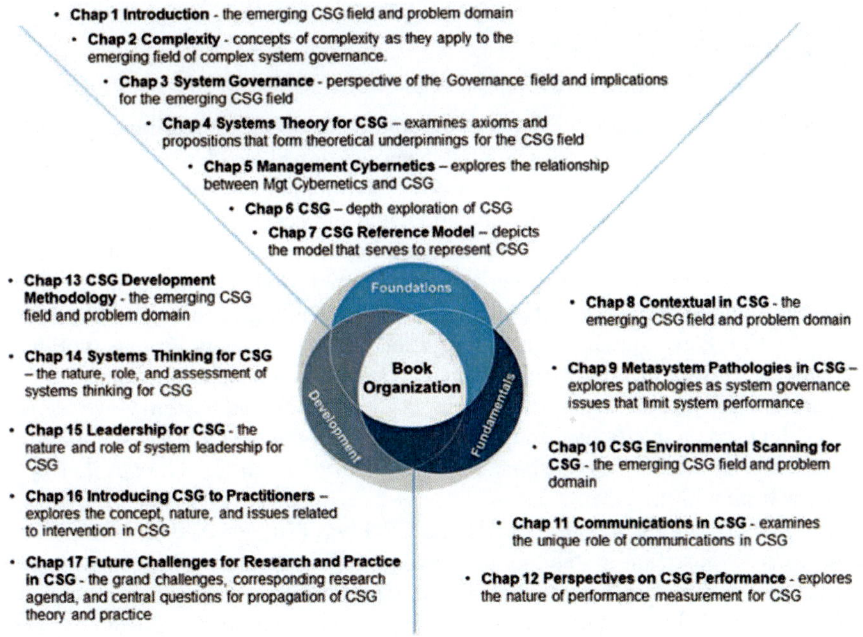

Fig. 7 Organization of the book

"Context in Complex Systems Governance", "Metasystem Pathologies in Complex System Governance, "Environmental Scanning for Complex System Governance", "Communications for Complex Systems Governance", and "Performance Perspective for Complex System Governance" are targeted to provide a treatment of the most fundamental aspects that serve to distinguish and lay the foundations for the CSG field. Emphasis is placed on exploring CSG in a manner such that the utility for practitioners is apparent. The conditions are established to demonstrate the utility CSG can offer practitioners for alternative decision, action, and interpretations to advance the state of complex systems. Third, chapters "Complex System Governance Development Methodology", "Systems Thinking for Complex System Governance", "Leadership for Complex System Governance", and "Introducing Complex Systems Governance to Practitioners", address several *development and application issues* central to further propagation of CSG. While this set is not presented as 'complete,' nevertheless they offer a sound starting point to better understand the nature and substance of CSG deployment. The final chapter "Future Challenges for Complex System Governance Research and Practice" closes the work with the grand challenges, corresponding research agenda, and the central questions that must be answered for CSG to maintain a trajectory that assures viability for the future. Ultimately, CSG is positioned as an emerging field with strong theoretical grounding and significant implications for improving practices and performance for modern complex systems and their problems.

6 Summary

In this chapter, we have provided a brief introduction to CSG. The examination identified the problem domain that CSG has been developed to address. This domain is characterized by high levels of *complexity* (high number of entities or variables, rich interconnections, dynamically changing, and subject to emergence), *ambiguity* (lack of clarity in definition of the system and its context), *context* (circumstances, factors, and conditions within which a system is embedded and enable or constrain the system), *uncertainty* (lack of cause-effect understanding), and *holism* (the spectrum of technology, human, social, managerial, organizational, policy, and political dimensions). Given this problem domain, CSG was introduced as a response at the intersection of Systems Theory, Management Cybernetics, and System Governance. CSG is focused on the *design, execution, and evolution of the metasystem functions necessary to provide control, communication, coordination, and integration of a complex system* [82].

The CSG paradigm was introduced and CSG was distinguished with respect to other more mainstream systems-based methodologies. Primarily, CSG was introduced as having an explicit grounding in Systems Theory, qualifying participants with respect to their systems thinking capacity, and establishing the state of CSG and the system context prior to engaging in development. Thus, unlike other systems-based methodologies, there is an extensive examination of the participants, context, and system of interest—all prior to full engagement. The current state of the CSG field was explored and the depth of work in the emerging field was established. Emphasis was placed on the holistic development of the field across conceptual/theoretical, methodological, model, methods, tools, and applications levels. While CSG has made inroads across each of these levels, there is much to be done to continue the development. Challenges to advance the CSG field were examined, for example, difficulties in confronting the 'status quo.' These challenges were presented as not insurmountable, but in need of being addressed if CSG is to continue a forward trajectory.

This book is presented as a waypoint for CSG development. It offers an appreciation of where CSG has evolved from, the current state of CSG, and projection into the future. The organization of the presentation of CSG is focused on three primary areas, including *foundations* (the basic conceptual and theoretical grounding for CSG), *fundamentals* (topics that establish the critical tenets for CSG), and *development and application issues* (several areas that have been selected as essential to the development of CSG).

Exercises

1. Discuss the need that the CSG field is emerging to address.
2. Briefly identify the contributions made by each constituent field (Systems Theory, Management Cybernetics, and System Governance) upon which CSG is based.
3. Explain the significance of the CSG paradigm and its implications for the design, execution, and development of complex systems.
4. Identify and discuss the importance of the distinctions that CSG makes in relationship to other systems-based approaches.
5. Discuss the importance of the CSG field making simultaneous advances across the spectrum of conceptual/theoretical, methodological, models, methods, tools, and applications.

References

1. Ackoff RL (1997) Systems, messes and interactive planning. Soc Engagem Soc Sci 3(1997):417–438
2. Adams KM, Hester PT, Bradley JM, Meyers TJ, Keating CB (2013) Systems theory as the foundation for understanding systems. Syst Eng 17(1):112–123. https://doi.org/10.1002/sys.21255
3. Adams KM, Keating CB (2011) Overview of the systems of systems engineering methodology. Int J Syst Syst Eng 2(2/3):112–119
4. Akers W, Keating CB, Gheorghe A, Sousa-Poza A (2015) The nature and behaviour of complex system archetypes. Int J Syst Syst Eng 6(4):302–326
5. Alfaqiri A, Hossain NUI, Jaradat R, Abutabenjeh S, Keating CB, Khasawneh MT, Pinto CA (2019) A systemic approach for disruption risk assessment in oil and gas supply chains. Int J Crit Infrastruct 15(3):230–259

6. Ashby WR (1958) Requisite variety and its implications for the control of complex systems. Cursos Congr Univ Santiago De Compostela 1(2):83–99
7. Arbesman S (2017) Overcomplicated: Technology at the limits of comprehension. Penguin
8. Baker W, Castelle K, Bradley J, Marino L (2020) Design for sustainment: governance engineering in major acquisition programs. In: Proceedings of the seventeenth annual acquisition research symposium. Monterey, California, pp 1–18
9. Baugh D (2015) Environmental scanning implications in the governance of complex systems. Int J Syst Syst Eng 6(1–2):127–143
10. Beer S (1979) The heart of the enterprise. Wiley, New York, NY
11. Beer S (1984) The viable system model: its provenance, development, methodology and pathology. J Oper Res Soc: 7–25
12. Beer S (1985) Diagnosing the system for organizations. England: Oxford University Press, Oxford
13. Bradley JM, Keating CB (2014) Systems theory for a competency model framework. In: 2014 9th International Conference on System of Systems Engineering (SOSE), pp 61–66. IEEE
14. Bradley JM, Soule RT (2018) Information technology governance through the complex system governance lens. In: INCOSE international symposium, vol 28, no 1, pp 1237–1249
15. Bradley JM, Joiner KF, Efatmaneshnik M, Keating CB (2017) Evaluating Australia's most complex system-of-systems, the future submarine: a case for using new Complex Systems Governance. In: INCOSE international symposium, vol 27, no 1, pp 187–199
16. Bradley JM, Unal R, Pinto CA, Cavin ES (2015) Competencies for governance of complex systems of systems. Int J Syst Syst Eng 6(1–2):71–89
17. Bradley J, Katina P, Keating C (2016) Complex system governance for acquisition. Leading Change, LLC Augusta United States
18. Calida BY (2013) System governance analysis of complex systems. Doctor of Philosophy (PhD) dissertation. Engineering Management and Systems Engineering. Old Dominion University
19. Calida B (2013) System governance analysis of complex systems. Unpublished doctoral dissertation, Old Dominion University
20. Calida BY (2016) Complex system governance: moving diverse theory to practice. Int J Syst Syst Eng 7(1–3):22–42
21. Calida BY, Jaradat RM, Abutabenjeh S, Keating CB (2016) Governance in systems of systems: a systems-based model. Int J Syst Syst Eng 7(4):235–257
22. Calida BY, Keating CB (2014) System governance: emergence of practical perspectives across the disciplines. Gheorghe AV, Masera M, Katina PF (eds) Infranomics: sustainability, engineering design, and governance. Springer, New York, pp 269–296
23. Carter B (2015) A metasystem perspective and implications for governance. Int J Syst Syst Eng 6(1–2):90–100
24. Carter B, Moorthy S, Walters D (2016) Enterprise architecture view of complex system governance. Int J Syst Syst Eng 7(1–3):95–108
25. Chesterman Jr CW (2018) Contextual framework of communications functions supporting complex system governance. Old Dominion University
26. Crownover WM (2005) Complex system contextual framework (CSCF): a grounded-theory construction for the articulation of system context in addressing complex systems problems. Old Dominion University
27. Dery D (1984) Problem definition in policy analysis. University Press of Kansas, Lawrence: KS
28. Espejo R, Harnden R (eds) (1989) The viable system model: interpretations and applications of Stafford Beer's VSM. Wiley, Chichester
29. Flood RL, Jackson MC (1991) Creative problem solving: total systems intervention, vol 831397533. Wiley, Chichester
30. Georgescu A, Gheorghe AV, Piso MI, Katina PF (2019) CSI—a complex system governance approach. In: Critical space infrastructures. Springer, Cham, pp 281–320

31. Georgescu A, Gheorghe AV, Piso MI, Katina PF (2019) Establishing governance for CSI: an MCDA approach. In: Critical space infrastructures. Springer, Cham, pp 249–279
32. Georgescu A, Gheorghe AV, Piso MI, Katina PF (2019) CSG: Towards CSI research. In: Critical space infrastructures. Springer, Cham, pp 321–343
33. Gheorghe AV, Vamanu DV, Katina PF, Pulfer R (2018) Governance vulnerability facets. In: Critical infrastructures, key resources, key assets. Springer, Cham, pp 39–79
34. Gheorghe AV, Vamanu DV, Katina PF, Pulfer R (2018) System of systems governance. In: Critical infrastructures, key resources, key assets. Springer, Cham, pp 93–130
35. Hester PT, Meyers TJ (2010) The role of measurement in complex system governance
36. Hodge RJ, Craig S, Bradley JM, Keating CB (2019) Systems engineering and complex systems governance–lessons for better integration. In: INCOSE international symposium, vol 29, no 1, pp 421–433
37. Ireland V, Keating C (2016) Introduction to special issue on complex system governance. Int J Syst Syst Eng 7(1/2):3
38. Jackson MC (1991) Critical systems thinking. In Systems Methodology for the Management Sciences pp 183–212. Springer, Boston, MA
39. Jackson MC (2003) Deeper complementarism: a brief response to Ulrich. J Opera Res Soc 54(11):1225–1226
40. Jackson MC (2019) Critical systems thinking and the management of complexity. Wiley
41. Jaradat RM (2015) Complex system governance requires systems thinking-how to find systems thinkers. Int J Syst Syst Eng 6(1–2):53–70
42. Jaradat RM (2017) Systemic intervention for complex system governance development. In: Proceedings of the international annual conference of the American Society for Engineering Management. American Society for Engineering Management (ASEM), pp 1–10
43. Jaradat R, Keating C (2014) Fragility of oil as a complex critical infrastructure problem. Int J Crit Infrastruct Protect 7(2):86–99
44. Jaradat RM, Keating CB (2016) Systems thinking capacity: implications and challenges for complex system governance development. Int J Syst Syst Eng 7(1–3):75–94
45. Katina PF (2015) Systems theory-based construct for identifying metasystem pathologies for complex system governance. Old Dominion University
46. Katina PF (2015) Emerging systems theory–based pathologies for governance of complex systems. Int J Syst Syst Eng 6(1–2):144–159
47. Katina PF (2015) Systems theory-based construct for identifying metasystem pathologies for complex system governance. Doctoral dissertation, Doctoral thesis, Old Dominion University, Norfolk, VA, USA
48. Katina PF (2016) Systems theory as a foundation for discovery of pathologies for complex system problem formulation. In: Applications of systems thinking and soft operations research in managing complexity. Springer, Cham, pp 227–267
49. Katina PF (2016) Metasystem pathologies (M-Path) method: phases and procedures. J Manag Dev 35(10):1287–1301
50. Katina P, Bradley JM (2016) Towards a systems theory-based curriculum for complex systems governance. In: 2016 ASEE annual conference & exposition
51. Katina PF, Calida BY (2017) Complex system governance: implications and research directions. National Academies of Sciences, Engineering, and Medicine
52. Katina PF, Keating CB (2014) Metasystem pathologies: towards a systems-based construct for complex system deficiencies. In: Proceedings of the international annual conference of the American Society for Engineering Management. American Society for Engineering Management (ASEM), p 1
53. Katina P, Keating C (2016) Metasystem pathologies: towards discovering of impediments to system performance. In: Proceedings of the 2016 industrial and systems engineering research conference, pp 1–6
54. Katina PF, Keating CB (2018) Cyber-physical systems governance: a framework for (meta) cybersecurity design. In: Security by design. Springer, Cham, pp 137–169

55. Katina PF, Keskin OF (2021) Complex system governance as a foundation for enhancing the cybersecurity of cyber-physical systems. Int J Cyber Warf Terror (IJCWT) 11(3):1–14
56. Katina PF, Despotou G, Calida BY, Kholodkov T, Keating CB (2014) Sustainability of systems of systems. Int J Syst Syst Eng 5(2):93–113. https://doi.org/10.1504/IJSSE.2014.064833
57. Katina PF, Keating CB, Jaradat RM (2014) System requirements engineering in complex situations. Requir Eng 19(1):45–62. https://doi.org/10.1007/s00766-012-0157-0
58. Katina PF, Keating CB, Bradley JM (2016) The role of 'Metasystem' in engineering a system of systems. In: Proceedings of the 2016 industrial and systems engineering research conference, pp 1–6
59. Katina PF, Keating CB, Gheorghe AV (2016) Cyber-physical systems: complex system governance as an integrating construct. In: Proceedings of the 2016 industrial and systems engineering research conference. IISE, Anaheim, CA
60. Katina PF, Gheorghe AV, Keating CB (2018) Pathology-informed approach in vulnerability assessment methods. In: Probabilistic safety assessment and management PSAM 14, September 2018, Los Angeles, CA
61. Katina PF, Keating CB, Bobo JA, Toland TS (2019) A governance perspective for system-of-systems. Systems 7(4):54
62. Katina PF, Calida BY, Bobo JA (2020) Complex system governance: advancing prospects for critical space infrastructure applications. Space Infrastruct Risk Resil Gov 57:29
63. Katina PF, Tolk A, Keating CB, Joiner KF (2020) Modelling and simulation in complex system governance. Int J Syst Syst Eng 10(3):262–292
64. Katina PF, Pyne JC, Keating CB, Komljenovic D (2021) Complex system governance as a framework for asset management. Sustainability 13(15):8502
65. Keating CB (2000) A systems-based methodology for structural analysis of health care operations. J Manag Med 14(3/4):179–198
66. Keating CB (2005) Research foundations for system of systems engineering. In: 2005 IEEE international conference on systems, man and cybernetics, vol 3. Presented at the 2005 IEEE international conference on systems, man and cybernetics, pp 2720–2725. https://doi.org/10.1109/ICSMC.2005.1571561
67. Keating C, Hester P, Kady R, Calida B (2009) R&D governance: an emerging systems-based paradigm for public sector R&D. In: Proceedings of the 30th American Society for Engineering Management Conference, pp 14–17
68. Keating CB (2014) Governance implications for meeting challenges in the system of systems engineering field. In: 2014 9th international conference on system of systems engineering (SOSE). IEEE, pp 154–159
69. Keating CB (2014) Governance implications for meeting challenges in the system of systems engineering field. In: IEEE system of systems engineering conference, Adelaide, Australia June 9–14, 2014
70. Keating CB, Bradley JM (2015) Complex system governance reference model. Int J Syst Syst Eng 6(1–2):33–52
71. Keating CB (2015) Complex system governance: theory to practice challenges for system of systems engineering. In: 2015 10th system of systems engineering conference (SoSE). IEEE, pp 226–231
72. Keating CB, Bradley JM (2015) Complex system governance reference model. Int J Syst Syst Eng 6(1/2):33–52
73. Keating CB, Katina PF (2011) System of systems engineering: prospects and challenges for the emerging field. Int J Syst Syst Eng 2(2/3):234–256
74. Keating CB, Katina PF (2012) Prevalence of pathologies in systems of systems. Int J Syst Syst Eng 3(3/4):243–267. https://doi.org/10.1504/IJSSE.2012.052688
75. Keating CB, Katina PF (2015) Foundational perspectives for the emerging complex system governance field. Int J Syst Syst Eng 6(1/2):1
76. Keating CB, Katina PF (2016) Complex system governance development: a first generation methodology. Int J Syst Syst Eng 7(1–3):43–74

77. Keating CB, Katina PF (2018) Emergence in the context of system of systems. In: Engineering emergence. CRC Press, pp 491–522
78. Keating CB, Katina PF (2019) Complex system governance: concept, utility, and challenges. Syst Res Behav Sci 36(5):687–705
79. Keating CB, Katina PF (2019) Enterprise governance toolset. In: Evolving toolbox for complex project management. Auerbach Publications, pp 153–181
80. Keating CB, Morin M (2001) An approach for systems analysis of patient care operations. J Nurs Adm 31(7/8):355–363
81. Keating CB, Katina PF, Bradley JM (2014) Complex system governance: concept, challenges, and emerging research. Int J Syst Syst Eng 5(3):263–288
82. Keating CB, Katina PF, Bradley JM (2015) Challenges for developing complex system governance. In: IIE annual conference. Proceedings. Institute of Industrial and Systems Engineers (IISE), p 2943
83. Keating CB, Katina PF, Bradley JM (2016) Complex system governance: failure mode effects and criticality analysis application. In: Presented at the proceedings of the 2016 industrial and systems engineering research conference, ISERC, Anaheim, CA
84. Keating CB, Calida BY, Jaradat R, Katina PF (2016) Systems thinking. In: Farr JV, Gandhi SJ, Merino DN (eds) The engineering management handbook, 2nd edn. ASEM Press, Huntsville, AL, pp 281–316
85. Keating CB, Bradley JM, Katina PF (2016) Systemic analysis of complex system governance for acquisition. In: Proceedings of the international annual conference of the American Society for Engineering Management. Huntsville, (2016)
86. Keating CB, Bradley JM, Katina PF, Jaradat REM (2017) A systems theoretic-based framework to discover pathologies in acquisition system governance
87. Keating CB, Katina PF, Gheorghe AV, Jaradat R (2017) Complex system governance: advancing prospects for system of systems engineering applications
88. Keating C, Rogers R, Unal R, Dryer D, Sousa-Poza A, Safford R, Rabadi G (2003) System of systems engineering. Eng Manage J 15(3):36–45
89. Keating CB, Katina PF, Bradley JM, Jaradat RE, Hodge R (2018) Systemic intervention for complex system governance. In: INCOSE international symposium, 28(1):1534–1548
90. Keating CB, Katina PF, Jaradat RE, Bradley JM, Gheorghe AV (2017) Acquisition system development: a complex system governance perspective. In: INCOSE international symposium, vol 27, no 1, pp 811–825
91. Keating CB, Katina PF, Bradley JM, Jaradat RE, Hodge R (2018) Systemic intervention for complex system governance. In: INCOSE international symposium, vol 28, no 1, pp 1534–1548
92. Keating CB, Katina PF, Joiner KF, Bradley JM, Jaradat RM (2018) A method for identification, representation, and assessment of complex system pathologies in acquisition programs
93. Marek C, Bradley J, McGowan J, Cui Q (2021) A governance model and safety management system framework for industrial fire safety during naval ship maintenance availabilities. In: Proceedings of the eighteenth annual acquisition research symposium. Monterey, California, pp 6–25
94. Pyne JC, Keating CB, Katina PF (2016) Enhancing utility manager's capability for dealing with complex issues. Proc Water Environ Fed 2016(8):4207–4232
95. Pyne JC, Keating CB, Katina PF, Bradley JM (2018) Systemic intervention methods supporting complex system governance initiatives. Int J Syst Syst Eng 8(3):285–309
96. Rhodes RA (2007) Understanding governance: ten years on. Organ Stud 28(8):1243–1264
97. Rittel HW, Webber MM (1973) Dilemmas in a general theory of planning. Policy Sci 4(2):155–169
98. Skyttner L (2005) General systems theory: Problems, perspectives, practice. World scientific
99. Sousa-Poza A, Kovacic S, Keating C (2008) System of systems engineering: an emerging multidiscipline. Int J Syst Syst Eng 1(1/2):1–17
100. Shauger JJ (2016) Introduction to service system governance. Int J Syst Syst Eng 7(1–3):189–206

101. Tatar U (2020) Critical space infrastructure: contributions of complex system governance. Space Infrastruct Risk Resil Gov 57:60
102. The I Ching: or book of changes (1968) (Wilhelm R, trans), vol 19. Taylor & Francis
103. von Bertalanffy L (1968) General systems theory. New York
104. Walters D (2016) Leadership issues in governance of complex systems. Int J Syst Syst Eng 7(1–3):130–142
105. Walters D (2018) Leadership and foresight in complex system governance. Int J Syst Syst Eng 8(3):249–267
106. Walters D, Moorthy S, Carter B (2014) System of systems engineering and enterprise architecture: implications for governance of complex systems. Int J Syst Syst Eng 5(3):248–262
107. Willke H (2007) Smart governance: governing the global knowledge society. Campus Verlag GmbH, Frankfurt, Germany
108. Whitney K, Bradley JM, Baugh DE, Jr CWC (2015) Systems theory as a foundation for governance of complex systems. Int J Syst Syst Eng 6(1–2):15–32
109. Yesudas R, Efatmaneshnik M, Joiner K (2019) Identifying focal points in IT project governance using a synthetic and systems thinking approach. In: International conference on complex systems design & management. Springer, Cham, pp 79–92
110. Yuchnovicz DE (2016) Understanding system structural tensions to support complex system governance development methodology. Int J Syst Syst Eng 7(1–3):109–129

Foundations

Complexity

Charles W. Chesterman Jr., Charles B. Keating, and Vernon Ireland

Abstract Complexity is an essential and fundamental concept in complex systems. The most rudimentary perspective of complexity suggests a large number of entities/variables in rich interaction, not totally "knowable," subject to emergence, and dynamically changing over time. However, for complex system governance (CSG), complexity has much more profound ramifications than the rudimentary perspective. Thus, the purpose of this chapter is to explore in-depth the nature, role, and implications of complexity for CSG. Three central themes of complexity are explored. First, the many different variations of complexity are synthesized into a set of cogent themes to provide a grounded perspective to inform CSG. Second, the role that complexity holds for the emerging CSG field is explored. Additionally, insights into the themes are provided in relation to CSG. Third, a set of implications of complexity for the design, deployment, and development aspects of CSG are examined. These implications are examined considering both field development as well as practice for CSG. The chapter closes with complexity-related challenges for CSG field development along with theoretical, methodological, and practice implications.

Keywords Systems theory · Complex system governance · Complexity · Emergence

1 Introduction

The historical background associated with "what is a system" is a rich reading of philosophical writings as well as the evolution of science, engineering, and social studies. Warren Weaver in writing about science and the multiple results that have

C. W. Chesterman Jr. (✉) · C. B. Keating
Norfolk, Virginia, USA
e-mail: cches008@odu.edu

C. B. Keating
e-mail: ckeating@odu.edu

V. Ireland
Adelaide, Australia

C. B. Keating et al. (eds.), *Complex System Governance*, Topics in Safety, Risk, Reliability and Quality 40, https://doi.org/10.1007/978-3-030-93852-9_2

affected the lives of mankind, some of the limited impact and of trivial consequences and some of profound impact and benefit, then posed two questions "How can we get a view of the function that science should have in the developing future of man? How can we appreciate what science really is and, equally important, what science is not? He goes on to posit [88, p. 1]:

> Subsequent to 1900 and actually earlier, if one includes heroic pioneers such as Josiah Willard Gibbs, the physical sciences developed an attack on nature of an essentially and dramatically new kind. Rather than study problems which involved two variables or at most three or four, some imaginative minds went to the other extreme, and said: "Let us develop analytical methods which can deal with two billion variables." That is to say, the physical scientists, with the mathematicians often in the vanguard, developed powerful techniques of probability theory and of statistical mechanics to deal with what he (Gibbs) called problems of *disorganized complexity* [88, p. 1].

The rest of Weavers article discusses the "problems of simplicity," "problems of disorganized complexity," "problems of organized complexity," and finally, the "boundaries of science" where he imparts the overall lesson that while there have been great advances, it is still incumbent on mankind to understand what science really is. These early writings on what the future is capable of and that the term *complexity* is an integral part of the future, proceeding many aspects of system theory.

Complex system theory has its roots in the general systems theory that Karl Ludwig von Bertalanffy published in the 1930s. This contribution is significant as it moved studies from the classical closed model of systems in accordance with the second law of thermodynamics to open systems. This movement to an open model allowed for the study of many classes of phenomena that were inadequately described.

Systems theory supports many fields of science that study the nature of complex systems in science, nature, society, and technology. Etymologically, complexity comes from the Latin *plexus*, which means interwoven. Thus, something complex is difficult to separate. This means that its entities are interdependent, i.e., their future is partly determined by their *interactions* [28]. Thus, studying the components in isolation—as reductionist approaches attempt—is not sufficient to describe the dynamics of complex systems. Complex systems theory is helpful in establishing an understanding of observed behavior of complex systems. Associated with complex systems are several observations and mechanisms associated with complex systems. Table 1 describes the observations and mechanisms in complexity.

Herbert Simon, following Gibbs and Weaver, pioneered much of the literature on complexity and social action and noted: "Everything is connected. Some things more than others." [81, p. 7]. This work was followed by Robert Jervis's landmark study on systems effects demonstrating the wide range of unanticipated and unintended consequences and often perverse, localized effects of action filtered through systemic level variables [46]. It is now understood that the challenges with many of today's problems are that they are complex across multiple domains, such as time horizons that vary both within and across the various issues and that many attributes associated with an entity can interact in a nonlinear manner as well as some of the interactions that are not anticipated [44]. The world is full of issues interacting with other problems

Table 1 Observations and mechanisms (adapted from Prokopenko [74] and Fernandez et al. [27])

Observation and mechanism	Description
Equilibrium	Complex systems can have periods of equilibrium, order, and stability that appear to be an established behavior, and then there can be short periods of instability and disorder
Open	Complex systems are open systems, meaning that they actively interact with their environment (receive a regular supply of energy, information, and/or matter) and have the observed behavior of constantly evolving and changing over time
Entity (component//agents)	Complex systems can be described as having entities that are not necessarily the same and as the system is what it does, means that there are interactions between the entity or components. The number of them is large, but not too large
Interactions	The scope of the interactions is best described as nonlinear, meaning that the interaction between two similar entities is not equal, such as in one instance the reaction or stimulus experienced by the receiving entity has a significant effect and, at other times, no effect or a minimal effect. This nonlinear interaction is also true for the reverse stimulus between the two entities The ensemble of individual entities interact in a nontrivial fashion and as such studying the system via statistical mechanics would miss important properties brought about by interactions The nontrivial interactions result in internal constraints, leading to symmetry breaking down in the behavior of the individual entities, from which *coordinated* global behavior arises
Feedback	Complex systems also have the mechanism of feedback, and the signal contained in this feedback loop can be described as either negative (damping) and positive (amplifying)
State history/Path dependence	Complex systems have a history where the prior state can have an influence on present states, and this can influence future states. This is path dependence, whereby current and future states can depend on the path of previous states
Levels of organization structure	Complex systems are nested, and the organization of the entities may be at various levels The system is now more organized than it was before because neither central director nor any explicit instruction template was followed. It is said that the system has "*self-organized*." Self-organization coordination can express itself as *patterns* detectable by an external observer or as structures that convey new properties to the systems itself. New behaviors "*emerge*" from the system

(continued)

Table 1 (continued)

Observation and mechanism	Description
Emergence/Adaptation	The state of a complex system is driven by the varied interactions of the entities, and as the level structures of the entities will change, the interactions will change The properties of a system are emergent if they are not present in their entities. In other words, global properties which are produced by local interactions are emergent Coordination and emergent properties may arise from specific response to environmental pressure; the system displays *adaptation*; adaptation occurs across generations at a population level; the system *evolved* Coordinated emergent properties give rise to effects at a *scale* larger than the individual entities. These interdependent sets of entities with emergent properties can be observed as coherent entities at lower *resolution* than is needed to observe the entity. The system can be identified as a novel unit of its own and can interact with other systems/processes expressing themselves at the same scale. This becomes a building block for new iterations, and the cycle can repeat itself
Attractors	An attractor is a preferred steady system's state set, to which a complex system evolves after a long enough time

creating a system of problems—or, similar to what Russel Ackoff simply called "messes" [1].

Complexity became widely recognized by the 1960s where cybernetics and open systems theory helped focus attention on intertwined problems and the complex arrays of social forces by which they are recognized and addressed [22]. Social science research into the structure and fabric of their field of interest recognized that theirs was a study of complex systems that are composed of a very large number of entities where the component to component interaction is iterative and recursive (i.e., nonlinear). These interactions can be described as direct and indirect feedback loops. The system or systems present the viewer with an unintentional emergent form of order due to the internal self-organizing operation and that, for the most part, is dynamic and adaptive. Importantly, the holistic perspective was recognized that the whole is not only more than the sum of its parts but most interesting that the whole was different from the sum of the parts.

Some of the most influential developments in systems thinking about complexity (emergence and self-organization) in open systems came [66, p. 11]:

> …from the physical and life sciences: Prigogine's work on dissipative structures in chemical systems along with Eigen's hypercycles and Haken's articulation of Synergistics, Maturana and Varela's concept of autopoiesis in living systems, and the articulation of evolutionary dynamics in artificial life and ecosystems. All of these highlight that self-organization is not the result of a priori design, it surfaces from the interaction of system and the environment and the local interactions between the system's components. This capacity for the spontaneous creation of order through intrinsically generated structures is captured in Stuart Kauffman's [50] expression 'order for free', in the notion of Prigogine's dissipative structures (Prigogine,

1967), Haken's Synergistics [36], Eigen's hypercycles [24] and in Maturana and Varela's theory of autopoiesis [85].

Concerning social systems and complex systems from the writings of Hayek [40, 41], Moroni points out:

> The fact that a certain structure has intrinsic complexity means that its characteristics do not depend solely on the properties of the individual entities of which it is composed – nor on the relative frequency with which these properties appear – but above all on the way in which these discrete elements interrelate with each other, creating emergent patterns. [69, pp. 250–251].

Additionally, on complex systems and the structure of their composition, it is important to note that:

> A complex system is a structure of phenomena, not a mass of phenomena: its very characteristics are those of a 'general order' whose specific elements are perpetually changing. Self-organization is certainly one of the most characteristic features of a complex system: this kind of system spontaneously seeks out some form of order, with articulated structures being created randomly; no one deliberately imposes order on its numerous entities – the system spontaneously exhibits synchrony. The crucial point is that when we are dealing with structures of essential complexity (such as social phenomena), it is not feasible to provide explanations of detail, but only an explanation of the principle." [69, pp. 250–251].

Complexity has also been described as "the array of concepts, methods and intuitions that emerged piecemeal from an engagement with specific nonlinear, adaptive, emergent and/or self-organizing phenomena/problems that revealed the limitations of existing scientific approaches." [18, pp. 1–2]. This description directs the focus of the defining character of a complex system directly to complexity.

Complexity science continues to evolve and provide the theoretical methodological framework to study problems that were not amenable to classic scientific methods as discussed by Weaver. Social science literature [67, 87] extensive work on the potential contribution toward the understanding of the micro–macro link (i.e., the relationship between different levels of analysis) where the use of agent-based modeling has been advanced to take advantage of both semantic and syntactic flexibility that computer languages as well as computer processing advances to provide an insight into micro–macro transition [31] is an example of advancement in methods of investigation.

Currently, the term complex adaptive systems (CASs) have emerged as a special area of study of complex systems [76]. CAS was initially described at the interdisciplinary Santa Fe Institute (SFI), by John H. Holland, Murray Gell-Mann, and others. This loosely organized area of study has more than one theoretical framework. Complex adaptive systems study is focused on the properties and features like self-similarity, complexity, emergence and self-organization, adaptation (or homeostasis), communication, cooperation, specialization, spatial and temporal organization, and reproduction. These systems are adaptive as they express a capability and capacity to change and as they learn from experience. This can be called a survival or flexibility capability enabling a priori to respond to and or adjust themselves to changes in the environment. A complex adaptive system is most interesting as it is constantly adapting nonlinear relationships.

Accordingly, complexity is not difficult to grasp and is used in everyday language to describe a host of situations, conditions, and systems. With this usage, the reader has surely been confronted with the reality that there is no single "universal" definition that is agreed to, applicable, or can be assumed to be above all others. Instead, the reader hopefully will accept that definitions or perspectives encountered are the ones that have been developed by experience and as such are neither correct nor incorrect. Rather the developed definitions can be most useful for a particular context or circumstances and that as such, it meets the needs of those who use them. However, from these developed definitions or perspectives, there are some common characteristics that can help develop a broad and rich understanding of the complexity and assist in framing a "useful perspective" of complexity. Therefore, a set of consolidated features of complexity and complex systems includes the following:

1. Large number of entities
2. Dynamically interconnected
3. Rich interactions and mutual influence among entities
4. Causality and feedback
5. Open systems, subject to transport of energy, information, or material across boundaries
6. Contextually embedded
7. Operate in conditions far from equilibrium
8. Have a "historical" context as a reference context
9. Individual entities operate without necessarily understanding the behavior of the whole system (of which they are a part).

The following section places emphasis on self-organization, emergence, nonlinearity, and adaptive before examining the implications for CSG in the next section.

2 Variations of Complexity

First, the many different complexity variations are synthesized into a set of compelling themes to provide a grounded perspective to inform CSG.

Self-Organization:

One key concept within complexity science is that of self-organization. Self-organization is used to refer to the emergence of stable patterns through autonomous and self-reinforcing dynamics at the micro-level [50, 78]. Simple studies such as [79] model of segregation at the macro-level self-organization appear as an unintended consequence of actions taken by entities. Axelrod's [10] research on an iterated prisoner's dilemma shows that cooperation is developed as an adaptive strategy because the traditional equilibrium of a one-off game works differently for an iterated version. This work is important to understand the dynamics of social science where there is a connection between individual action and the observed results at a particular population level. In previous work using traditional research methods, the connections between the entities were either unknown or improperly described.

Self-organization has been more robustly articulated since its inclusion within the complexity framework. Importantly, self-organization is not a subordinate concept to complexity, rather it should be acknowledged as in existence and viewed as one of cross-fertilization [6]. Most current linkage to self-organization refers to Ashby's [7] work in cybernetics as the contemporary precursor. Although there are other authors in the natural, biological, and social sciences focused on self-organizing dynamics, Ashby is acknowledged as the first one using explicitly the concept "self-organization" in a somewhat similar manner to contemporary accounts [6] primarily due to the important character of self-organizing in complex systems.

While there is not an agreed upon definition of self-organization, [32] suggest that there are four factors that are common across definitions: pattern formation, autonomy, robustness and resilience, and dynamics.

- Pattern formation is directly related to the observed process of self-organization, that there are observable types of self-organization. According to Gilbert et al. these patterns, within the context of social science are "patterns of interest are usually designated by nominalized verbs (e.g., cooperation, segregation, stratification, normalization, etc.)" [32, p. 529]. Observational criteria such as the geometric distribution may work for simple systems. Complicated or complex systems need criteria that are more robust where the definition is consistent and well understood and does not need further qualification. As an example, self-organization from the literature has been explained as

> an increase in coordination, a reduction in entropy, a movement from simple to complex structures, or a decrease in degrees of freedom. It is not yet clear to what extent these criteria can be translated from their original domains to social science and if they can provide any insightful information about traditionally studied social processes, such as normalization, stratification, specialization, etc. [32, p. 529].

There still remains the issue of whether a change in self-organization can be described and is this observable for social science. Gilbert et al. point out:

> Measurement in many cases relies on qualitative or stylized perceptions of the system. The observer's subjective point of view and interpretation is important. Comparative studies have shown, for example, that traditional societies have subsystems with high levels of organization and complexity, and, yet, because of the past observers' biases and prejudices, for a long time these societies were understood to be simple and wrongly labeled 'primitive. [32, p. 529]

- Autonomy deals with the controlling force or mechanism that is behind the process. Autonomy covers the "self" in self-organization as change in a self-organized system must come from inside the system. It must be originated by any of the entities, or an interaction between two or more entities, or by an internal entity reacting to an external event. Importantly, the external event is not a controlling event, as controlling events are only internal to a system. According to Gilbert et al. [32]

> Social systems pose two particular challenges for this feature of self-organizing systems. First, there are asymmetric power relations between and within levels; second, the boundaries of social systems and subsystems are difficult to draw. Regarding the former, humans are intentional actors and can have designs on controlling or influencing a system through individual or institutional action, for example, by the implementation of public policy. Regarding the latter, setting the system and subsystem boundaries is crucial. The outside and inside of the system need to be clearly identifiable in order to make sure any controlling force is not ignored via undue manipulation of the boundary of the system. [32, p. 529].

- Robustness and resilience offer that self-organizing systems display a level of stability over time and space that makes their identification possible. Robustness refers to a system's ability to resist change. Resilience implies that there is change, but that the system has internally the inherent ability to accommodate the change. Again, for the social sciences, [32] suggest that robustness and resilience require subjective criteria so as to accommodate observed intervening factors. They provide the example of a political system that experiences a coup where robustness and resilience may be used to describe the political system's ability to accommodate potential civil unrest or the terms associated with the perpetrators of the coup and their ability to replace the prior government and there is no change to the normative framework. Finally, it must be understood that robustness and resilience do not need to match. As in the example of the coup, if it is successful then the political system did lack robustness. If, after the coup, if the political system remains, then it was resilient to a disruptive change in power/leadership.
- Dynamics refers to the properties or structure of systems that appear to be constant over time. As such, there are going to be changes and the changes are reflected "by the non-static character of the constitutive elements and relations" [32, p. 529] and can be attributed to the results of path-dependent processes as the future states are directly dependent upon the historical states. Accordingly, the concept of dynamics is dependent on the conceptualization of self-organization, and the analysis of the system considers the combined variability rather than the individual states.

It hopefully is clear that self-organization is a process that occurs with a complex system. Under this continuous process, overall order forms both from the interactions between the entities of a complex system as well as interactions that the entities have with the environment, and critically, it is this process that brings order, where before there was disorder, to the complex system. Thus, self-organization reflects the ability of the complex system to develop a new system structure as a result of the system's internal structure and critically not the result of external management [72]. The concept of organization is related to an increase in the structure or order of the system's behavior.

The concepts of self-organization and emergence while related to each other are different and do occur together. Self-organizing systems usually display emergence, but that is not always the case. Self-organization exists without emergence, and emergence exists without self-organization. As a complex system will have many types of adaptation or change, this new organization or variation of a prior organization is internal to the system and the system's entities. This creation of variety either in creation of new entities or new relationships, either internal or external, is one of the cardinal attributes of complex systems in terms of creating new entities and relations.

Emergence:

Historically, the use of emergent and hence emergence was coined in 1875 by the philosopher Lewes when discussing the changing nature of causality:

> … although each effect is the resultant of its components, we cannot always trace the steps of the process, so as to see in the product the mode of operation of each factor. In the latter case, I propose to call the effect emergent. It arises out of the combined agencies, but in a form which does not display the agents in action [Lewes 59, pp. 368–369].

The common understanding of the term emergence was proposed as a supplement or a substitution to the mechanistic and incrementalist view proposed in Darwin's theory of evolution [33]. It held that emergence enabled the viewer to switch between mechanism and vitalism. With the emergence of complexity theory, the construct of emergence was suggested as an alternative way that systems change without the imposition from a command or control hierarchy. This helps explain how innovations in a system functioning at the macro-level are out of connectivity at the micro-level. As these innovations are not the result of imposition, it is believed that they will exhibit the charter of creative solutions, and thus, there is an increased sense of ownership that can be associated with empowerment. Emergence, for many, is a crucial property and is the core of complex systems [71, 84].

Goldstein has synthesized six prototypes of emergent phenomena based upon a review of complex systems research of emergence that are presented in Table 2 below.

Although emergent phenomena cover a wide range of diverse disciplines and various typologies have been suggested for them, common characteristics include radically novel macro-level entities and properties with respect to a micro-level substrate: ostensibly in the sense of unpredictability and non-deducibility; integrated coordination characterizing the macro-level; and dynamical in the sense of coming to be over time [33]. The Nobel laureate Laughlin [57] holds that society is now leaving the Age of Reductionism and entering the Age of Emergence.

One of the dominant theoretical underpinnings shared by many of these researchers employing the idea of emergence is a "dissipative structures" model derived predominantly from the approaches found in the first three prototypes of emergent phenomena listed above, particularly that of number two, namely, the work on self-organizing physical systems pioneered by Prigogine and Haken and, to a lesser extent, the nonlinear dynamical systems prototype of number one, the latter the main perspective, for example, taken by Guastello [35].

As emergence is associated with the occurrence of unique and complete structures, patterns, and properties during the process of self-organization in complex systems

Table 2 Emergence prototypical conceptualizations (adapted from Goldstein, [33], p. 2)

Prototypical conceptualizations	Author	Characteristic theoretical underpinnings
Phase transitions, "quantum protectorates" and similar critical phenomena in condensed matter physics	Anderson [5], Laughlin [57], Batterman [11]	Symmetry-breaking; order parameters: renormalization group: universality; and criticization
Self-organizing physical systems	Haken [37], Allen and McGlade [4], Nicolis and Prigogine [70], Haken [38]	(Dissipative structures: far-from-equilibrium conditions; order parameters; "enslaved" variables; and self-organization
Mathematical emergence in dynamical systems	May [64], Cohen and Stewart [83], Martelli [61], Scott (2007)	Nonlinearity; phase space; bifurcations; attractors; and chaos
Computational emergence	Langton [55, 56], Adami [2], Crutchfield [20, 21], Holland [42, 43], Marinaro and Tagiaferri [60], Griffeath and Moore [34]	Artificial life; neural nets; Game of Life; and computational mechanics
Social emergence including virtual social networks of the Internet	Addis [3], Johnson [47], [77]	Collectivity; social networks; artificial societies; and cooperation
Biological emergence	Goodwin and Sole [82], Reid [75], McShea [65]	New speciation; morphogenesis; symbiogenesis; and hierarchical constructions

[33], this property is reflected in the autonomous properties at a higher (macro) level that cannot be understood by reduction to lower (micro) levels [77]. This is also applicable to the emergent properties of a group of entities having varying properties and showing deviant behavior at a higher structure level than the individual entities at a lower structural level.

Non-linearity:

Ireland points out that many nonlinear systems, "when studied mathematically, show another odd behavior in their otherwise chaotic conduct. Certain states may never occur even though the system seems to move unpredictably through space. And even more surprising: possible states intermingle with not possible ones at any scale" [45, p. 9]. It also has been observed "that complex systems may stabilize themselves in fairly stable patterns as attracted by an odd force" [15, p. 5].

This follows in line with the contribution to complexity thinking provided by Ross Ashby in 1956 in which he named *requisite variety*, in which he stated that the variety of an organization (system) must match the variety of the environment if the organization (system) is to exert a degree of control over the environment. With respect to organizations, Boisit and McKelvey following Ashby propose that organizations/systems are always in one of the following states; stable equilibrium, partial equilibrium, or chaos [17 p. 61].

When in a chaotic state, the impact of change has an unpredictable long-term effect.

Therefore, within the context of Ashby's law:

> Increasing tension often increases the level and strength of connectivity between the two unconnected phenomena, thus transforming what could ordinarily appear to be production of regularities into scale-free ones. Tiny initiating events (TIES) will then propagate more rapidly and easily through a system, getting amplified in the process to produce a unified, linear, and possibly extreme, outcome. To illustrate: imagine a fishing net crumpled in a pile. The net between the two nodes and the rest of the net will remain undisturbed and the effects of the cut will remain strictly local. Now place the net under tension by stretching it taught. If the net is taught enough then a simple cut would initiated a tear that would instantaneously spread from one end of the net to the other." [17, p. 290]

Complexity Vignette—Electrical Blackouts

A simple dynamic underlying the power blackouts that occasionally occurred in a fully-fledged New England power grid is an excellent example of Non-Linearity and Complexity.

The scenario is that utility companies may temporarily shut down one overloaded station to reduce the load on the entire grid. The result can be a cascade of further blackouts throughout the Northeastern US. Given tension plus connectivity, then, what starts off as a Tiny Initiating Event (TIE), can rapidly propagate throughout the network, growing in severity as it does so, resulting in an extreme outcome.

So nonlinearity of complex systems implies that the effect resulting from a cause will not be proportional to the stimulus. This effect is critical to understanding the behavior of such systems. While complex systems exhibit the imprint of initial conditions, the states that evolve because of a perturbation may not be as predicted. With feedback, the effects of a perturbation without regard to its size, either major or minor, may not be proportional as an entity may absorb a major perturbation and have no effect. Hence, moving beyond a reductionist perspective is critical to understanding complex systems where the observed nonlinear behaviors reflect outputs that are not proportional to input.

The nonlinear equations that describe the system's dynamics have a number of possible solutions, and the path that the system takes will depend on the system's history and the prevailing environmental conditions at that precise moment. As the system is in a constant state of flux, the combination of system state and environmental conditions is unique for each dissipative structure, and this means that over the longer term it is impossible to predict what the next system state will be. It is this impossibility of prediction that distinguishes complex adaptive systems from chaotic systems. The term "chaos" has been popularized in the managerial literature on dynamism, innovation, and creativity, and is often used to refer to a state of disorder and randomness out of which arises a new order. However, technically, a chaotic system is a deterministic system that has parts of its trajectory that are not stable, so that its future is very sensitive to its precise path and current state. In practice, the degree of accuracy (of measurement of start conditions) needed in order to predict an outcome is likely to be impossible to obtain. Chaotic systems share properties with complex systems, including their sensitivity to initial conditions. However, in the study of chaotic systems, the systems' dynamics are generally described by a small number of variables interacting in a non-linear fashion, while complex systems have many degrees of freedom.

Adaptive:

The adaptive nature associated with complex systems is the ability to adapt and evolve in the interaction with the environment, and, by the nature of the environment, it is always in a dynamic state. Adaptation occurs at the macro level (the "whole" system), and this can be characterized by emergence and self-organization based on the local adaptive behavior of the system's entities. Holland points to simplicity arising from the aggregated behavior of interdependent adaptive agents driven by a set of rules [42]. Agents/entities following rules adapt to each other and create an emergent order. A given set of rules governing interaction, a specified number of interacting agents, and random events seem to uncover a "hidden" process. Once in motion, the process follows a route toward a stable end, periodicity, or even apparent randomness.

Ever since Conway's Game of Life (1976), successful and more comprehensive computer applications have developed in the field of organization studies including the study of emergent social behavior [25], organizational adaptation [19], culture formation [39], and stock market evolution [58]. In these models, emergent system level behaviors are achieved through the interaction of agents. Stable patterns are not just the outcome of random encounters. Internal dynamics and random events coalesce to produce different orders by following multiple paths. Questions thus arise concerning the role of causation and the nature of the order that emerges, and in which chance plays a key role. The nature of the interaction itself, however, is deterministic; rules are fixed by choice or by nature. Once chance has opened up opportunities for viable alternative combinations, the system evolves toward an un-programmed, emergent order.

Ireland in reviewing complexity and specifically the adaptive character of complex systems noted that [17] viewed that adaptive character is tension defined as an "edge of order" which is a way of exploiting the environment. This does not apply to all complex systems representing organizations as they can lose their capacity to adapt and die. There is an analogy with respect to the edge of order that was made by [17] with respect to heat that is being applied to a container of water, in which energy exchange shifts from conduction to convection. This shift significantly reduces the imposed energy differential. Thus, for adaptive character, complex adaptive systems may become increasingly efficient and exploitative of their environment.

Ireland in writing about complex systems also noted that complex systems as the entities/agents interact with each other and adapt themselves to other agents and changing conditions was described by Rotmans and Loorbac described complex adaptive systems having "unique features, such as co-evolution, emergence, and self-organization" [76, p. 186]. Co-evolution is an interesting property between two non-interbreeding populations that share the same ecological environment such as entities in a complex system. The property can cause a series of reciprocal changes in two or more non-interbreeding populations. Within the context of complex systems, co-evolution helps to describe the change of the entities with respect to each other, "leading to irreversible patterns of change within each of the systems" [76, p. 186]. This single direction or irreversibility means that the "complex system evolves with its environment or that there are interdependencies and positive feedbacks between the complex system and its environment" [76, p. 186].

Homeostasis:

The concept of homeostasis was developed to describe internal and physiological regulation of "bodily functions." The concept of regulation of the human internal environment was described by French physiologist Claude Bernard in 1849, and the word *homeostasis* was coined by Walter Bradford Cannon in 1926 in his work on *Physiological regulation of normal states: some tentative postulates concerning biological homeostatics"* where homeostasis from the Greek h´omoios (similar) and stasis (standing still). The use of the term homeostasis describes the observed natural resistance to change when already in the optimal conditions and equilibrium is maintained by many regulatory mechanisms. The term does not imply that the system is immobile or a stagnant state, rather there are mechanisms that are tending toward equilibrium.

The British cybernetician William R. Ashby proposed the use of the term homeostasis within a systems theory perspective, which implies an adaptive reaction to maintain "essential variables" within a range [8, 9]. Cybernetically control systems such as thermostats function as homeostatic mechanisms. Ashby also contributed by linking the concepts of ultrastability and homeostatic adaptation where ultrastability refers to the normal operation of the system within a "viability zone" to deal with environmental changes. The viability zone is defined by the lower and upper bounds of the essential variables. If for example the value of variables crosses the limits of its viability zone, the system has a chance of finding new parameters that make the challenged variables return to their viability zone.

Homeostasis can be seen as reflecting the dynamic process associated with self-regulation and adaptation. Ashby indicated that a dynamic system has a high capacity for stability due to the homeostatic capacity to maintain its dynamics close to a certain state or states (attractors). This construct implies that when perturbations or environmental changes occur, the system adapts to face the changes within the viability zone, that is, without the system "breaking" [8].

Autopoiesis:

Autopoiesis comes from the Greek *auto* (self) and *poiesis* (creation, production) and was originally as a system description to define and explain the nature of living systems. A canonical example of an autopoietic system is the biological cell. The cell is made of various biochemical entities and has defined structures such as the nucleus, organelles, a cell membrane, and cytoskeleton. These structures, based on an internal flow of molecules and energy, *produce* the entities which, in turn, continue to maintain the organized bounded structure that gives rise to these entities [85, 62].

[54] building upon the work of Maturana and Varela described that a autopoietic system is to be contrasted with an allopoietic system such as a factory that uses recycled or new materials (components) to fabricate and assemble a car (an organized structure) which is something other than itself (the factory). However, if the boundary of the allopoietic system is extended from the factory to include components in the factory's environment such as supply chains from raw material through the lifecycle of the car, then as a total viable system it could be considered to be autopoietic.

Maturana noticed that living beings, as dynamic systems, were in continuous change where interactions between entities of an autopoietic system regulate the production and regeneration of the system's entities, having the potential to develop, preserve, and produce their own organization [86]. This leads to what can also be described as autopoiesis and is closely related to autonomy. Autonomy is always limited in open systems, as their states depend on environmental interactions. However, differences in autonomy can be clearly identified, just like in the previous example.

Fernandez describes autopoiesis as the ratio between the complexity of systems and the complexity of their environment [26]. In this generalized view of autopoiesis, systems are self-producing but not in terms of their physical entity, but in terms of their organization (self-organization), which can be measured in terms of information and complexity. In other words, an autopoietic system is producing more of its own complexity internally than a non-autopoietic system can produce by its environment [29].

With this background in the character of complexity and the set of cogent themes as a grounded perspective, the role of complexity as well as the implications of complexity in CSG will now be covered in the next two sections of this chapter.

3 Role of Complexity in CSG

In this section, the role that complexity plays for the emerging CSG field is explored. This exploration will examine the themes of complexity and how they inform a robust and important underpinning perspective of CSG. Additionally, insights of the themes are provided in relation to CSG, focusing on the implications that complexity holds for the emerging field and continuing practice implications. But firstly, there is a necessary exploration of CSG, so that practitioners can have a better foundation of the CSG field prior to examination of complexity implications for the field.

The essence of CSG is found in the current definition of "Design, execution, and evolution of the [nine] metasystem functions necessary to provide control, communication, coordination, and integration of a complex system." [51, p. 3]. As CSG continues to evolve, there are several essential distinctions (Table 3) that serve to distinguish CSG from existing systems-based approaches. Arguably, many proponents of other systems-based approaches may claim that several of these delineations are in fact embedded in the approaches, either tacitly assumed or superficially expounded.

There are several points of emphasis for this depiction of CSG. First, "design accentuates the purposeful and proactive engagement in the creation of the governance system." [52, p. 2]. While it should appear that the idea of design could be taken for granted, from a truly purposeful, holistic, and comprehensive perspective, design of governing systems represents the exceptional case rather than the norm. Based

Table 3 CSG distinctions (adapted from Keating and Katina [52])

Distinction	Implications
1. Deep and explicit grounding in the underlying systems theory upon which it has been developed	Systems theory is an essential aspect of CSG. The field has made the explicit grounding in systems theory detailed, rigorous, and explicit. This is in contrast to the "bulk" of the systems based approaches that either assume this grounding or provide a superficial treatment
2. Qualification and acceptance that the level of "systems thinking capacity" held by individuals/organizations are critical to proper deployment of CSG	CSG makes deployment contingent on the state of systems thinking capacity held by participants and the aggregate. This is distinct from other systems based approaches and seeks to (1) ensure that participants are sufficiently sophisticated in systemic thinking levels necessary to engage the CSG methodology, and (2) ensure that the participants do not attempt to engage in activities in support of CSG development for which they lack systemic thinking sophistication to have a likelihood of success
3. Permitting the tailoring of the approach and the tempering of expectations based on the unique context, system in focus, implementing entity, and support infrastructure for deployment	CSG does not assume a standard "fit" to a generalized approach. On the contrary, CSG is explicit in adjusting to the uniqueness of the specific context and system of interest for which CSG is being deployed. The constraining and enabling aspects of both the problem system and associated context are critical to deployment of CSG
4. Holistic system examination across the spectrum of technical, organizational, managerial, human, social, policy, and political dimensions of complex systems and problems	CSG resists the temptation to narrowly bound a complex system/problem such that assumptions necessary for reductionist approaches can be fulfilled. Instead, CSG emphasizes the necessity to consider a system/problem in its entirety. This includes consideration of the enabling and constraining aspects that cross beyond technology, to include the human/social, organizational/managerial, and political/policy dimensions of a system/problem. Thus, instead of focusing on the linear, part, property aspects of a system, CSG defers to the nonlinear, whole, relationship nature of complex systems

(continued)

Table 3 (continued)

Distinction	Implications
5. Emphasis on the discovery, classification, and engagement of "deep system issues" (pathologies) that limit system performance	CSG operates beyond the surface, objective, and observable aspects of system problem manifestations. Instead, CSG examines problems as outputs of an underlying system that awaits discovery to unfold the "deep system" sources responsible for the production of undesirable performance or behavior. Deep system issues are evidenced as superficial outputs. These issues are rooted in violations of underlying systems theory propositions and termed "pathologies"
6. Purposeful system development that prioritizes and simultaneously targets individual, system, organizational, and support infrastructure for improvement	CSG focuses on holistic development of a complex system. This development extends beyond the strict limits of the system of interest. Included in CSG development are those aspects/entities that act to enable or constrain the system of interest. Among these, external aspects are included individual participants, higher level systems that depend, or are dependent upon the system of interest, the larger organization, and the support infrastructure influencing the system of interest. CSG system development must appreciate the totality of these different aspects to facilitate holistic systems-based development
7. Focus on functions already being performed by all systems and the pathologies being experienced in the design, execution, or development of those functions	The language and approach to development of CSG may be new and novel. However, all complex systems that remain viable already perform mechanisms that serve to achieve CSG functions at a level that supports viability (continued existence). Therefore, although the language, functions, and systems perspective may be new, the actual performance of the functions already exists. The focus of CSG is to rigorously examine the performance of the functions to identify pathologies. This sets the stage to target system development to the greatest system developmental needs, consistent with the feasibility to successfully address the identified pathologies

on issues propagating all manner and form of our "manmade" complex systems, the anecdotal evidence suggests that our systems are not sufficiently serving the needs or expectations intended to enhance societal well-being. In addition, irrespective of purposeful/purposeless design, execution embodies the notion that a design without effective deployment offers little more than good intention. Execution is where a design meets the harsh realities of the "real world," which is fraught with complexity and emergent conditions that test the most thoughtful system designs. Design can also be flawed because as the design is being applied to a model of the real-world it is, in truth, being applied to an abstraction of the real-world. This acknowledges that complexity can be neither totally captured nor completely understood. The third leg of CSG, evolution, recognizes that systems, as well as their environments, are in constant flux. Therefore, governance must also be able to flex (evolve) in response to internal and external changes impacting the system over time. Evolution by its very nature suggests that the developmental emphasis is on long-term sustainability, irrespective of the need to operate a system in real time. In effect, governance must be capable of absorbing, processing, and responding to external turbulence and internal system flux. This can ensure the system remains viable (continues to exist) in both the short-term operational sense that delineates current system existence and the long-term evolutionary sense that positions the system for the future.

The second aspect of the CSG definition lies in the articulation of the metasystem as the set of nine interrelated functions that produce governance for a complex system. These functions find their basis in, and offer an extension of, Beer's metasystem concept in the viable systems model [12–14] as well as three additional communication channels following the work of [53].

The metasystem for CSG is the set of nine interrelated functions that act to provide governance for a complex system; see Table 4.

In effect, these metasystem functions provide for the absorption of variety (a measure of complexity) generated both externally and internally to the system of interest. The nature of complexity for CSG functions includes several important points, including: (1) the metasystem functions are performed by "mechanisms" (vehicles that implement the functions), (2) the mechanisms performing functions "absorb" variety for the system and thus are the means of maintain viability (continued existence) for the system through the design and execution of function supporting mechanisms, (3) the degree to which the mechanisms effectively dispose of variety determine the state of the system and are outwardly observed as "performance," (4) effectiveness of variety absorption is evidenced by the absence or presence of pathologies in a complex system—meaning that pathologies are evidence of excessive complexity (variety) not being absorbed by the system mechanisms in support of CSG functions, (5) system development of CSG is focused on dealing with unabsorbed variety (complexity)—accomplished through the purposeful modification of complex system design and execution. Thus, the relationships between mechanism and functions of CSG, and their purposeful development, delineate the specific relationship of complexity to CSG.

A third primary aspect of the metasystem construct is found in the communication channels that provide for the flow of information between system entities as they perform functions. These channels support the flow of information for decision and action as well as produce consistency in interpretation for exchanges within the metasystem and between the metasystem and external entities. With respect to complexity, the communication channels play a critical role in the absorption of variety (complexity) in the system. Communication channel design, operation,

Table 4 CSG metasystems (adapted from Keating and Katina [52])

Metasystem Five (M5)	Policy and identity	Focused on overall steering and trajectory for the system and maintains identity and balance between current and future focus
Metasystem Five Star (M5*)	System context	Focused on the specific context within which the metasystem is embedded. Context is the set of circumstances, factors, conditions, or patterns that enable or constrain the execution of the system
Metasystem Five Prime (M5')	Strategic system monitoring	Focused on oversight of the system performance indicators at a strategic level, identifying performance that exceeds or fails to meet established expectations
Metasystem Four (M4)	System development	Maintains the models of the current and future system, concentrating on the long-range development of the system to ensure future viability
Metasystem Four Star (M4*)	Learning and transformation	Focused on the facilitation of learning based on correction of design errors in the metasystem functions and planning for the transformation of the metasystem
Metasystem Four Prime (M4')	Environmental scanning	Designs, deploys, and monitors sensing of the environment for trends, patterns, or events with implications for both present and future system viability
Metasystem Three (M3)	System operations	Focused on the day to day execution of the metasystem to ensure that the overall system maintains established performance levels
Metasystem Three Star (M3*)	Operational performance	Monitors system performance to identify and assess aberrant conditions, exceeded thresholds, or anomalies

(continued)

Table 4 (continued)

Metasystem Five (M5)	Policy and identity	Focused on overall steering and trajectory for the system and maintains identity and balance between current and future focus
Metasystem Two (M2)	Information and communications	Designs, establishes, and maintains the flow of information and consistent interpretation of exchanges (communication channels) necessary to execute metasystem functions

maintenance, and development are essential to maintenance of viability through the disposition of variety generated internally and externally to the system. The ten communication channels are adapted from the work of [12–14] and extensions of [53]. A short summary of these communication channels is provided in Table 5.

From Table 5, we recognize the critical role played by complexity in CSG. Complexity is evidenced as unabsorbed variety in a system. The design and execution of system functions, achieved through mechanisms, act to produce a level of unabsorbed variety. This unabsorbed variety (complexity) must be absorbed to a degree necessary to maintain system viability. CSG mechanisms, functions, and communications channels act to absorb and match the variety generated from internal flux and external turbulence. The unabsorbed variety produces a level of system pathologies evidenced as performance issues in system design or execution. The pathologies represent violations of underlying system theory propositions that are accumulating as unabsorbed variety. To deal with this unabsorbed variety, CSG is focused on the purposeful development of functions and mechanisms to adjust system design and execution and better match variety (complexity).

The final part of the definition of CSG is focused on the elements of control, communication, coordination, and integration as determinants of system performance. These terms and their basis emanate from management cybernetics (communication and control) and systems theory (coordination and integration). With respect to complexity, these elements must be designed, executed, and developed such that they permit system variety to be absorbed (matched) to maintain system viability. Here are the extended perspectives for each of these elements as they relate to inform CSG:

- ***Control—constraints that provide regulation necessary to ensure consistent performance and future system trajectory***. In the formulation of control, it is important to look to a more informed system view for guidance. This view suggests that control is not a pejorative term, to be scorned as a form of domination over a particular venue, activity, or entity. On the contrary, from a CSG systems view, it is suggested that control is essential to ensure that the system stays on a trajectory that will provide future viability in response to changing conditions and circumstances.

Table 5 Communication channels in the VSM

Communication channel	Primary functions	Complexity perspective
Command	• Provides direction to operational units • Dissemination of non-negotiable direction to the system	Amplifies variety through constraining commands to system elements. Attenuates variety by provision of filtering instructions. Provides transduction to permit consistency in interpretation of events
Resource bargain/Accountability	• Provides/determines the resources (manpower, material, money, information, support) for operational units • Defines performance levels to which operational units will be held responsible • Determines how operational units will interface for performance reporting and accountability	Engages complexity by the allocation of resources necessary to support mechanisms essential to amplification, attenuation, and transduction of variety within the system. Accountability provides for the balance between autonomy and integration in the midst of variety generation
Operations	• Provides for the routine interface between operational system entities and from the metasystem to operational units	Permits the system to deal with the present and emergent variety being generated by internal flux and external turbulence
Coordination	• Provides for system balance and stability by ensuring that information concerning decisions and actions necessary to prevent disturbances are shared among operational units	Provides for variety absorption to prevent unnecessary fluctuations in the system. Necessary to amplify and match variety generated by interactions between different system entities
Audit	• Provides routine and sporadic feedback on the performance of system operations • Investigates and reports on areas on problematic areas	Identifies unabsorbed variety in the system that creates performance issues. Acts to amplify variety by identification of the sources contributing to unabsorbed variety
Algedonic	• Provides instant alert to crisis or potentially catastrophic situations occurring in the system • Bypasses routine communications channels and structure to identify system threats	Provides a variety"relief valve" to identify potential viability issues stemming from unabsorbed variety and threatening system viability. Variety amplification in response to sources of unabsorbed variety

(continued)

Table 5 (continued)

Communication channel	Primary functions	Complexity perspective
Dialog	• Provides examination and interpretation of organizational decisions, actions, and events • Seeks alignment of perspectives and shared understanding of organizational decisions and actions in light of system purpose and identity	Acts to absorb variety through the establishment of amplification achieved through strengthened identity. Identity acts to provide consistency in decision, action, and interpretation in the face of internal flux and external turbulence
System learning	• Provides detection and correction of system errors, testing of assumptions, and identification of system design deficiencies • Ensures that the system continually questions the adequacy of its design	Absorbs variety through the correction of system error. Correction follows detection of unabsorbed variety and seeking to resolve that variety through variety amplification. Targets purposeful development in response to issues in design and execution
Informing	• Provide routine transmission of information throughout the system • Routes information that is not appropriate for other channels for accessibility throughout the system	Provides for variety absorption through the redundancy of information. Redundancy acts to amplify variety to match unabsorbed variety existing in the system

This is achieved by providing the greatest degree of autonomy (freedom and independence of decision, action, and interpretation) possible while still maintaining the system at desired levels of performance and behavior. In effect, this suggests that excessive constraint of a system wastes resources (constraint is not free), limits system initiative/creativity/innovation, and unnecessarily diverts important metasystem resources to lower levels of the system (inefficiency). In contrast, insufficient constraint in a system risks the system underperforming to expectations resulting from insufficient integration of the system elements. Achieving the minimal constraint necessary to continue to produce desirable performance is a primary objective for CSG.

- ***Communication—flow and processing of information necessary to support consistent decision, action, and interpretation across the system.*** Communication is essential to governance and operation of the metasystem. Communications include not only the exchange of information but also the interpretative schemas that permeate the system. These interpretative schemas are necessary to provide coherence in making, understanding, and interpreting the myriad of exchanges in a system. Communications may range from formal to informal, explicit to tacit, and patterned to emergent. There is not an optimal configuration for communication in a system, and the arrangements are certainly subject to shifts over time and emergent conditions. However, from a CSG perspective, communications are something that would be better off not left to chance self-organization. Instead, purposeful design and evolution of communications within a system is more likely to produce and support desirable results. With respect to complexity, the design and execution of communications mechanisms are essential to absorb variety to the degree necessary to support system viability.

- ***Coordination—providing for effective interaction among different entities within the system, and external to the system, to prevent unnecessary fluctuations***. Certainly, coordination is an essential aspect to ensure that a system provides sufficient interaction among different elements to maintain consistency. Quite possibly, the most important aspect of coordination is the damping of unnecessary fluctuations as the system operates. In effect, this implies that there must be sufficient standardization to provide routine interface as well as a sufficiently robust design to absorb emergent conditions that could not have been known in advance. The design and execution of coordination in a complex system permits the absorption of variety emanating from internal flux. This internal flux is generated from the interactions between the different entities and their mechanisms in a system. In essence, coordination is necessary to support continuing system viability by "damping" fluctuations from unabsorbed variety. Although original work in management cybernetics focused on coordination as an internal function, one should also consider the necessity for coordination with entities external to the system.
- ***Integration—design for system unity with common goals, accountability, and balance between individual autonomy and system level interests***. The primary focus of integration is to insure that the system achieves desirable levels of performance while (a) providing the maximum level of autonomy to constituents, (b) invoking the minimal constraint necessary for the system to function as a unity in achieving the intended purpose, and (c) strategically shifting the balance point between autonomy and integration based on changes in contextual factors, environmental shifts, and system performance levels. Integration is necessary to provide the appropriate balance between autonomy of elements and system level integration. This balance requires the absorption of variety generated between the system and constituent elements. The variety generated from the tension between entity levels and system levels must be absorbed to the degree necessary to assure continued viability. Integration is not achieved through serendipity but rather by active design and continuous purposeful evolution (development).

The value accrued by CSG in dealing with complexity stems from engagement designed to absorb the variety (complexity) generated in a system, including:

- Scanning of the capacity an organization (entity) to engage in a level of systems thinking compatible with the complexity demands of the system environment.
- Exploration of the design and execution of essential governance functions.

- Identification and prioritization of system performance constraints tracked to problematic governance functions and violations of systems laws.
- Establishment of developmental strategies across multiple levels (e.g., entities, units, staff teams, departments) essential to enhancing CSG to improve system performance across the enterprise to better engage complex systems and problems.
- Examination and development of support infrastructure (processes, procedures, technologies, systems) for compatibility with system governance design, execution, and development.
- Identification of context and development considerations for unique circumstances, factors, and conditions that influence (constrain or enable) achievement of system governance functions and system performance (e.g., stakeholders, regulatory requirements, staff, leadership style, etc.). Providing identification of impediments to system performance rooted in specific deficiencies in design, execution, and development of governance functions and corresponding system laws.

Complexity is critical to understanding and implementation of CSG. The thinking, decision, actions, and interpretations of CSG are driven from an underlying grounding in complexity and the systemic perspective that grounding provokes. Having set the basis for the role of complexity in CSG, we shift focus to examine the implications of complexity for CSG.

4 Implications of Complexity in CSG Development

In this section, we develop a set of implications of complexity for the design, deployment, and development of CSG. These implications have been developed to consider both CSG field development as well as enhancing practices for deployment of CSG.

The primary role of complexity in CSG is found in relationship to absorption of variety in a complex system. This variety absorption is most aptly identified as "variety engineering" for a complex system. At a basic level, variety is a measure of complexity in a system, which is determined by the number of states that a system can occupy. Very quickly, a simple calculation of variety for a simple non-trivial system can be found in the equation:

$$V = z^n \tag{1}$$

where

V is the measure of variety that is a representation of complexity for a system

z is the number of states for an entity of a system, and

n is the number of entities in a system.

Clearly, the calculation of variety grows exponentially, as the number of states and entities grows for a system. In fact, for all intents and purposes, the calculated variety for a system escalates to infinity rapidly. Thus, for complex systems, the engineering of variety represents a critical step in mastering complexity. The engineering of variety involves those efforts to absorb (match) variety. Variety absorption occurs in a multitude of different forms targeted to permit a system to "match" variety being generated from internal flux or external turbulence. Among these, forms of variety matching are:

- *Variety Attenuation*—the absorption of variety through the setting of filters that limit the variety which must be "dealt with" for a system. This approach to variety matching engineers the mechanisms that reduce the variety that is presented to the system for disposition. Therefore, the potential for "unabsorbed variety" is reduced. An example of attenuation in CSG is the filtering of information generated from the environment into categories that can be more effectively analyzed and disseminated to the aspects of the system that are potentially influenced by the information.
- *Variety Amplification*—the absorption of variety through creation of vehicles that serve to better match variety being generated internally or externally for a system. In contrast to attenuation, amplification seeks to increase the variety of the system in response to externally/internally generated variety. Thus, system variety is increased to match variety, rather than limiting the variety that reaches the system. The engineering of variety amplification requires development of mechanisms that will more effectively match variety being imposed on the system. For example, the design of mechanisms targeted to rapidly identify, process, respond, and evaluate response to emergent events would be an attempt to amplify system variety.
- *Variety Transduction*—the absorption of variety through provisions for consistency in interpretation of information, resources, or energy that crosses a boundary. Expectations for direct, complete, and consistent interpretation of variety generated by boundary crossing information for complex systems is somewhat naïve. Transduction matches variety by providing an interpretative framework that can make the information intelligible for a system. In effect, this represents an appropriate translation of information (matching variety) through mechanisms. For example, an excessive data stream might be transduced through an AI system that discerns patterns in the information as it crosses into the system.

For CSG, the execution of a system design produces variety which must be maintained below threshold levels to support continued system viability. Variety, and unabsorbed variety, are part of the landscape for complex systems. However, CSG suggests that variety is something that can be engineered to more effectively deal with complexity. While all variety will never be eliminated for a complex system, CSG suggests that the focus on systems which are more robust (have a wider range of perturbations over which the design can compensate), resilient (be capable of returning to the system state that was occupied prior to a disturbance), and antifragile (modifications of a system that support the system "thriving" in response to perturbations). These three aspects of CSG are intended to match variety and create conditions for enhanced viability (continued existence) of a system in the presence of increasingly frequent and disturbing perturbations.

The engineering of variety for CSG is focused on the purposeful development of metasystem functions, performed by mechanisms, to absorb (match) system variety generated from internal flux and external turbulence. Variety engineering was an essential aspect of the viable system model (management cybernetics) that has carried forward to CSG. The essence of variety engineering is in the purposeful development of the design of functions and mechanisms directed to enhance CSG. Purposeful development also invokes the necessity to be proactive, explicit, and continuous in the development of CSG. It is shortsighted to leave CSG development (i.e., variety engineering) to achievement by accretion (adding pieces and parts in an ad hoc manner) or self-organizing (permitting relationships to form without constraint). Either of these approaches to CSG development is not likely to produce desirable results. Variety is not something that should be left to develop unconstrained. Instead, CSG development should be focused on the purposeful engineering of variety to support continued system viability.

Unabsorbed variety produces pathologies, degrades system performance, and lessens system viability. Pathologies are an indicator of unabsorbed system variety (complexity). Minimally, if variety is not absorbed (matched) by design, execution, and development for a complex system, performance degradation is inevitable. In fact, if unabsorbed system variety exceeds threshold levels, a system can experience a catastrophic failure.

Thus, the purposeful development of CSG is focused on "variety engineering" to manage complexity more effectively. Figure 1 below captures the relationships of complexity to purposeful development of CSG.

5 Conclusions

Complexity is fundamental to CSG. Complexity is routinely understood as stemming from a high number of variables/entities, rich interconnections, dynamic nature, and subject to emergence. However, for CSG, the development and implications of complexity are much deeper and wide ranging. To expound on the nature and role of complexity for CSG, in this section, we summarize eight key themes and offer several challenges for the future development of complexity for CSG.

THEME 1: Complexity is a defining characteristic of design, execution, and development aspects of CSG.

Complexity is an attribute that spans the entirety of design, execution, and development of CSG. The structural definition (design) of a complex system will elaborate over time, with new structural configuration emerging. The initial design can hardly be expected to survive new knowledge/understanding, shifts in environment/context, or reformulation based on reframing in the face of complexity. In addition, execution in complex systems will undoubtedly produce emergent patterns, behaviors, and new structure. The development aspects of CSG will also find it necessary to confront complexity. The modifications in design (second order change) or modifications in

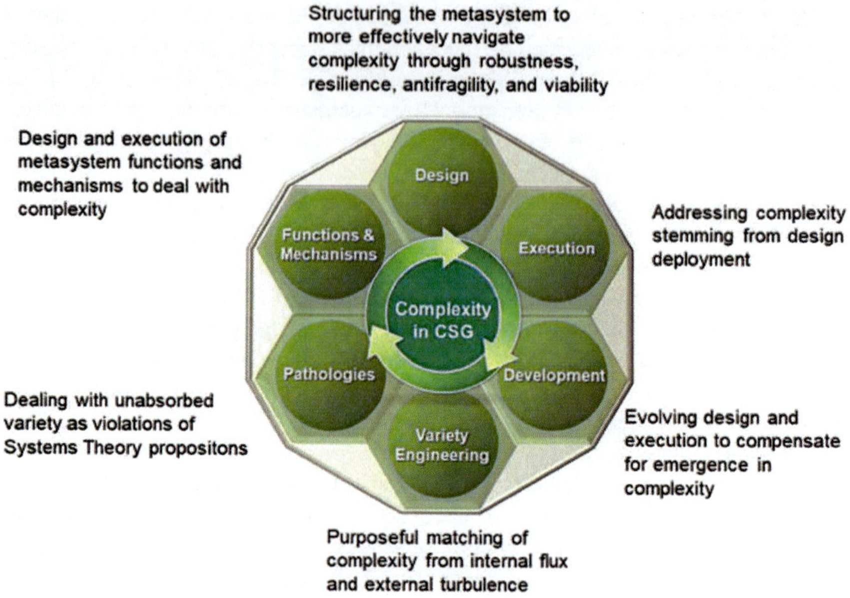

Fig. 1 Relationships of complexity to performance of CSG

execution (first order change) will experience complexity as new initiatives or modifications are installed. It would be naïve to assume that complexity will not be a mainstay of CSG across design, execution, and development. This does not suggest that "nothing" can be done to confront complexity. On the contrary, complexity presents a particular set of attributes that should be taken into account in performance of CSG,

THEME 2: Complexity impacts all nine metasystem functions as well as the ten communication channels for CSG.

Complexity is indigenous to the metasystem functions and communication channels. There is no escape from the reach of complexity in CSG. Each of the nine metasystem functions must be designed, executed, and developed against the backdrop of complexity. Failure to take complexity into account is tantamount to engaging CSG from a simple system perspective. The simplifying assumptions and logic necessary to execute CSG require a complexity perspective. Short this perspective is naïve, inconsistent with the formulation of CSG, and unlikely to produce the results expected from engagement of CSG. Unfortunately, complexity becomes somewhat of a mindset first that must be translated to an operational setting to become actionable. Beyond the concept of CSG lies the harsh realities that await in operational deployment. Although a "complexity mindset" cannot guarantee success in CSG endeavors, it forms a set of necessary underlying perspectives to appropriately engage CSG.

THEME 3: Dealing effectively with complexity requires "variety engineering" to develop the mechanisms capable of matching variety

Complexity can be measured by calculating variety for a complex system. In reality, while calculation of variety is interesting, the crux is that variety quickly becomes unmanageable. Variety of a complex system begins to approach infinity rapidly as the number of system entities and potential states they can occupy escalates as complexity increases exponentially. The importance of complexity (variety) for CSG is found in the necessity to actively engage in "variety engineering" across design, execution, and development of CSG. This involves the purposeful design of appropriate mechanisms to attenuate (filter), amplify (enlarge), or transduce (convert) variety and thus "match" variety encountered by a system. Complexity suggests that variety can emanate from internal flux or external turbulence. Irrespective of the source, variety (complexity) must be absorbed by the design of the system or execution of the system. Unabsorbed variety is a source of system dysfunction and can degrade system performance in the best case or result in catastrophic failure in the worst case. Thus, it falls on CSG to foster variety engineering to address exploding complexity in complex systems.

THEME 4: Design for complexity embodies robustness, resilience, and antifragility to support continuing system viability

The CSG response to complexity is to engineer variety such that the attributes of robustness, resilience, and antifragility are embedded across design, execution, and development for a complex system. In this sense, variety engineering and achievement of the attributes of variety engineering are all focused on enhancing the prospects for system viability (continued existence). Robustness is focused on the range of disturbances over which a system has been designed to absorb. Resilience targets the degree to which a system can recover from disturbances that are unexpected and emergent. Antifragility is focused on the degree to which a system can thrive in the wake of disturbances that are unexpected (stress, shocks, failures). Design and execution for robustness, resilience, and antifragility enable system viability (the maintenance of existence of a system in the wake of internal flux and environmental turbulence). CSG offers a path forward to deal with complexity by enhancing system viability through the proactive design, execution, and development of a system.

THEME 5: Pathologies are indicative of a system that is producing unabsorbed variety

Regardless of the effectiveness of design, execution, and development of a complex system, there will be pathologies. We consider pathologies to be a mainstay of complexity in CSG. Pathologies are indicative of unabsorbed variety and serve to diminish system performance. In this respect, pathologies are simply a "fact of life" for complex systems. The development aspect of CSG is focused on addressing pathologies through one of two development modes. First, the correction of pathologies stemming from execution of a complex system represents the "first-order" of "single-loop" learning. "First-order" system development adjusts execution aspects of CSG. In contrast, "second-order" development is focused on adjusting the system design to absorb variety stemming from internal flux or external turbulence.

THEME 6: Systems theory provides the basis for a worldview that can cope with increasing complexity

Complexity requires a mindset that can guide design, execution, and development for CSG. This mindset for CSG is informed from systems theory. Systems theory exists as the set of axioms (taken for granted assumptions) and associated propositions (the collection of laws, principles, and concepts that define the behavior and performance of complex systems). The mindset informed by systems theory can be called a systemic worldview. This worldview provides a path forward for understanding and responding to complexity.

THEME 7: Purposeful system development is targeted to proactively match unabsorbed variety from internal flux and external turbulence.

A product of complexity in complex systems is unabsorbed variety. CSG is focused on resolution of the unabsorbed variety that exists in a complex system. The purposeful design, execution, and development of the functions and mechanisms of CSG for a complex system addresses complexity. CSG holds that purposeful development can adjust the system to compensate for unabsorbed variety through adjustments to design or execution of the complex system. Thus, CSG is focused on providing "variety matching" to deal with complexity in a system. It is important to note that adjustments to a system are not a one-time event. On the contrary, adjustments must be continuous, as the environment and context of a system will shift over time.

THEME 8: Complexity is always a hallmark of complex systems and CSG; however, it is dynamic in degree, impact, and manifestation form.

Complexity is inescapable in complex systems. However, while it is known that complexity will be present, the precise form and impact will not be revealed until complexity manifests itself. Emergence is a central characteristic of complexity and suggests that the structural and behavioral patterns develop as a system operates—irrespective of the intention of design or execution. Understanding that complexity cannot be specifically known in advance of the occurrence suggests the importance of CSG design and execution. However, perhaps more important is development in CSG. Development is where the adjustments to design or execution can be made to compensate for complexity.

Given that complexity (as measured by variety) is continuing to increase, mastery of complexity is essential for the viability of systems. CSG is an attempt to master complexity through the purposeful development of systems. Purposeful development invokes modifications to design or execution. Although CSG has been developed to address complexity, there are additional developmental contributions across theoretical, methodological, and practice dimensions. Table 6 succinctly captures the future directions for development of complexity for CSG. This research uses a select set of characteristics associated with complexity. However, a case could be made for a.

Table 6 Theoretical, methodological, and practice challenges for CSG

Development area	Implications
Theoretical	Need to use a comprehensive set of laws, principles, and theorems related to complexity and general systems theory (e.g., see Katina [48, 49]), albeit, with the context in mind. Such an approach would solidify the range of utility of applicability of present research
Methodological	Although complexity is considered in CSG, extension of methodological considerations can be a significant focal area for future development. This is particularly the case in dealing with emergence through design, execution, and development
Practice	There are significant aspects related to practice that can be enhanced with respect to complexity. Specifically, further development of systemic worldview and its influence on ability to deploy CSG is an area that could use additional development

In this chapter, we have explored complexity and its implications for CSG. Complexity touches all aspects of CSG, including implications for design, execution, and development. Complexity was explored with respect to CSG metasystem functions and their corresponding mechanisms. Understanding, incorporation, and integration of complexity into the CSG landscape are essential to further development of the CSG field and ultimately incorporation into operational practice.

6 Exercises

1. Identify the key attributes for complexity.
2. Discuss the implications of complexity for Complex System Governance across design, execution, and development.
3. Recommend strategies that might be pursued to deal with the inescapable complexity characteristic of complex systems and their problems.
4. Discuss the themes of complexity and their implications for dealing with complexity in CSG.

References

1. Ackoff R (1974) Redesigning the Future. Wiley, New York
2. Adami C (1998) Introduction to artificial life. Springer Science & Business Media
3. Addis L (1975) The logic of society: a philosophical study, vol 7. U of Minnesota Press
4. Allen P, McGlade J (1987) Modelling complex human systems: a fisheries example. Eur J Oper Res 30(2):147–167
5. Anderson PW (1972) More is different. Science 177(4047):393–396
6. Anzola D, Barbrook-Johnson P, Cano JI (2017) Self-organization and social science. Comput Math Organ Theory 23(2):221–257

7. Ashby W (1947) Principles of the self-organizing dynamic system. J Gen Psychol 37:25–128
8. Ashby WR (1947) The nervous system as physical machine: with special reference to the origin of adaptive behavior. Mind 56 (221) (January):44–59. http://tinyurl.com/aqcmdy
9. Ashby WR (1960) Design for a brain: the origin of adaptive behaviour, 2nd ed. Chapman & Hall, London. https://doi.org/10.1037/11592-000
10. Axelrod R (1984) The evolution of cooperation. Basic Books, New York
11. Batterman RW (2009) On the explanatory role of mathematics in empirical science. Br J Philos Sci 61(1):1–25
12. Beer S (1979) Heart of enterprise. Wiley
13. Beer S (1981) Brain of the Firm. Wiley
14. Beer S (1985) Diagnosing the system for organizations. Oxford University Press
15. Bertelsen S (2002) Bridging the gap–towards a comprehensive understanding of lean construction. In: IGLC-10, Gramado, Brazil
16. Boisot M, McKelvey B (2010) Integrating modernist and postmodernist perspectives on organizations: a complexity science bridge. Acad Manag Rev 35(3):415–433
17. Boisot M, McKelvey B (2011) Complexity and organisation—environment relations: revisiting Ashby's law of requisite variety. In: Allen P, Maguire S, McKelvey B (eds) The SAGE book of complexity and management. London, pp 279–298
18. Bousquet A, Curtis S (2011) Beyond models and metaphors: complexity theory, systems thinking and international relations. Camb Rev Int Aff 24(1):43–62
19. Carley KM, Hill V (2001) Structural change and learning within organizations. In: Lomi A, Larsen ER (eds) Dynamics of organizational societies: computational modelling and organization theories. Cambridge, MA: MIT Press, pp 63–92
20. Crutchfield JP, Young K (1993) Computation at the edge of chaos. In: Complexity, entropy and the physics of information: SFI studies in the sciences of complexity, pp 223–269
21. Crutchfield JP, Young K (1994) What lies between order and chaos. In: The sciences. New York Academy of Sciences
22. Deutsch KW (1963) The nerves of government: models of political communication and control. The Free Press, New York
23. Eigen M, Schuster P (1977) A principle of natural self-organization. Naturwissenschaften 64(11):541–565
24. Eigen S, Schuster P (1979) The hypercycle [electronic resource] a principle of natural self-organization
25. Epstein JM, Axtell R (1996) Growing artificial societies: social science from the bottom up. MIT Press, Cambridge, MA
26. Fernández N, Gershenson C (2013) Measuring complexity in an aquatic ecosystem. arXiv preprint arXiv:1305.5413
27. Fernández A, Gómez C, Hornero R, López-Ibor JJ (2013) Complexity and schizophrenia. Prog Neuro-Psychopharmacol Biol Psychiatry 45:267–276
28. Gershenson C (2013) The implications of interactions for science and philosophy. Foundations of Science Early View. http://arxiv.org/abs/1105.2827
29. Gershenson C (2014) Harnessing the complexity of education with information technology. arXiv preprint arXiv:1402.2827
30. Gershenson C (2015) Requisite variety, autopoiesis, and self-organization. Kybernetes
31. Gilbert N (2008) Agent-based models. Sage, London
32. Gilbert N, Anzola D, Johnson P, Elsenbroich C, Balke T, Dilaver O (2015) Self-organizing dynamical systems. In: Wright JD (ed) International encyclopedia of the social & behavioral sciences. Elsevier, London
33. Goldstein J (1999) Emergence as a construct: History and issues. Emergence 1(1):49–72
34. Griffeath D, Moore C (eds) (2003) New constructions in cellular automata. Oxford University Press on Demand
35. Guastello SJ (2002) Managing emergent phenomena: nonlinear dynamics in work organizations. Lawrence Erlbaum Associates

36. Haken W (1973) Connections between topological and group theoretical decision problems. In: Studies in logic and the foundations of mathematics, vol 71. Elsevier, pp 427–441
37. Haken H (1981) Chaos and order in nature. In: Chaos and order in nature. Springer, Berlin, Heidelberg, pp 2–11
38. Haken H (2008) Self-organization of brain function. Scholarpedia 3:2555. https://doi.org/10.4249/Scholarpedia.2555
39. Harrison JR, Carroll GR (1991) Keeping the faith: a model of cultural transmission in formal organizations. Adm Sci Q 36(4):552–582. https://doi.org/10.2307/2393274
40. Hayek FA (1967) The theory of complex phenomena. In: Hayek FA (ed) Studies in philosophy, politics and economics. Routledge, London
41. Hayek FA (1978) The pretence of knowledge. In: Hayek FA (ed) New studies in philosophy, politics, economics and the history of ideas. Routledge, London
42. Holland JH (1988) The global economy as an adaptive system. In: Anderson PW, Arrow KJ, Pines D (eds) The economy as an evolving complex system. Addison-Wesley, Reading, MA, pp 117–124
43. Holland JH (1994) Echoing emergence: objectives, rough definitions, and speculations for ECHO-class models. In: Cowan GA, Pines D, Meltzer D (eds) Complexity: metaphors, models, and reality. Addison-Wesley
44. Homer-Dixon T (2015) Synchronous Failure. Ecol Soc 20(3)
45. Ireland V, Gorod A (2016) Contribution of complex systems to entrepreneurship. Entrep Res J 6(1):1–41
46. Jervis R (1997) System Effects. Princeton University Press. Kaski, T, Princeton
47. Johnson S (2002) Emergence: the connected lives of ants, brains, cities, and software. Simon and Schuster
48. Katina PF (2015) Emerging systems theory–based pathologies for governance of complex systems. Int J Syst Syst Eng 6(1–2):144–159
49. Katina PF (2016) Systems theory as a foundation for discovery of pathologies for complex system problem formulation. In: Applications of systems thinking and soft operations research in managing complexity. Springer, Cham, pp 227–267
50. Kauffman S (1995) At home at the universe. Oxford University Press, Oxford
51. Keating CB, Katina PF, Bradley JM (2015) Challenges for developing complex system governance. Paper presented at the Proceedings of the 2015 industrial and systems engineering research conference
52. Keating CB, Katina PF (2019) Complex system governance: concept, utility, and challenges. Syst Res Behav Sci 36(5):687–705
53. Keating CB, Morin M (2001) An approach for systems analysis of patient care operations. J Nurs Adm 31(7/8):355–363. https://doi.org/10.1097/00005110-200107000-00007
54. Koskinen KU (2013) Processual autopoietic knowledge production in organizations. In: Knowledge production in organization. Springer, Heidelberg, pp 1–5
55. Langton CG (1986) Studying artificial life with cellular automata. Physica D: Nonlinear Phenom 22(1–3):120–149
56. Langton CG (1996) SFI studies in the sciences of complexity, vol XVII. Addison-Wesley
57. Laughlin RB (2005) A different universe: reinventing physics from the bottom down. Basic Books (AZ)
58. LeBaron B (2000) Empirical regularities from interacting long- and short-memory investors in an agent-based stock market. IEEE Trans Evol Comput 5(5):442–455. https://doi.org/10.1109/4235.956709
59. Lewes GH (1875) On actors and the art of acting, vol 1533. B. Tauchnitz
60. Marinaro M, Tagliaferri R (2002) Neural nets. Springer-Verlag
61. Martelli M (1999) Introduction to discrete dynamical systems and chaos. Wiley
62. Maturana V, Varela FJ (1980) Autopoiesis and cognition the realization of the living. In: Boston studies in the philosophy of science, vol 42. D. Reidel Pub, Dordrecht, Holland; Boston
63. Maturana HR, Varela FJ (2012) Autopoiesis and cognition: the realization of the living, vol 42. Springer Science & Business Media

64. May RM, Oster GF (1976) Bifurcations and dynamic complexity in simple ecological models. Am Nat 110(974):573–599
65. McShea DW (2000) Functional complexity in organisms: parts as proxies. Biol Philos 15(5):641–668
66. Merali Y, Allen P (2011) Complexity and systems thinking. In: The SAGE handbook of complexity and management, 31–52
67. Miller J, Page S (2007) Complex adaptive systems. Princeton University Press, New Jersey
68. Morin E (2007) Introduction to complex thinking. Barcelona: Gedisa: Barcelona. PMI (Project Management Institute). (2014). Navigating complexity: a practice guide. pp. 1–113
69. Moroni S (2015) Complexity and the inherent limits of explanation and prediction: urban codes for self-organising cities. Plan Theory 14(3):248–267
70. Nicolis P, Prigogine I (1989) Exploring complexity: an introduction. W.H. Freeman, New York
71. Orsini A, Le Prestre P, Haas PM, Brosig M, Pattberg P, Widerberg O, Gomez-Mera L, Morin J-F, Harrison NE, Geyer R, Chandler D (2019) Forum: complex systems and international governance. Int Stud Rev, 1–31
72. Prigogine I, Stengers I (1984) Order out of chaos: man's new dialogue with nature. Boulder, CO: New Science Library
73. Prigogine I, Stengers I (2018) Order out of chaos: man's new dialogue with nature. Verso Books
74. Prokopenko M, Boschetti F, Ryan AJ (2009) An information-theoretic primer on complexity, self-organization, and emergence. Complexity 15(1):11–28
75. Reid G (2007) The foundations of small business enterprise. Taylor and Francis
76. Rotmans J, Loorbach D (2009) Complexity and transition management. J Ind Ecol 13(2):184–196
77. Sawyer RK (2005) Social emergence: societies as complex systems. Cambridge University Press, Cambridge, UK
78. Skar J (2003) Introduction: self-organization as an actual theme. Philos Trans Ser A Math Phys Eng Sci 361(1807):1049–1056
79. Schelling T (1971) Dynamic models of segregation. J Math Sociol 1(2):143–186
80. Simon H (1962) The architecture of complexity. Proc Am Philos Soc 106(6):467–482
81. Simon H (1981) The Sciences of the Artificial, 2nd edn. MIT Press, Cambridge
82. Solé R, Goodwin B (2000) How complexity pervades biology. Basic Books
83. Stewart I, Cohen J (1994) Why are there simple rules in a complicated universe? Futures 26(6):648–664
84. Teisman G, Gerrits L (2014) The emergence of complexity in the art and science of governance. Complex, Gov Netw 1(1):17–28
85. Varela F, Maturana H (1972) Mechanism and biological explanation. Philos Sci 39(3):378–382
86. Varela FG, Maturana HR, Uribe R (1974) Autopoiesis: the organization of living systems, its characterization and a model. Biosystems 5(4):187–196. https://doi.org/10.1016/0303-2647(74)90031-8
87. Urry J (2003) Global complexity. Blackwell, London
88. Weaver W (1948) Science and Complexity. Am Sci 36:536

System Governance

Behnido Y. Calida

Abstract This chapter offers a synopsis mapping of the recent and significant advances in the research of *systems* and *governance* concepts; highlighting any conceptual synergies of one to the other, and altogether strengthening any other obvious emergent themes resulting from the confluence of ideas across several disciplinary fields and/or problem domains. This synthesis is designed to establish the current state of the field, to provide a scholarly critique of the literature and to present relevant research gaps in need of further exploration, elaboration, or confirmation. An additional goal of this chapter was to establish the position and fit of the current research within the larger body of knowledge for which it will become an original contribution.

Keywords Complex systems · Governance · Systems theory · Systems thinking

1 Introduction

The present chapter highlights the state-of-the-art research and practice involving both systems and governance concepts. There are three primary objectives of this review. First, a synthesis of the literature related to system governance is designed to establish the current state of the field. Particular attention was given to works that feature the multidisciplinary nature of system governance. The second objective was to provide a scholarly critique of the literature to identify the strengths and limitations of the state of the topic. Third, in conjunction with the critique, relevant gaps in need of further exploration, elaboration, or confirmation were established. The overarching goal for these primary objectives was to clearly establish the position and fit of the current research within the larger body of knowledge for which it will become an original contribution. The chapter is organized to first provide an overview of the body of knowledge scope. This provided a boundary for the literature and the scope of the effort to cross multidisciplinary lines. Next, the chapter explores the state

B. Y. Calida (✉)
Agylabs, LLC, Virginia Beach, Virginia, USA
e-mail: behnido.calida@agylabs.com

C. B. Keating et al. (eds.), *Complex System Governance*, Topics in Safety, Risk, Reliability and Quality 40, https://doi.org/10.1007/978-3-030-93852-9_3

of literature for systems philosophy and the systems-based approach. This establishes the nature of "systems" as the basis for establishing the analytic framework for governance. Following the examination of the systems literature, the literature with respect to governance is elaborated. This examination is truly multidisciplinary, as it is expansive across several disciplinary fields and the corresponding sets of literature. The literature review then provides a synthesis of the general themes that have emerged from the review. Care is taken to establish the basis for the themes that run through the literature as well as the absence of thematic areas that are ripe for research exploration. This is used to position the current research within the body of knowledge as elaborated by the literature review.

2 Generating Science Overlay Maps

To begin an informed foray into system governance across different disciplinary knowledge domains, the literature review process initiated with a search query through ISI Web of Knowledge Social Science Citation Index (SSCI) and Science Citation Index (SCI) as the database of record since it is the most comprehensive database of peer-reviewed research work for both the social sciences and sciences, respectively. The resulting search records served as a starting point to initiate the literature review process although the entirety of the reviewed literature was extended to sources from outside those initially identified from the primary indexes. Mainly, this established a coarse research context (mainly by setting main disciplinary and seminal works sources) which was then used to narrow down previous works that were deemed to be relevant to this research.

Using a science overlay map [19, 73], a visual interdisciplinary knowledge domain representation of the resulting search records can be visualized like those shown in Fig. 1 thru 3 below. These representations provided "simple and quick" visualizations of the disciplinary diversity of governance-related research context without the need for sophisticated combined indices.

The below mappings gave a better appreciation of the existing intellectual diversity of governance research. Intellectual diversity as represented by (1) the variety of disciplines involved directly or indirectly in governance research, (2) the balance of how each of the disciplines has contributed to pushing the envelope of "governance" research thus far, and (3) the disparity conveyed by how accounts of "governance" from different disciplines are proximally located on a cognitive spatial map.

For instance, as one interpretation from the set of retrieved data, a cognitive knowledge space mapping of mainstream "governance" research is predominantly contextualized from specific disciplines. There were also dispersed weak accounts of "governance" research that are indicative of emergent research on associated conceptual ideas and applications of "governance." Also, from the collection of literature sources, it was useful to bear in mind how possibly each conceptual account of governance evolved from the diverse philosophical (axiomatic, epistemological,

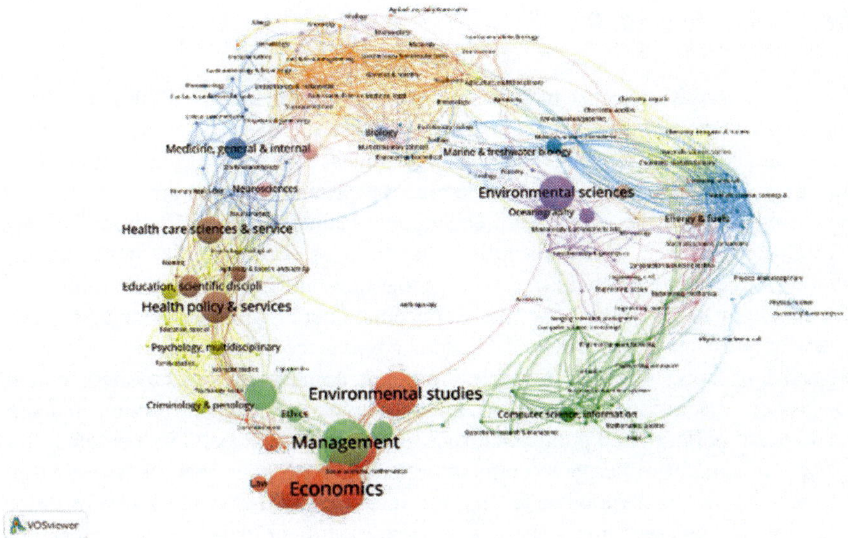

Fig. 1 A science overlay map of governance-related research: Year 2000 and prior years

and ontological) orientations and methodological choices that were inherent in the domain under which different strands of governance research were explored. We have partitioned our observations across primarily three epochs.

2.1 Prior to 2000

Figure 1 shows a glimpse of "system governance" research from prior years up to the year 2000. A high-level assessment of the diversity of research for system governance from research work up to year 2000 can be seen as already active research in the domains of many disciplines (as high variety), where several of the governance-research treatments were expected to be arguably qualitative in nature coming from subjectivist disciplinary paradigms (one way of interpreting research balance), and being significantly largely framed within economics, management (highly dense disciplinary nodes in mentioned areas as an indicator of low disparity), political science, and regional urban planning. From an engineering management and systems engineering standpoint, quantifiable research on systems governance was practically nonexistent, if not limited to concepts associated with computer and information science fields.

2.2 2001 Thru 2010

Next, Fig. 2 shows a glimpse of "system governance" research from 2001 thru 2010. There was even more diversity of research for system governance continuing from the momentum from the same singular disciplines. There are arguably more varieties of problem contexts. Research interest purporting "system governance" sees increasing focus in the mainstream economics, management, and political science disciplines. However, related disciplines and domains of practice like public administration, environmental studies/sciences, geography, and international relations have also caught on to the suitability of these concepts. There is also a sense that problems getting tackled as the purview of "system governance" encompass slightly more expansive in scope typical of "complex" problem domains that start to involve more than just one discipline like in ecology, laws, urban planning, water resource management, health policies, and public environment occupational health to mention a few. Albeit an indicator of things to come, more and more researchers are investigating research theories and frameworks that are increasingly useful research constructs beyond their own disciplines. There is cross-cultivation of ideas, and we find similar concepts and theories getting explored as alternative approaches to new problem domains. Even still focusing on the engineering practice field, quantifiable research on systems governance have slowly started to gain traction.

Figure 3 brings us to the current snapshot and state of "system governance" research from 2011 thru 2021. If anything, there is a lot more research activity that are more multidisciplinary in nature. Problem domains are no longer tackled within the

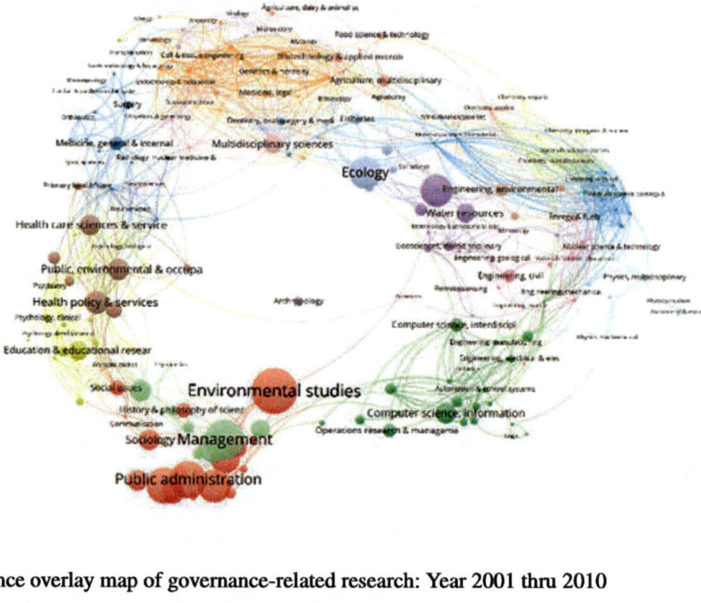

Fig. 2 A science overlay map of governance-related research: Year 2001 thru 2010

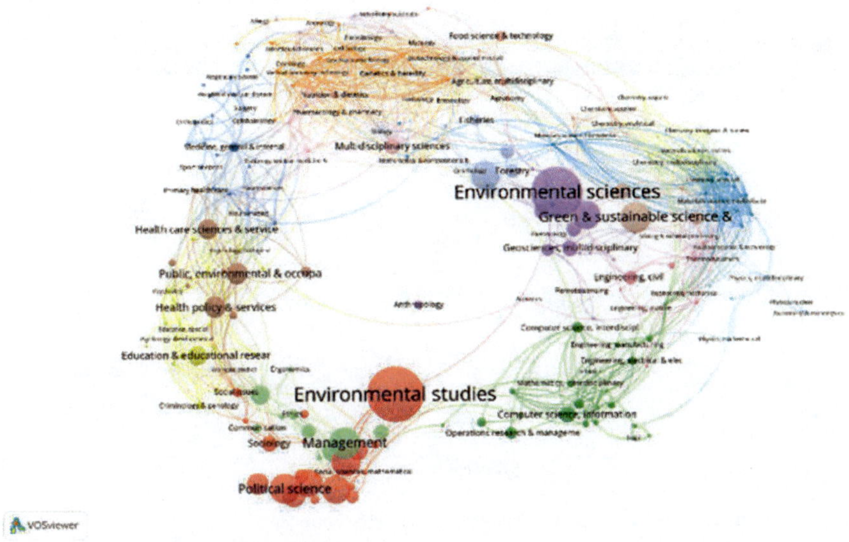

Fig. 3 A science overlay map of governance-related research: Year 2011 thru 2021

confines of a single discipline. Instead, we can easily pick out research that supports solutions originating from more multidisciplinary approaches. For example, during this past decade, where communities and cities are increasingly sounding the alarm if not directly feeling the effect of climate change, large research initiatives spanning multiple disciplines, countries, and expertise are leveraged altogether to address the systemic roots of the problem.

2.3 *2011 Thru 2021*

This resulted in the sudden emergence of sustainability, green technologies, environmental engineering, ethics, resiliency building, and various risk mitigating approaches to the forefront of the climate change challenge. Also, coincidentally, the contributions from engineering practice are slowly becoming more relevant to this problem domain.

To demonstrate further, a "funnel down" mapping of the relevant literature on system governance, the research frame initialized by disciplines and communities of practice familiar with the bodies of knowledge investigating associated phenomena. System governance had for its root components *systems* and *governance* which were separately cultivated from specific disciplines or observed from particular application or problem-focused communities. The literature review shown in Fig. 4 resulted in several informative articles. However, one can easily cast doubt concerning their cross-concept consistencies, more specifically on the development of the concepts

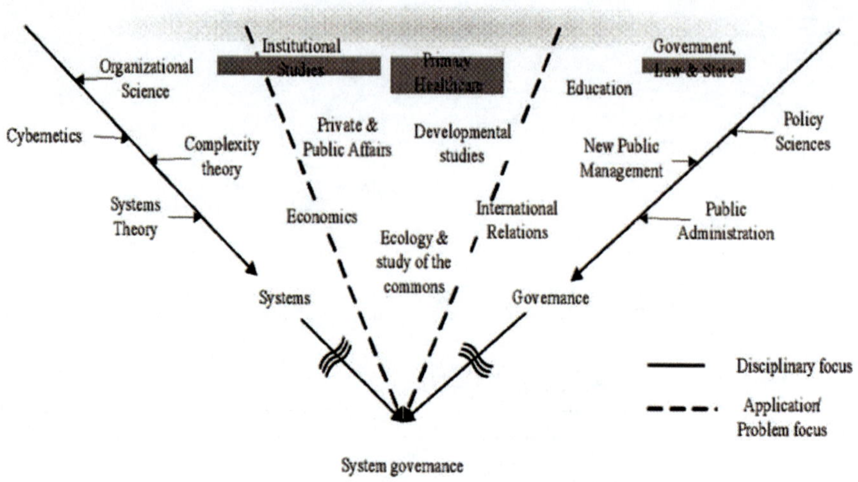

Fig. 4 Multidisciplinary evolution of "system governance" concepts

and theories themselves as opposed to more superficial treatment of the phenomena associated with system governance. While versions of "systems theories" and "governance theories" abound, a "system governance concept or theory was not available and was not explicitly articulated. Though studies on "system" or "governance" have progressed, a "system governance" research thread was not determined to have been approached from an integrative perspective—that is appreciative of the purview of disciplines investigating systems or governance, nor from those from practitioner communities engaged in "governance" application or problem domains.

The different highlights from each disciplinary research line are presented in the following sections. In particular, the next section discusses the state of the literature in systems and systems approaches which were closely followed by the state of the literature for governance research mostly from more predominantly "governance" focused disciplines.

3 System Philosophy Highlights

The main highlights to be covered in this section focused on the state-of-the-literature in systems research, including an articulation of its philosophy (e.g., systems philosophy) and its approach (e.g., systems approach) as reflected from investigations in recent systems research.

The modern systems movement has grown in prominence over the years since Von Bertalanffy [6] first posited his theory on open systems that became the basis of the renowned general systems theory or simply GST [11]. Resulting from these seminal works, the body of knowledge or BoK has been enriched by several closely woven research threads in complex systems [5, 48, 79], systems analysis [22, 37], second-order cybernetics [87], system dynamics [28, 78], soft systems methodology [21], critical systems thinking [40, 41, 85], systems architecting [65], systems engineering [34], and systems of systems [1, 42, 50, 51]. While a complete and exhaustive account was pertinent in understanding the history of the systems movement, it is beyond the scope of this research. One may, however, endeavor a more in-depth look at any of those seminal works mentioned above. What is pertinent to the current research was the articulation of the underlying system philosophy that enabled us to draw a clear understanding of a "system" that was consistent with the contemporary understanding of the systems approach and directly relevant to this research with respect to system governance.

The main philosophical strands that are brought into focus in this study make a distinction between the traditional reductionist philosophies, which support a traditionally mechanistic view from the natural sciences, versus the emergentist philosophies now being embraced by modern-day interdisciplinary science [70, 89]. Using these ideas, many of the key developments in traditional disciplines of science promote what is now considered a mechanistic science worldview that promoted mostly mechanical properties of things as primary, in contrast to the derivative and secondary properties divulged in other sciences. Due to the unprecedented success of the scientific method, its philosophy that proved so successful in resolving vexing problems of physical phenomena continued to slowly find its way outside of the natural sciences. However, there was a rejection of the appropriateness of the approach beyond the successes found in the natural sciences. According to Checkland [21], this paved the way to realizing that Cartesian reductionist philosophy, when applied to the social science domain, is seriously constrained to explain problems of complexity (e.g., emergence), problems of social science (e.g., rational behavioral capacity) and problems of management (e.g., problem uniqueness). Similarly, Casti [20] also noted the same limitations of scientific modeling when indiscriminately applied to the modeling of processes in the social and behavioral sciences. He contended that fundamental aspects that allow classical scientific modeling to work flawlessly, such as the existence of fundamental "laws" that are either absent or unknown, are characteristically indeterminable for systems that demonstrate complexity, manmade structures, and several possible social interactions. Based on this premise, an alternate philosophy is being argued that would consider the possibility of considering the absence of laws and of operational forms of key concepts in the social sciences [70].

Several significant contributions of the science-based philosophy emanating from the natural sciences shaped the present disciplines of physics, chemistry, and biology among many others. Furthermore, several scholarly advances in the sciences and social sciences have pushed for an alternative way of thinking based this time on systems philosophy. This systems philosophy, according to Checkland, can be attributed to mainly the following two sets of ideas: (i) emergence and hierarchy, originating in organismic biology and generalized in GST; and (ii) communication and control, originating in communication engineering and generalized in cybernetics. As a main distinction that makes it broader than traditional disciplines, these sets of ideas support a systems approach that is fundamentally interdisciplinary.

Separately, Bunge [18] articulated system philosophy or simply *systemism* as distinct from the reductionist/mechanistic philosophy of *atomism* and *individualism* (or micro-views) but also likewise different from ideas of *holism* (or macro-views) that is often conflated by some to mean one and the same as systems philosophy. He clarifies that while the holistic approach supposes to accept only the idea that a whole is more than a mere aggregation of its parts: it also maintains also that wholes must be taken at prima facie value, understood by them, not through analysis. Below is his reasoning as to why systemism should be considered as different from holism:

> Because the holistic approach rejects the possibility of analysis, it relies upon the method of intuition, not rational explanation or empirical experiment. While the systems approach recognizes the existence of emergent properties, it nevertheless seeks to explain them in terms of how their constituent parts are organized. Where holism is satisfied with a non-rational apprehension of unanalyzed wholes, systems aims to demystify emergent properties by providing scientific understanding that utilizes analysis as well as synthesis. Therefore, it is equally important that the systems approach be distinguished from holism as from mechanism [18].

Having recognized that both macro- and micro- entities and their processes are at best partial contributors toward complete understanding, systems require a full set of linkages for purposes of theorizing. In other words, systems philosophy, and the systems approach views systems as a function of its composition, environment, and structure, with the appreciation of the necessary linkages or mechanisms that specify its functional form. Bunge posits that the systems philosophy is the adoption of a worldview that is underpinned by the following postulates:

1. Everything, whether concrete or abstract, is a system or an actual or potential component of a system.
2. Systems have systemic (emergent) features that their components lack, whence
3. All problems should be approached in a systemic rather than in a sectoral fashion.
4. All ideas should be put together into systems (theories); and
5. The testing of anything, whether idea or artifact, assumes the validity of other items, which are taken as benchmarks, at least for the time being.

Based on the above postulates, the system notion adopted in this research closely follows Bunge's characterization of systems in terms of its composition, environment, structure, and mechanisms or simply called the CESM model (through substitution using each the initials of the key concepts). Composition is the collection of all the parts of the system. The environment is a collection of items, other than those composing the system, that act on or are acted upon by some or all components of the system. Structure is the collection of relations, in particular the linkages, among which components of the system interact with themselves or with their environment. Mechanisms are those collections of processes in the system that explain why the system behaves the way it does or more specifically, these are the processes or entities that mediate between the observable inputs and outputs of a system.

Fig. 5 Systemic research paradigm

Following from the earlier discussion, and specifically on Bunge's updated notion of the systemic view, the distinction in different interrelated classes of philosophical considerations are important foundations for the research. As depicted in Fig. 5, these may fall under the following several classes: (i) epistemological, (ii) ontological, (iii) methodological, (iv) axiological, and (v) ethical.

By epistemological, these refer to the starting assumptions of knowledge, or in this case the manner in which "system governance" constructs are formed. Epistemology is about how we came to know? According to Bunge, this is an elaboration on the roles of observation and speculation, intuition and reason, discovery, and invention. Johannessen and Olaisen [45] add that it also concerns the distinction behind intention and behavior. For instance, the interpretation of meaning becomes an important part of the intention aspect while explanation and prediction become an important part of the behavior aspect. These provide an important consideration for systemic research where Johannessen and Olaisen [45, 46] state:

In the systemic research model, the mental (emic) does not precede the behavioral (etic), but constitute different knowledge domains to be studied, together or separately. Sometimes the one may be the case of the other, and, at other times, vice versa. Constructs from both domains are used on the condition that workable indicators can be developed. Further, it should be noted that according to the systemic approach, all adequate explanations in social science are pluralistic, i.e., they are related to the model of the human being and the social systems we use, and it is therefore only partial truths…Much of the existing confusion in social science emanates according to systemic thinking, from a lack of distinction between intention and behavior [45].

Meanwhile, ontological considerations pertain to the nature of reality that is reflected in the constructs. In basic philosophy, ontology is the study of what is said to exist. In the case of system governance, by its adherence to systemic precepts, it views the world as a system consisting of subsystems. It would entail an examination of the nature of system governance in society, the kinds of social processes, actions, events, and artifacts involved in governance, as well as the different levels affected by this governance. It would also be concerned with questions like: What precisely are the systems being governed, and who are those responsible for governing? What type of relationships exists with the greater environment? What are the engines of governance: a system of values, norms, laws, culture, politics, economics, or some combination of all these? Do these systems refer to entire social systems, or only aggregate or only individuals? What are the macro–micro relationships that need to be considered? In systems terms, what by-products of system governance may be considered as *emergent*? Emergence takes place as something new emerges which previously did not exist at a lower system level. Emergence, an important systems concept, is crucial in establishing the exact nature of the relation between micro and macro processes. Systemic thinking is based on the premise that society is a concrete system of interrelated individuals, and that some properties are aggregates of individual properties, while others are "global" and emerge because of relations between the individuals. The emergent properties must be studied at different levels in a system, and the relations between the levels must also be studied.

Next, there are methodological considerations, or just simply the methodology, which pertains to anything related to general method or technique. From a systemic view, the methodology helps to maintain the interconnections, both in terms of concrete things, ideas, and knowledge of the problems or phenomena under study. In general, methodology looks at the nature of this data—its meaning, how it should be interpreted, possible means of validation among others. However, Guba and Lincoln [33] suggest that methodology is constrained by earlier epistemological and ontological assertions. Take, for instance, the role of the observer/inquirer, where the observer's conception of social systems would influence their actions regardless of whether their conceptions are justified to be right or wrong. A systemic methodological consideration should therefore start "from individuals embedded in a society that pre-exists them and watch how their actions affect society and alter it" [17]. Johannessen and Olaisen [46] further added that a systemic approach must reasonably always include actors, observers, and social systems. The methodology should investigate the mental model actors have about their social system. An observer attempts to disclose the system's composition, environment, and structure. Social systems themselves have inherently specific processes and mechanisms that need to be disclosed. From all these, the methodology reflects the researcher's decision as to what needs to be analyzed (i.e., unit of analysis like individual, aggregate, organization, enterprise, and society). Thinking in terms of systems, this unit of analysis should be viewed considering its relationships with a larger system where it is a part of, and how it is involved with the lower-level system.

Lastly, there is axiology and ethics to enhance the systemic research paradigm. Although each has their specific place in philosophy, both will be discussed together in this section. Axiology is also known as value philosophy that refers to a philosophical school of thought "that examines the common ground for various forms of evaluations' [46]. Ethics, on the other hand, established the code of conduct of researchers. Specifically, ethics asks: "What is the role of moral norms in the development of theories, frameworks, and models?" Both axiology and ethics have objective and subjective elements that need to be made explicit given a specific situation or research purpose. Therefore, axiology and ethics as applied to considerations for a systemic research paradigm deal, among other things, with the question of the role of values/ethics in the research. Research based on a presumed value and ethical philosophy, specifically from a systems standpoint, will allow for an assessment of effectiveness in the eventual outcome of the research. Some research situations or purposes call for a concerted effort to address or study social phenomena or problems. These types of problems may be properly addressed if addressed by interdisciplinary/multidisciplinary teams that have similar axiology and ethical foundations. What is important for a systemic research paradigm is to allow axiology and ethics to achieve their defined goal while reflecting the objective needs and subjective wishes of actors at multiple levels of the system.

These include the key system tenets of system boundary, multiple perspectives, the notion of a system paradigm, and emergence. Adams [2] succinctly summarized these tenets among many others. These systems tenets are discussed below to draw out some underlying system foundations that may be relevant for system governance:

- *Systems boundary*—The notion of system should be understood as a representation of an entity as a complex whole open to exchange or feedback from its environment. Adhering to this tenet is crucial as it dictates a proper framing to problems of complexity (e.g., emergence), problems of social science (e.g., rational behavioral capacity), and problems of management (e.g., problem uniqueness) that are not comprehensively addressed by reductionist thinking.
- *Multiple perspectives*—The existence of macro- and micro- entities and their processes each can only provide at best partial contributions toward complete understanding. Any problem that uses the systems approach requires a full set of linkages for purposes of theorizing. The value of adopting a systems approach is drawn from the critical examination of simplifying assumptions. This helps to make explicit the limits of applicability, such that transformation of the relevant assumptions can possibly extend the application of scientific model building.
- *System paradigm*—Systems philosophy and the systems approach view systems as a function of their composition, environment, and structure, with the appreciation of the necessary linkages or mechanisms that specify their functional form. When presented with a problem, one must reflect on how to make explicit distinct but different interrelationships of the nature of the problem in terms of epistemological, ontological, methodological, axiological, and ethical considerations.
- *Emergence*—In systems, it is an instantiation of a transformation of something new which previously did not exist at a lower system level. Emergence is crucial in establishing the exact nature of the relation between micro and macro processes.

> The transformations apply in general to reductionist assumptions that wholes do not have properties apart from the properties of their components and to linear thinking about causation, composition, and control. In general, the premise of emergence is the revelation of interrelations of certain entities that have properties that are not simply aggregates of individual properties, or in other cases may be "global" because of relations between themselves. The emergent properties must be studied at different levels in a system, and the relations between the levels must also be studied.

In summary, by enriching our understanding of its history leading to what is now referred to as system philosophy and its approach, we can draw a rich context of important system tenets which will be foundational for the research. Up next is a review of the various research highlights related to the other key concepts on governance.

4 Governance Highlights

Like the last on systems, this section highlights the state-of-the-literature in governance research including an enumeration of the different ways "governance" has been understood in different disciplines and areas of practice, and to make a distinction between two broad categories, namely (1) the rationalist approaches and (2) the empirical school of thought on governance research.

4.1 A Litany of "Governance" Concepts

The meaning of governance is undergoing transformation and is far from offering any semblance of a generally accepted definition, perspective, or related practices. At first glance, studies have noted that there is an ambiguity between the concept and the practice of "governance" [88]. Walters further adds that beyond mere asymmetry of concepts and practice, the problem is deeper, going back to the actual presupposition roots and commitments in the implementation of "governance." Indeed, uncovering the history of governance over the years reveals the interestingly arbitrary deviations of the concept. There have been accounts that governance was originally first used by Plato himself. Historically, the origin of the word governance can be traced to the Greek verb "kubernân" or its Latin roots "gubernare." As early as a passage in Plato's classical work *Republic*, Plato himself used it metaphorically to indicate the fact of controlling men in the context of *steering* or *piloting* a ship [56]. Rosenau [76] emphasizes the value of recognizing governance as distinct but related to the concepts of command and control. He clarifies that governance is more expansive than the concept of command mechanisms which implies hierarchy and government. Governance most certainly is not limited to hierarchical processes of "framing goals, issuing directives, and the pursuit of policies" (p. 146). Instead, governance is closely related to the mechanisms relevant to control or steering. This highlights the purposeful nature of governance such that it may still evolve without any involvement of a hierarchy in place. He further promotes an idea of governance that is consistent with the concept of control which consists of relational phenomena that may comprise systems of rule that are used by the system to steer itself. By its relational nature, the dynamics of communication and control are important keys to the overall process of governance that are easily amenable to integration with system-based approaches. These are reflected in several of the definitions including governance purported in various works.

In another work, Eric Voegelin, a German political philosopher [86] regarded "governance" as *Herrschaft* (closely related to "governing" as *Herrschen*) and further acknowledged it to be a richly nuanced word and highly context dependent. That is easily interchangeable with ideas like *dominion, domination,* and *rule.* A lot has changed in the history of man and his social systems, but the notion of governance persists albeit in different forms and varying levels of articulation. Table below presents a sampling of some recent well-articulated meanings of "governance". From what the previous table has suggested, there are innumerable notions of governance (Table 1).

Tables 2 and 3 provide many more perspectives on the streams of governance one may encounter when examining the literature. Underlying these notions of governance, one may ponder what ideas or concepts reinforce each notion.

Vignette Water and Governance?

*At first glance, the terms water and governance may seem incompatible. However, the terms 'water governance' convey the political, social, economic, and administrative systems in place that influence water use and management. Essentially, it refers to who gets what water, when and how, and who has the right to water and related services and their benefits. It determines the equity and efficiency in water resource and services allocation and distribution and balances water use between socio-economic activities and ecosystems. Governing water includes formulating, establishing, and implementing water policies, legislation, and institutions and clarifying the roles and responsibilities of government, civil society, and the private sector concerning water resources and services. The outcomes depend on how the stakeholders act concerning the rules and roles that have been taken or assigned to them. The water sector is a part of broader social, political, and economic developments and is thus also affected by decisions by actors outside of the water sector (*https://www.watergovernance.org/governance/*what-is-water-governance/).*

4.2 *Rationalist "Governance"*

Rationalist approaches have afforded the formulation of knowledge utilizing base sets of theories, models, and ideas to provide an explanation for "governance." These rationalizations provide either a descriptive or prescriptive account of governance constructs. The logical starting points are sets of theories, propositions,

Table 1 Survey of "governance" from discipline and practice

Type	Definition/Description	Sources
General		
Process-centric	"A governing arrangement where one or more public agencies directly engage non-state stakeholders in a collective decision-making process that is formal, consensus-oriented, and deliberative and that aims to make or implement public policy or manage public programs or assets."	[3]
	"Social turbulence kept within bounds, and change steered in desired directions… preserves order and continuity, but not necessarily the maintenance of the status quo."	[25]
	"…the totality of conceptual ideas about these interactions" (these in relation to the act of governing)	[58]
Structure-centric	"…the activity of coordinating communications in order to achieve collective goals through collaboration."	[93]
	"…the reflexive self-organization of independent actors involved in complex relations of reciprocal interdependence, with such self-organization being based on continuing dialogue and resource-sharing to develop mutually beneficial joint projects and to manage the contradictions and dilemmas inevitably involved in such situations."	[44]
Hybrid	"…interdependence between organizations… continuing interactions between network members, caused by the need to exchange resources and negotiate shared purposes, … game-like interactions, rooted in trust and regulated by rules of the game negotiated and agreed by network participants,a significant degree of autonomy; they are self-organizing."	[75]
	"…the system of checks and balances, both internal and external to companies, which ensures that companies discharge their accountability to all their stakeholders and act in a socially responsible way in all areas of their business activity."	[13]
Restrictive corporate governance	"…the means for achieving direction, control, and coordination of wholly or partially autonomous individuals or organizations on behalf of interests to which they jointly contribute."	[64]
	"…the ways in which stakeholders interact with each other in order to influence the outcomes of public policies."	[10]
New public management	"…the processes and institutions, both formal and informal, that guide and restrain the collective activities of a group."	[53]

(continued)

Table 1 (continued)

Type	Definition/Description	Sources
Public policy	"...the emergence and recognition of principles, norms, rules and behavior that both provide standards of acceptable public behavior and that are followed sufficiently to produce behavioral regularities."	[52]
International security	Governance denotes the structures and processes which enable a set of public and private actors to coordinate their interdependent needs and interests through the making and implementation of binding policy decisions in the absence of a central political authority	[60]
Social and political	"...arrangements in which public as well as private actors aim at solving societal problems or create societal opportunities, and aim at the care for the societal institutions within which these governing activities take place."	[57]

Table 2 Core usages (Part 1): Governance "IS"

Governance "IS"	References
The act, process, or power of governing; government: The state of being governed	[32]
The activity of coordinating communications in order to achieve collective goals through collaboration	[93]
Mainly concerned with creating conditions for ordered rule and collective action	[82]
Stewardship of formal and informal political rules. Rules refer to measures that involve setting the rules for the exercise of power and settling conflicts over such rules	[39]
Emergence and recognition of principles, norms, rules, and behavior that both provide standards of acceptable public behavior and that are followed sufficiently to produce behavioral regularities	[53]
Entirety of interactions instigated to solve societal problems and to create societal opportunities; including the formulation and application of principles guiding those interactions and care for institutions that enable or control them	[59]

and/or principles that aim to provide an explanation for the process of governance (~descriptive) and how governance should be (~prescriptive). For instance, for a descriptive-rationalist overview, Buchinger [16] relates how the biological concept of "autopoiesis" and the philosophically oriented concept of "meaning" may be adapted to provide an explanation for governance in modern societies. Nicolescu [68] likewise suggests how different theories (such as agency theory, resource dependency theory, stakeholder theory, and stewardship theory) as well as varying organizational models (corporate, consensual, and shared organizational models) should be adopted to make sense of "governance" irregularities that plague the *system*. Then, there are rationalist-prescriptive accounts that characteristically show the use of specific concepts and trace them back to a specific problem domain or discipline practice

Table 3 Core usages (Part 2): Governance "AS"

Governance "AS"	Usage context
Corporate governance	How businesses should be directed and controlled. Posit openness (disclosure of information), integrity (straightforward dealing and completeness), and accountability (holding individuals responsible for their actions)
New public management	The introduction of corporate management techniques to the public sector (performance measures, managing by results, value for money, etc.) or marketization (introduction of incentive structures into public service); steering as a synonym for governance
Good governance	Government reform that encompasses systemic, political, and administrative dimensions (key concepts include distribution of power, promoting legitimacy and authority, accountable, and audited public service)
International interdependence	Multilevel governance
A socio-cybernetic system	Interdependence among social-political-administrative actors; shared goals; blurred boundaries between private, public and volunteer sectors; new forms of action, intervention, and control
New political economy	Interrelationships of the economy to civil society, the state, and the market economy
Networks	Self-organizing, autonomous, inter-organizational entities as an alternative to indirectly and imperfectly steer networks

like those by [14] for international relations, environmental development [27, 38] as well as primary clinical practice [83].

While there is a distinct set of literature constructs that mainly report on governance challenges in practice (see for instance [9, 61, 84]), a rationalist-prescriptive account posits the alternative use of other concepts such as polycentricity, participation, legitimacy, social capital, effectiveness, leadership, teamwork, and communication in relation to governance. The "rationalist" account, by way of minimizing the effort in scoping the examination of available literature of this nature, helped to critically examine the general themes of governance as they apply to this research.

4.3 Empirical "Governance"

Alternatively, another thrust of accumulated knowledge reflecting "system governance" may be found in studies that are empirical in nature. Due to the wide range of experience that may be considered as empirical, there are also several different configurations for empirical claims about governance. This diversity is expected across different disciplines but surprisingly, empirical evidence may also be divergent even within a single discipline. Consider the discipline of Public Administration, Rhodes [74] enumerates several diverse usages of governance as given in Table 3.

With the range of "governance" phenomena, one would assume a level of consistency within a single discipline. However, there is too much variation in the manner empirical evidence which is collected and the corresponding interpretations of that evidence. Kersbergen and Waarden [54] recently suggested that part of the difficulty lies in the problem of empirical identification which touches on the extent one is still able to sensibly describe new empirical phenomena using traditional conceptual tools (p. 164). Therefore, research in governance must consider that empirical data reflects the phenomena purported as governance may represent a shift in the phenomena itself, a shift in the causes confronting it, or even a shift in consequences or effects of the governance phenomena. Available empirical studies on governance only serve as supporting evidence for a particular account of governance from the perspective of one discipline [64].

In many of the above use cases, governance, as traditionally defined, is something related to government. Clearly over the years, it is now referred to as something broader than government as some of the above definitions imply. Where can we attribute the diversity of evidence constituting "system governance"? Part of the reason for such diverse accounts is because the identified "governance" concept is instantiated from a specific level with the involvement of users, approving bodies, sponsors, etc. [31, 36, 80, 91], mode—in terms of economic firms or assets, public, or private markets [24, 26, 35, 59], or order of governance—in terms of day-today affairs, institutional arrangements, or the general incorporation to practice of basic sets of values, norms, and principles [59]. Like Kooiman and Jentoft [59], who provided a conceptual framework to form the empirical logic of governance systems, there were also integrative governance studies that lie somewhere within the rationalist and empirical spectrum such as those by Brown et al. [15] and Garcia-Meca and Sanchez-Ballesta [29]. In these studies, new developments from other disciplines not traditionally associated with the practice of governance, such as risk management and earnings management, were incorporated. These types of research revealed some form of empirical coupling evident across different conceptual levels, modes, or order.

5 Synthesis of Themes for System Governance

It will not be surprising that the scope of governance literature just about covers any problem as a problem of governance. For instance, one account of the problem of governance in modern society suggests that it is a problem of adaptation, capacity, and scale [55]. Under the paradoxical reality of globalization and devolution, terms used to refer to the simultaneous *internationalization* and in parallel *localization* of traditionally government-centered decision processes, the agenda for modern governance must find ways to address these problems. The problem of adaptation, specifically in government, refers the need for non-traditional structured and staffed bureaucracies to support newer strategies and tactics, suggesting the role as "fitting traditional vertical systems to the new challenges of globalization and devolution, and integrating new horizontal systems to the traditional vertical ones" (p. 495) The problem of capacity is a call for effective management and accountability as enhancing government's ability to govern and manage effectively in this transformed environment. This is uncharted territory not accounted for in traditional intellectual foundations supporting hierarchical authority, bureaucratic exchange mechanisms, and delegation of power practices.

Closely related to the problems of adaptation and capacity, there is also the problem of scale that makes issues harder to address, as it remains unclear as to which levels of governance are best suited or best fit to address it. In other words, the problem of scale implies sorting out the functions of different levels of governance and finding better alternatives of channeling available capabilities rather than relying on ad hoc mechanisms most of the time.

Though examples were found in very distinctly different disciplines and problem domains, the rhetoric sounds all too familiar and almost resounding very similar themes. The next few sections in this chapter will espouse the general themes that this research has highlighted.

5.1 Need for a Systems Perspective

Theorizing system governance would imply an attempt at formulating an acceptable multilevel abstraction of the system. This allows for the accommodation of underlying worldviews to be made explicit and perceived *governance* situations to be accurately depicted. To help confront this issue, a systems-based approach is the primary study lens where perceived systems of interests will provide the focus to study generalizable aspects of governance situations. The process of governance and the system of interest themselves exist as independent societal entities and are embedded within the society at large. As such, they are easily captured conceptually as complex systems, as system-of-systems (SoS), or just simply, as systems. Motivated by several system-based principles, certain anticipated paradoxical divergences of perspectives help in resolving the practical difficulties in theorizing about governance. Keating [50], like Baldwin et al. [4], promoted the use of system-based articulations of context and its associated boundaries as the key tools in resolving such paradoxical perspectives. Whereas several definitions were available, Lycan [63] suggests a definition of paradox as "an inconsistent set of propositions, each of which is very plausible" where its resolution is a matter of deciding, on principled ground, which of the propositions are to be abandoned. This is the usual case and the domain of complex system governance. Paradoxes can be traced to propositional inconsistencies arising from philosophical, methodological, axiological, axiomatic, and even application logical levels of divergence [50]. Without a way to study these paradoxes, it would be impossible to even begin to understand how to design or embark on development of a *system governance* platform that would make sense with the vast array of other relevant theories and/or frameworks. Any resemblance to replicable governance phenomena, though interesting and novel, is coincidental and, at best, existential in the context of time, place and prevailing logic of someone else's decisions and actions. In other words, while there are examples of the utility in examining accounts of governance, the main argument in this work is toward an attempt for a well-articulated universal governance concept. It is a grand and complicated effort, but it should be attempted nonetheless because of its greater relevance to resolving paradoxical dead ends that confound day-to-day practice related to governance.

Hence, moving forward it would be convenient to explore the notion of the concept of governance in greater depth. Current understanding of governance is either conceived too broadly or too narrowly, limiting the recognition of the paradoxical phenomena that carries over to conflicting approaches of implementation.

5.2 *Diverse Notions of Governance*

The literature is replete with studies that are about governance but are totally standing on very dissimilar conceptual bases. To date, there is still no comprehensive conceptual account of "governance" [47, 56]. This does not imply a shortage of well-thought rigorous scholarly studies at all. In fact, several works on the usual "what" question have been articulated quite sufficiently and extensively [58, 71, 82]. Multidisciplinary literature would reveal two prevailing perspectives in the practice of "governance." Either governance is deployed supposedly for a *system of interest* for purposes of (i) maintaining its operation despite any recurring problem, and/or (ii) adapting its capabilities in anticipation of future challenges. While it is the contention in this study that existing governance systems were predominantly designed toward either one of the previously mentioned perspectives, new and existing governance systems will benefit from analysis that reaches back to basic concepts and approaches supporting such perspectives. Most governance systems will have to merge both perspectives given their underlying purposes. Such an appreciation is starting to emerge as evidenced by many studies about governance within the specific topical contexts of the Internet [67], urban culture [69], knowledge [81], enterprise information systems [66], networks [72], resilience and vulnerability [30] to name a few.

In some general sense, all these initiatives seem to converge on governance as either the last resort solution or as the ultimate cause of failure. There are several successful realizations where resulting outcomes can be evaluated against some theoretical backdrop of "governance." In each of those instantiations, however, the claims will not allow for enough comparison to suggest similar conceptualizations of "governance." In some instances, one implicitly assumes that "governance" is viewed not as the problem but the solution. Conversely, the problem perspective is stated in terms of the "lack of" where new efforts toward correct "governance" will progress toward improvement. There is also the difficulty to clearly draw out what is being governed and to what end. Presumably, a system is assumed at the receiving end where governance reflects the effort to realize a system's purpose. Each unique system state often invariably requires its own unique kind of governance which was also identified as a gap in the literature. The current state is described by an internal differentiation of dynamics and complexity residing within the system in relation to its environment [62]. There are of course several available ways to reveal the state of a system by way of systematic classifications or typologies [1, 11, 12, 79, 90]. These have been instrumental in advancing understanding that are useful for application in real-life complex systems. Therefore, the rich diversity of interpretations for governance brings to light a key systems concept, specifically the notion of multiple perspectives. This consideration has implications for anyone responsible for the design, development, or transformation of governance systems. They will have to utilize these perspectives to comprehensively allow the system to accomplish their purpose.

5.3 *Irresolvable Conflicts of Perspectives*

Several reasons for conflicts in perspectives on governance are traceable to the multiple "levels" and roles of different actors and their associated interests in implementing governance. Because each perspective held by every actor is important in the actual implementation of governance, blurring of traditional "functional" boundaries (i.e., political, administrative, public, private, etc.) is inevitable. Having no clear delineation presiding over practice, the active "governance" concept is a tenuous implementation of overlapping and often conflicting hierarchical and network/collaborative paradigms. We can draw perspectives based on both assumptions from a single very recent real-life example—the US financial market collapses that triggered damaging effects throughout the global economy. Depending on how an individual's epistemological stance or knowledge boundaries are drawn, one can make a good case either way that some form of governance already exists or was in fact absent. Before the financial collapse, the financial market is a good case example of sophisticated layers of governance. Governance in the financial market can be described as a dizzying array of regulations, policies, laws, and standards through a complex interaction between public, private, and government sectors [93]. Shortly after the collapse, everyone was insisting on better governance as a pressing concern since taxpayers' money was used for bailout or stimulus money. However, if one is a keen fan of Adam Smith's genius, the financial market as it was conceived was one that can function without any individual's awareness of obvious governance, whether minimal or if any at all. Hence since then, free markets are famous for the "Invisible Hand" metaphor [92]. This shows that no matter which assumption is held, governance is perceived sometimes as a solution and sometimes as the problem.

5.4 Uncovering Underlying Philosophical Debates

Undoubtedly, there are much larger philosophical roots underlying the debates that feature these differing perspectives. This goes back to the great debates between philosophers like Plato, Aristotle, and much more recently Kant regarding the very nature of existence, of reality, of knowledge and of truth, of wholes, and of entities [77]. It is not the intent of this work to offer a resolution to these debates as they are expected to persist irrespective of any ongoing scholarly deliberation of governance. Instead, it is supposed that to have a good foundational understanding of governance, an integrative philosophy should be adopted that is appreciative of the different ontological, epistemological, and axiological perspectives found in the literature. While governance can mean very different things based on which philosophical strand dominantly persists, it will be helpful to establish the preliminary conceptual boundaries before going any further in this study.

6 Critique of the Literature

The focus of this critique revolved around (i) the conceptual ambiguities underlying theories of "systems" and "governance and (ii) the absence of a specific set of criteria to be able to compare and assess existing and new theories related to governance of complex systems.

6.1 Need to Address Conceptual Ambiguities

Jessop [43] notes that "governance" according to its usage in the social sciences may often be considered as still "'pre-theoretical' and eclectic; and lay usages are just as diverse and contrary." Further, Jessop observed that the conceptual interest in governance clearly has "precursors of the current interest in governance in various disciplines" (p. 31). These precursors call out a distinct set of assumptions, models, and theories that bring about a concept of governance characterized by *heterarchy*, understood as 'self-organization across different levels. Walters [88] likewise observed that despite the growing prominence of governance and its use in policy circles, that "(T)here is still a striking imbalance between the exponential growth of literature applying governance to particular cases and areas, and research that critically examines the foundation assumptions and political implications of governance (p. 27)." He also noted that "there are also continuities, certain core ideas, assumptions, propositions which attach to the term as it moves from one locale to the next." These comments, however, are still made within the purview of a single discipline—political science. There is yet a reconceptualization that marries insights from different disciplines, although there are already applications across different problem domains. Therefore, there is a need to formulate a theory of "governance" that adequately analyzes the various conceptual underpinnings or presuppositions. Hence, as alluded to by joining the term "systems," what should be attempted here is a reconceptualization that synthesizes "governance" in terms of more general *systems*.

6.2 Lack of a Criteria Set for Theorizing and Practice

Meanwhile, due to the diversity of theorizing practices, there is also a need to establish an agreed set of criteria as a basis for theorizing and practicing normative concepts of governance. Four different categories of criteria will be presented. These different criteria cover ontological, theoretical, pragmatic, and axiological grounds. These different area categories are summarized in the following vignette.

Criteria set for 'theorizing' on system governance

We suggest that governance should be seen through the lens of ***ontology****, where concerns of scope and simplicity (e.g., parsimony) address the principal question of "What can be said to exist?". It should also focus on the* ***theoretical*** *implications that embody a degree of testability given presented evidence and conservatism when compared with other related theories. It needs to be* ***pragmatic*** *through the judgment of a posited theory by its usefulness. Finally, offer* ***axiological*** *implications, where the suggested theory tracks the "truth" based on some measure of value, worth, and quality.*

An *ontological* criterion, in the case of system governance, should consider treatment of ontological issues concerning the "levels of analysis" and the "status of entities" that are posited in the theories. The scope of the suggested theory should be able to arrive at the same level of resolution as to the type of questions we expect governance to answer. Simplicity refers to the use of a generic set of forces and entities for as broad a scope of "governance" phenomena. A *theoretical* criterion implies that any scientific explanatory theory on governance should be responsive to evidence, in the sense that it is able to accommodate a wide range of evidence (does not mean insulate itself from possible counterexamples). Another theoretical criterion is that the posited theory should fit with nearby theories (conservatism or principle of theoretical unification). *Pragmatic* criteria have two routes to applying this either through (i) its theoretical merit and/or (ii) its methodological merit. Theoretical merit asks a predetermined set of relevant "why" questions. The methodological aspect refers to how a good theory often also offers indications of the right level of resolution (unit of analysis) and techniques to manipulate the phenomena under investigation. Lastly, an *axiological* criterion is mostly important to be able to drive the other earlier suggested criteria. This is what sets apart normative theories from descriptive theories. A good theory tracks the "truth" if it makes good predictions and generally fits the data, as a basis for setting a baseline to pursue action/intervention.

Having understood how these different criteria can be applied; suggested theories related to "governance" can be assessed, clarified, and dismissed from consideration, or to be used in support of development of a better conceptual definition for governance. Any indication of a good theory on governance or for any theory on any phenomena for that matter should be assessed based on some acceptable criteria set. In the case of governance, any theory posed is reviewed against ontological, theoretical, pragmatic, and axiomatic grounds. One such exemplar articulation is a growing body of knowledge related to complex system governance (CSG) that supports the view of governance from the perspective of cybernetics and systems theory. [49] suggest an evolved definition of CSG as the "Design, execution, and evolution of the [nine] metasystem functions necessary to provide control, communication, coordination, and integration of a complex system.". A quick reframing of this definition aligns the notion of a "metasystem" as an appropriate focus and analytic resolution for its ontological and theoretical basis. As part of its metasystem construct, we are presented a set of nine interrelated functions that each focus on different aspects of the performance of a system, including (i) policy and identity, (ii) system context, (iii) strategic system monitoring, (iv) system development, (v) learning and transformation, (vi) environmental scanning, (vii) system operations, (viii) operational performance, and (ix) information and communications. With these nine functions, there are now pathways to a more pragmatic as well as an axiological grounded framework toward a purposeful, "holistic" and comprehensive approach when addressing problems pertaining to the design, execution, and evolution to sustain and evolve system performance. In CSG, it becomes evident that while systems may be different (e.g., operational, tactical, managerial), there can be a great degree of interconnectedness tied together by the meta-aspects of the system. Once again, the metasystem idea highlights consideration of the whole, including interactions, complexity, emergence, and ambiguity traversing the boundaries within and external to a single system.

In practice, systems and the analysis of any governing capacity can benefit from insights drawn from systems principles and the perspective they invoke as in the CSG framework. The CSG framework draws from system principles like emergence, holism, and complementarity that forms a basis of a language to inform governance thinking from a systems theoretic perspective. To fully design an existing system, geared toward system performance, it is crucial to recognize that governance encompasses both the control and informational linkages of the system. These linkages originate across different metasystem functions of governance and involve various implementing practitioners tasked with executing the purpose of the system. In addition, these interactions, and in turn their input/output exchange relationships, exist within the contextual environment. The governance challenge recognizes that inevitably practitioners face the difficulties of dealing with systems, events, and contexts that cannot be fully grasped. Rightfully so, the concern needs to shift instead to how best to implement governance while representing the "true" situation building on advances in our toolkit of frameworks, theories, concepts, and methods.

7 Summary

In summary, the literature review showed that several disciplines advanced certain versions of systems and governance without regard for a wider multidisciplinary perspective of system governance. Adopting a multidisciplinary purview as the primary impetus, the challenge was to investigate the ambiguous nature of relevant ideas for a more precise articulation of system governance. These entailed a thorough investigation at the conceptual and empirical level of governance-related situations that reflect the mental images, memories, concepts, propositions, theories, inferences, problems, and many more. This resulted from a deep investigation of the state-of-the-art in diverse research in systems theory and in governance practice.

As such, the body of knowledge introduced here highlights the multidisciplinary lens to investigate system governance. Having implemented a thorough literature review process, an overview of the body of knowledge (BoK) was produced to help narrow down the key literature boundary themes on system governance. Both systems and governance are well studied terms with each having undergone advanced conceptual development and a long history from the purview of multiple independent disciplines and practice domains [7, 10]. System governance, however, is not an easy transition from both key ideas (e.g., systems and governance), although there were already a few recent studies which used the compound notion of "system governance" [8]. The difficulty was in the heterogeneous paradigms and plurality of conceptions expected when associated *ideas* were cultivated from the diverse world of traditional disciplines and practice [23, 54]. These were evidenced by a set of systemic themes emerging from the literature. Finally, the chapter concluded by presenting a critique of the literature. The focus of the given critique revolved around (i) the conceptual ambiguities underlying theories of *systems* and *governance* and (ii) the absence of specific criteria set to be able to compare and assess existing and new theories.

8 Exercises

The following exercises provide an opportunity to examine the concepts presented in this chapter through several questions.

1. Imagine the 2021 wildfires in the West Coast that affected families, communities, threatening livelihoods, and entire economies as our "hypothetical" problem context. Find out the definition of a *system* that is relevant to this context. Write up (i) the specific system purpose, (ii) the system boundaries, and (iii) the system(s) elements as well as interrelationships that are meaningful for such a *system*.
2. With the same problem context laid out in the previous question, what does *governance* look like from the perspective of (i) the homeowners, (ii) the insurance industry, (iii) the local and state leadership?

3. Give an example of a system and discuss how management is the same/ or differs from governance.

References

1. Ackoff RL (1971) Towards a system of systems concepts. Manage Sci 17(11):661–671
2. Adams KM (2011) Systems principles: foundation for the SoSE methodology. Int J Syst Syst Eng 2(2):120–155
3. Ansell C, Gash A (2007) Collaborative governance in theory and practice. J Public Adm Theory 18:547–571
4. Baldwin W, Sauser B, Boardman J, John L (2010) A typology of systems paradoxes. Inf, Knowl, Syst Manag 9(1):1–15
5. Bar-Yam Y (1997) Dynamics of complex systems. Westview Press Boulder, Colorado
6. Bertalanffy L (1950) The theory of open systems in physics and biology. Science 111(2872):23–29
7. Bevir M (2004) Governance and interpretation: What are the implications of postfoundationalism? Public Administration 82(3):605–625
8. Bevir M (2006) Democratic governance: systems and radical perspectives. Public Adm Rev 66(3):426–436
9. Biermann F, Pattberg P (2008) Global environmental governance: taking stock, moving forward. Annu Rev Environ Resour 33:277–294
10. Bovaird T (2005) Public governance: balancing stakeholder power in a network society. Int Rev Adm Sci 71(2):217–228
11. Boulding KE (1956) General systems theory-the skeleton of science. Manage Sci 2(3):197–208
12. Boulding K (1985) The world as a total system. Sage Publications.
13. Brennan NM, Solomon J (2008) Corporate governance, accountability and mechanisms of accountability: an overview. Account, Audit & Account J 21(7):885–906
14. Brinkerhoff D (2005) Rebuilding governance in failed states and post-conflict societies: core concepts and cross-cutting themes. Public Adm Dev 25(1):3–14
15. Brown I, Steen A, and Foreman J (2009) Risk management in corporate governance: a review and proposal. Corporate governance—an international review 17(5):546–558.
16. Buchinger E (2006) The sociological concept of autopoiesis: biological and philosophical basics and governance relevance. Kybernetes 35(3/4):360–374
17. Bunge MA (1996) Finding philosophy in social science. Yale University Press
18. Bunge MA (2000) Systemism: the alternative to individualism and holism. J Socio-Econ 29:147–157
19. Carley S, Porter A, Rafols I, Leydesdorff L (2017) Visualization of disciplinary profiles: enhanced science overlay maps. J Data Inf Sci 2(3):68–111. https://doi.org/10.1515/jdis-2017-001
20. Casti J (1981) Systemism, system theory and social system modeling. Reg Sci Urban Econ 11:405–424
21. Checkland P (1981) Systems thinking, systems practice. John Wiley and Sons, Chichester, UK
22. Digby J (1989) Operations Research and Systems Analysis at RAND. Tech Rep N-2936-RC: 1948–1967. The RAND Corporation.
23. Dixon J, Dogan R (2003) Analyzing global governance failure: A philosophical framework. J Comp Policy Anal: Res Pract 5(2–3):209–226
24. Driver C (2008) Varieties of governance. Recherches Economiques De Louvain- Louvain Economic Review 74(4): 425-+.
25. Dunsire A (1990) Holistic governance. Public Policy Adm 5(1):4–19.
26. Fligstein N, Choo J (2005) Law and corporate governance. Annu Rev Law Soc Sci 1:61–84

27. Folke C, Hahn T, Olsson P, Norberg J (2005) Adaptive governance of social-ecological systems. Annu Rev Environ Resour 30:441–473
28. Forrester J (1961) Industrial dynamics. Pegasus Communications, Waltham
29. Garcia-Meca E, and Sanchez-Ballesta J (2009) Corporate governance and earnings management: a meta-analysis. Corporate governance—an international review 17(5):594–610.
30. Gheorghe AV (2004) Risks, vulnerability, sustainability and governance: a new landscape for critical infrastructures. Int J Crit Infrastruct 1(1):118–124
31. Gideonse H (1993) The governance of teacher-education and systemic reform. Educ Policy 7(4):395–426
32. "Governance" (2004) The American Heritage (c) Dictionary of the English Language, 4th edn. (Houghton Mifflin Company). http://dictionary.reference.com/browse/governance
33. Guba E, Lincoln Y (1994) Competing paradigms in qualitative research. In: Denzin NK, Lincoln YS (eds) Handbook of qualitative research. Sage, Thousand Oaks, CA, pp 105–117
34. Hall A (1965) Systems engineering from an engineering viewpoint. IEEE Trans Syst Sci Cybern 1(1):4–8
35. Hawley J, Williams A (2005) Shifting ground: emerging global corporate-governance standards and the rise of fiduciary capitalism. Environ Plan A 37(11):1995–2013
36. Hill C, Lynn L (2005) Is hierarchical governance in decline? Evidence from empirical research. J Public Adm Res Theory 15(2):173–195
37. Hitch C (1955) An appreciation of systems analysis. J Oper Res Soc Am 3(4):466–481
38. Huitema D, Mostert E, Egas W, Moellenkamp S, Pahl-Wostl C, and Yalcin R (2009) Adaptive water governance: assessing the institutional prescriptions of adaptive (co-) management from a governance perspective and defining a research agenda. Ecol Soc 14(1): 26. http://www.ecologyandsociety.org/vol14/iss1/art26/
39. Hyden G (1999) Governance and the reconstitution of political order. In: Joseph R State, Conflict, and Democracy in Africa. Rienner, Boulder
40. Jackson MC (1985) Social systems theory and practice: the need for a critical approach. Int J Gen Syst 10:135–151
41. Jackson MC (1991) The origins and nature of critical systems thinking. Syst Pract 4(2):131–149
42. Jackson MC, Keys P (1984) Towards a system of systems methodologies. J Oper Res Soc 35(6):473–486
43. Jessop B (1998) The rise of governance and the risks of failure: the case of economic development. Int Soc Sci J 50(155):29–45
44. Jessop B (2003) Governance and metagovernance: On reflexivity, requisite variety, and requisite. In: Bang HP (ed) Governance, as Social and Political Communication. Manchester University Press, Manchester, pp 142–172
45. Johannessen J, Olaisen J (2005) Systemic philosophy and the philosophy of social science—Part I: transcendence of the naturalistic and the anti-naturalistic position in the philosophy of social science. Kybernetes 34(7/8):1261–1277
46. Johannessen J, Olaisen J (2005) Systemic philosophy and the philosophy of social science: Part II: the systemic position. Kybernetes 34(9/10):1570–1586
47. Jose J (2009) Conceptualising 'Governance': Discourse, theory and ontology. In: Paper presented at the annual meeting of the WPSA Annual Meeting "Ideas, Interests and Institutions", Vancouver, BC, Canada Online. http://www.allacademic.com/meta/p316878_index.html
48. Kauffman S (1993) The origins of order: self-organization and selection in evolution. Oxford University Press, Oxford
49. Keating CB, and Katina PF (2019) Complex system governance: Concept, utility, and challenges. Syst Res Behav Sci 36: 1–1. doi: https://doi.org/10.1002/sres.2621
50. Keating CB (2005) Research foundations for system of systems engineering. In: 2005 IEEE international conference on Systems, Man and Cybernetics, vol 3, pp 2720–2725 https://doi.org/10.1109/ICSMC.2005.1571561.
51. Keating CB, Katina PF (2011) System of systems engineering: prospects and challenges for the emerging field. Int J Syst Syst Eng 2(2/3):234–256

52. Keohane R, Nye J (1989) Power and interdependence. Harper Collins, New York
53. Keohane R, and Nye J (2000) Governance in a globalizing world. In: Nye J and Donahue J (eds) Governance in a globalizing world. Brookings Institution, Washington DC
54. Kersbergen K, Waarden F (2004) 'Governance' as a bridge between disciplines: cross-disciplinary inspiration regarding shifts in governance and problems of governability, accountability and legitimacy. Eur J Polit Res 43:143–171
55. Kettl D (2000) The transformation of governance: globalization, devolution, and the role of government. Public Adm Rev 60(6):488–497
56. Kjær A (2004) Governance: Key concepts. Polity Press, Malden, MA, USA
57. Kooiman J (2000) Societal governance: levels, models and orders of social-political interaction. In: Pierre J (ed) Debating governance. Authority, steering and democracy. Oxford University Press, UK, pp 138–166
58. Kooiman J (2003) Governing as governance. Sage, London
59. Kooiman J, Jentoft S (2009) Meta-governance: values, norms and principles, and the making of hard choices. Public Adm 87(4):818–836
60. Krahmann E (2003) Conceptualizing security governance. Coop Confl 38(1):5–26
61. Lemos M, Agrawal A (2006) Environmental governance. Annu Rev Environ Resour 31:297–325
62. Luhmann N (1977) Differentiation of society. Can J Sociol/Cah Can Sociol 2(1):29–53
63. Lycan WG (2010) What, exactly, is a paradox? Analysis 70(4):615–622
64. Lynn L, Heinrich C, Hill C (2000) Studying governance and public management: challenges and prospects. J Public Adm Res Theory 10(2):233–261
65. Maier MW (1998) Architecting principles for systems-of-systems. Syst Eng 1(4):267–284
66. Marks E (2008) Service-oriented architecture governance for the services driven enterprise. John Wiley & Sons, Hoboken, New Jersey, USA
67. Mathiason J (2009) Internet governance: the new frontier of global institutions. Taylor & Francis, New York, USA
68. Nicolescu L (2010) Governance in higher education: theories and practices applied in the metallurgical industry. Metal Int 15:201–205
69. Ostrom V (2008) Constitutional foundations for a theory of system comparisons. In: Sproule-Jones M, Allen B, Sabetti F (eds) The struggle to constitute and sustain productive orders: Vincent Ostrom's quest to understand human affairs. Lexington Books, Plymouth, UK, pp 11–26
70. Pickel A (2007) Rethinking systems theory: a programmatic introduction. Philos Soc Sci 37(4):391–407
71. Pierre J (ed) (2000) Debating governance: Authority, steering, and democracy. Oxford University Press, New York, NY, USA
72. Provan K, Kenis P (2008) Modes of network governance: structure, management, and effectiveness. J Public Adm Res Theory 18(2):229–252
73. Rafols I, Porter AL, Leydesdorff L (2010) Science overlay maps: a new tool for research policy and library management. J Am Soc Inform Sci Technol 61(9):1871–1887
74. Rhodes RA (2000) The governance narrative: key findings and lessons from the ESRC's Whitehall programmes. Public Administration 78(2):345–363
75. Rhodes RA (2007) Understanding governance: ten years on. Organ Stud 28(8):1243–1264
76. Rosenau JN (1997) Along the domestic-foreign frontier: Exploring governance in a turbulent world. Cambridge University Press, Cambridge, UK
77. Santas G (2001) Goodness and justice: Plato, Aristotle, and the moderns. Malden, MA, USA: Wiley-Blackwell.
78. Senge PM (1980) A system dynamics approach to investment-function formulation and testing. Socioecon Plann Sci 14(6):269–280
79. Simon H (1962) The architecture of complexity. Proc Am Philos Soc 106(6):467–482
80. Slowther A, Boynton P, Shaw S (2006) Research governance: ethical issues. J R Soc Med 99(2):65–72

81. Stehr N (2004) A world made of knowledge. In: Stehr N The Governance of Knowledge. Transaction Publishers, New Brunswick, pp 9–26
82. Stoker G (1998) Governance as theory: Five propositions. Int Soc Sci J 50(155):17–28
83. Tait A (2004) Clinical governance in primary care: A literature review. J Clin Nurs 13(6):723–730
84. Tickell C (1997) Security, environment, and global governance. Interdisciplinary Sci Rev 22(3):246–250
85. Ulrich W (1983) Critical heuristics of social planning: a new approach to practical philosophy. Haupt, Berne
86. Voegelin E (2003) The theory of governance and other miscellaneous papers, 1921–1938. University of Missouri Press
87. Von Foerster H (1979) Cybernetics of cybernetics. In: Krippendork K (ed) Communication and control in society. Gordon and Breach, New York
88. Walters W (2004) Some critical notes on 'governance.' Stud Polit Econ 73:27–46
89. Wan PY (2011) Emergence à la systems theory: epistemological total ausschluss or ontological novelty? Philos Soc Sci 41(2):178–210
90. Weaver W (1948) Science and complexity. Am Sci 36:536–544
91. Whitehead M (2003) "In the shadow of hierarchy": meta-governance, policy reform and urban regeneration in the West Midlands. Area 35(1):6–14
92. Williamson OE (1994) Visible and invisible governance. Am Econ Rev 84(2):323–326
93. Willke H (2007) Smart governance: governing the global knowledge society. Campus Verlag GmbH, Frankfurt, Germany

Systems Theory for Complex System Governance

Kaitlynn Castelle, Joseph M. Bradley, and Charles W. Chesterman Jr.

Abstract This chapter introduces a refinement on the conceptual framework for the placement of systems theory propositions relevant to complex systems and provides a perspective for understanding their linkages through systems theory axioms. An overview of the evolution of the framework to its current state is provided. The expanded framework offers a taxonomy of axioms and related concepts to support complex systems analysis from a governance perspective. A view of a complex system through this framework supports an enriched view of the total system. The logical interrelations between the identified axioms may be beneficial in understanding of different aspects of a complex system and provide a referential foundation from which to evolve our systems thinking capacity. Use of the framework supports complex systems analysis through articulation of complementary perspectives and relation to systems theory propositions of the complex system to enhance decision-making and governance.

Keywords Systems theory · Complex system governance

1 Introduction

The historical background and academic literature associated with the definition of "system" and its associated properties is a rich reading of philosophical writings as well as the evolution of science, engineering, and social studies. Systems literature

K. Castelle (✉)
Department of Engineering Management and Systems Engineering, Old Dominion University, Norfolk, USA
e-mail: kcastell@odu.edu

J. M. Bradley
National Centers for System of Systems Engineering, Old Dominion University, Norfolk, USA
e-mail: josephbradley@leading-change.org

C. W. Chesterman Jr.
Department of Engineering Management and Systems Engineering, Old Dominion University, Norfolk, USA
e-mail: cwjstellami@aol.com

C. B. Keating et al. (eds.), *Complex System Governance*, Topics in Safety, Risk, Reliability and Quality 40, https://doi.org/10.1007/978-3-030-93852-9_4

explains the properties and behaviors of physical, chemical, biological, social, and economic systems, as well as others. Jackson states "a system is a complex whole the functioning of which depends on its parts and the interactions between those parts" [1] (p. 3). There have been two perspectives when observing or studying systems.

> The traditional, scientific method for studying such systems is known as reductionism. Reductionism sees the parts as paramount and seeks to identify the parts, understand the parts, and work up from an understanding of the parts to an understanding of the whole [1] (p. 3).

An alternative to reductionism is,

> Holism considers systems to be more than the sum of their parts. It is of course interested in the parts and particularly the networks of relationships between the parts, but primarily in terms of how they give rise to and sustain in existence the new entity that is the whole [1] (p. 4).

While the understanding of a system may follow the description provided by Jackson or as attributed to Stafford Beer, a system is what it does, and an underlying intent of the provided conceptual framework is that a system is: "Thus, a system may be identified as such if it exhibits and can be understood within this set of axioms. Conversely, any entity that exhibits these seven axioms is, by definition, a system." [2] (p. 120). Accordingly, a universally agreed-upon definition for systems theory does not exist at present, though the term is ubiquitous in systems literature.

Adams et al. [2] proposed a systems theory construct, resting upon an axiomatic set supported by a set of cited propositions from systems theory literature. Whitney et al. [3] revised the construct based on additional research and constructive feedback from the community. This construct was developed by use of the axiomatic method that will be described. This resulting construct affords both practitioners and theoreticians a prescriptive set of axioms by which a system must operate; conversely, any entity defined as a system will be characterized by a set of seven (7) axioms: contextual axiom, purpose axiom, design axiom, operational axiom, centrality axiom, information axiom, and viability axiom. These axioms are presently organized to conform to the discoverers' induction as proposed by William Whewell, where knowledge can be constructed through the union of sensations and ideas [4]. The use of this inductive inference methodology provided insight of the common themes integrated among systems theory propositions in order to produce a set of axioms that describe systems.

2 Systems History

Between WWI and WWII, a multidisciplinary problem-solving research effort began that incorporated a decomposition of the problem system into individual problems that were related to the respective fields in which they applied. These disparate problems were then to be solved independently of each other, and the independent solutions were later aggregated. As can be anticipated, this approach was later realized as ineffective. Ackoff [5] notes that "different terms are used to refer to the same thing, and the same term is used to refer to different things. This state is aggravated by the fact that the literature of systems research is widely dispersed and is, therefore, difficult to track. Researchers in a wide variety of disciplines and interdisciplinary are contributing to the conceptual development of the systems sciences, but these contributions are not as interactive and additive as they might be" [5] (p. 661).

Thus interdisciplinary research began, in which representatives from different disciplines confronted problem complexes together to solve them collaboratively. The growth of systems theories commenced from immense pressure to develop theories capable of interdisciplinary application. In 1954, biologist von Bertalanffy, economist Kenneth Boulding, physiologist Ralph Gerard, and mathematician Anatol Rapoport collaborated at the Palo Alto Center for advanced study in behavioral sciences, where they discovered the wide applicability of their convergent thoughts stemming from their different fields of study [6]. They soon formed the original bylaws for the foundation of the Society for General Systems Research (SGSR) to: (i) investigate the isomorphy of concepts, laws, and models from various fields and to help in useful transfers from one field to another, (ii) to encourage development of adequate theoretical models in fields which lack them to minimize the duplication of theoretical effort in different fields, and (iii) to promote the unity of science through improving communications among specialists [7] (pp. 435–436).

Then, von Bertalanffy [6] continued writing on the subject throughout his career, recognizing the gravitation toward integrated natural and social sciences, centered in systems theory. He noted that by unifying principles expressed in dissonant fields, the effort could eventually lead to a "much-needed integration in scientific education" [6] (p. 37). Biologist Paul A. Weiss declared that this conceptual integration would "render the map of knowledge more complete and more consistently coherent" [8] (p. 159).

There has not been a full adoption of a generally accepted canon of systems theory within the discipline, albeit the potential for systems theory has been realized in theory or practice, as noted by Checkland [9]. Still, practitioners can greatly benefit from the body of knowledge that does exist, which certainly provides necessary propositions that are relevant for common practice.

3 Discoverers' Induction Methodology and Criteria for Inclusion

This section will discuss the use of discoverers' induction as proposed by William Whewell where knowledge can be constructed, in particular the use of this inductive inference methodology provided insight of the common themes integrated among systems theory propositions in order to produce a set of axioms that describes systems. The axioms as they are currently organized conform to the discoverers' induction as proposed by William Whewell where knowledge can be constructed through the union of sensations and ideas [4]. The use of this inductive inference methodology provided insight of the common themes integrated among systems theory propositions enabling the formulation of a set of axioms that describes systems. There are two steps to discoverer's induction, as follows [4]: First, colligate known members of a class by the use of an idea or conception and second, generalize this concept over the complete class, including its unknown members.

Colligation, as defined by Snyder, is "the mental operation of bringing together a number of empirical facts by "super inducing" upon them some idea or conception that unites the facts and renders them capable of being expressed by a general law" [4] (p. 585). This new knowledge adds to the current body of facts, causing them to be seen in a new light. With respect to systems theory, there is an elucidation of the generalized properties of systems.

Generalizable knowledge projected onto unknown members of a class (i.e., unidentified propositions that would support the development of axioms) suggests that the listing of proposed axioms may be incomplete or omitting some aspect of absolute truth.

3.1 Axioms

The purpose of a systems theory construct is to unify the large set of systems theory concepts related to systems studied in academic literature and broad field of systems research, to develop an organizing construct for understanding and studying systems. Axioms capture irrevocable truths that can be universally accepted for the sake of studying systems, in that they have been regarded as established, legitimate, and accepted without further demands for justification:

- Publication as organizing construct in multiple venues for systems literature
- Acceptable to experts/scholars in the field.

Axioms are at the core of the systems theory construct formulated for complex system governance (CSG) and convey themes about systems as supported by the systems theory propositions.

3.2 Proposition

A proposition is a principle, law, or concept presented for consideration as it pertains to the inherent nature of a system by providing insight about the qualities or tendencies of systems, as articulated in empirical research in a variety of disciplines that discuss systems. Propositions reflect the current state of knowledge, without assuming fundamental, universal *truth* about a system. They reflect a widely accepted set of concepts proposed about systems, through empirical research, and discussed in the body of systems theory literature. The following captures the criteria for inclusion as a proposition:

- Multiple citation in the systems literature or capture in seminal work
- Acceptable to experts/scholars in the field
- Provides explanatory or predictive power for system behavior, structure, or performance
- Assignable to an existing axiom or foster creation of a new axiom.

Ontology is the study of what is and Epistemology is the study of knowledge and justified belief. This chapter does not limit the system theory concepts to an objective ontological and epistemological perspective basis. Table 1 below, drawn from [10], lays out a topology of thinking and the developed views from the subjective to the objective approaches to social science.

4 Framework

This section introduces the systems theory framework that provides a basis for complex systems governance. The inductive analysis resulted in the following set of supporting propositions mapped to axioms. The systems theory axioms are provided below. Section 5 will expand on each of the axioms for systems theory with descriptions and their primary proponents in systems literature.

Table 1 Network of basic assumptions characterizing the subjective–objective debate within social science. Adapted from [10] (p. 492)

Subjectivist approaches to social science ⟵⟶ Objectivist approaches to social science

Research methods	Exploration of pure subjectivity	Hermeneutics	Symbolic analysis	Contextual analysis of Gestalten	Historical analysis	Lab experiments, surveys
Core ontological assumptions	Reality as a projection of human imagination	Reality as a social construction	Reality as a realm of symbolic discourse	Reality as a contextual field of information	Reality as a concrete process	Reality as a concrete structure
Basic epistemological stance	To obtain phenomenological insight, revelation	To understand how social reality is created	To understand patterns of symbolic discourse	To map contexts	To study systems, process, change	To construct a positivist science
Assumptions about human nature	Man as a spirit, consciousness, being	Man as a social constructor, the symbol creator	Man as an actor, the symbol user	Man as an information processor	Man as an adapter	Man aa a responder
Research methods	Exploration of pure subjectivity	Hermeneutics	Symbolic analysis	Contextual analysis of Gestalten	Historic analysis	Lab experiments, surveys

(continued)

Table 1 (continued)

Subjectivist approaches to social science ⟷ Objectivist approaches to social science

Core ontological assumptions	Reality as a projection of human imagination	Reality as a social construction	Reality as a realm of symbolic discourse	Reality as a contextual field of information	Reality as a concrete process	Reality as a concrete structure
Basic epistemological	To obtain phenomenological insight	To understand how social reality is constructed	To understand patterns of symbolic discourse	To map context	To study systems, processes change	To construct a positivist science
Assumptions about human nature	Man as pure spirit, consciousness, being	Man as a social constructor, the symbol center	Man a an actor, the symbol user	Man as information processor	Man as an adaptor	Man as a responder
Research methods	Exploration of pure subjectivity	Hermeneutics	Symbolic analysis	Contextual analysis of Gestalten	Historic analysis	Life experiments, surveys

1. **Contextual Axiom**: The contextual axiom states that *system meaning is informed by the circumstances and factors that surround the system*. The contextual axiom's propositions are those which bound the system by providing guidance that enable an investigator to understand the set of external circumstances or factors that enable or constrain a particular system.
2. **Purpose Axiom**: The purpose axiom states that *systems achieve specific goals through purposeful behavior using pathways and means*. The goal axiom's propositions address the pathways and means for implementing systems that are capable of achieving a specific purpose.
3. **Design Axiom**: The design axiom states that *system design is a purposeful imbalance of resources and relationships*. Resources and relationships are never in balance because there are never sufficient resources to satisfy all of the relationships in a system's design. The design axiom provides guidance on how a system is planned, instantiated, and evolved in a purposive manner.
4. **Operational Axiom**: The operational axiom states that *systems must be addressed in situ, where the system is exhibiting purposeful behavior*. The operational axiom's propositions provide guidance to those that must address the system in situ, where the system is functioning to produce behavior and performance.
5. **Centrality Axiom**: The centrality axiom states that *central to all systems are two pairs of propositions; emergence and hierarchy and communication and control*. The centrality axiom's propositions describe the system by focusing on (1) a system's hierarchy and its demarcation of levels based on emergence arising from sub-levels and (2) systems control which requires feedback of operational properties through communication of information.
6. **Information Axiom**: The information axiom states that *systems create, possess, transfer, and modify information*. The information axiom provides understanding of how information affects systems.

7. **Viability Axiom**: The viability axiom states that *key parameters in a system must be controlled to ensure continued existence*. The viability axiom addresses how to design a system so that changes in the operational environment may be detected and affected to ensure continued existence.

5 Framework Use and Identification of Anticipated Outcomes

This section will describe how the framework can be used and provides to the user what can be anticipated outcomes with the use of the framework.

5.1 Contextual Axiom

Contextual axiom states that *system meaning is informed by the circumstances and factors that surround the system*. The contextual axiom's propositions are those which bound the system by providing guidance that enable an investigator to understand the set of external circumstances or factors that enable or constrain a particular system.

Proposition and primary proponent	Description of proposition
• Complementarity [11]	Two different perspectives or models about a system will reveal truths regarding the system that are neither entirely independent nor entirely compatible
• Boundary [6, 12]	The abstract, semi-permeable perimeter of the system defines the components that make up the system, segregating them from environmental factors, and may prevent or permit entry of matter, energy, and information
• Incompressibility [13, 14]	Each element in the system is ignorant of the behavior of the system as a whole and only responds to information that is available to it locally. As such, the best representation of a complex system is the system itself and that any representation other than the system itself will necessarily misrepresent certain aspects of the original system
• Holism [15]	A system must be considered as a whole, rather than a sum of its parts

The way we interpret systems is dependent upon the perspective of the observer and the boundary drawn around the open system, which determines what is included and excluded to inform the interpretation of system throughput and system environment. No two vantage points are identical.

Imagine you are in a restaurant with two of your friends. The menu has recently changed. What caused the menu to change? Your friends are disappointed and begin a debate, speculating reasons for the new menu. One decides that the restaurant is trying to increase profits by changing the portion sizes and limiting the amount of ingredients in inventory. The other says the changes must be as a result of a change in management.

You are not disappointed because the menu changed, because you understand that the needs the restaurant fulfills extend beyond the preferences of your friends. You reason that the menu change could be due to one or many factors, including that the menu selections may be governed by, for example, time of year for seasonal items, tourism fluctuations, internal change of staff, locals' preferences, market trends, and others. As the conversation drifts towards your ongoing software modernization project, you realize the debate between use of one automation tasking tool versus another also largely depends on the context surrounding the system's use case and desired outcomes.

System actors and observers are limited by their perspectives, and the more complex the system, the more challenging it is for the represented system to be "compressed" as perception is limited to the only available information. The best representation of the system is the system itself, with explicit and shared understanding of the intended system purposes, as the set of contextual elements is rarely fully exhaustive and objectively interpreted. Understanding the nature of the intended outcomes of the system stakeholders and use cases becomes a basis to inform the development of a solution to address a system need. Through any transformation, the solution system with selection of changes based on contextual considerations (with or without appropriate appreciation of context) can have intended or unintended outcomes as new component interactions take place that change the system definition.

When we think about systems, we must think of them as integrated wholes, as they are more than a collection of interacting parts decoupled from other systems and their environment. Their combined interaction transcends our ability to model the total system's behavior; thus, we acknowledge incompressibility. Holism is a metaphysical ideal, defined by Smuts [16] as "the ultimate synthetic, ordering, organizing, regulative activity in the universe which accounts for all the structural groupings and syntheses in it, from the atom and the physic-chemical structures, through the cell and organisms, through mind in animals, to personality in man" (p. 314).

As our view of the complex system is limited, similarly, our interpretation of system purpose is also limited. As such, Stafford Beer conceded that the observer of the system is the one that recognizes the purpose of the system; i.e., what the system does [17]. What the system does, and whether it meets the intended needs, is largely influenced by the system design choices and their fit for the application context.

5.2 *Purpose Axiom*

Purpose axiom states that *systems achieve specific goals through purposeful behavior using pathways and means*. The goal axiom's propositions address the pathways and means for implementing systems that are capable of achieving a specific purpose.

Proposition and primary proponent	Description of proposition
• Equifinality [18]	If a steady state is reached in an open system, it is independent of the initial conditions and determined by the system parameters, e.g., rates of reaction and transport
• Multifinality [19]	Radically different end states are possible from the same initial conditions
• Purposive behavior [20]	Purposeful behavior is meant to denote that the act or behavior may be interpreted as directed to the attainment of a goal, i.e., to a final condition in which the behaving object reaches a definite correlation in time or in space with respect to another object or event
• Satisficing [21, 22]	The decision-making process whereby one chooses an option that is, while perhaps not the best, good enough

Now, consider the perspective of a franchise restaurant owner. Equifinality holds that a given outcome (Y), for example, increased year-over-year net profit of 15%, can be reached from a number of different strategies or development paths (X_1, X_2, ..., X_n), including a range of possible menu changes. Although it may be that the paths are not equal, rather, other interacting factors in the environment may contribute to the achievement of the goal.

Conversely, the vast contextual factors enhancing or restricting system performance increase the difficulty in proving a causal relationship from a chosen path. Multifinality reduces our confidence that a restaurant franchise's market strategy X will lead to predicted success Y, as even franchises with similar initial conditions will have dissimilar outcomes (Y_1, Y_2,... Y_n). To establish a casual relationships between strategy X and outcome Y, it is required that: (i) the two variables covary, such that changes in the change in the strategy correlate with change in the profit; (ii) the change in the variable assumed to be the cause, in this case, the strategy precedes in time the observed change in the resulting profit; and (iii) alternative explanations for the rise in profits have been ruled out (e.g., overall food costs, improved efficiency).

Consider another case study: The LCS class Littoral Combat Ship was allowed two different design philosophies for which the Navy would later down select to a single design. Simultaneously, the Freedom (displacement hull) class and the Independence (trimaran) class were developed. Both ships designs met the operational requirements established by the US Navy and achieved them through different approaches, demonstrating equifinality. To demonstrate multifinality, consider the program's significant cost and schedule overruns due to changing requirements (among other reasons), eventually leading to contract cancelations for the first two contractors. A range of cost and schedule variances in either direction are possible when deviating from an initial set of requirements. Still, entirely different outcomes could have occurred, and tracing the specific contextual factors leading to outcomes and the point in time in which their individual and combined contribution to the inevitable outcome is difficult to draw absolute conclusions. The Virginia Class submarine provides another example, as it fulfilled its purpose of meeting delivery cost and schedule demands, but carries lessons forward for improving other program outcomes in future classes, such as supply chain growth and sustainment costs.

Achieving a specific purpose is not accomplished without a vision, philosophical assumptions, stategic plan, and feedback process to inform the governance scheme. The execution of the design will be discussed in the next section as well as the effort involved with the evolution of the design.

5.3 Design Axiom

Design axiom states that *system design is a purposeful imbalance of resources and relationships.* Resources and relationships are never in balance because there are never sufficient resources to satisfy all of the relationships in a systems design. The design axiom implies that the system viability is influenced by the governing framework by which a system is planned, instantiated, and evolved in a purposive manner.

Proposition and primary proponent	Description of proposition
• Minimal critical specification [23, 24]	This proposition has two aspects: negative and positive. The negative simply states that no more should be specified than is absolutely essential; the positive requires that we identify what is essential
• Power law [25]	The probability of measuring a particular value of some quantity varies inversely as a power of that value
• Requisite parsimony [26, 27]	The capacity of human short-term recall is no greater than seven plus or minus two items
• Requisite saliency [28]	The factors that will be considered in a system design are seldom of equal importance. Instead, there is an underlying logic awaiting discovery in each system design that will reveal the significance of these factors
• Requisite hierarchy [29]	The weaker in average are the regulatory abilities and the larger the uncertainties of available regulators, the more hierarchy is needed in the organization of regulation and control to attain the same result, if possible at all

A system under observation by the observer can be considered as in existence, undergoing change, or the observer is part of a team/group that has been tasked with creating something new. Any of these observations of where a system is in its evolution does not detract from the contribution of system design and the various propositions. Associated with system design, most are familiar with the organization or structure of formal elements in the terms: requirements, intentions, synopsis of intent, and specification. Each of these terms helps bring forward to the system design an understanding of what the system is to be doing. The practicality of these terms is to reduce the instruction so that the end is fully achieved. So as part of observing a system, matching system construct and mechanisms to specifications lends itself to determining how well the system has been organized.

The ability to fully observe and understand a system under observation may not be fully achievable, especially due to the size of the system under observation and the capacity of the observer. Hence, the parsing of the observation to a team centered on solving a specific problem and a minimum viable product can be effective, remembering that the team likewise will need a design, and in many respects, the propositions associated with system design are most applicable to the creation and tasking of a team. For example, software development team's shift from "waterfall"- to "agile" development will find their team members focused more distinctly on the features and properties most salient to the end users, based on their use case and needs. This recognition goes beyond small teams that are collocated, but rather is thought to be adaptable to any size and dispersion of groups of people.

There are three interrelated concepts that can help with the creation and tasking of a team for system design or observation of a system under observation. Where the system under observation appears to be large and complex, the needs of the hierarchy will be large as well, but with a degree of purposeful design for how a team of teams will work together to achieve outcomes, and how they will account for learning. Observations of what to build or how to build it are not all equal in importance, and with continual observation, the actual system design will materialize, and with this emergence, the significant factors more easily identifiable. Lastly, as the human has limited capacity, a team of teams must be organized in such a way that they may focus their energy on a minimum viable products with a common understanding of goals. This supports the team's ability to maintain momentum in a sustainable manner, and reduces the frequency of context switching and the need to re-orient within the system. At scale, teams must be brought back together to observe what has been captured, adjust for learning and prioritization, and then sent with new tasking The selected items to accomplish within a prescribed timebox are meant to be parsimonious in nature, and allow for natural evolution of a robust and viable system that may be flexibly integrated.

The observation of the system in situ, the understanding of its capabilities as it performs operational functions while maintaining viability will be discussed in the next section.

5.4 Operational Axiom

Operational axiom states that *systems must be addressed in situ, where the system is exhibiting purposeful behavior*. The operational axiom's propositions provide guidance to those that must address the system in situ, where the system is functioning to produce behavior and performance.

Proposition and primary proponent	Description of proposition
• Dynamic equilibrium [6, 30]	An entity exists as expressions of a pattern of processes of an ordered system of forces, undergoing fluxes and continuing flows of matter, energy, and information in an equilibrium that is not static
• Homeorhesis [31, 32]	The concept encompassing dynamical systems that return to an acceptable trajectory through adjustments in dynamic equilibrium controlled by interrelated regulation mechanisms
• Homeostasis [33]	The property of an open system to regulate its internal environment so as to maintain a stable condition, by means of multiple dynamic equilibrium adjustments controlled by interrelated regulation mechanisms
• Redundancy [34]	Means of increasing both the safety and reliability of systems by providing superfluous or excess resources
• Relaxation time [35, 36]	Systems need adequate time to recover from disorder that disturbs its equilibrium, at which point characteristic behavior resumes
• Self-organization [37]	The spontaneous emergence of order out of the local interactions between initially independent components
• Sub-optimization [38]	If each subsystem, regarded separately, is made to operate with maximum efficiency, the system as a whole will not operate with utmost efficiency

Where the system design provides guidance on how a system is planned created, established, and that with time it has modified itself in a fashion that is reflective of a purposive manner that is directly related to the system design, the propositions associated with this section deal with the guidance on the system functioning to produce behavior and performance. Whether one is involved with a small system or only associated with a small portion of a large system, one with little time can recognize that even small systems are composed of small entities that form a whole.

The abstraction of this can be that government is made up of various hierarchies from local up to the federal level and beyond. Each type of government has the tendency to operate within their level in this hierarchy. There are several observations with respect to government that can be made:

- If each level of government is allowed to operate with maximum efficiency, the whole system as a whole will not operate with utmost efficiency (sub-optimization).
- Means of increasing both the safety and reliability of government is by providing superfluous or excess resources (redundancy).
- Government needs adequate time to recover from disorder that disturbs its equilibrium (earthquakes, extensive fire, hurricanes) at which point characteristic behavior resumes (relaxation time).

These observations while not always universal nor necessarily globally applicable, hopefully, they will convey some of the characteristics of where human guidance does produce a behavior and performance of a system. For the observer of a system that is a commercial enterprise, human characteristics will be recognizable.

Taking the observations and overlaying them on the system design, it is possible for there to be a one-for-one match between the observations and the design. In fact, one would expect that there may be a one-to-many match between design and observations where the differences found between the many observed identify where there is more than one process requiring more observation and evaluation. Where the observation leads one to conclude that the system does not appear to be static but is not undergoing a wide range of radical changes, this reflects the system design exercising regulation of its internal environment so as to maintain a stable condition, by means of multiple dynamic equilibrium adjustments controlled by interrelated regulation mechanisms.

A Systems Theory—Operational Axiom in CSG Vignette—Sub-optimization
Behaviors expected from systems should be described by the axioms proposed in this chapter. As an example, one should expect that any system should exhibit sub-optimization. For a system as complex as the Boeing 747, this means that there had to be trade-offs made, so for increased cargo-carrying capacity, there was an associated maximum airspeed. For a system such as a laptop computer that there may need to be a minimum temperature for optimum operation of the faster processing chip, hence, the use of the laptop in the artic may not be advisable. These examples hopefully illustrate that the use of one of the propositions described in the book, the axioms, and associated propositions provides to the reader insight and hopefully understanding of the internal system behavior. Gaining this insight affords all how system theory affords a more significant overall system understanding.

5.5 Centrality Axiom

Centrality axiom states that *central to all systems are two pairs of propositions: emergence and hierarchy and communication and control*. The centrality axiom's propositions describe the system by focusing on (1) a system's hierarchy and its demarcation of levels based on emergence arising from sub-levels and (2) systems control which requires feedback of operational properties through communication of information.

Proposition and primary proponent	Description of proposition
• Communication [12, 39, 40]	Communication is a transaction between the information source terminal and the destination terminal, with the sole aim of generation and reproduction of symbols. Information is transmitted as a selection along possible alternative states
• Control [9]	The process by means of which a whole entity retains its identity and/or performance under changing circumstances

(continued)

(continued)

Proposition and primary proponent	Description of proposition
• Emergence [9, 41, 42]	Whole entities exhibit properties and patterns that are meaningful only when they are attributed to the whole, not its parts
• Hierarchy [9, 43]	Entities meaningfully treated as wholes are built up of smaller entities, which are themselves, wholes. In a hierarchy, emergent properties denote the levels

The axioms and propositions up to this point have described the various areas to consider with respect to a system with the exception of the boundary that demarks the system and the environment. The centrality axiom is a focus on the vital, critical, and important aspect of the condition of being central to a system. It is this pairing of two sets of propositions that describes that of being central.

It has previously been described that systems exhibit properties and patterns that only when they are considered as whole entity, they exhibit meaning. The first pairing of propositions deals with structure of the entities or parts of a system. While there may be character to some of the parts, it is not recognized as complete till it is all assembled. An excellent example is a train, where it is more than just an engine, it has one or more cars, and one of any of the cars being in the furthest position away from the engine is considered the caboose or end of the train. This combination of engine and car/s exhibits properties that when considering just an engine and/or a car/s, none of these individual units can emulate a complete train. Additionally, as there is a hierarchy or combination of the smaller entities, which all would agree are whole themselves, then one can understand that as the train exists in a hierarchy, that each of the entities can have emergent properties that are different from the whole train.

The second pairing of communication and control brings forth identity order. The train example will be used to continue the discussion. The movement of a train is limited by various forms of communication to its operators as well as the design and material condition of the track as well as weather. The overall effect of communication and control is to have a train pass from one geographic place to another safely and on schedule. In the design of the communication for the train, safety is a paramount factor even with ever-changing weather conditions. Communications provide a foundation for control of the train as well as informing entities external to the system. The execution of control ensures that the identity and performance of the train are within the design.

The centrality of a system has been discussed, and the next section will develop how information is involved in a systems operation.

5.6 *Information Axiom*

Information axiom states that *systems create, possess, transfer, and modify information.* The information axiom provides understanding of how information affects systems.

Proposition and primary proponent	Description of proposition
• Information redundancy [44]	The number of bits used to transmit a message minus the number of bits of actual information in the message
• Redundancy of potential command [45]	Effective action is achieved by an adequate concatenation of information
• Conway's law [46]	The basic thesis … is that organizations which design systems … are constrained to produce designs which are copies of the communication structures of these organizations

The use of this axiom leads to an understanding of how information (data/information) affects a system of interest. Specifically, the information is created, it is retained/stored, it is moved from one location/individual to another, and the information is not static, but changes. Chapter "Perspectives on Complex System Governance Performance" will go in much greater detail on the mechanisms of communication, but the reader when looking at a system of interest can start to question, observe, and articulate answers to the following questions; how information is created, where is it possessed, how can it be changed (it will be), and how it has moved from its initial starting point.

The secondary level of questions/observations with respect to information is to determine the "who" whether internal to the system under observation or the external environment—coupled with the "who" is the rate of information or volume of information created. Additionally, when the information is received, does it result in the accomplishment of the intended action? Does the set of observations begin to develop answers, relative to the information, a result of normal system actions, or is it the observation the instigator of actions in the system under observation? Also, where the results are anticipated? And, was there an increase or decrease of variety?

This set of efforts develops a mapping of information (from-to), the reason for the instigation of information movement, and the effect of the information on the recipient. It starts to layout part of the foundation of the system under observation identity as well as defining the roles of the participants, either internal or external. This set of efforts must also take Conway's law into account when designing the organization, as it will also affect the products of that organization.

A Systems Theory—Information Axiom in CSG Vignette—Conway's Law
Mel Conway proposed the following idea in a paper from 1968, now famously known as Conway's Law:"Any organization that designs a system (defined broadly) will produce a design whose structure is a copy of the organization's communication structure." The hypothesis orginated from Conway's observations that software with components that function well together were developed by teams that worked closely and communicated often. What this tells us is that when segregated teams are building parts of a system, the design of the subsystem or components may be based on uncommunicated and differing assumptions, leading to locally optimized design choices, potential inconsistencies and dependencies, integration risk, and requiring additional communication. Frequent iteration is also less likely, and reliant on formal mechanisms that constrain the team's ability to self-organize.

5.7 *Viability Axiom*

Viability axiom states that *key parameters in a system must be controlled to ensure continued existence.* The viability axiom addresses systems that remain in continued existence do so by adequately adapting to changes in their environment.

Proposition and primary proponent	Description of proposition
• Circular causality [47]	An effect becomes a causative factor for the future "effects," influencing them in a manner particularly subtle, variable, flexible, and of an endless number of possibilities
• Feedback [48]	All purposeful behavior may be considered to require negative feedback. If a goal is to be attained, some signals from the goal are necessary at some time to direct the behavior
• Recursion [17]	The fundamental laws governing the processes at one level are also present at the next higher level. Recursive Systems: The fundamental laws governing the processes, functions, and structure at one level are also present at the next higher level. In a recursive organizational structure, any viable system contains and is contained in a viable system
• Requisite variety	Control can be obtained only if the variety of the controller is at least as great as the variety of the situation to be controlled

Ashby's work on the Law of Requisite Variety postulated that for a system to remain viable, the variety of the environment must be matched by the variety of the system Variety represents complexity or the number of potential states of a system or environment: The more possible states, the more complexity is present [49]. Attenuation describes a system's feedback mechanism which allows regulation of key parameters by transduction or filtering of variety from the environment, a necessary ability for a viable system [17] to independently detect, respond, or adapt to challenges in the environment [50].

Viability is the ability of a system to continually maintain function and structure within a certain environment [51] at a system's level of recursion: "In a recursive organizational structure, any viable system contains, and is contained in, a viable system" [17] (p. 118). Stafford Beer's work on the Viable System Model (VSM) is built on the work of Ashby and defined necessary and sufficient conditions for a system to remain viable at any level of recursion [52]. "Recursion" refers to a concept of viable systems existing within each other and is applicable to any organization, regardless of size, sector, scope, or purpose.

The basis of viability is founded on adequate system governance: adequate regulation, control, communication, and coordination [53]. Systems attenuate variety from the environment through generation of requisite variety resulting from:

- Events within the system environment that become a causative factor for the future environmental effects
- Interaction with other systems
- Interaction of inter-system components.

Consider the As We Flourish You Lose theorem (AWFUL) [54]: We live in a zero-sum resource world, without biophysical limits on growth and expansion, yet competition prevents everyone from winning, and the success of any species necessarily requires comparative disadvantage of others in the exploitation of finite resources. Laszlo [55] states that one of the main challenges to humanity at this point in our collective history is to,

> find systemic alternatives to either adapting the world to us to the point of overload or adapting ourselves to the world to the point of evanescence. The options in this third direction must promote systemic sustainability, that is, integral approaches to human relationships between ourselves and co-adaptation—strategies for adapting with the world, rather than either adapting ourselves to it or forcibly adapting it to us (p. 165).

6 Conclusion

This chapter built upon the reader's understanding of the systems theory framework to enable it to be the basis of informed design and decision-making concerning governance functions. The theoretical basis of systems theory increases one's understanding of real-world systems and provides for improved interpretation while supplying the fundamental underpinning for analyzing complex systems. The construct for systems theory presented in this chapter provides a foundation for understanding multidisciplinary systems by improving the ability to explain and predict the behavior derived from the natural order of systems, thereby enabling holistic analysis and problem-solving. An associated language of systems is enabled in the assimilation of systems theory, which becomes a "lens" to facilitate the interpretation of complex systems and related problems by allowing the grounding of observations in a theoretical-based foundation. Systems theory is also multidisciplinary in application, as it is removed from traditional disciplinary problem-solving approaches. As such, it provides an ideal groundwork for the consideration of governance in complex systems.

The authors believe that building upon the propositions associated with systems theory as presented here enables the reader to develop an important foundation to navigate through issues related to systems. Practitioners can especially use this chapter and the reading of other chapters to develop the appropriate perspective to use as a lens when viewing multidisciplinary systems and their associated issues and problems. This lends itself to decision-making that is informed by systems theory allowing for informed considerations by the user. Specifically, these sets of seven axioms with supporting propositions cover the vital arena of systems theory and inspire confidence in understanding issues that one encounters. We suggest that the use of the presented well-developed foundation based upon the theory will increase confidence in systems theory-based decision-making.

7 Exercise

1. Provided is a table of axioms and propositions that are relative to the metasystem functions. From the reading, determine which axioms and relative propositions are appropriate for the metasystem functions.

Axiom	Principle	Information and Communications-(M2)	System Operations-(M3)	Operational Performance-(M3*)	System Development-(M4)	Environmental Scanning-(M4')	Learning and Transformation-(M4*)	Policy and Identity-(M5)	Strategic System Monitoring-(M5')	System Context-(M5*)
Contextual Axiom	Complementarity									
	Incompressibility									
	Holism									
	Boundary									
Goal Axiom	Equifinality									
	Multifinality									
	Purposive Behavior									
	Satisficing									
Design Axiom	Minimal Critical Specification									
	Power Law									
	Requisite Parsimony									
	Requisite Saliency									
Operational Axiom	Dynamic Equilibrium									
	Homeorhesis									
	Homeostasis									
	Redundancy									
	Relaxation Time									
	Self-Organization									
	Sub-Optimizations									

		Information and Communications-(M2)	System Operations-(M3)	Operational Performance-(M3*)	System Development-(M4)	Environmental Scanning-(M4')	Learning and Transformation-(M4*)	Policy and Identity-(M5)	Strategic System Monitoring-(M5')	System Context-(M5*)
Centrality Axiom	Communication									
	Control									
	Emergence									
	Hierarchy									
Information Axiom	Information redundancy									
	Redundancy of Potential Command									
Viability Axiom	Circular causality									
	Feedback									
	Recursion									
	Requisite Hierarchy									
	Requisite Variety									

References

1. Jackson MC (2003) Systems thinking: creative holism for managers. Wiley, Chichester, West Sussex, Hoboken, N.J
2. Adams KM, Hester PT, Bradley JM, Meyers TJ, Keating CB (2014) Systems theory as the foundation for understanding systems. Syst Eng 17(1):112–123
3. Whitney K, Bradley JM, Baugh DE, Chesterman CW Jr (2015) Systems theory as a foundation for governance of complex systems. Int J Syst Syst Eng 6(1):15–32
4. Snyder LJ (1997) Discoverers' Induction. Philos Sci 64(4):580–604
5. Ackoff RL (1971) Toward a system of systems concepts. Manage Sci 17(11):661–671
6. Bertalanffy LV (1968) General system theory: foundations, development, applications, Rev. George Braziller, New York
7. Hammond D (2002) Exploring the genealogy of systems thinking. Syst Res Behav Sci 19(5):429–439. https://doi.org/10.1002/sres.499
8. Laszlo E (1969) System, structure, and experience: toward a scientific theory of mind, vol 1. Taylor & Francis, Florida
9. Checkland P (1993) Systems thinking, systems practice. Wiley, New York
10. Morgan G, Smircich L (1980) The case for qualitative research. Acad Manag Rev 5(4):491–500. http://www.jstor.org/stable/257453
11. Bohr N (1937) Causality and complementarity. Philos Sci 4(3):289–298
12. Skyttner L (2005) General systems theory: problems, perspectives, practice, 2nd edn. World Scientific Publishing Co., Pte. Ltd., Singapore

13. Richardson KA (2004) Systems theory and complexity: part 1. Emerg Complex Organ 6(3):75–79
14. Cilliers P (1998) Complexity and postmodernism: understanding complex systems. Routledge
15. Smuts JC (1961) Holism and evolution. Viking Press, New York
16. Smuts J (1926) Holism and evolution. Greenwood Press
17. Beer S (1979) The heart of enterprise. Wiley, New York
18. Bertalanffy LV (1950) An outline of general systems theory. Br J Philos Sci 1(2):134–165
19. Buckley W (1967) Sociology and modern systems theory. Prentice-Hall, Englewood Cliffs
20. Rosenblueth A, Wiener N, Bigelow J (1943) Behavior, purpose and teleology. Philos Sci 10(1):18–24
21. Simon HA (1955) A behavioral model of rational choice. Quart J Econ 69(1):99–118
22. Simon HA (1956) Rational choice and the structure of the environment. Psychol Rev 63(2):129–138
23. Cherns A (1976) The Principles of Sociotechnical Design. Human Relations 29(8):783–792
24. Cherns A (1987) The principles of sociotechnical design revisited. Human Relat 40(3):153–161
25. Newman ME (2005) Power laws, Pareto distributions and Zipf's law. Contemp Phys 46(5):323–351
26. Miller GA (1956) The magical number seven, plus or minus two: some limits on our capability for processing information. Psychol Rev 63(2):81–97
27. Simon HA (1974) How big is a chunk? Science 183(4124):482–488
28. Boulding K (1966) The impact of social sciences. Rutgers University Press, New Brunswick, NJ
29. Aulin-Ahmavaara A (1979) The law of requisite hierarchy. Kybernetes 8(4):259–266
30. Miller JG (1978) Living systems. McGraw Hill, New York
31. Waddington CH (1957) The strategy of genes: a discussion of some aspects of theoretical biology. George Allen & Unwin, London
32. Waddington CH (1968) Towards a theoretical biology. Nature 218(5141):525–527
33. Cannon WB (1929) Organization for physiological homeostasis. Physiol Rev 9(3):399–431
34. Pahl G, Beitz W, Feldhusen J, Grote KH (2011) In: Wallace K, Blessing LTM (Trans) Engineering design: a systematic approach, 3rd edn. Springer, Darmstadt
35. Clemson B (1984) Cybernetics: a new management tool. Abacus Press, Kent
36. Holling CS (1996) Engineering resilience versus ecological resilience. In: Schulze P (ed) Engineering within ecological constraints. National Academies Press, Washington, DC, pp 31–43
37. Ashby WR (1947) Principles of the self-organizing dynamic system. J Gen Psychol 37(1):125–128
38. Hitch CJ (1953) Sub-optimization in operations problems. J Oper Res Soc Am 1(3):87–99
39. Shannon CE (1948a) A mathematical theory of communication, Part 1. Bell Syst Tech J 27(3):379–423
40. Shannon CE (1948) A mathematical theory of communication, Part 2. Bell Syst Tech J 27(4):623–656
41. Aristotle (2002) Metaphysics, Book H—form and being at work (Sachs J, Trans), 2nd edn. Green Lion Press, Sante Fe
42. Keating CB, Katina PF (2018) Emergence in the context of system of systems. In: Engineering emergence. CRC Press, Boca Raton, pp 491–522
43. Pattee HH (ed) (1973) Hierarchy theory: the challenge of complex systems. George Braziller, New York
44. Shannon CE, Weaver W (1949) A mathematical model of communication. University of Illinois Press, Urbana, IL, p 11
45. McCulloch WS (1959) Embodiments of mind. MIT Press, Cambridge, MA
46. Conway ME (1968) How do committees invent. Datamation 14(4):28–31

47. Korzybski A (1933) Science and sanity: an introduction to non-Aristotelian systems and general semantics, 1st edn. International Non-Aristotelian Library Pub. Co., Lancaster, Pa
48. Wiener N (1948) Cybernetics: or control and communication in the animal and the machine. MIT Press, Cambridge, MA
49. Ashby R (1956) An introduction to cybernetics. Chapman & Hall, Ltd.
50. Ríos JP (2010) Models of organizational cybernetics for diagnosis and design. Kybernetes 39(9/10)
51. Nechansky H (2011) The cybernetics of viability: an overview. Int J Gen Syst 40(7):679–700
52. Beer S (1985) Diagnosing the system for organizations. John Wiley & Sons Inc
53. Beer S (1994) Cybernetics of national development evolved from work in Chile. In: How many grapes went into the Wine–Stafford Beer on the art and science of holistic management. John Wiley and Sons, Chichester
54. Patten BC (1998) Ecology's AWFUL theorem: sustaining sustainability. Ecol Model 108
55. Laszlo A (2015) Living systems, seeing systems, being systems: learning to be the system that we wish to see in the world. Spanda J Syst Chang 6(1):165–173

Management Cybernetics

Joseph Sisti

Abstract The purpose of this chapter is to examine Management Cybernetics as a primary underpinning for Complex System Governance (CSG). The origins of Management Cybernetics and how the Viable System Model (VSM) can be used to model systems (organizations) as a means of understanding control and governance within an organization are suggested. The central tenets of the Management Cybernetics field are surveyed. The essential background for the Viable System Model (VSM) is provided as a critical foundation for CSG. This background includes the historical basis of the VSM, basic laws of cybernetics, the characteristics of the VSM, and the relationship between cybernetics and control for the VSM. The approach to system modeling with the VSM is provided. The five systems of the VSM are presented in detail with respect to their unique role within the model. Additionally, interactions within the VSM are examined. The communication channels within the VSM are explained. The chapter closes with a set of exercises.

Keywords Management cybernetics · Viable System Model (VSM)

1 Introduction

1.1 Management Cybernetics

Management Cybernetics is the "science of control"; cybernetics can be management's "profession of control" [10]. Cybernetics gets its roots from Norbert Wiener, an American mathematician (1894–1964), who studied the control and communications associated with living organisms and organization operations. Cybernetics is "concerned with general patterns, laws, and principles of behavior that characterize complex, dynamic, probabilistic, integral, and open systems" [15, p. 19]. Cybernetics highlights the existence of circular causality (feedback) and the concept of systems

J. Sisti (✉)
National Centers for System of Systems Engineering, Old Dominion University, Norfolk, VA 23,529, United States of America
e-mail: jsisti@odu.edu

C. B. Keating et al. (eds.), *Complex System Governance*, Topics in Safety, Risk, Reliability and Quality 40, https://doi.org/10.1007/978-3-030-93852-9_5

having a "holistic" behavior. The holistic behavior is described as belonging to the system and not the individual parts [9, 22]. Beer [9] states that a system "consists of a group of elements dynamically related in time according to some coherent pattern" [9]. The observer of the system is the one that recognizes the purpose of the system; i.e., what the system does [9]. The characteristics of a system emerged from the interaction of the parts, actions from whose individual parts, together created reactions not otherwise understood by looking at the individual parts separately [15]. Stafford Beer's *The Brain of the Firm* proposed the use of a nuerocybernetic model to be used as the model of a viable system for any organization. The underlying theoretical foundation for the VSM is based on cybernetics. It is here that Stafford Beer suggested that the human nervous system stipulates the rules whereby an organization is survival worthy and that it is regulated, learns, adapts, and evolves [9].

1.2 Three Basic Laws of Cybernetics

The laws of cybernetics are founded around three basic laws: (1) The Self-Organizing Systems Law; (2) Feedback; and (3) The Law of Requisite Variety.

- The Self-Organizing Systems—The Self-Organizing System Law states *complex systems organize themselves; the characteristic structural and behavior patterns in a complex system are primarily a result of the interactions among the system parts.* [15, p. 26]. Within this realm is a sub-law or subordinate that "complex systems have basins of stability separated by thresholds of instability" [15, p. 27]. "The mechanism through which complex systems organize themselves is, to a large extent, through sets of interlocking feedback loops. Parts A interacts with Part B and Part B affects Part A and they tend to continue to interact with each in some region of stability under the conditions provided by the other" [15, p. 40].
- Feedback—The Feedback Law states: *The output of a complex system is dominated by the feedback and, within limits, the input is irrelevant.* [15, p. 24]. Within this realm is a sub-law that states "All outputs that are important to the system will have associated feedback loops" [15, p. 30].
- The Law of Requisite Variety—The Law of Requisite Variety states: *Given a system and some regulator of that system, the amount of regulation attainable is absolutely limited by the variety of the regulator*" [15, p. 36]. The Law of Requisite Variety highlights the importance of continuous interactions between the system and the regulator. Variety is the technical expression for complexity of the systems or the number of states a system may have. Ashby's Law of Requisite Variety: "control can be obtained only when the variety of the controller (and in this case of all the parts of the controller) is at least as great as the variety of the situation to be controlled" [10, p. 41].

> The paradigm conflicts somewhat with our traditional images of science and ways of thinking about complex phenomena such as organizations. The cybernetic paradigm developed herein

> builds and broadens our image of what constitutes science and thereby provides powerful new ways of dealing with extreme complexity [15, pp. 44–45].

The measure of complexity is "variety" and Beer [9] refers to "variety" as the measure of the "number of possible states of whatever it is whose complexity we want to measure" [9, p. 23]. Ashby's Law describes the conditions under which a complex system can be externally controlled [16]. Understanding these conditions under which complex systems can be controlled is an underpinning for the understanding of how the VSM works.

1.3 VSM in Terms of Systems View: A Brief Perspective on System's View

Within an organization, governance of complex systems is needed to navigate the business world. Understanding what a system is paramount to the organization's ability to govern itself. There is a way of looking at creation which emphasizes the relationships between things equally with the things themselves. A brief perspective of a "system's view" is described by [16] below:

1. A system is a bounded collection of three types of entities: elements, attributes of elements, and relationships among elements and attributes. Both attributes and relationships are characterized by functions called "variables," which include the familiar quantifiable variety as well as the non-numerical types described by Warfield and Christakis (1987). The "state" of a system at any time is the set of values held by its variables at that time.
2. The values of certain variables of the system must remain within physiological determined limits for the system to continue in existence as the system; these are called "essential" variables [2, p. 41] of the system; examples are blood pressure and temperature in human systems and cash flow and net income in the firm.
3. Many system variables display equilibrium; that is, a tendency toward a single or small range of values, and when displaced form these values, a tendency to return. This quality, exhibited by all living systems, is known in teleological or goal-seeking behavior.
4. Within the category of living goal-seeking system is the class of systems whose goals and reasons for existence are consciously set by man, called "purposive" [3] or "purposeful" (Ackoff and Emery 1972) systems.
5. Most natural systems are "complex," which means that their possible states are so numerous that they cannot be counted in real time. The unit of complexity is "variety." The variety of a dynamic system is the number of distinguishable states that it can occupy. The essential quality of a complex system is that its variety is so great that it cannot be controlled or managed by any method that depends on enumerating or dealing sequentially with its states.

6. Ashby's Law of Requisite Variety states that to control a complex system, the controlling system must generate at least as much variety as the system being controlled: "Only variety in the control mechanism can deal successfully with variety in the system controlled" [3, p. 50].
7. The concept of systemic "control" operates at two levels. First is physiological control, required to allow the system to continue in existence (see 3 above); the values of all the essential variables are held within physiologically set tolerances. If physiological control fails, the system dies.
8. The second level is operational control, or the control of one system by another. This also requires the presence of physiological control, but in addition requires the maintenance of the value of a set of variables (essential or otherwise), chosen by the controlling system, according to its purpose for existence (see 5 above and 9 below), within tolerances set by the controlling system. If operational control fails, the system can still live, but (by definition) it fails to accomplish its purpose. Ashby's law governs both types of control.
9. An "organization" is a complex purposive system that man brings into being (or maintains in being) for the purpose of creating some desired change in the environment (i.e., society, organization, etc.). In order to accomplish its societal purpose, the organization must have the ability and power to influence and cause change in other organizations and the other complex natural systems that make up its environment. The organization must operationally "control" some part of the environment, which requires (Ashby's Law) that it must possess—contrary to normal expectations—at least as much variety as the societal systems it strives to control [10].
10. In classical cybernetics, there are only three methods that an organization (or any system intent on operationally controlling another complex system) can use to establish the variety surplus it needs: it can amplify its own variety beyond that of the system to be controlled; it can exactly match its variety to that of the system to be controlled (a special case); or it can reduce the variety of the system to be controlled to less than its own.

Cybernetics as a "science of control" examines the "holistic" system verses just its individual parts [10]. The cybernetic basic laws and the law of Requisite Variety described above form the foundations used for the VSM. The variety and complexity of describing organizations using the systems view was articulated by Beer [16] and described in the previous ten points as the emphasis of the relationship between things equally with the things themselves; things being the components of the system.

2 Characteristics of the Viable System Model (VSM)

The Viable System Model (VSM) is a model of the organizational structure of a viable system developed by Stafford Beer [7, 9–12]. Beer [9, 10] has explained how management manages a process within an environment and how the interactions of these processes reflect the two-way communications between those components of these processes. Organizations can use this model as a framework for Complex System Governance. Beer [9, 10] explains the levels of communication between the components as being "variety" (the measure of complexity). Variety is seen as the number of possible states of the system. Beer [9, 10] further describes the organization as having multiple operations that require management.

The five systems of the model are shown to communicate with each other in the Viable System Model and work to balance the system to ensure that variety generated within the system is absorbed. A system "consists of a group of elements dynamically related in time according to some coherent pattern" [10, p. 7]. A Viable System Model can be seen in Fig. 1 to highlight the systems and their interactions organization.

The VSM can be used to develop a model of a complex organization (or project) to clearly show how this organization functions as compared to the way the organization may be perceived to be functioning. Once developed, the model can be used to identify areas where changes could be made to improve the governance of the organization. These changes may be for streamlining the organization or to make it more effective in its working environment [10]. The Viable System Model is intended as a diagnostic tool [10]. The diagram is setup to have logical not organizational implications [10]. Beer further states that a researcher can "map the exact organization onto the model, and then ask whether the parts are functioning in accordance with the criteria of viability, as these have been set forth in neourocybernetic language" [10, p. 7]. Mapping will be described in the characteristics area for systems and channels in Sects. 6 and 7. The mapping does not create an organizational chart, but rather focuses on the process and communication aspects of the organization [10]. The processes are not assigned to one person as in a hierarchical chart, but are seen to be spread out throughout the organization. Following these processes and the communication associated with these interactions help define the underlying aspects of the VSM. The variety of roles required of the viable system is spread throughout the activity. The VSM, when modeling a branch within an organization, similarly follows the same conventions when describing the divisions above or when describing the project operations below the branch level of organizations. "The whole of the chart is reproduced within each circle representing a division, and of course this means in turn that (if we could write or read that small) the whole chart would be reproduced in each division of each division—which is to say in each little circle within every big circle" [10, p. 156]. This makes this a "competent chart for any organization" [10, p. 156]. The hierarchical chart is referred to as the "machine for apportioning blame" that the organization chart comprises [9].

Beer discusses in *Decision and Control* [7] the concepts and the three essential characteristics of a viable system:

1. "Viable systems have the ability to make a response to a stimulus which was not included in the list of anticipated stimuli when the system was designed. They can learn from repeated experience what the optimal response to that stimulus is. Viable systems grow. They renew themselves—by, for example,

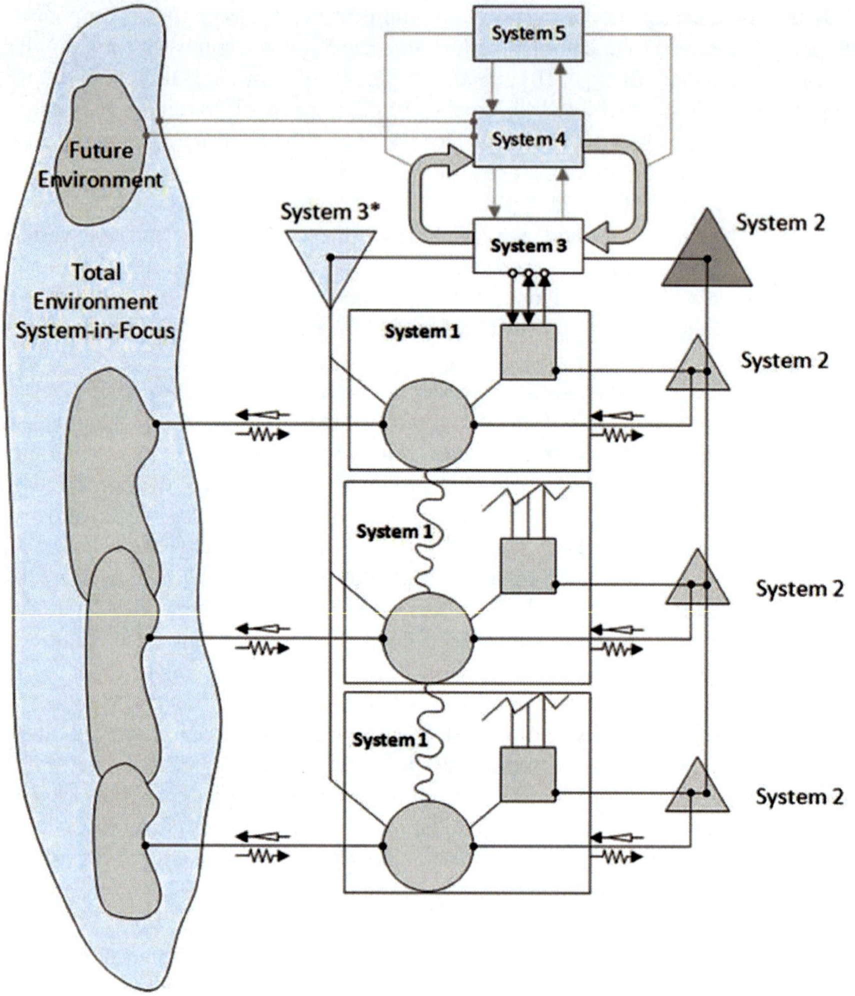

Fig. 1 Viable System Model {Adapted from [18, p. 49]}

self-production. They are robust against internal breakdown and error. Above all, they continuously adapt to a changing environment, and by this means survive—quite possibly in conditions which had not been entirely foreseen by the designer" [7, p. 256].

2. "Viable systems maintain equilibria behavior only by multiple contact with whatever lies outside themselves" [7, p. 257].
3. "It is characteristic of a viable system that all its parts may interact; not indeed to the extent that all possible permutations of all possible parts with all other

possible parts must manifest themselves, but to the extent that subtle kinds of interaction drawn from all these permutations can and do take place" [7, p. 257].

Beer summarizes these three attributes of a viable system as the systems innate complexity, complexity of interaction with the environment, and complexity of internal connectivity [7].

3 Understanding the VSM: A Discussion of Cybernetics and Controls

The Viable System Model (VSM) developed by Stafford Beer is explained by describing the conceptual components that make up the model and the relationship to how these components form the model. As modern management has developed so too has the complexity of the organizations that need to be managed [10]. Complex System Governance as an emerging field can use the VSM as a framework of analysis. The desire to gather and maintain all the data in one huge database to be used by managers to make the best decisions is often perceived as the way to manage [10]. What is really needed is a control system for change where the manager is the instrument of change [10]. The study of control science is the basis of cybernetics which is the science of communication and control through which management makes decisions [10, 11]. Management Cybernetics is the science of effective organization [11]. With the increase in available data, the interface between man and machine (computers for example) has become more complex. Cybernetics offers a managerial methodology for the management of complex control requirements within an organization [10]. Management is the profession of regulation, "and therefore of effective organization, of which cybernetics is the science" [11]. To understand the concepts of cybernetics and the modeling accomplished by using the VSM, one must understand the language that describes the decision-making process. The principle of control requires that the controller is part of the system that is being controlled [10, 11]. The controller is part of the system as it is and develops within the system as it evolves; it is not something that is attached to the systems, but rather part of the system architecture [10]. The control of the system are through the channels of communications between the systems. With a VSM, the communication channels link to the systems allowing communication and control between the systems. This can be seen in Fig. 1.

Within the VSM, an understanding of how the system is stimulated, and how the system is made aware of this stimulation, is important in describing how the system is to be controlled. Stimulation of the system is how the operation of the system is changed; whether the system accepts the stimulation for the better or rejects it due to its disruptive behavior are both important aspects for the manager to be able to be aware of, and in control of, within the system [10]. The mechanisms to allow the manager to be aware of changes and the effects within the organization are important aspects of the control of the system [10]. "Control is what facilitates the existence and the operation of a system" [10, p. 27]. The control of the system affects the internal stability of the system [10]. The manager needs to have a control system that has "a way of measuring its own internal tendency to depart from stability, and a set of rules for experimenting with responses which will end back to an internal equilibrium" [10, p. 27]. The stability pertains to not only known stimuli but the unknown events that occur to the organization as well [10].

The system design should be designed to allow the system to maintain stability in a complex environment where not all variables are known. In cybernetic terms, ultra-stability is when a system can survive in arbitrary and un-forecasted interference [10]. Anything within a system that can register and classify the existence of a stimulus is known as a sensorium [10]. Within this area, a decision is made that compares the outcomes of making a choice against its criterion of stability [10]. This is where there must be a mechanism that registers something has happened and is able to translate it into terms that have meaning to the control, so that it understands the stimulus and can react accordingly [10]. This detection is made within the system as this device is part of the system, not the stimulus itself [10]. The "bringing across" of the stimulus into the system is defined as the transducer [10]. The Sensory Input Channel (SIC) is the channel along which this information flows to bring the information into the system [10]. The Motor Output Channel (MOC) refers to the effects (output) caused by the stimulus [10]. It is this function of input and output that reflects the balance of input and output. When large numbers of input stimulus and the associated outputs are produced, they are often grouped together; as each individual input–output is too complex and exponential in number to describe [10]. This network or area of inputs/outputs within a system can be called reticulum, and the variety of reticulum in cybernetics is called anastomotic [10]. Anastomotic refers to the fact that many branches of the network intermingle to such purpose that it is no longer possible to sort out quite how the messages traverse the reticulum [10]. The idea is similar to understanding that if you add a bucket of water to the tub, you know that the tub has more water in it than before the water was added, but you do not know exactly where it is in the tub, nor is it deemed important to the overall description as to the amount of water in the tub [10]. Another analogy is the understanding of our heart within our own body. We know our heart is there but we do not consciously control it, but we know it is being controlled by our body.

Stability of a system is to be designed into the system [10]. Stability is "a self-regulating mechanism which does not rely on understanding causes of disturbances but deals reliably with their effects" [10, p. 34]. This begins to help describe the term feedback which is an adjustment to the input, so that the existing transfer function determines a corrected output within the system [10]. The pattern of the output as described by a plot of all the inputs over the range is this transfer function. Beer stated that "negative feedback corrects output in relation to fluctuating inputs from any cause. It does not matter what noise gets into the system, how great it is compared to the input signal, how unsystematic it is, nor why it arose. It tends to disappear" [10, p. 36].

There are three fundamental components of the control system: an input setup, an output setup, and the network that connects the two together [10]. An input arrangement may be a set of receptors which transmits information about some external situation into the affective channels, and concludes with a sensory register (or sensorium) on which this information is collected [10]. The capacity to distinguish detail at each end of the input arrangement should be equivalent in efficient systems [10]. The capacity to transmit the information between receptors and sensorium must be sufficient to take the traffic [10]. This needs to occur for the output arrangement—the second component of the control system [10]. The third part is the anastomotic reticulum which connects the sensory to the motor plate [10]. This means that there needs to be the same capacity to generate the inputs as there is on the output area for the outputs to go [10]. This balancing of the control systems creates the desirable stability the manager seeks; it is the management of complexity [11]. In cybernetics, the number of distinguishable items is called the "variety" [10, p. 41]. "Variety is a measure of complexity, because it counts the number of possible states of a system" [11, p. 41]. In cybernetics terms, then the input variety of the system as a whole must equal the output variety of the system as a whole to maintain a state of stability. This is an application of Ashby's Law of Requisite Variety which states "that control can be obtained only if the variety of the controller (...range of the controller) is at least as great as the variety of the situation to be controlled" [10, p. 41]. To understand the importance of variety, one must understand the scale to which variety can proliferate within a system; it often is exponential [10].

The scale of variety within the system and from nature can be enormous, but managers still need to choose effective solutions and reduce the variety for decision making [10]. "We may devise variety-generators in control mechanisms, just as nature disposes of variety-proliferators in proposing problems of control" [10, p.45]. Variety that is reduced to a set of possible states is referred to as attenuated variety [11]. "The real problem of control, the problems which a brain is needed to solve, is the problem of connecting an input pattern to an output pattern by means of an anastomotic reticulum" [10, p. 46]. We must understand that there is a fundamental degree of uncertainty in nature already [10]. This added to needed decision making by managers contributes to the complexity of managing an organization.

"There's a capability inherent in natural systems to self-organize the anastomotic reticulum in ways in which we do not properly understand" [10, p. 52]. To help distinguish these two terms they needed to be defined: algorithm and heuristic. "An algorithm is a technique, or a mechanism, which prescribes how to reach a fully specified goal" [10, p. 52]. Examples include a flight path for pilots, a math formula for calculation area, and the program a programmer has set up on a computer. "An heuristic specifies a method of behaving which will tend towards a goal which cannot be precisely specified because we know what it is but not where it is" [10, p. 52]. "These two notions are very important in cybernetics, for in dealing with unthinkable systems it is normally impossible to give a full specification of a goal, and therefore impossible to prescribe an algorithm. But it is not usually too difficult to prescribe a class of goals, so that moving in some general direction will leave you better off (by some criterion) than you were before. Instead of trying to organize it in full detail, you organize it only somewhat; you then ride on the dynamics of the system in the direction you want to go" [10, p. 53]. "These two techniques for controlling a system are dissimilar...we tend to live our lives by heuristics and try to control them by algorithms" [10, p. 53]. It is like making plans to a destination and then trying to get there. Beer points out 13 points to be made about heuristic controls [10, pp. 54–57]:

1. An heuristic will take us to a goal we can specify but do not know, and perhaps cannot even recognize when we reach it.
2. If we give a computer the algorithm which operates the heuristic, and wait for it to evolve a strategy, we may find that the computer has invented a strategy beyond our own ability to understand.
3. This being the case, it is time to start recognizing the sense in which man has invented a machine "more intelligent" than he is himself.
4. "Computers can do only what they are told" is correct, but highly misleading.
5. The argument that the output of a computer is only as good as its input, summed up in the phrase "garbage in, garbage out….is true for algorithms specifying algorithms, but not for algorithm specifying heuristics.
6. The mechanism we are using is precisely the old servomechanism discussed much earlier, in which error-correcting feedback is derived by a comparator from actual outcomes contrasted with ideal outcomes. But the outcome is measured, not in terms of the input data transformed by a transfer function, but in terms of the whole system's capacity to improve on its results as measured in another language.
7. The servomechanism's feedback does not operate on the forward transfer function as such. It operates on the organization of the black box which houses the transfer function. It experiments with the connectivity of the anastomotic reticulum. As effective structure emerges, this is what cuts down the capacity to proliferate variety.
8. Feedback dominating the outcome still holds. Hence, everything depends on the other- language criteria which the system is given to decide what to learn and what to unlearn.
9. There must be another control system, using the output of the first system as input, and operating in another plane. This higher-order, other language system would experiment with the fluctuating outputs of the first system, and produce new outputs in the other plane. Feedback from there (compared with some other-plane criteria) would establish the meaning of "better" or "worse" for the first system.
10. The second system needs a third system to evaluate its outputs in a higher-order language, and to say what counts as more or less profitable. This third system would experiment heuristically with the time-base of the second system's economic evaluations.
11. This argument continues until the hierarchy of systems, and the levels of language that go with them, reach some sort of ultimate criterion. It can only be survival.
12. And what is true of the firm in this generation of management, and true of this man, son of his father, becomes true of the firm as a continuing entity in perpetuity, and of all men, fathers of their sons. The training process for here and now is the evolutionary process for the epochs ahead.

13. So when we said that a heuristic organizes a system to learn by trying out a new variation in its operation control strategy, we might equally have said that a heuristic organizes a family of systems to evolve, by trying out a new mutation in its genetic control strategy. The aim of adaptation is identical.

The controls described above sets up a meta-language—a language of a higher order in which propositions written in a lower order language can be discussed [10]. Virtually any language must contain propositions whose truth or falsity cannot be settled within the framework of that language of which logical paradoxes are the familiar example [10]. These propositions will then have to be discussed in the meta-language, at which level we understand what is paradoxical about them [10]. "Activities can create an algedonic mode of communication between two systems which do not speak each other's language" [10, p. 59]. This is used to translate between the two systems. Errors in communication occur. The vital point is that mutation in the outcome is not the absolute enemy we have been taught to think, it is a precondition of survival [10]. The flirtation with errors keeps the algedonic feedback toned up and ready to recognize the need for change [10]. The systems' errors are wasted as progenitors of change, and change itself is rarely recognized as required [10]. "All the managerial emphasis is bestowed on error-correction rather than error-exploitation" [10, p. 62]. Errors themselves are reiterated and are deemed as being essentially bad. "Thus it follows that when change is really understood to be necessary, people resist the need, because to attempt to change is automatically to increase the error rate for a time, while the mutations are under test" [10, p. 62]. "We use organizational charts that are really devices for apportioning blame when something goes wrong. They specify 'responsibility' and the 'chain of command', instead of the machinery that makes the firm tick" [10, p. 75]. "Models are more than analogies, they are meant to disclose the key structure of the, system of study" [10, p. 75]. If we want to understand the principles of viability, we had better use a known-to-be-viable system as a model. It turns out our body is a familiar analogy to the model and will be used in describing the VSM [10, p. 76]. "Once the issues are properly understood, there will be no real need to remember the details" [10, p. 77].

It still holds true today that control in a business "has to do with the information of an extent and complexity beyond the capacities of those senior people to absorb and interpret it. It has to do with the structure of the information flows, with the method of information handling, with the techniques for information reduction, and so forth. All these features of information's role used to be determined by the cerebral capacities of the senior staff' [10, p. 80]. "There exists today a capacity to cope with information vastly in excess of the human capacity, with the result that the manager is no longer the arbiter of sophistication in control. He must delegate this role to the electronic computer" (or the information available and presented) [10, p. 80]. The manager has to organize the team and information flow. The need for a new language to be used with the VSM differs from the hierarchical models and languages often used in representing organizations [10]. The language associated with the VSM differs and hence enables better articulation of the model proposed as opposed to using the language associated with the hierarchical model. "We are constrained by our own experience as well as informed by it" [10, p. 82]. "We have a managerial culture in which some things, distinctively modern, cannot be expressed although we know them" [10, p. 82].

4 VSM: Modeling Systems

The purpose of modeling has different perspectives from different people [10]. "A model's scaling down to transfer the functions to a more manageable size allows workability in describing an organization that is complex" [10, p. 83]. "A good model is one that is appropriate and one is able to learn something about the thing that is being modeled" [10, p. 84]. Beer presents that the self-reproduction of a viable, system is usually thought of as the outstanding characteristic of that viable system, but it is continuous and regenerative self-production that is an underlying characteristic of its identity [11]. These are the characteristics of a learning organization.

> "Models are more than analogies; they are meant to disclose the key structure of the system under study" [10, p. 75]. Beer [10] suggests we look at the body as a model of a system where we have subsystems such as the heart and lungs. We have a body and we understand it, but not necessarily the "how it happens" part of things [10]. The importance of the model is to allow the reader to understand how the project works as opposed to how the project is said to work [10]. To reiterate, the VSM is intended as a diagnostic tool that can "map the exact organization onto the model, and then ask whether the parts are functioning in accordance with the criteria of viability, as these have been set forth in neourocybernetic language" [10, p. 7]. The mapping does not create an organizational chart for the project, but a framework of analysis of the viable functionality of the project as a whole. The variety of roles required of the viable system is now seen spread throughout the activity as compared to a hierarchical model. The VSM can be used to map the project or organization into Five Systems and six primary communication channels.

"The criticism of the organization chart as a model of a firm is that it is not appropriate as modeling those aspects of the firm we most wish to understand—which have to do with control" [10, p. 84]. The organizational chart was never intended for control anyway [10]. If you want to look how control is accomplished in an organization, it makes sense to use a control system as a model [10]. Control systems are the topic of study of the science of cybernetics [10]. "The trouble is that control systems of sufficient complexity to serve as adequate models of the firm are themselves so complicated that cybernetics does not fully understand them—except through models" [10, p. 84]. "Cybernetics is actually done by comparing models of complex systems with each other and seeks the control features which appear common to them all" [10, p. 84]. The VSM seeks to learn about the structure of control in complex systems. "That would mean deriving a model of a complex system in which control was already recognized as highly successful. Such a system could teach us about structure, provided that the rules of the modeling were followed carefully [10, p. 85]. "Scaling down, transferring, and investigating workability in an appropriate description would be essential, but the cybernetician is used to doing this job" [10, p. 85].

The VSM is based on a nuerocybernetic model with similarities of the way an organization is controlled [10]. The modeling after the human nervous system is also very familiar to many. “A useful model must be able to handle the differences in scale, transference, workability, and appropriateness in convincing style” [10, p. 87]. The “Nuerocybernetic model pursues and hunts down organizational invariances in large, complex, probabilistic systems within the methodology of model-building” [10, p. 87]. Invariance is when one thing is invariant with respect to something else,does not change as the other thing changes [10, 11]. Invariant in this case is a factor in a complicated situation that is not affected by the changes surrounding it [11]. “There are invariant rules governing such a system, which is derived from the theory of probability and expressed mathematically. It does not matter whether we are dealing with a brain or a firm” [10, p. 87]. Within the VSM, information within the model needs to be inspected to see whether the information coming up is appropriately dealt with at specific levels [10]. A modification of the information is passed on and upwards according to the rule sets instilled into the organization [10]. There is a filtering of information within a model as the variety or amount of information must be reduced or amplified to adequately manage the levels within the model of this organization [10, p. 93]. A filter is a variety reducer, which acts as an attenuator for variety [10, p. 94]. “There has to be a central command axis, and specialized controllers have to be integral to it—even if they are operating in a different mode…they all have their tasks to be performed” [10, pp. 95–96]:

1. Testing incoming data and recognizing any on which command action should be taken; taking the action, and sending on the original information, suitably modified.
2. Test and recognize any data which have to be filtered at this level, compressing, facilitating, and inhibiting the ascending path (handling the data at this level).
3. Store a record of these transactions in case details have to be retrieved.

We are confronting what seems to be a five-level hierarchy of systems contained within a major computer configuration…..five being somewhat arbitrary [10, p. 98]. “All five systems are serially arranged along the vertical command axis of the firm, and they model the somatic nervous system of the body” [10, p. 98]. “The middle three of the five are divided out of the cord and the brain stem” [10, p. 98]. “The cord itself is at the lowest level, the medulla and pons are grouped together next” [10, p. 98]. The third of the three echelons is the diencephalon along with the thalami and basal ganglia [10]. You see two subsystems when looking at the outer part of the five subsystems: the lateral axis which mediates afferent and efferent information and the cerebral cortex itself [10]. The upper level creates a homeostasis of stability of its system one’s environment, despite each of the systems having to cope with the unpredictable external environment [11]. “What matters to the firm’s top management is not so much the ‘facts’ as ‘the facts as presented’, and the presentation chosen can govern the outcome of even the most important and well considered decision” [10, p. 98]. “Just as the cerebral cortex is not in direct touch with peripheral events at all, but receives only such data and in such form as the subordinate echelons pass on, so top management should be presumed to be isolated from actual events” [10, p. 98].

"The exteroceptors are looking outward at captured information from the outside world" [10, p. 100]. "Telereceptors work at a distance to see whatever functions are responsible for example: examining markets, economic conditions, and the credit-worthiness of customers" [10, p. 100]. There are chemical and cutaneous receptors as well that are all analogous to any kind of data-logging signal in a distant production plant [10]. The receptors are there to detect delicate situations that may be arising [10]. The idea of this is to describe how information is detected and retrieved at the lowest level within the VSM and analogous to the human nervous system; this information is collected and disseminated along the lateral axis [10]. "The cortex, we said, has to do with intellect,it is the seat of consciousness. Its functions are incredibly complex, but they seem concerned with one thing: pattern" [10, p. 102].

"Large areas of complex organizations should be autonomous" [10, p. 103]. Autonomous means that the branch or function indicated is "responsible for its own regulation" [10, p. 103]. "The autonomic function is essentially to maintain a stable internal environment" [10, p. 103]. "Autonomic control must correct imbalances to the internal environment; the first necessity is to detect the change; receptors then alter their state, transducing the change into efferent impulses which then go to the control center" [10, p. 103]. "The impulses are then computed and associated adjustments are made through the motor part of the system (the autonomic reflex)" [10, p. 104]. Hierarchical control is "not the only dimension of control" [10, p. 105]. "The main pathways up and down the central command axis are used to inter-relate the activities of the different departments and functions within the total plan" [10, p. 105]. "If the managers in the line kept everyone fully informed with details, the major planning networks would become overloaded" [10, p. 107]. "There is a complete society of peripheral management, which operates for the most part at the social level, and whose control language is not hierarchical in the sense of the line command, but informational" [10, p. 107]. The internal balance within the organization has a goal of a general 10homeostasis [10, 11]. There can be checks and counter-checks to maintain stability and the conscious and unconscious processes are put in place for stability [10]. "For the management scientist, the model provides the bridge between practical problems of control in the enterprise, and apparently too simple, too analytic, too demanding computable models of servomechanisms" [10, p. 113]. "In autonomic control, a basic operational system and a basic set of instructions are taken for granted and then proceeds to keep what is happening in balance and in economic health. Of course consciousness can take control when it wishes" [10, pp. 116–117].

5 Application Areas of the VSM

The VSM as developed by Stafford Beer [10, 11] has been used extensively in many different application areas around the world. Applications have centered on organization structures and how to diagnose, develop, or reorganize from a cybernetics perspective. In the following development, examples of the global application of the VSM area are discussed.

Designing a Viable Organization [14] talks about the usefulness of the VSM as "a tool for anticipating, planning for, and implementing large-scale organizational change" [14, p. 49]. The model was used "as part of a research and consultancy intervention with Telecom (NZ) Limited during a period of extensive reorganization and downsizing" [14, p. 49]. The authors determined that the " VSM framework provides a useful tool for thinking about the workings of any system, particularly business organizations" and "provide a pictorial representation" to organizational questions [14, p. 51]. The authors summarize and state the VSM "provides a common framework that allows one to capture organizational idiosyncrasies, each organization's systemic strengths, and unique weakness" [14, p. 51].

"*Designing Freedom, Regulating a Nation: Socialist Cybernetics in Allende's Chile*" [21] examines the history of "Project Cybersyn." This was a project that developed "an early computer network…in Chile … to regulate the growing social property area and manage the transition of Chile's economy from capitalism to socialism" [21, p. 571]. Medina points out that "Beer recognized that his cybernetic toolbox could create a computer system capable of increasing capitalistic wealth or enforcing fascist control" [21, p. 599]. This is an example where the cybernetic use of the VSM could be used as a political tool for monitoring and controlling a nation.

Another unique article, "Design for viable organizations: The diagnostic power of the viable system model" by Markus Schwaninger [23] set out to document five applications of the VSM. The five cases were:

1. Transformation of a Swiss insurance company.
2. Redesign of a meta-system for Aditora Abirl—a company famous for journals, magazines, and travel/cultural books.
3. Enhancing a small chemical corporation, Togo, from three separate companies into one.
4. Developing a strategy for health Services Company: Kur- und Klinikverwaltung Bad Rappenau.
5. Examining the corporate ethos of the national auditing institution of the Republic of Colombia: Contralia de la Republica.

The interesting significance of this article was that they were using case studies at the organizational level as their research method. The author states "VSM has proved to be an extraordinary instrument. It not only enables a better understanding of the cases under study, but it facilitates the work enormously" [23, p. 965].

And finally there is an example of VSM being applied to the healthcare services area. "Improving Practice: A systems-based methodology for structural analysis of healthcare operations" by Keating [18]. This article introduces a systems-based methodology for conducting analysis of organizational structure for healthcare operations. The methodology enlightened higher orders of learning through structural inquiry. Several contributions to this methodology included a better method of understanding the organizations identity, an analysis that supports establishing priorities for structural improvements, decision support for better utilization of resources, and identification of its use across a wide range of applicability for structural analysis of other organizations within context [19].

The preceding examination demonstrates how the VSM has been used as an organizational analysis tool in a variety of applications areas to include: organizational structural change within corporations, government organizational reform, insurance services industries, chemical corporations, auditing institutions, and healthcare service industries. The following sections explain the systems and channels integral to the VSM.

6 Characteristics of the VSM's Systems

6.1 System One

The System One (the productive function) as described by the VSM is related to the operational units of the organization that deliver the product or service that the organization is built around. An element of control in this area centers on the detection of patterns of achievement that can be reported through System Two (coordination) to the organization [10, pp. 171–172]. System One is embedded in a meta-system, which is in fact an operational element of another system at a higher level of recursion [10, 11]. The set of embedded productive functions is known as the System One of the System-in Focus [11].

"System One must produce itself. This is the one criterion of viability that everyone seems to accept. It means that the existing enterprise has to go on being itself....the investment required to enable System One to produce itself is mandatory" [9, p. 254].

Figure 2 shows the VSM with Operational units of System One identified. The meta-system is highlighted to focus on operations and management areas.

System One is responsible for the production and delivery of organizational goods and services to the environment [18]. System One is made of operational organizational units (each of which is a complete viable system), each of which is responsible for an activity or product [18]. The other units play a supportive role and are non-viable regulatory units; that is to say they are unable to exist independently outside of the organization, unlike System One units [18]. The following describes the relationship between System One and the other units [18]:

1. With corporate management (System 3) via the three kinds of fundamental relations represented by "receiving instructions and guidelines," "accountability," and "resource bargaining."
2. With its specific environment comprising, among others, its market or the addresses of the services offered by the unit.
3. With its regulatory unit (System Two).
4. With the auditing function (System 3*: Specific information channel).
5. With the other operational units (System One components).
6. With the management of the various operational units.

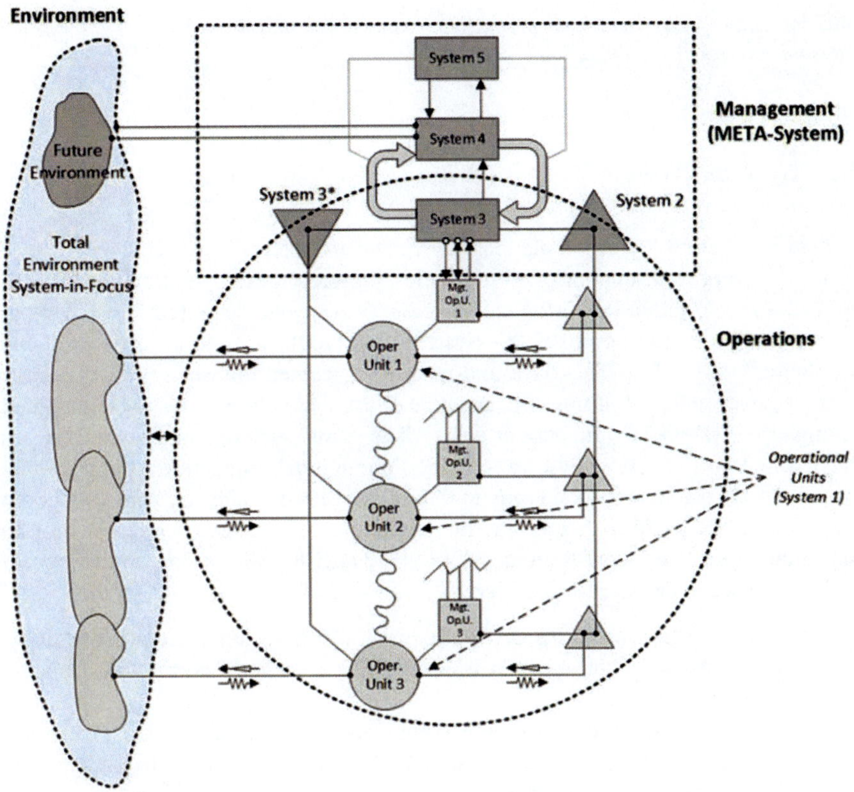

Fig. 2 VSM with operational units noted {Adapted from [18, p. 26]}

7. With the metasystem via algedonic channel.

System One controls execution in response to policy directives and overriding instructions from above in response to the environment and other divisional needs [10, p. 167]. The metasystem (in its role as an operational element of the next level of recursion) may know something affecting oscillatory behavior of our System One that is not seen by System One [9, p. 182]. System One is seen as the operational level of a project.

A Management Cybernetics Vignette—The Black Box

When modeling System One's with the VSM, logical groupings that reflect an autonomous grouping allow development of communications within other systems within the model to be clearly visualized. Modeling a shipbuilding organization, for example, may have the pipefitters, electricians, welders, accounting departments, etc. each described as a unique System. Each has unique roles and responsibilities and have unique controls and communications within their group. This 'black box' approach would be described by its input and output communications as seen within the organization while its internal characteristics describe it's autonomous structure of operation.

6.2 System Two

System Two acts as "an elaborate interface between Systems One and Three" whose purpose is to prevent uncontrolled oscillation between these operational areas [10, pp. 172–173]. "System Two is logically necessary to any viable system, since without it System One would be unstable—System Two would go into an uncontrollable oscillation" [9, p. 177]. This back-and-forth disagreement between operation units over resources and procedures is an example of this oscillation that is to be mitigated through the System Two functional areas. "The viable system engages the services of System Two to cut down the variety of its operational interaction insofar as they are inherently oscillatory—and *only* to that extent" [9, p. 177]. "System Two is not dedicated to the performance of routine procedures of whatever kind, but only to those routines that are anti-oscillatory" [9, p. 184]. This is important to distinguish as System Two is cybernetic discovery [9]:

1. Although every enterprise dedicates much effort to anti-oscillatory activity, under all manner of guises, there is no orthodox managerial correlation available to match it.
2. System Two failures are extremely common—to be corrected it must be understood that this whole question of oscillatory behavior is endemic to System One, and of System Two as an antidote.

Viability is the ability of a system to maintain a separate existence and depends on a number of necessary conditions [9]. System Two's main role can be seen to prevent oscillation within the System One—System Three areas. It is also an amplifier of the self-regulating capacity of the units themselves [18]. Examples of System Two are [18]:

1. Information systems.
2. Production planning or task programming tools.
3. Knowledge basis.
4. Accounting procedures.
5. Diverse types of operational norms intended to provide behavior standards.
6. Activities associated with personnel policies, accounting policies, the programming of production and operations, and legal requirements.

The System Two mechanism deals with the transmission of information which is taken from the operational units and once filtered, forwarded by the central regulatory unit to System Three [18]. System Three will then decide whether or not to act as a function of the information provided from System Two [18]. The System One's communicate with their associated System Two to update the upward channels of their operational status, its System Two collective role is to filter and forward to System Three the needs and balance the System Ones.

Figure 3 shows the System Two portion of the VSM. It is here where the anti-oscillatory actions occur between the System One's.

A Management Cybernetics Vignette—Part 2

The System Two coordination within the system can be described similarly with the shipbuilding example given previously. The electricians and pipefitters may have IT resources that need to be balanced across the organization. Warehousing, material, and space access to ship working areas critical to each of the systems needs to be coordinated to ensure meeting scheduling and performance goals. Elements of safety between the trades require the coordination of resources as each group can affect other system members as they work in the same shipboard environment as an example.

Fig. 3 System two (S2) {Adapted from [10, p. 173]}

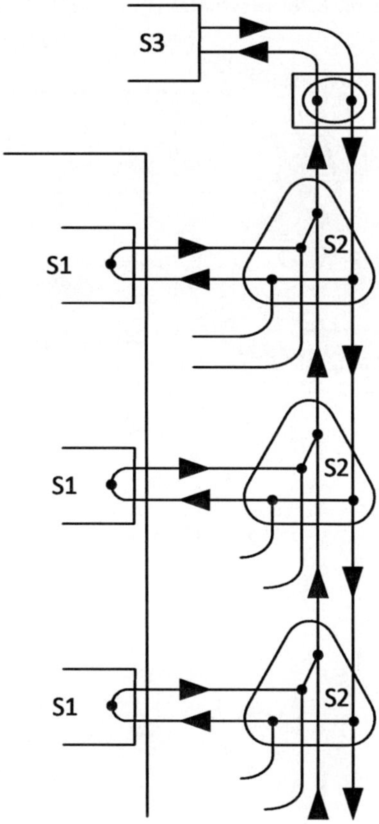

6.3 System Three

System Three is "the highest level of autonomic management and the lowest level of corporate management" whose purpose is to "govern the stability of the internal environment of the organization" [10, pp. 175–176]. It is here in System Three where routine information about the internal regulation is available to System Four. Systems Three characteristics include the following [9, p. 202]:

1. It surveys the total activity of the operational elements of the enterprise.
2. It is aware of what is going on inside of the firm in the current state.
3. Direct links with all managerial units – real time.
4. It is aware of System Two—its own subsystem.

Figure 4 highlights Systems Three, Three* (Star), Four, and Five.

System Three is usually handled by corporate executives since they are positioned to have the time to overview without the operational concerns of the working division level personnel [9, p. 203]. "Common services that contribute to synergy are

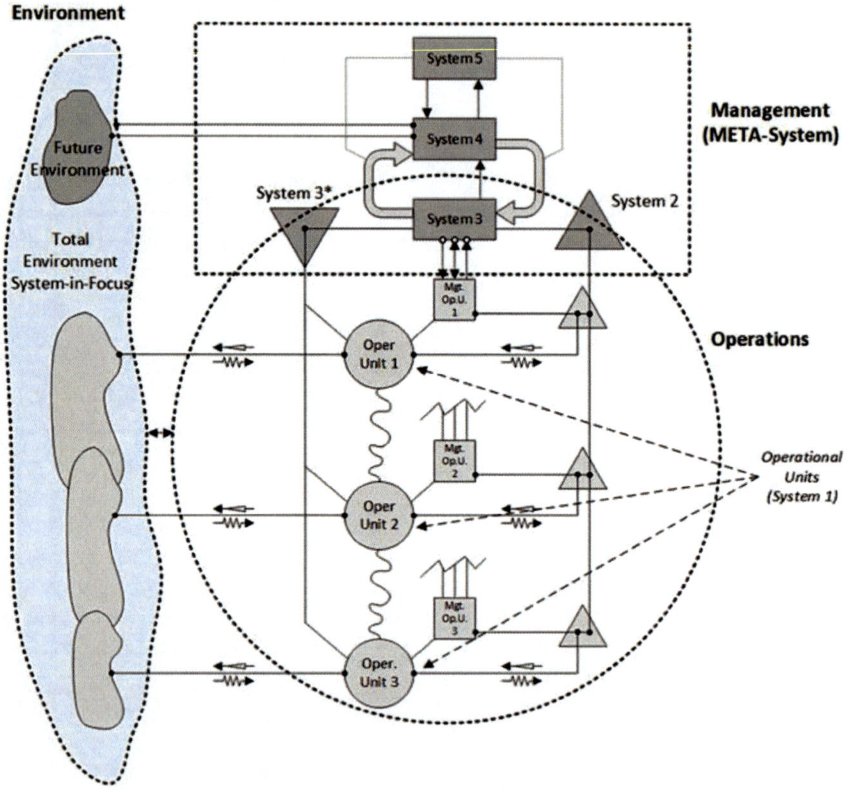

Fig. 4 VSM highlighting systems 3, 3* (Star), 4, and 5 {Adapted from [18, p. 26]}

always System Three functions" [9, p. 204]. System Three has the task of managing the set of operational units comprising System One sometimes being referred to as the "Operational Management" of the organization [18]. System Three is fundamentally interested in the "here and now" [18]. It should always be remembered that the direct involvement by the vertical line of authority has to be limited to special circumstances so as not to jeopardize the autonomy of the operational units which need this autonomy to directly absorb most of the variety generated in their specific environments [18]. Functions may include [18, pp. 32–35]:

1. Transmitting information from "management" on aspects related to the organization's aim or purpose.
2. Information concerning the policies of the organization and operational instructions to the operational units.
3. Receives information on the organization's internal situation (includes the algedonic signals that give warning of extreme risk).
4. Modifying goals.
5. Changes needed in System One as suggested by System Four.
6. Negotiation of resources.
7. Should have fluid communication with System Four on functioning and opportunities/difficulties of modifying System One.

6.4 System Three * (Star)

System Three * (Star) is a support system for System Three getting information of the status of System One; information that does not follow the normal direct channel of communication [18]. The purpose of System Three * (Star) is to ensure that the information between System One and System Three is complete [18]. Information and activities include [18, pp. 35–39]:

1. Quality audits.
2. Opinion surveys.
3. Compliance with accounting procedures.
4. Work studies.
5. Operational research.
6. Surveys.
7. Special studies.
8. Information gathering techniques.

6.5 System Four

"System Four can be described as the "development directorate of the firm" [10, p. 181]. "System Four provides all the information to System Five, the highest level of decision making within the organizational unit" [10, p. 183]. "System Four demonstrates recursive logic as it mirrors or maps the totality it serves by self-duplication" [10, p. 192]. System Four's principal responsibility is connected with the future and the external environment of the organization [18]. System Four is seen to expand variety by "contemplating rather than creating alternatives" and is able to reduce variety by "mental elimination of those alternatives" [9, p. 230]. "We hope to acquire the degrees of freedom needed to promote mutation, learning, adaptation, and evolution (in a word survival-worthiness, or in another word VIABILITY) by *stimulating* the amplification and attenuation of variety" [9, p. 230]. System Four activities may include research and development, market research, corporate planning, and economic forecasting [9]. These areas are constantly changing and in need of continuous attention.

"It's quite normal, in a large enterprise, for the elements of System Four to have virtually no knowledge of each other's activity" [9, p. 232] because: (1) each member is part of the staff of some other director or vice president, and (2) top people believe they are affecting the integration themselves. "The "integration" of System Four entails an involvement between its elements at the level of their own variety generation" [9, p. 233]. "Every regulator mechanism must contain a model of that system which is being regulated" [9, p. 234]. Beer proposed using the model as a "screen," to obtain the "focus" that would manifest "integration," exemplifying sound cybernetic underpinnings [9]. System Four can be considered the "outside and then" level [9]. System Four performs the following actions to achieve its task or functions [18, pp. 39–46]:

1. Make use of prospective study tools (example Delphi studies).
2. Scenario analysis.
3. Sensitivity analysis.
4. Simulation modeling.
5. Operational room to make strategic and operational decisions.
6. Looking at the past, present, future, and real-time data.
7. Development and innovation.
8. Market research; other research.
9. Prospective studies; projects.
10. Financial innovations.
11. Analysis of relations with the environment.

"System Four must be ready to handle the variety input generated by System Three and to design the attenuation filter that conveys that variety to System Five" [9, p. 238]. "System Four is the innovation generator that uses "existing channels and transducers through which to stimulate and interrogate the problematic environment" [9, p. 238]. The unique design of the return channel is the difference in organizations. "Innovators devise new attenuating filters and new transducers, in order to understand the novelties which (by definition) they are not aware of in advance" referred to as feedback [9, p. 239].

System Four is designed to handle the regulation of the System Three environment of the System One operations environment and the larger organizational environment. An organization needs to invest in itself to ensure its own viability [9]. System Four develops these areas where investments are advised. Investments in time, talent, care, and attention are needed [9]. As most resources go to the System One areas, the balance is divided primarily to System Three and System Four; again an area of resource competition. System Four uses its resources to expand its ability to absorb System Three variety by contemplating versus creating alternatives [9]. System Four reduces variety here by the mental absorption of alternatives [9]. Some elements of System Four that allow for the variety changes are from functions such as [9, pp. 230–231]:

1. Research and development.
2. Market research.
3. Corporate planning.
4. Economic forecasting.
5. Market development.

These functional areas are typically dispersed among different areas of the organization and not centralized to one specific area [9]. System Four's goal is to focus the goals for each of the functional areas to the goals of the desired organization [9]. System Four then is able to have a model of the organization as it is "now" and how the organization should strategically be "then". By comparing the elements of the models, System Four is able to make recommendations for changes [9]. It is here where [9] says that every regulator must contain a model of that which is to be regulated. When two different models converge into one, learning is said to have occurred [9]. System Four's goal is to make recommendations based on the functional inputs that would allow their individual models of the organization's goals to be merged into one organizational model to be called the corporate strategic model [9].

System Four has to manage the functional elements in their normal interactions with their environment as well as the larger environment [9]. The focus area is called the kernel. "An Operations Room, considered as the physical manifestation of our focus—in which in particular the kernel of the System Four model of itself is displayed—might take on any form. But outstandingly it must be an ergonomically viable locale" [9, p. 243]. System Four consists of people who spend the money that is made in System Three, the resource area [9]. Beer states that synergistic behavior derives from the recognition of mutual support between the operational elements [9]. Synergy as the sum is greater than the aggregate productivity of constituents [9].

6.6 System Five

System Five is the highest decision point within the organization unit and forms the policy for the rest of the organizational unit [10]. The power to balance the natural tension that exists between Systems Three—System Four resides in the equation of variety between System Three and System Four [10]. System Five can delegate power, if the machinery associated with System Four is in place. Beer [9, 10] reiterates that variety absorbs variety. All that remains for System Five to do is monitor the regulatory machinery—to ensure that it does not embark on an uncontrolled oscillation [9]. Recursiveness embraces the notion of local closure at any given level of recursion [9]. Within any one viable system, System Five is the metasystemic administrator of Ashby's law [9]. System Five is then seen to absorb the residual variety of the System Three—System Four interaction [9, p. 263]. System Five representatives can be representatives of management, shareholders, investors, unions, potential workers, and project managers. System Five represents the identity of the project or organization. Responsibilities of System Five would include [18, pp. 46–49]:

1. Determining the vision, mission, and strategic goals of the organization.
2. Monitoring organizations stability and internal equilibrium.
3. Ensure organization such that identity is maintained.
4. Manage stakeholders.

The four responsibilities are the major areas that System Five must perform as part of the defining identity of the system (project).

Figure 5 also shows the recursive nature of the VSM as noted by the embedded VSM within the operations area.

A Management Cybernetics Vignette—Part 3

System Five can be seen as the organization's owners and board of directors. They are the face to the customers and give vision to the internal management teams. With the shipbuilding organization example, reputation for quality work is paramount for future work for this organization. Managing budgets and customers stakeholders expectations and concerns occurs here. The board of directors provides answers to the external world, and the image of the organization is projected from here.

7 System Interactions Within the VSM

When developing the foundations of the model, three divisions of management will be recognized to suggest that the "large part of their activity, perhaps eighty percent of it, is purely anti-oscillatory" [9, p. 180] as below:

1. Interventions on the vertical line from the metasystem to System One which constrain horizontal variety for legal reasons.
2. Interventions on the vertical line from the metasystem to System One which constrain horizontal variety for the sake of institutional cohesiveness, as judged from the *purpose* of the institution.
3. System Two activities, which are purely anti-oscillatory.

"The second proposal is that all documentation dealing with the accounting functions (1) and (2) should be distributed uniquely as a sign that they relate to mandatory interventions on elemental variety" [9, p. 181]. "Without a System Four clearly in

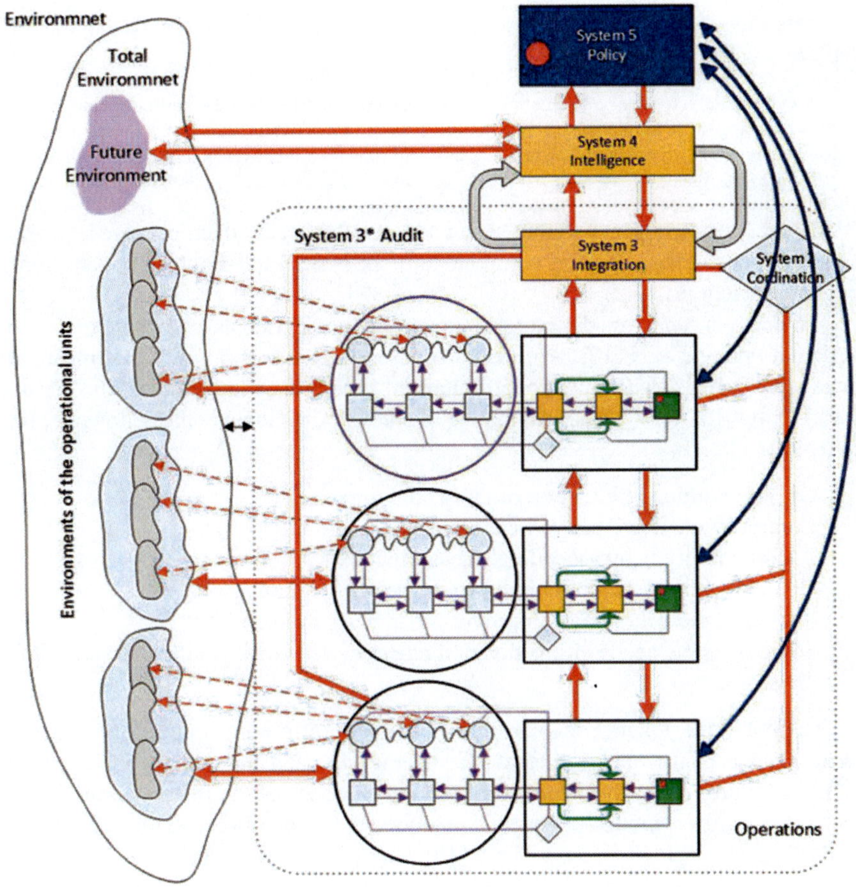

Fig. 5 VSM showing system five {Adapted From [18, p. 60]}

place, and with a System Five whose very nature is ambiguous, there is no System Three - System Four interaction, and no System Five monitoring of that interaction" [9, p. 181]. In this case, the whole metasystem collapses into System Three. "The operation of the first three principles must be cyclically maintained through time, and without hiatus or lags" [9, p. 258]. This is instantiated with the concept of an Operations Room where "System Three and System Four would exhibit themselves to each other, in a continuous mode, and absorb each other's variety" [9, p. 258]. System Five will monitor the balancing operation between Systems Three and System Four. Systems "Three-Two-One plus Three-Four-Five is a viable system - where the second group is metasystemic to the first" [9, p. 259]. "What is beyond System Five is the next level of recursion, of which this fivefold viable system is an operational element" [9, p. 259]. The "boss" within System Five supplies closure.

Beer has identified the necessary interactive elements of the viable systems as he states below [9, p. 261]:

> Our cybernetic enquires ... have elicited Six interactive elements in the vertical plane, all of which appear to be necessary to a viable system, all of which can be identified with logical precision, all of which can be measured in terms of variety exchanges under the three principles of organizations

All are present in every viable system; normally five of them are not formally recognized or studied as vertical components of the system and should be to determine requisite variety [9].

A division is run by its directorate, shown on the diagram as a box square on the vertical command axis [10]. A division is essentially autonomous. "That means it 'does what it likes' within just one limitation: it continues to belong to the organism" [10, pp. 158–159]. Practical managerial constraints include the following [10, pp. 159–161]:

1. Operate within the intention of the whole organism.
2. Communicate down the vertical command chain.
3. Accountability....by ascending lines in that axis.
4. Operate within the Coordinating framework of System Two.
5. Submit to the Automatic Control of System Three itself.
6. Sometimes the needs of one division must be sacrificed...to the needs of other divisions.

The first three managerial constraints are the variety-interconnections in the vertical plane of the environmental, the operational, and the managerial domains [10]. The fourth managerial constraint are the channels of the metasystemic intervention, the anti-oscillation channels that innervate System Two, and the operational monitoring channels of System Three [10]. The last three are "there to contain the residual variety not absorbed by the first three, given the *purposes* of the enterprise as a corporate entity" [10, p. 260]. Beer suggests that the first three variety absorbers just happen (but must be recognized) and the second three must be recognized and then designed [9, p. 261]. First Axiom of Management states:

> The sum of horizontal variety disposed by *n* operational elements = the sum of. vertical variety disposed on the six vertical components of corporate cohesion (Beer 1970, p. 261).

It is a question of creating a language that will discuss a viable system and then using this language to describe how enterprises actually *are* run" [9, p. 225]. "To use this work, in short, it is VITAL to know at all times at exactly which level of recursion one is operating. And since many managers operate at different levels of recursion, in different roles, confusion often occurs" [9, p. 226]. The environment of the viable system is the environment that has to be considered as an operational element of the metasystem (a level of recursion higher) [9]. The use of the VSM necessitates the understanding of the system boundaries chosen and their relationship to the boundaries established at the next higher level of recursion.

8 Characteristics of the VSM's Channels

Communication paths exist within the elements of the VSM [9]. "From the standard organizational chart, one would think communication would be one vertical channel up and down the chart and would be called the "command channel where authority is delegated downwards and in return the acceptance of responsibility and accountability would flow upwards" [9, p. 216]. Beer had identified six primary channels that operate along the vertical plane and handle the channel variety associated with the viable system [9]. The first three primary communication channels Beer describes are the "variety-interconnections in the vertical plane of the ENVIRONMENTAL, the OPERATIONAL, and the MANAGERIAL domains" [9, p. 216]. Beer describes these as:

> "Proliferating variety is absorbed by the interactions of elemental units among themselves. Environments can never be disconnected. Operations are invariably connected, although their interactions may be strong or weak – and therefore may absorb much or little of each other's variety. In the vertical managerial domain, managers necessarily curtail the variety of their colleagues as the stamp of their own personalities on the behavior of the elemental units becomes manifest, and as each learns to tolerate the resulting performance profile of adjacent units is a willing spirit of teamwork" [9, p. 216].

The second three primary communication channels Beer describes are the channels of "METASYSTEMIC INTERVENTION (normally confused with inherited 'chain of command'), and the ANTI-OSCILLATION CHANNELS that innervate System Two, and the OPERATIONAL MONITORING CHANNELS of System Three" [9, p. 216]. Beer describes these as:

> "These are all management activities that result from the embedding of System One in a metasystem. Unlike the first three variety absorbers, which are given in the nature of the enterprise for that particular System One, these three variety absorbers are subsystems of the metasystem itself. They are there to contain the residual not absorbed by the first three, given the purposes of the enterprise as a corporate entity. The first three variety absorbers just happen, but must be recognized. The second three must be recognized, and then designed" [9, p. 216].

The communication channels in the VSM are the elements that connect both the diverse functions specified in the VSM and the organization with its environment(s) [18]. The channels provide the equilibrium, balance, or homeostasis of the internal environment of the system in view. The six primary channels and one additional channel of the VSM can be characterized as follows [18, p. 61]:

1. Channel One–C1—Channel connecting and absorbing variety between the environments of each elementary operational unit.
2. Channel Two–C2—Channel connecting the various elemental operations (operational units making up System One).
3. Channel Three–C3—Corporate intervention channel (System Three-System One).
4. Channel Four–C4—Resources bargaining channel (System Three – System One).

5. Channel Five–C5—Anti-oscillatory channels (Coordination) (System Two).
6. Channel Six–C6—Monitor channel (Auditor).
7. Algedonic Channel—Transmits alert signals concerning any event or circumstance that could jeopardize the organization. Travels straight to the top through existing links.

The primary VSM communication channels can be seen in Fig. 6.

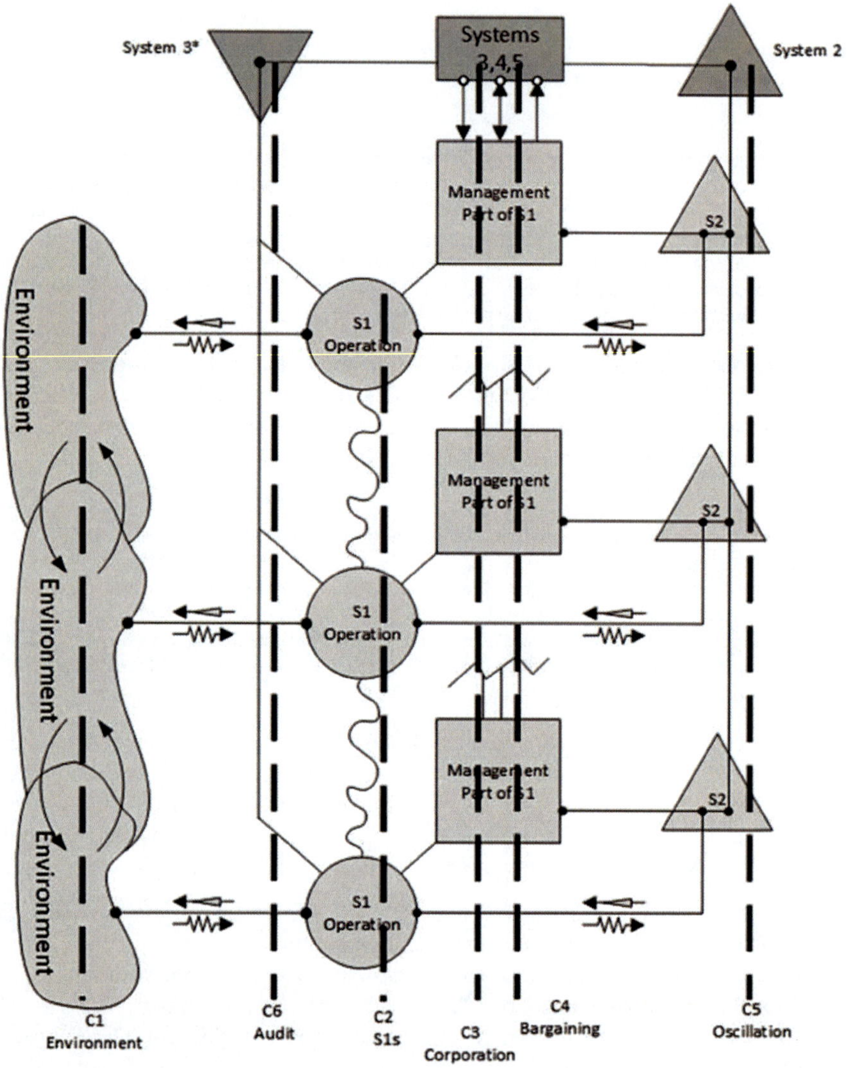

Fig. 6 VSM six primary channels {Adapted from [18, p. 61]}

The communication channels include those between the environment and the systems called C1. The C2 channels are between the S1's. The C3 cooperation channels are between the management portion of the S1's up and including the management portion of S3. The C4 channels provide the bargaining that goes on between the S1's and managed by the S3. The C5 channel monitors and controls oscillation between the S2's. The C6 channel that provides the auditing function of the S1's using unfiltered data and managed as a S3* (Star) function. The Algedonic channel provides the emergency channel directly to the top without filtering from the lower systems.

The systems and channels of the VSM were described above in the previous paragraphs. These systems and channels are the elements of the model that are used in the VSM lenses into the system of interest for the framework.

9 Summary

This chapter has introduced the field of Management Cybernetics and the Viable System Model (VSM). The origins of Management Cybernetics were established, and the field was anchored in cybernetics and the concepts of control and Requisite Variety. The VSM was introduced as the primary instantiation of Management Cybernetics. The background for the VSM was examined, and the model was anchored in cybernetics and dealing with complexity in systems. Management Cybernetics is one of the three fields, along with systems theory and governance, that are intersected to inform the conceptual/theoretical foundations for CSG. The five constituent systems of the VSM were examined in detail. The utility of the VSM to support modeling of complex systems was established. Additionally, the communication channels role in the VSM were examined. These channels provide for the flow and interpretation of information within the viable system as well as between the system and the environment.

Exercises

1. Think of your own organization as a system of interest. Choose an area within the organization (branch, project or overall organization itself) and identify the Five Systems that would make up a VSM representing your organization.
2. Identify how your "systems" maintain control from anti-oscillation "forces" between themselves.

Glossary of Terms

Algorithm A comprehensive set of instructions for reaching a known goal [10, p. 401].

Anastomotic the variety of reticulum expected to see in cybernetics; refers to the fact that the many branches of the network intermingle to such purpose that it is no longer possible to sort out quite how the messages traverse the reticulum [10, p. 30].

Autonomous A law onto itself; function indicated is responsible for its own regulation [10, p. 103].

Cybernetics concerned with the general patterns, laws and principles of behavior that characterize complex, dynamic, probabilistic, integral, and open systems [15, p. 19] about the manner of control, all kinds of structure, all sorts of systems [17].

Feedback The return of part of a system's output to its input, which is thereby changed. Positive feedback takes an increase in output back to increase the input; negative feedback takes back an output increase to decrease the input—and is therefore stabilizing in principle [10, p. 402].

Feedback Law "The output of a complex system is dominated by the feedback and, within limits, the input is irrelevant" [15, p. 28].

Filter A variety reducer [10, p. 94].

Heuristic Serving to find out; specifies a method of behaving which will tend towards a goal which cannot be precisely specified because we know what it is but not where it is [10, p. 52].

Holistic Systems Systems whose important characteristics are not ascertainable from the properties of the system components [15, p. 26].

Homeostasis Wherever one system impinges on the other, it recognizes a match which is normal to their coexistence [10, p. 145].

Invariant A mathematical term; one thing is invariant with respect to something else; it doesn't change as the other thing changes [10, p. 87].

Models More than analogies; they are meant to disclose the key structure of the system under study; a model is good if it is appropriate [10, p. 75, 84].

Regulation to select certain results from those that are possible [15, p. 70].

Requisite Variety Law Given a system and some regulator of that system, the amount of regulation attainable is absolutely limited by the variety of the regulator" [15, p. 36].

Self-Organizing Systems Principle Complex systems organize themselves; the characteristic structural and behavior patterns in a complex system are primarily a result of the interactions among the system parts" [15, p. 26].

Sensorium anything within a system that can register and classify the existence of a stimulus [10, p. 28].

SIC Sensory Input Channel.

State of the system is defined as a particular allocation of forms to events, given a particular configuration of events [10, p. 144].

Variety The total number of possible states of a system, or an element of a system [10, p. 403]. The measure of the "number of possible states of whatever it is whose complexity we want to measure" [9, p. 23]. The technical expression for complexity of the systems or the number of states a system may have.

ViabilityPrinciple ViabilityThe ability of a system to maintain a separate existence and depends on a number of necessary conditions [9, p. 199].

References

1. Ashby RW (1956) An introduction to cybernetics. Chapman & Hall Ltd., London
2. Ashby RW (1960) Design for a brain: the origin of adaptive behavior. John Wiley and Sons
3. Beer S (1959) Cybernetics and management. John Wiley and Sons
4. Beer S (1974) Designing freedom. John Wiley and Sons
5. Beer S (1978) Platform for change. John Wiley and Sons
6. Beer S (1965) The world, the flesh and the metal*: the prerogatives of systems. Nature Publishing Group, vol 4968, pp 223–231
7. Beer S (1966) Decision & control, the meaning of operational research & management cybernetics. John Wiley and Sons
8. Beer S (1967) Management science: the business use of operations research. Doubleday and Company, Inc
9. Beer S (1979) The managerial cybernetics of organizations, the heart of enterprise. John Wiley and Sons
10. Beer S (1981) The managerial cybernetics of organizations, brain of The firm, 2nd edn. John Wiley and Sons
11. Beer S (1985) Diagnosing the system for organizations. John Wiley and Sons
12. Beer S (1994) The managerial cybernetics of organizations, beyond dispute the invention of team syntegrity. John Wiley and Sons
13. Beer S (2000) Ten Pints of Beer: The Rationale of Stafford Beer's cybernetic books 1959–04), Kybernetes, MCB University Press, 29(5/6), 558–572
14. Brocklesby J, Cummings (1996) Long Range Planning, Elsevier Science Ltd, 29(1):49–57
15. Clemson B (1984) Cybernetics: a new management tool. Abacus Press
16. Espejo R, Harnden R (1989) The viable system model, interpretations and applications of Stafford Beer's VSM. John Wiley and Sons
17. Harnden R, Leonard A (1994) How Many Grapes Went Into the Wine: stafford beer on the art and science of holistic management, John Wiley & Sons, Chichester, ISBN 0-471-94296-0
18. Jose PR (2012) Design and diagnosis for sustainable organizations: the viable system method. Springer-Verlag, Berlin, Heidelberg
19. Keating C, Varela M (2002) Project management systems. In: Proceeding of the 23rd Annual american society for engineering management. Tampa, Florida, pp 1–11
20. Keating CB (2000) Improving practice: a systems-based methodology for structural analysis of health care operation. J Manag Med 14(¾):179–198
21. Medina E (2006) Designing Freedom, Regulating a Nation: Socialist Cybernetics in Allende's Chile, Cambridge University Press 38:571–606
22. Patton MQ (2002) Qualitative Research & Evaluation Methods, Sage Publications, (3rd ed.)
23. Schwaninger M (2006) Design for viable organizations: the diagnostic power of the viable system model Kybernetes, 35(7/8):pp 955–966
24. Wiener N (1950) The human use of human beings: cybernetics and society. The Riverside Press, Cambridge
25. Wiener N (1948) Cybernetics, or control and communication in the animal and the machine. John Wiley and Sons

Complex System Governance

Charles B. Keating

Abstract This chapter examines the definition, nature, and role of complex system governance (CSG). We begin by introducing the need and utility for the CSG field. Next, we address five primary elements for elucidation of CSG. First, the context for CSG is examined. This is achieved by exploring the underpinnings of CSG and acknowledgment of the conceptual foundations upon which the field is grounded. Second, the nature and definition of CSG are explored. This examination includes the underlying philosophical, conceptual, and practical utility foundations for the emerging field. Particular emphasis is placed on the underlying CSG paradigm and supporting systems worldview upon which it is based. Third, the applicability of CSG to the design, execution, and development of governance functions is explored. This exploration is conducted through the examination of several vignettes and scenarios that serve to demonstrate the utility and contributions offered by CSG. As part of this exploration, advantages, limitations, and challenges brought by CSG to practitioners and the practices for governing complex systems are suggested. Fourth, the implications of CSG development to enhance practice are examined. Specific suggestions of the utility and contributions that CSG can make to both practices and practitioners who must navigate complex systems and their problems are explored. The chapter closes with some concluding thoughts and several exercises that serve to underscore central concepts from the chapter.

Keywords Complex systems · Systems thinking · Systems development · Management cybernetics

1 Introduction

The problems facing practitioners in modern systems appear to be intractable given the apparent ineffectiveness of the responses provided to address them. These problems continue to proliferate into all aspects of human endeavor and the systems

C. B. Keating (✉)
Old Dominion University, Virginia, USA
e-mail: ckeating@odu.edu

C. B. Keating et al. (eds.), *Complex System Governance*, Topics in Safety, Risk, Reliability and Quality 40, https://doi.org/10.1007/978-3-030-93852-9_6

designed to orchestrate those endeavors. They are not the privilege, or curse, of any particular field or sector (energy, utilities, health care, transportation, commerce, defense, security, services), as none are immune to the effects of this problem domain. Problems stemming from this domain do not have a precise cause-effect relationship that would make understanding and resolution easy. In fact, they are more likely products of a 'circular causality', where the precise singular determination of cause is doubtful [28, 35] Instead, these problems are consistent with the notion of [1] 'messes' (interrelated sets of problems that are not well formulated, understood, or easily resolved) and Rittel and Webber's [32] 'wicked problems' (problems that are intractable with current levels of thinking, decision, action, and interpretation). This problem domain is likely to continue and perhaps accelerate, as we continue to grapple with twenty-first-century complex systems and their problems.

Arguably, complex systems and their associated problems have been in existence long before the twenty-first century. However, the landscape for modern systems has changed appreciably into a much more 'complex problem space'. This problem space is marked by difficulties encountered across the holistic range of technical, organizational, managerial, human, social, information, political, and policy issues. The different aspects of this 'new normal' complex problem space have been previously established [17, 19, 20, 23] as being characterized by conditions identified in Fig. 1.

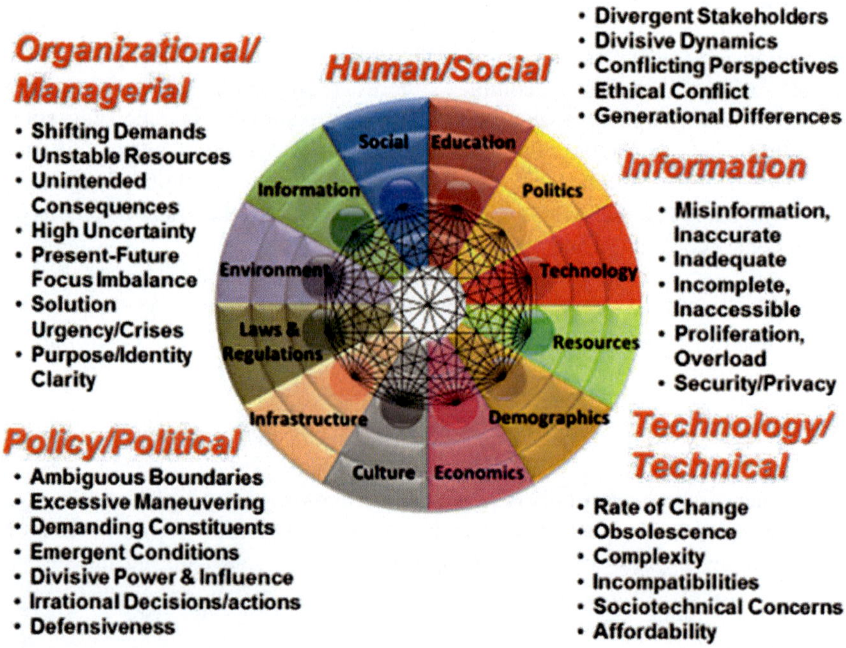

Fig. 1 Challenges for practitioners in the complex system problem domain

While this problem listing is not exhaustive, it illustrates two important points. First, the issues emanating from the complex system domain continue without consistent resolution methods. Thus, there is certainly room for new thinking and derivative approaches to address this domain. Second, the challenges identified are not likely to recede in the future. In essence, this domain represents the 'new normal' for the practitioners dealing with complex systems.

The problems emanating from this domain appear to be intractable. In any cursory look at present systems and their problems, it is easily concluded that we have not, and are not, mounting an effective solution. Given the current level of ineffectiveness in dealing with complex system problems that have proliferated into all aspects of human endeavor, CSG has been introduced. It is in the domain presented above that CSG is being postured to impact practitioner capabilities to more effectively address growing concerns. CSG is primarily based in general systems theory [2, 8, 33, 34] and management cybernetics [4–6] and has been built upon their philosophical, theoretical, and methodological underpinnings. At this point in our development, we introduce CSG as it has been previously defined as the design, execution, and evolution of the metasystem functions necessary to provide control, communication, coordination, and integration of a complex system. This chapter will focus on the elucidation of CSG as a response to the problem domain identified in Fig. 1.

In many cases, our systems have developed over time through processes of accretion or self-organization. Accretion is a process whereby elements are added in a piecemeal fashion until the whole system appears fragmented and no longer makes sense. Self-organization involves letting system structure and resulting behavior develop with minimal design oversight. This can produce results that may or may not be consistent with expectations or desirable performance. The result of either of these system development processes, accretion or self-organization, can and often do result in systems that fail to meet performance expectations. In effect, system development is not purposeful, resulting in a condition we refer to as 'system drift'. Just as a powerless ship drifts along its intended course subject to uncontrollable currents, so too can our systems experience drift resulting from development by accretion or self-organization. System drift symbolizes a system that is subject to the unintended consequences that accrue in the absence of a purposefully executed design. In the end, system drift describes a condition all too familiar to practitioners who must navigate systems through the increasingly complex environment, while confronting seemingly intractable issues on a daily basis. CSG is a coherent response to system drift.

CSG is one of many systems-based approaches [16] designed to better deal with complexity and what we referred to earlier as 'system drift'. System drift denotes systems that, irrespective of the noblest intentions, have either which is never been properly designed or whose execution continually fails to meet desired performance expectations. In short, these 'drifting' systems fall short of delivering minimal value expected, much less producing high performance. We do not need to look far to see examples of drifting systems. In fact, it would be a rare day that we would not be impacted by systems in drift. Consider the following examples: (1) launching of a new Enterprise Resource Planning initiative that collapses due to emergent incompatibilities with existing systems, (2) a costly crisis from discovery of non-compliance to a regulatory requirement that has been in existence for several years but never identified, or (3) introduction of a new purchasing policy that achieves intended reductions in supplier costs but increases overall costs due to resulting schedule delays. Unfortunately, the impacts of system drift are not limited to increased costs. These drifting systems have considerable associated human cost. These human costs are borne by those that must suffer through these drifting systems by compensating for their ineffectiveness. CSG supports thinking, decision, and action to proactively and purposefully address system drift. Ultimately, CSG is intended to reduce the high human costs characteristic of these systems in drift.

Systems-based approaches, such as CSG, and the systems thinking upon which they are founded, are certainly not 'new' in trying to address what we described as system drift. The foundations of systems thinking have been traced as far back as the ancient Chinese work The I Ching (translated as Book of Change dated prior to 400 B.C.) that noted the dynamic nature of changing relationships among elements. Additionally, the central philosophical tenet of systems thinking, holism, can be traced back to the writings of Aristotle, who suggested that 'the whole is more than the sum of its parts'. Thus, approaches based in systems thinking and 'holism' are not new and have historically represented a significant step toward dealing with system drift. However, what is new in bringing CSG-applied research to the problem domain is the fusion of general systems theory and management cybernetics to provide practitioners with perspective, supporting methods, and tools to confront drifting systems. This practitioner-focused CSG research seeks to increase capabilities for better understanding, decision, and action in dealing with complex systems and their associated problems. In essence, CSG seeks to increase effectiveness in dealing with system drift.

CSG is focused on providing practitioners with perspective, methods, and tools to better understand and deal with complexities they must routinely confront. In essence, CSG helps avoid system drift through purposeful design, similar to a ship changing heading or speed to compensate for the effects of wind or current. Figure 2 below depicts five critical realities that practitioners responsible for modern complex systems must face. The ability to effectively respond to these realities will separate the high-performance systems from the 'also ran' systems in the future. We might hope that this situation would only be a temporary aberration from normal. Unfortunately, these conditions are not likely to subside in the near or distant future. Instead, they are more likely to intensify. Practitioners responsible for systems must adjust to thrive in this 'new normal' reality. Those who do not shift the level of decision, action, and understanding in response, in the best case scenario, will likely be experiencing system drift firsthand. In the worst case scenario, they are likely to experience outright failure and system collapse.

Effectiveness in dealing with these problem domains beckons for individuals and organizations capable of engaging in a different level of thinking, decision, and action to produce alternative paths forward. As one response, CSG is proposed as an emerging field to enable practitioners to build capabilities to better diagnose and effectively respond to deeper level systemic issues that impede system performance.

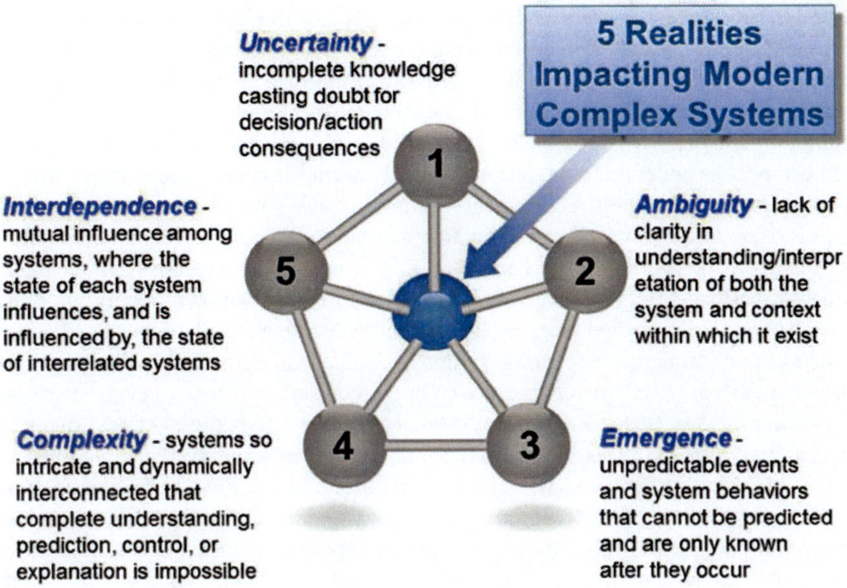

Fig. 2 Realities for practitioners in modern complex systems

Thus, CSG seeks to identify and 'design through' fundamental system issues such as those identified in Fig. 1.

A Complex System Governance Vignette—Water Utilities.

The water utilities industry provides an excellent demonstration of the pervasive nature of the complex system problem domain. Multiple sources indicate a challenge for water utilities as the industry tries to navigate the twenty-first century [3, 12, 31]. Among these recognized challenges are economic and financial uncertainty, resilience of operations, aging infrastructure, new and emerging contaminants in water supplies, an aging workforce and requirements for new skills in a future workforce, uncertainties in water resource demands and adequacy of current supplies, instantaneous access to information and public perceptions of performance, proliferation of information and advanced technologies, regulatory changes, the uncertainty of climate impacts, and the scarcity in resources as demands for efficiency increase. These conditions are not going to resolve or reside in the near future. It is also evident that the thinking and approaches to address these issues will not rest in those that have brought the industry to the present state. Instead, there is a clarion call that different thinking and approaches will be necessary even to maintain industry performance much less improve performance. Time continues to run short as industry crises loom eerily on the horizon. The water industry does not stand alone in these challenges. We only need to look to other industries, enterprises, and sectors facing similar circumstances (e.g., energy, transportation, health care, education, defense, security, infrastructure, etc.).

Unfortunately, these issues exist at deep tacit levels and appear only as symptomatic at the surface. Thus, efforts to address the problems at the surface level, although providing temporary 'fixes', continually fail to resolve the deeper fundamental system issues. This deeper fundamental system-level resolution is necessary to preclude recurrence of the symptomatic issue in another superficial form. Continual treatment of symptomatic conditions contributes to 'system drift' by focusing on temporary correction of deficiencies at a superficial level. Unfortunately, this correction behavior is endemic to modern systems, fostering 'system superheroes'. These 'system superheroes' are recognizable as individuals who resolve surface symptoms (crises) through brute force and knowing how to navigate problematic systems. However, this behavior for error correction fails to address underlying systemic inadequacies, instead opting to reactively focus on apparent resolution that only serves to mask deeper systemic inadequacies. This is not to disparage the hard work and noble efforts of practitioners who become skilled at compensating for poorly designed and executed systems (system superheroes). On the contrary, we seek to draw attention to the liabilities of dependence of 'system superheroes' to resolve 'crises' invoked by faulty systems. We should ask three important questions of systems that operate in the 'system superhero' reactive problem resolution mode. First, is the existence of 'system superhero' behavior masking more fundamental deficiencies in the underlying system? Second, is reliance on 'system superheros' unsustainable, creating conditions for an eventual system collapse? And third, what happens when the 'system superhero's get overwhelmed, tired, retire, or just leave? While CSG cannot claim to eliminate the existence of system superheroes, it does provide an opportunity to address underlying systemic deficiencies that this behavior masks. And perhaps, if not making them obsolete, at least reducing reliance on superheros for system performance.

CSG is certainly not portrayed as a 'panacea' to singularly guarantee success with the present and future twenty-first-century problems facing organizations and their systems. However, CSG does offer a compelling argument as an approach to generate alternative thinking, decision, and action to address system problems. In addition, CSG can foster enhanced collaboration and partnerships across a system. This includes supporting: (1) a 'total systems view' based in a holistic perspective, (2) effective communication with multiple stakeholders through more explicit system understanding and system representations, (3) development of systems-based leadership skills that enhance capabilities for dealing with increasingly complex systems, and (4) increasing the likelihood of achieving expected performance. Again, while CSG is not a singular remedy to produce better-performing systems, it does provide a solid complementary set of methods, tools, and thinking to enhance practice.

The reminder of this chapter is organized as follows: Sect. 2 explores the context for CSG. This provides a brief background into the initial formulation of CSG to explain the particular genesis and contributing fields that inform CSG. Section 3 provides a detailed definition and development of CSG. This provides an articulation of the precise definition of CSG and the essential context necessary to grasp the essence of an emerging field. In Sect. 4, the applicability of CSG to modern complex systems and their problems is examined, where the emphasis is placed in the particular role that CSG might play in addressing a class of problems that appears intractable given present paradigms and approaches. Section 5 explores the implications that

CSG holds for addressing present and future complex systems and their problems. The implications for practitioners are also examined, with the intent to demonstrate the potential this field holds for advancing capabilities for dealing with complex system problems. Finally, the chapter concludes several exercises to consolidate thinking.

2 The Context for Complex System Governance

In this section, we examine the foundations for CSG, including the background of the initial formulation. Our examination is focused on development of the three primary fields informing the development of CSG. CSG lies at the intersection of three fields, including *general systems theory, management cybernetics,* and *governance* (Fig. 3). In broad terms, general systems theory provides the set of propositions (laws, principles, concepts) that defines the behavior and performance of all complex systems. For CSG, [general] systems theory provides the basis for integration and coordination. Management cybernetics (the science of effective system organization) complements general systems theory by identifying the essential functions performed by all complex systems to remain viable (continue to exist). Governance is concerned with the providing for direction, oversight, and accountability for system(s). Each of

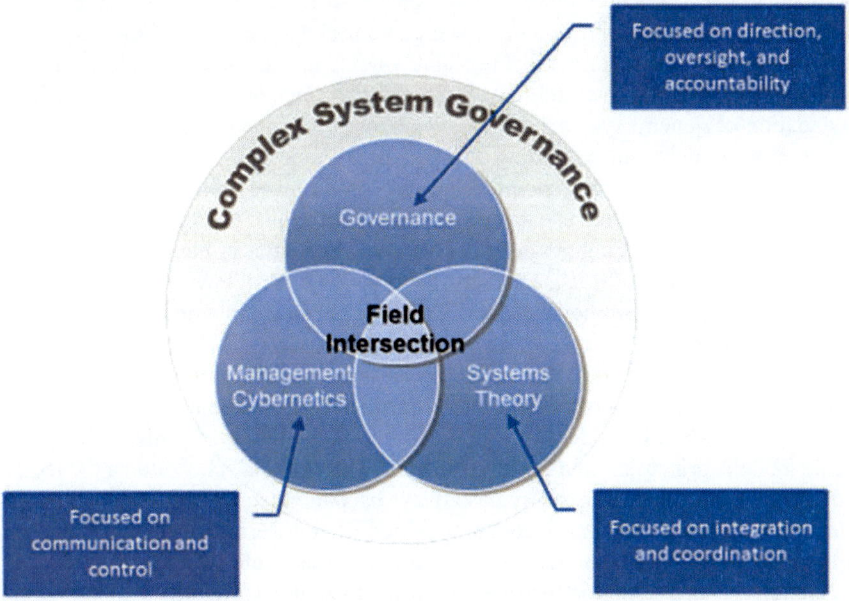

Fig. 3 CSG at the intersection of three fields

these fields will be examined for their unique contributions as the conceptual basis for CSG.

2.1 Contributions of General Systems Theory to CSG

General systems theory cannot be depicted by a common definition that is accepted by a preponderance of those scholars and practitioners for which the field has significance. In fact, the foundations of systems thinking, upon which general systems theory and CSG have been built, can be traced as far back as the ancient Chinese work *The I Ching* (translated as Book of Change dated prior to 400 B.C.). From the earliest beginnings of mankind, the struggle with increasingly complex and troublesome systems and the continually evolving general systems theory has endured. The early Chinese work noted the dynamic nature of changing relationships among elements—a condition that has not changed in well over two thousand years since it emerged. Additionally, the central philosophical tenet of systems thinking, *holism*, can be traced back to the writings of Aristotle, who suggested that 'the whole is more than the sum of its parts'. The more recent depictions of general systems theory are frequently attributed to Anatol Rapoport, Norbert Wiener, Karl Ludwig von Bertalanffy, and Ross Ashby [27, 30], having emerged in the 1940s in response to the inabilities of 'reductionist' approaches to adequately account for behavior of more complex systems. Reductionism depicts a particular intellectual stance rooted in the knowledge that is objective and understandable from the behavior of the parts, relationships that can be precisely and repeatably defined, and a close coupling with the tenets of the scientific method [8, 14]. In contrast, *holism emerged* as the driving foundation of general systems theory suggesting that knowledge is subjective and observer dependent, understanding of behavior is found in the relationships among parts, and that behavior in (complex) systems is not necessarily capable of being completely understood or repeatable [9]. They kept re-discovering the Aristotelian dictum of the whole being greater than the sun of its parts in biology, psychology, sociology, and physics [8, 29]. This sets in motion a different level of thinking, based in understanding systems behavior/performance not being explained from traditional reductionist thinking.

The genesis of general systems theory is thus found in pursuit of the goal to find a common platform of understanding the behavior/performance for all systems and thus provide a basis for a common frame of reference for universally applicable models, principles, and laws that help explain 'system' phenomena [7, 15, 29, 30]. Thus, general systems theory has always been targeted to discovery and understanding of 'universally' applicable propositions that govern the behavior, function, and performance of all systems, be they natural or manmade.

General systems theory provides a strong theoretical grounding for complex system governance. General systems theory has been identified as a set of axioms and associated propositions (principles, concepts, and laws) that seek to describe the behavior of systems, either natural or manmade [2, 38]. A full development of general systems theory is beyond the scope of this chapter. However, following the development of Whiteny et al. [38] and adapted from the earlier work of Keating [20], general systems theory is provided as a set of seven systems axioms and their implications for CSG:

1. *Centrality Axiom:* Central to all systems are emergence and hierarchy and communication and control. This implies that there should be consideration for flexibility in design for uncertainty, minimal constraint on constituents within a system, and the flow of information by design.
2. *Contextual Axiom:* Meaning in systems is derived from the circumstances and factors that surround them. This implies the necessity to account for influence of system context and the holistic consideration of the range of socio-technical-political aspects of the domain within which a system is embedded.
3. *Goal Axiom:* Systems achieve specific goals through purposeful behavior using pathways and means. This implies that there must be clarity in system purpose as well as the pathways, strategies, and resources necessary to achieve those purposes.
4. *Operational Axiom:* Systems must be addressed in situ, where the system is exhibiting purposeful behavior. This implies that system performance must be monitored and balanced to alleviate variability and provide for integration of constituent elements in their operational setting.
5. *Viability Axiom:* Key parameters in a system must be controlled to ensure continued existence. This implies that external perturbations and internal flux must be managed to maintain viability consistent with the continuing identity of the system.
6. *Design Axiom*: Purposeful imbalance of resources and relationships. This implies that there must be responsive system reconfiguration through trade-offs consistent with the identity of the system. Also that, there is a rebalancing of constituent autonomy with system-level integration considerations as well as resource allocation balancing.
7. *Information Axiom:* Systems create, process, transfer, and modify information. This implies that information necessary to support consistency in decision, action, and interpretation on behalf of the system must be by purposeful design. Also, sufficient redundancy in information must be available to ensure continuity of the system.

In effect, general systems theory provides a theoretical grounding for CSG such that integration and coordination necessary to ensure continuation of a system can be achieved.

2.2 Contributions of the Governance Field to CSG

Governance provides a critical set of grounding insights for CSG. There is an abundance of perspectives on governance stemming from the literature. However, tailoring this work for CSG, the following developments based on the work of Calida [10] and subsequently Calida and Keating [11], provides discovery of the multitude of perspectives that permeates the governance field. We offer three that are influential in providing a grounding perspective of governance for CSG:

1. *Process-centric:* collective decision making processes that are based in formal, consensus seeking, and deliberative execution in nature. The aim is to provide effective processes that enable the act(s) of governance to be performed.
2. *Structure-centric:* emphasis on the formulation and execution of structures that preserve order/continuity and steer the system in desired directions. The aim is to install sufficient structure that provides and maintains the trajectory of a system toward desired ends.
3. *Policy-centric:* emphasis on the formulation of policies that act to inculcate the principles, norms, rules, and behaviors that produce sufficient regularity in performance. The aim is to invoke policies with sufficient capacity to direct/control aspects essential to achieve/maintain system performance.

In addition, it is important in the development of CSG to make a distinction between 'governance' and 'management' perspectives. Based on the work of Keating [22], Table 1 identifies the management–governance critical distinctions.

Based on this spectrum of governance perspectives suggested by Calida [10], we can draw several important themes, which serve to inform a systems perspective of governance from the literature. For CSG, we suggest that governance embodies continuous achievement of: (1) *Direction:* sustaining a coherent identity and vision that support consistent decision, action, interpretation, and strategic priorities, (2) *Oversight Design:* providing control and integration of the system and corresponding initiatives, and (3) *Accountability:* ensuring efficient resource utilization, performance monitoring, and exploration of aberrant conditions.

2.3 Contributions of Management Cybernetics to CSG

Management cybernetics has been described by its founder as the science of effective organization [4]. Management cybernetics provides a critical contribution to the emerging paradigm of complex system governance. Beer [4, 6] introduced the concept of the 'metasystem' as a set of functions that must be performed by any viable (continuing to exist) system. The metasystem acts to provide the integration and coordination necessary to ensure that a system continues to produce the products or services that allow it to meet performance levels necessary to continue to operate (exist). Failure of any of the metasystem functions would jeopardize the overall system. Beer's formulation of the metasystem provides five essential functions for continued system viability. These functions are summarized below:

1. *Coordination function:* provides for system stability by preventing unnecessary oscillations within the set of systems being integrated by the metasystem.

Table 1 Differences between management and governance

Characteristic	Management	Governance	Implications for CSG
Emphasis	Outputs (tangible, objective, short term)	Outcomes (less tangible, subjective, long term)	Determination of governance 'goodness' is not simple or straightforward
Central questions of concern	What? And How?	Why?	Governance exists at a higher logical level of performance—emphasizing purpose
Focus	Near-term demonstrable results	Long-term future focused trajectory	The focus of governance is expansive, entertaining long view questions of strategic rather than operational significance
Determinants of success	Easily defined, measured, and tracked	Difficult to define and measure	While governance measures might be developed, they necessarily lack precision
Time horizon	Short term	Long term	The nature of governance invokes a much longer time horizon
Action-response proximity	Close separation between action and system response	Tenuous separation and relationship between action and response	Instabilities in understanding, knowledge, and magnitude create separation between action-response certainty
Uncertainty	Local uncertainty concerns	Global uncertainty concerns	Governance has a more global level of uncertainty and its resolution
Stability and emergence	Local proximity stability, local-level emergence	Global proximity stability, global-level emergence	Global focus of governance questions assumptions of long range or time stabilities

Promotes operational system performance by ensuring sufficient integration within the system. Acts to harmonize the system such that the system acts in unison. Without the coordination function, the system would be subject to unnecessary turbulence, decreasing both efficiency as well as effectiveness.

2. *Operational control function:* maintains operational performance on a day-to-day basis. Provides for the execution of policy, distribution of resources, and accountability within the system. Governance must provide a focus that allows near-term achievement to be balanced with longer term system shifts necessary to maintain viability.
3. *Audit and accountability:* provides monitoring of the system to identify aberrations and invoke necessary explorations to determine the source of the aberrant behavior or unexpected variance. Essential to understanding the nature of variance and focus actions to resolve variance.

4. *Development function:* scans and captures information from the environment and assesses that information for strategic implications and system- level impacts. Models the future and strategic evolution of the system. Critical to governance since the early indicators of strategic system threat are identified and interpreted.
5. *Policy function:* provides for the strategic decisions and direction that maintain the identity of the system. Monitors and maintains a balance between the inherent tension between the long-term external focus and the short-term internal focus of the system. For governance, this function is essential to ensure that the system maintains itself on a trajectory consistent with the desirable future.

In development of CSG, management cybernetics brings three important contributions. First, the strong grounding in cybernetics provides a strong theoretical foundation for CSG. Cybernetics, at a most basic level, is concerned with communication and control—in effect deriving from the Greek notion of 'steering'. This is consistent with the function of governance as providing the direction and monitoring the movement of the system along that trajectory. With respect to control, taking a cybernetic viewpoint allows inclusion of the more expansive perspective of control. This perspective is consistent with providing the highest degree of autonomy within a system, while preserving integration necessary to maintain system performance. Second, the work of Beer [4, 6] provides a model (viable system model) which includes functions (metasystem) consistent with achievement of governance for a system. This reference model, identified by the functions above, provides CSG with an established frame of reference upon which to build. The management cybernetics foundation provides a strong systemic/cybernetic set of underpinnings, is logically consistent with CSG articulation from a systems perspective, and offers a field which has withstood several decades of scrutiny. Since its development in the 1970s, management cybernetics has been successfully applied for over five decades. It has maintained a sustainable footing, even with the arrival and departure of a multitude of other methods or approaches that have ceased to exist in any formidable fashion.

A CSG Vignette—No Way Out of the Crisis Mode.

The mantra is frequently heard in the corporate halls, 'all we do is continually deal with crises, moving from one fire to the next.' One executive, tired of the continual 'firefighting mode', decided to examine their system from a 'governance' perspective. This entailed structured accounting for the design, execution, and development of the system. The discovery from an introspective examination suggested that while people were working harder, they were masking system deficiencies that created inefficiencies, inconsistencies, and a seeming endless engagement in the 'status quo' operation in continual crisis mode. Engaging in a protracted 'CSG study' of their system focused on identification and assessment of critical systemic issues (governance) in the design and execution of their system. The result of the 'guided' self-study identified and prioritized multiple design issues based on their impact on system performance. For example, one design issue was the near absence of coordinated efforts with respect to scanning the environment for trends, patterns, and shifts. This absence precluded early identification, analysis, and response planning—prior to the inevitable crises that would eventually erupt upon their 'too late' discovery. The stage was set to establish and integrate appropriate mechanisms for more effective environmental scanning.

3 Defining Complex System Governance

In this section, we examine the nature and definition of CSG. We begin with introduction and amplification of a definition of CSG. This is followed by examination of the systems-based paradigm that captures the essence of CSG. We close this section by the examination of five fundamental aspects that capture the essence of CSG.

3.1 Defining CSG

There is a growing body of knowledge related to CSG. The essence of CSG lies in the current state of the definition captured as the '*design, execution, and evolution of the metasystem functions necessary to provide control, communication, coordination, and integration of a complex system*' [24]. This depiction of CSG allows for several points of emphasis.

First, *design* accentuates the necessity to purposely and proactively engage in the creation of the governance system. While this seems as though it should be a taken for granted proposition, we suggest that truly purposeful, holistic, and comprehensive design of governing systems represents the rare case. While we might argue the merits of this conclusion, at this point, it suffices to say that based on the current level of performance and issues propagating all manner and form of our 'manmade' complex systems, the anecdotal evidence suggests that what we are doing with respect to our systems is not working.

Irrespective of purposeful/purposeless design, *execution* embodies the notion that a design without deployment offers little more than good intention. Execution is where a design meets the harsh realities of the 'real world' which is fraught with complexity and emergent conditions that are sure to test the most thoughtful designs. We should note that the need to adjust a system during execution is not indicative of poor design, but rather recognition that all designs are flawed. They must be flawed because they are abstractions of real-world complexity that can be neither totally captured nor completely understood.

The term *evolution* recognizes that systems as well as their environments are in constant flux. Therefore, governance must also be able to flex (evolve) in response to internal and external changes impacting the system. *Evolution* by its very nature suggests that the emphasis is on long-term sustainability, notwithstanding the need to operate a system in real time. In effect, governance must be capable of absorbing, processing, and responding to external turbulence and internal system flux to ensure the system remains viable (continues to exist). Viability must be maintained in both the short-term operational sense that delineates current system existence as well as the long-term evolutionary sense that positions the system for the future.

The CSG definition with the articulation of *design, execution, and evolution* focuses attention on the second part of the definition, *metasystem* as the set of functions that produces governance for a complex system. Nine (9) interrelated functions serve to capture the essence of CSG [24, 25]. These functions find their basis in and offer an extension of Beer's metasystem concept in the viable system model [4–6]. The metasystem for CSG is the set of 9 interrelated functions that acts to provide governance for a complex system. These functions include:

- ***Metasystem Five (M5)—Policy and Identity***—focused on overall steering and trajectory for the system. Maintains identity and balance between current and future focus.
- ***Metasystem Five Star (M5*)—System Context***—focused on the specific context within which the metasystem is embedded. Context is the set of circumstances, factors, conditions, or patterns that enables or constrains execution of the system.
- ***Metasystem Five Prime (M5')—Strategic System Monitoring***—focused on oversight of the system performance indicators at a strategic level, identifying performance that exceeds or fails to meet established expectations.
- ***Metasystem Four (M4)—System Development***—maintains the models of the current and future system, concentrating on the long-range development of the system to ensure future viability.
- ***Metasystem Four Star (M4*)—Learning and Transformation***—focused on facilitation of learning based on correction of design errors in the metasystem functions and planning for transformation of the metasystem.
- ***Metasystem Four Prime (M4')—Environmental Scanning***—designs, deploys, and monitors sensing of the environment for trends, patterns, or events with implications for both present and future system viability.
- ***Metasystem Three (M3)—System Operations***—focused on the day-to-day execution of the metasystem to ensure that the overall system maintains established operational performance levels.
- ***Metasystem Three Star (M3*)—Operational Performance***—monitors system performance to identify and assess aberrant conditions, exceeded thresholds, or anomalies.
- ***Metasystem Two (M2)—Information and Communications***—designs, establishes, and maintains the flow of information and consistent interpretation of exchanges (through communication channels) necessary to execute metasystem functions.

The means for executing the set of 9 interrelated CSG functions providing governance is found in the *metasystem communication channels* that provide for the flow of information between system entities as they perform functions. These channels support the flow of information for decision and action as well as produce consistency in interpretation for exchanges within the metasystem and between the metasystem and external entities. The ten CSG communication channels are adapted from the work of Beer [4–6] and extensions of Keating and Morin [26]. Table 2 below provides a concise listing of the communication channels, their primary CSG metasystem function responsibility, and the particular role they play in CSG metasystem execution.

The final part of the definition of CSG is focused on the elements of *control, communication, coordination, and integration*. These terms, and their basis, emanate from management cybernetics (communication, control) and general systems theory

Table 2 Communication channels of the metasystem for CSG

Communications channel and responsibility	CSG metasystem role
Command (Metasystem 5)	• Provides non-negotiable direction to the metasystem and governed systems • Primarily flows from the Metasystem 5 and disseminated throughout the system
Resource bargain/accountability (Metasystem 3)	• Determines and allocates the resources (manpower, material, money, methods, time, information, support) to governed systems • Defines performance levels (productivity), responsibilities, and accountability for governed systems Primarily an interface between Metasystem 3 to the governed systems
Operations (Metasystem 3)	• Provides for the routine interface concerned with near-term operational focus • Concentrated on providing direction for system production of value (products, services, processes, information) consumed external to the system • Primarily an interface between Metasystem 3 and governed systems
Coordination(Metasystem 2)	• Provides for metasystem and governed systems balance and stability • Ensures design and achievement (through execution) of design: (1) sharing of information within the system necessary to coordinate activities and (2) ensures decisions and actions necessary to prevent disturbances are shared within the metasystem and governed systems • Primarily a channel designed and executed by Metasystem 2
Audit (Metasystem 3*)	Provides routine and sporadic feedback concerning operational performance • Investigation and reporting on problematic performance issues within the system • Primarily a Metasystem 3* channel for communicating between Metasystem 3, the governed systems, and the metasystem concerning performance issues
Algedonic (Metasystem 5)	• Provides a 'bypass' of all channels when the integrity of the system is threatened • Compels instant alert to crisis or potentially catastrophic situations for the system • Directed to Metasystem 5 from anywhere in the metasystem or governed systems

(continued)

Table 2 (continued)

Communications channel and responsibility	CSG metasystem role
Environmental scanning(Metasystem 4')	• Provides design for sensing to monitor critical aspects of the external environment • Identifies environmental patterns, activities, or events with system implications • Provided for access throughout the metasystemMetasystem as well as governed systemsSystem by MetasystemMetasystem 4'
Dialog (Metasystem 5')	• Provides for examination of system decisions, actions, and interpretations for consistency with system purpose and identity • Directed to MetasystemMetasystem 5' from anywhere in the metasystem or governed systems
Learning(Metasystem 4*)	• Provides detection and correction of error within the metasystem as well as governed systems, focused on system design issues as opposed to execution issues • Directed to MetasystemMetasystem 4* from anywhere in the metasystemMetasystem or governed systems
Informing (Metasystem 2)	• Provides for flow and access to routine information within the metasystem or between the metasystem and governed systems • Access provided to entire metasystem and governed systems • Primarily designed by Metasystem 2 for utilization by all metasystem functions as well as governed systems

(coordination, integration). Here are the extended perspectives for each of these elements provided by CSG:

- *Control:* constraints necessary to ensure consistent performance and future system trajectory. In our formulation of control, we look to a more informed system view for guidance. This view suggests that control is not a pejorative term, to be scorned as a form of domination over a particular venue, activity, or entity. On the contrary, in the systems view we take, control is essential to ensure that the system stays on the trajectory that will provide future viability in response to changing conditions and circumstances. This is achieved by providing the greatest degree of autonomy (freedom and independence of decision, action, and interpretation) possible while still maintaining the system at desired levels of performance and behavior. In effect, this suggests that over-constraint of a system wastes resources (constraint is not free), limits system initiative/creativity, and diverts important emphases of a metasystem unnecessarily to lower levels of the system (inefficiency). However, underconstraint may sacrifice system level performance by providing excessive autonomy at the expense of integration necessary to maintain system level performance.

- *Communication: f*low and processing of information necessary to support consistent decision, action, and interpretation across the system. Communication is essential to governance and operation of the metasystem. Communications include not only the exchange of information, but also the interpretative schemas that permeate the system. These interpretative schemas are necessary to provide coherence in making, understanding, and interpreting the myriad of exchanges in a system. Communications may range from formal to informal, explicit to tacit, and patterned to emergent. There is not an optimal configuration for communication in a system, and the arrangements are certainly subject to shifts over time and emergent patterns. However, from a complex system governance perspective, communications are something that would be better off not left to chance self-organization. Instead, purposeful design and evolution of communications within a system are more likely to produce and maintain desirable results.
- *Coordination: p*roviding for effective interaction among different entities within the system, and external to the system, to prevent unnecessary fluctuations. Certainly, coordination is an essential aspect to ensure that a system provides sufficient interaction among different elements to maintain consistency. Quite possibly, the most important aspect of coordination is the damping of unnecessary fluctuations as the system operates. In effect, this implies that there must be sufficient standardization to provide routine interface as well as a sufficiently robust design to absorb emergent conditions that could not have been known in advance. While original work in management cybernetics focused on coordination as an internal function, we should also consider the necessity for coordination external to the system.
- *Integration:* design for system unity with common goals, accountability, and balance between individual constituent autonomy and system level interests. The primary focus of integration is to insure that the system achieves desirable levels of performance while (1) providing the maximum level of autonomy to constituents, (2) invoking the minimal constraint necessary for the system to function as a unity in achieving the intended purpose, and (3) strategically shifting the balance point between autonomy and integration based on changes in contextual factors and system performance levels. Integration is not achieved through serendipity, but rather by active design and continuous evolution.

The definition of CSG is incomplete without recognition of the underlying paradigm within which it is embedded. We now turn our attention to examine this paradigm and its importance to the deeper understanding and development of CSG.

A Complex System Governance Vignette—Where is the Metasystem?

Our systems continually act to disappoint by producing behavior, performance, and outcomes that are inconsistent with our intentions. An exemplar of this can be found in the case of an urban university, seeking to better understand their 'system' for bringing new students into the university system. Engaging in examination of the 'system' used for bringing on new students, several discoveries came to the forefront. Among these were the realization that the 'system' for

student entry was not designed, executed, maintained, or developed as a system at all. Instead, what was purported to be a system was a loose 'aggregate' of different processes, elements, and components. This aggregate had developed over time, without the benefit of a higher-level view concerning 'how it actually fits together' and made sense to effectively guide student entry. What was discovered was that there were individual units (e.g., finance, admissions, housing) that individually functioned very well. However, from a metasystem viewpoint, the individual units failed to function together. This was evidenced by the many problems that spanned multiple units and required their integrated efforts to address. Unfortunately, lacking a design or understanding as to how the 'metasystem' functioned, each issue that required multiple units for resolution generated inconsistent performance, emergent crises, and high human costs (frustration for administrators as well as students who had to navigate the system). The active examination from the 'lenses of the CSG metasystem helped to identify system design deficiencies and understand their source in systems principles/laws violated). Thus, a different path forward was made visible. A new decision, action, and performance interpretation space was opened to the system designers.

3.2 The CSG Paradigm

CSG is developing and exists in the early stages of emergence at the intersection of the governance, general systems theory, and management cybernetics fields as shown in Fig. 3. As such, it has the advantages of being tied to three fields with substantial substance, acceptance, and longevity. On the contrary, this intersection also invokes the criticisms and limitations of the fields as well as the potential for incompatibilities of the fields. Potential incompatibilities might exist across philosophical, theoretical, or methodological lines. This does not diminish the pursuit of CSG as an integrated field, but rather establishes a set of cautionary considerations in movement forward.

In the light of this caution, we have produced a succinct paradigm for CSG. This paradigm is related but distinct from each of the informing fields. As such, the paradigm exists as the particular way of thinking (worldview), which defines the grounding essence of the field. At this early stage, we would hesitate to suggest that CSG could be either considered a field or possessed a generally accepted paradigm. Instead, we have deliberately chosen to suggest an emerging paradigm for CSG—rooted in the governance, general systems theory, and management cybernetics, we have previously articulated. Although this does not preclude discovery or inclusion of other works or bodies of knowledge, it does offer a tenable starting point for further exploration. CSG could proceed absent a defining paradigm. However, this would be shortsighted, particularly given the CSG emphasis on enhancing the prospects for long term, sustainable systems, and solutions to their problems.

A paradigm offers a particular way of thinking (worldview). For CSG, we offer the following articulation of the paradigm:

From a systems theoretic foundation, a set of functions is enacted by mechanisms that invoke metasystem governance to produce the communication, control, coordination, and integration essential to continued system viability.

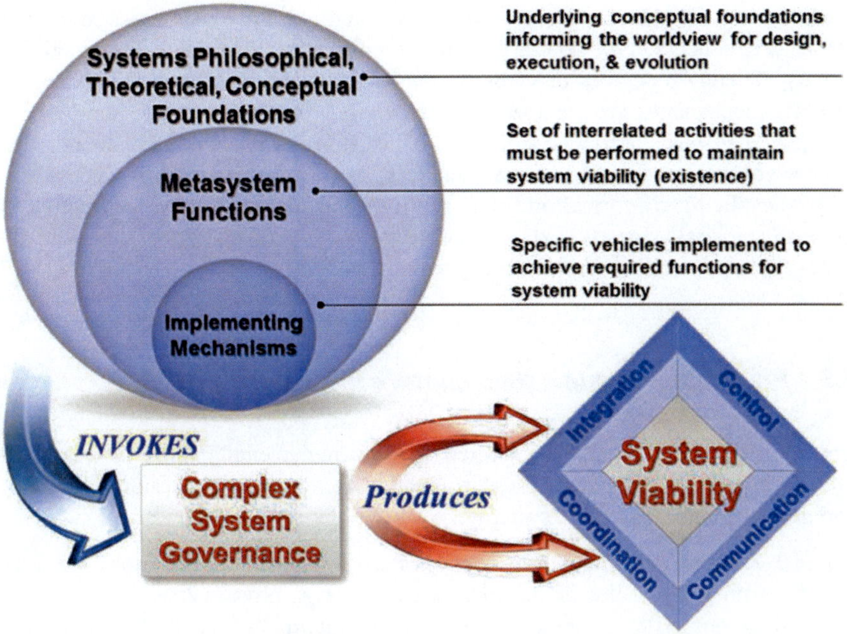

Fig. 4 Emerging paradigm for metasystem governance

Figure 4 below provides a pictorial representation of the emerging paradigm for CSG. It should be noted that the set of (metasystem) functions referred to in the paradigm as well as their development do not operate in isolation from one another. On the contrary, the functions themselves form an inseparable unity. The paradigm includes the relationship of three primary elements that serve as a triad for CSG. The first element consists of the *systems philosophical, theoretical, and conceptual foundations*. This foundation is rooted primarily in the general systems theory field, although foundational concepts from management cybernetics and governance are not excluded. The second element stems from the *metasystem functions* (specified above) that exist within the systems theoretic foundations and are subject to the laws, principles, and concepts that constitute general systems theory. *Implementing mechanisms* are the final element of the CSG triad, complementing conceptual foundations and metasystem functions. Implementing mechanisms are the 'vehicles' through which the metasystem functions are performed.

Conceptual foundations help to explain and understand 'why' systems behave and perform as they do, based on the laws and principles of general systems theory and management cybernetics. These laws and principles are immutable and cannot be negotiated away. The consequences for violation of the laws are real and will impact system viability. The metasystem functions identify 'what' must be achieved to ensure continued system viability. All systems must perform these functions at a minimal level to maintain viability. However, viability is not a 'guarantee' of performance excellence. On the contrary, viability only assures that the system continues to exist. There are degrees of viability, the minimal of which is existence. Implementing mechanisms are the specific vehicles (e.g., processes, procedures, activities, practices, plans, artifacts, values/beliefs, customs, more) that implement metasystem governance functions for a specific system of interest. These mechanisms may be explicit/tacit, formal/informal, routine/non-routine, effective/ineffective, or rational/irrational. However, all mechanisms can be articulated in relation to the metasystem governance functions they support.

3.3 Five Fundamentals that Capture the Essence of CSG

The essence of CSG can be captured in five fundamental points that serve to provide a succinct depiction of a very detailed approach to the design, execution, and development of complex systems.

1. ***All systems are subject to the laws of systems.*** Just as there are laws governing the nature of matter and energy (e.g., physics law of gravity), so too are our systems subject to laws. These system laws are always there, non-negotiable, unbiased, and explain system performance. *Practitioners must ask, 'do we understand systems laws and their impact on our system(s) design and performance?'*
2. ***All systems perform essential governance functions that determine system performance.*** Nine system governance functions are performed by all systems, regardless of sector, size, or purpose. These functions define 'what' must be achieved for governance of a system. Every system invokes a set of *unique implementing mechanisms* (means of achieving governance functions) that determines 'how' governance functions are accomplished. Mechanisms can be formal-informal, tacit-explicit, routine-sporadic, or limited-comprehensive in nature. CSG produces system performance which is a function of previously discussed *communication, control, integration, and coordination. Practitioners must ask, 'do we understand how our system performs essential governance functions to produce performance?'*
3. ***Governance functions can experience pathologies (deviations from 'healthy' system conditions) in performance of functions.*** There is no perfect system in execution. Regardless of the nobility of a system design, execution includes too many variabilities to 'guarantee' complete or absolute realization of design intentions. The effectiveness of governance is evident in the efficacy of identification, assessment, response, and evaluation to inevitable pathologies. Governance provides the degree of resilience and robustness to withstand and persevere in the middle of external turbulence and internal system flux. Good systems deal with pathologies as they occur—great systems continually design out pathologies before they escalate into crises. *Practitioners must ask, 'do*

we purposefully design and redesign our system to address and preclude pathologies?'

4. ***Violations of systems laws in performance of governance functions carry consequences.*** Irrespective of noble intentions, ignorance, or willful disregard, violation of system laws carries real consequences for system performance. In the best case, violations degrade performance. In the worst case, violation can escalate to cause catastrophic consequences or even eventual system collapse. *Practitioners must ask, 'do we understand problematic system performance in terms of violations of fundamental system laws?'*
5. ***System performance can be enhanced through development of governance functions.*** When system performance fails to meet expectations, deficiencies in governance functions can offer novel insights into the deeper sources of failure. Performance issues can be traced to governance function issues as well as violations of underlying system laws. Thus, system development can proceed in a more informed and purposeful mode. *Practitioners must ask, 'how might the roots of problematic performance be found in deeper system governance issues and violations of system laws, suggesting development directions?'*

4 Applicability of CSG

Organizations and practitioners must deal with increasingly complex systems and their inevitable problems. In essence, the complex system problem domain represents the *'new normal'* for the practitioners who must contend with complex systems and their associated problems. As a summary of this domain, following earlier work [17], we suggest that the domain is marked by the following characteristics:

- ***Uncertainty***—incomplete knowledge casting doubt for decision/action consequences as well as the appropriate approach(es) to proceed
- ***Ambiguity***—lack of clarity in understanding/interpretation of the system, environment, boundary conditions, context (circumstances, factors, conditions) within which it exists, and the nature of problems stemming from system operation
- ***Emergence***—occurrence of events and system behaviors that result from interactions, cannot be predicted, and are only known after they occur
- ***Complexity***—systems so intricate and dynamically interconnected that complete understanding, knowledge, prediction, control, or explanation is impossible
- ***Interdependence***—mutual influence among systems, where the state of each system influences, and is influenced by, the state of other interrelated systems

Complex systems, their associated problems, and the conditions that mark their problem domain are not going away. Practitioners (designers, owners, operators, performers) facing this domain are left in a precarious position. They must mount an effective response to develop systems and resolve problems within this domain, without the luxury of waiting for more effective support that lies 'just' beyond the horizon. CSG has applicability for practitioners who are interested in engaging complex systems at a different level of thinking.

At first glance, this reality is somewhat 'off-putting'. However, a closer examination of three questions is helpful for better understanding this current state of affairs in relation to the development of CSG in response.

How did our present day systems come to this reality? In many cases, our 'systems' have not been conceived, designed, or in fact executed as systems. Think of a problematic system—chances are it is like most of 'our systems', having come about through one of two primary means, *ad hoc* or *self-organized* design. An *ad hoc* system evolves by adding pieces and parts over time to respond to new requirements, never really being designed or evolved as an integrated whole. A fragmented 'system' emerges for which individual 'pieces' in the hodgepodge might make sense, but as a whole, the system becomes incomprehensible. Eventually, well-intended individual pieces detract from one another and degrade overall system performance. Examples of ad hoc systems are everywhere. Take for instance, a maintenance system intended to provide integrated and efficient maintenance operations across multiple entities and products. Over time, new maintenance programs, which all individually make sense and provide value, are added. However, although they individually might make great sense, collectively as a system, they comprise a 'hodgepodge' of fragmented pieces. This fragmented collection can actually detract from the primary purpose of the larger system intended to effectively integrate maintenance across the larger organization.

A second means of system development is *self-organization*, where the structure and functions of a system are permitted to develop 'on their own' without imposition of external constraints. This approach works great, as long as the system continues to produce expected behavior and desired performance levels. In effect, with *self-organized* system design, 'you get what you get', which may or may not continue to meet expectations given the present and future system realities. System design by *self-organization* might be great for low-stakes endeavors (e.g., a dinner party). However, for high-stakes complex systems, such as the maintenance system, exclusive reliance on self-organization is a recipe for disastrous system performance.

The third means of development is by *purposeful design*. This development involves the rigorous examination of a system through a set of systemic lenses. Although there are other systems- based approaches for applicability in system development, CSG is offered as a rigorously grounded systems-based approach to see underlying systemic issues and generate potential alternative paths forward.

If the situation of our systems is so 'dire', how do they continue to operate? Quite simply, systems continue to operate—in spite of poor designs—through 'brute force' execution. Without getting into an elaborate systems explanation, brute force can be recognized by such compensating activities as: (1) requiring excessive resources to overcome seemingly endless emerging issues, (2) simply living with the high cost (including human costs) of poorly designed/executed systems, or (3) reliance on *'system superheros 'to* sufficiently Band-Aid poor system designs to keep things working. Everyone has experienced system *superheroes* or might even be/have been one! *System superheroes* know 'how to get things done', 'can cut through the garbage', or 'know how to navigate the dark spaces of the system'. They are not bad people, however, system superheros frequently mask poorly designed and executed systems. And let us face it, even *superheroes* get tired, retire, or move on. In effect, it is 'us' who have let poor system designs evolve as they have and 'us' who have become so adept at accepting and 'compensating' for their poor performance—sometimes with incredible nimbleness. In effect, we frequently suffer and compensate for our poorly designed systems with execution that continues to mask system inadequacies.

Why is CSG not in the mainstream of system development approaches? CSG is an emerging field, with associated methods, applications, and technologies rapidly being developed for deployment. However, even though CSG might seem to 'make sense', engagement at any level of CSG development is not a casual decision. Underperforming systems do not appear overnight. They have gone through an evolutionary development (generally, *ad hoc* or *self-organized* as previously mentioned) and become entrenched in structure, strategies, support systems/processes, and even the identity of an organization (system). In essence, they have a large momentum based in the status quo. Thus, CSG exploration, analysis, and redesign can represent a 'sea change' to the 'status quo' within which a system exists. In other words, CSG is hard work, can be resource intensive, and can potentially discover fundamental system issues that may not be 'feasible' or 'palatable' to address given current circumstances. This does not diminish the value of CSG, but rather serves to establish more realistic expectations for CSG, or any approach that seeks to challenge entrenched systems, regardless of potential payoff.

CSG is not a 'silver bullet' or 'magic elixir' promising to cure all ills of modern systems. It requires hard work and commitment, but the payoff can be substantial. What is the payoff? Imagine having to navigate to a destination in the dark, without a map, having questionable directions, and no local knowledge of 'bad spots' to avoid. The result is very likely the 'trip from hell'. CSG provides practitioners with the equivalent of a real-time guide—providing directions, identifying impediments along the way, and tailoring the route to the capabilities of the vehicle (system) and practitioners making the trip. CSG is an invitation to generate a different experience in navigating complex systems and their problems. In effect, a *governance positioning system (GPS)* to provide directions to the future via more effective and compatible routes.

CSG has been developed as an alternative to *ad hoc* or *self-organized* system design, execution, and evolution. The CSG alternative is one of *'purposefully designed'* systems or p-systems. P-systems are focused on active design, execution, and evolution of governance functions in ways that are consistent with the laws (principles) of systems.

A Complex System Governance Vignette—Where is the Owner's Manual for this System?

We have all been in the situation where we are driving a rental car and cannot seem to find where a particular control is located and operates (e.g., heat, windshield wipers, gas tank release, trip mileage control, cruise control, radio, resetting clock). Since we do not 'own' the vehicle, there is a certain 'acceptance' of the annoyance, unless we breakdown and go to the owner's manual in the glovebox to learn how the function we desire is performed. However, for the vehicle we 'own', we have an owner's manual that provides guidance to make the intricate system adjustments we desire (e.g., Bluetooth settings). We would not purchase a complex vehicle without also receiving the owner's manual that tells us critical things about our car (system) such as maintenance intervals, troubleshooting problems, meaning of indicators, and performance of essential functions. Suggesting that a car is nowhere near the complexity we find in a modern organization, *why do we not have an integrated owner's manual that specifies the design, execution, and development for our organization (system)?* In many ways, we have pieces and parts—for example, processes, policies, and procedures provide some indicators of system execution. However, at the 'metasystem' level for governance, it is the rare case that we find an owner's manual equivalent for governance of an enterprise.

CSG is a system(s)-based approach to enable practitioners to better deal with complex systems and their problems.

CSG can provide value across several levels (Fig. 5), including:

- ***Practitioner:*** enhanced capacity of individual practitioners to engage in the level of systems thinking necessary to more effectively deal with the issues related to design, execution, and evolution of complex systems and their problems.
- ***Enterprise:*** provide competency development (knowledge, skills, abilities) for targeted entities (units, staff teams, departments) across the enterprise to better engage complex systems and problems.

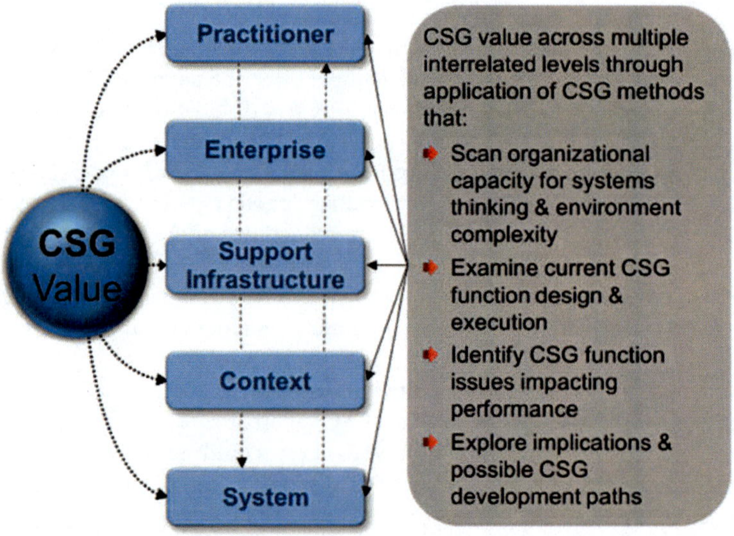

Fig. 5 Value-added at multiple levels from CSG

- ***Support Infrastructure:*** examination and development of support infrastructure (processes, technologies, systems) for compatibility with system governance design, execution, and development.
- ***Context:*** identification and development consideration for unique circumstances, factors, and conditions that influence (constrain or enable) achievement of system governance functions and system performance (e.g., stakeholders, regulatory requirements, staff, leadership style).
- ***System:*** providing identification of impediments to system performance rooted in specific deficiencies in design, execution, and development of governance functions and corresponding system laws.

The value accrued by CSG stems from: (1) scanning of the capacity an organization (entity) to engage in a level of systems thinking compatible with the complexity demands of the system environment, (2) exploration of the design and execution of essential governance functions, (3) identification and prioritization of system performance constraints tracked to problematic governance functions and violations of systems laws, and (4) establishment of developmental strategies across multiple levels essential to enhancing CSG to improve system performance.

To illuminate the applicability of CSG and potential contributions, we examine three scenarios of application.

4.1 SCENARIO 1: Workforce Capacity for System Thinking

Situation: A workforce is continually behind in producing innovative thinking to effectively respond to complexity demands of their environment—resulting in crises, surprises, or inefficiencies. The errors continue to mount with increasingly deficient performance, discontent in the workforce, and the seeming inability to effectively function in relationship to the demands of the complex environment within which the system and practitioners must function.

CSG Perspective Discussion: A critical element of CSG is the dependence on the capacity of the workforce to engage at a level of systems thinking necessary to realize the inherent value in CSG. Without the correct frame of reference (system thinking capacity), the results desired from CSG are not likely to be achieved. In essence, if the workforce does not have the necessary systemic thinking skills, then CSG is just another approach that an organization might grasp at for relief. Regardless of how dire the organization circumstances might be, there is no shortcut to having the requisite capacity in individuals to effectively engage any systems-based endeavor. There are two primary drivers for this situation. First, as mentioned, is the capacity of the workforce to think systemically. Second is the degree to which the environment demands systems thinking capacity. Performance will largely be determined by the degree that there is a sufficient 'match' between the systems thinking capacity that exists in the workforce to that demanded by the environment they must navigate.

CSG Response Discussion: Systems thinking capacity (ST-Cap) and environment complexity demand assessment instruments can be used to identify gaps between ST-Cap of the entity (team, department, organization) and the demands of their environment. 'Critical' areas for enhancing ST-Cap are identified. Figure 6 depicts this gap along the seven dimensions of systemic thinking. As can be seen by the diagram, there are gaps between what is demanded by the environment and what the workforce is capable of providing. For example, in flexibility the environment demands over 80 percent. However, the workforce is only operating at roughly 20 percent. This disparity, left unattended to, is a source of system dysfunction.

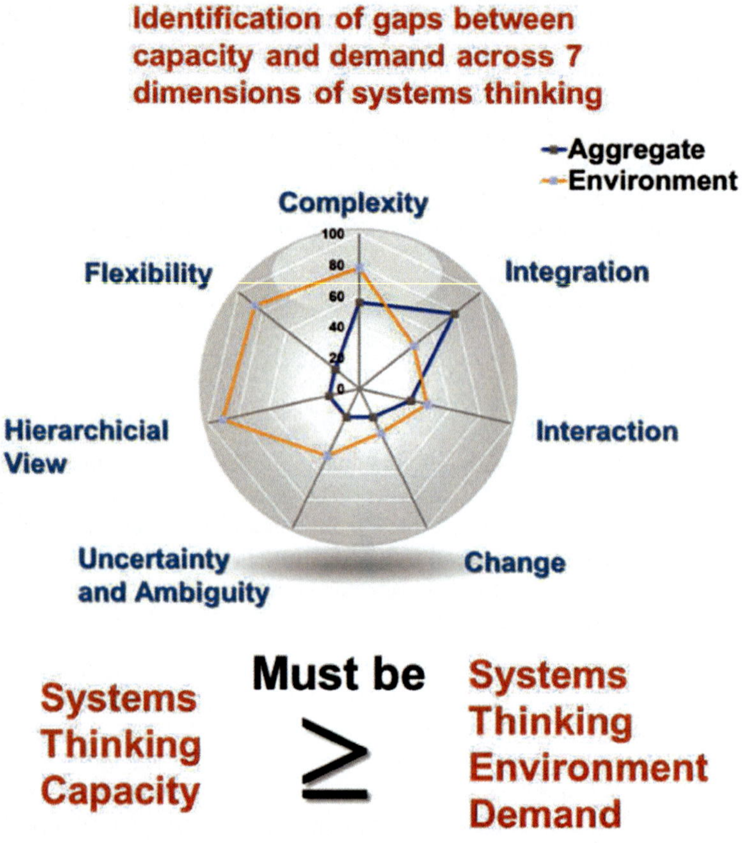

Fig. 6 Gaps between systems thinking capacity and environmental demand

4.2 *SCENARIO 2: System Governance Pathologies Identification*

Situation: A focal system is experiencing continual failures (e.g., cost overruns, schedule delays, missed performance targets) that are resistant to improvement efforts. The external manifestations of failures are evident in either product/service quality, missing milestones, required customer completion schedules, or conflicts in the adequacy, utilization, or outcomes achieved for resources consumed. There have been several failed attempts to locate the source of the deficiencies, but there does not appear to be a singular root cause to which failures can be attributed. The result is sagging customer confidence, resource scarcity, and a diminished workforce from the anxiety and frustration being experienced without an apparent path forward or end in sight.

CSG Perspective Discussion: It is quite easy to identify the results for violation of underlying systems principles (evidenced as pathologies). Pathologies are the outward manifestation of underlying system design, execution, or development issues. Being able to properly trace the systemic issue requires a 'deeper dive' into the actual system producing the performance issues. In essence, a system can only produce what it produces, nothing more and nothing less. If the system performance is not consistent with that we desire, we must understand the system that is producing the undesirable behavior/performance. Focusing only on the outward signs (symptoms) of the underlying systemic issues can at best provide a temporary fix. At worst, more damage than good might accrue from superficial treatment of symptoms of underlying system deficiencies (pathologies).

CSG Response Discussion: Focal group completes a system governance pathologies assessment instrument. Deep system pathologies (aberrations from healthy system conditions) across nine governance functions are identified, mapped, systemically explored, and prioritized for response. This approach provides an opportunity to discover the underlying source of deficiencies in a system. These are not necessarily observable from the inspection of their superficial deficiencies produced. Figure 7 below shows a mapping of one particular pathology (of 53 different possible pathologies) in a system.

4.3 *SCENARIO 3: System Governance Development*

Situation: An organization has difficulty in providing a clear, coherent, and accountable system innovation strategy to address persistent criticisms from oversight bodies. External forces are continually challenging the organization to provide information, performance indicators, and reasons for major decisions and strategies being pursued.

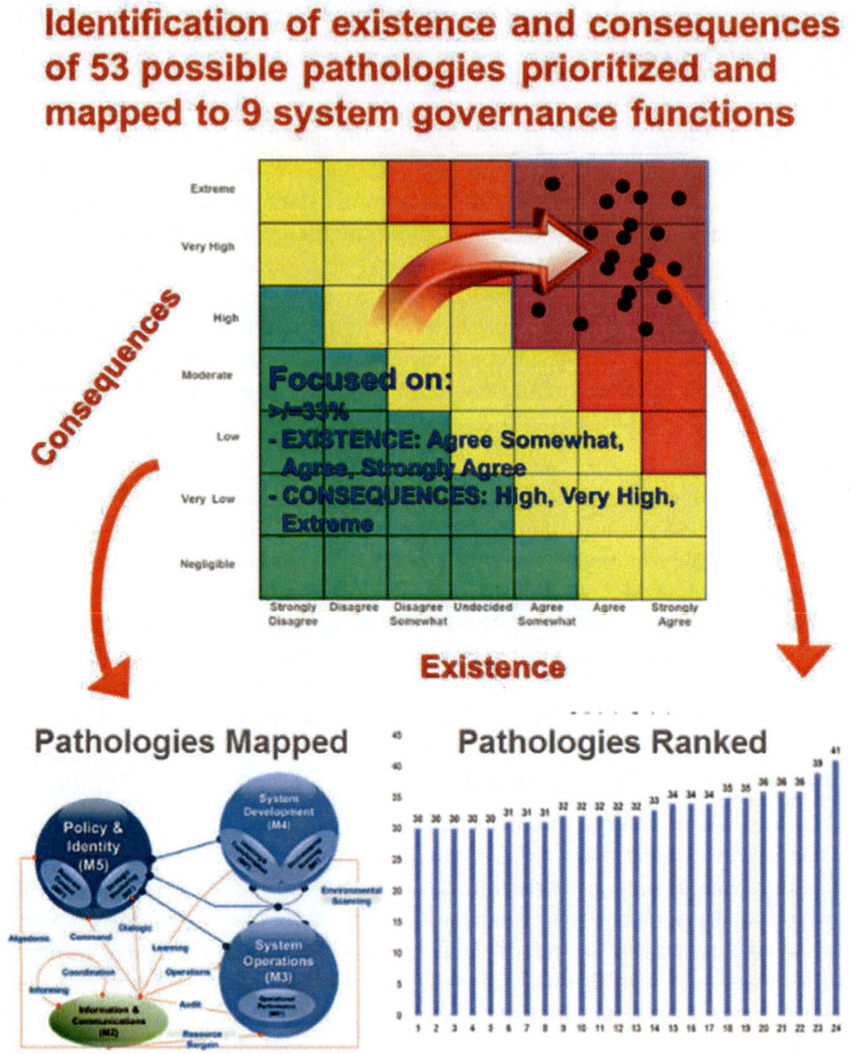

Fig. 7 Mapping deep-seated system pathologies in an organization

CSG Perspective Discussion: Although it is common to receive 'oversight' in the performance of the organizational mission, care must be taken to understand the degree to which the system is designed, executed, and developed such that oversight is not a burden but rather a welcomed opportunity to 'demonstrate' the system. As most systems are not purposefully designed, it is not uncommon to look at external 'hands' as an annoyance at best and at worst an impediment to performance. The need to constrain a system may in fact stem from inadequacies in the design or execution of a system to be commensurate to that which is demanded. Without a robust design against which to reference external perturbations, it is not likely that a system will generate sufficient resilience to effectively direct external 'meddling' in a system. System development should emphasize development of robustness in the design such that externally imposed 'reaching' can be better understood and responses can question the system design/execution for appropriateness to 'classes' of probing, not just individual cases.

CSG Response Discussion: Mapping of the CSG landscape provides visualization for analysis of the most critical challenges facing CSG development (peaks). Past, ongoing, and future planned system development initiatives are mapped against the existing governance landscape, pathologies, and system criticisms. 'Holistic' analysis provides clarity and focus for an integrated system development response strategy. Adjacent figures are representative of the current research, including the application of a 16-point CSG governance check. Figure 8 shows a mapping of a CSG landscape for a system.

5 Implications

With respect to development of CSG, there are two interrelated aspects. First, there is the development of active governing systems. This *governance development* is focused on identifying and engaging in a set of interrelated activities designed to establish, execute, and evolve the continuing development of the CSG metasystem functions. CSG development is always focused on identification and execution of feasible development activities consistent with initial assessments of the state of governance in an organization (system). Development involves purposeful improvement of the system of interest (context, pathologies, system). Ultimately, the purpose of governance development is to enhance system performance through the process of continual integrated activities to move the system to a more desirable, feasible, achievable, and sustainable level of performance. It would be shortsighted not to include the multiple aspects of development for CSG, including practitioners, organization, larger enterprise, support infrastructure, context, and system.

Table 3 below identifies the details of the five interrelated development activities that can be engaged to further *governance development*. These five elements include: (1) *Exploration*—examination of the performance of the metasystem functions, (2) *Innovation*—identification and prioritization of feasible decisions and actions to improve the metasystem functions, (3) *Transformation*—implementation of innovation strategies and initiative deployment planning to improve the metasystem functions, (4) *Evaluation*—continuous monitoring of the impact of strategies and initiatives undertaken to enhance metasystem performance, and (5) *Evolution*—monitoring development of system governance toward more desirable levels of performance and higher states of maturity.

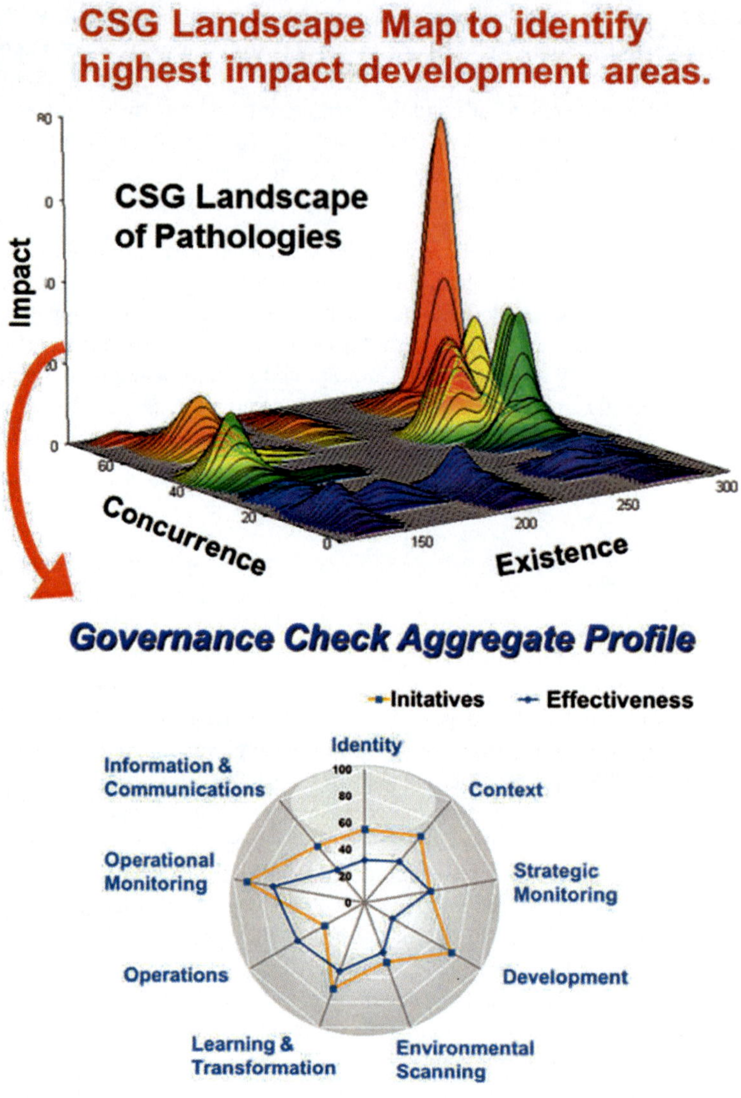

Fig. 8 Mapping CSG landscape to identify the highest priority development areas

There are three critical points of consideration for implications concerning CSG development. First, while the different governance development activities listed above are presented as separate, they are not independent or linear in execution. In fact, they are considered to be interrelated and overlapping. Therefore, the consideration and performance of the different activities are not mutually exclusive of one another. In essence, they set a frame of reference for a holistic and continuous

Table 3 Interrelated activities for CSG development

Governance development activity	Purpose	Objectives
Exploration	Holistic analysis and synthesis of metasystem context, design, execution, and pathologies	• Systemic investigation and self-study of the metasystem • Identification of completeness of the set of mechanisms performing CSG functions as well as the effectiveness of individual mechanisms • Conduct systemic inquiry to explore multiple perspectives and implications for the metasystem state • Identify, represent, and prioritize systemic meaning and implications of deficiencies, pathologies, and patterns • Define the metasystem current state and trajectory • Examine models of the current system, future system, environment, and context
Innovation	Definition of compatible and feasible metasystem development decisions, actions, priorities, strategies, and initiatives	• Develop the high-level strategy for systemic modifications to the metasystem • Identification, evaluation, and prioritization of compatible and feasible (contextually) first-order (correction within existing system) and second-order (correction by system redesign) initiatives to advance governance of the metasystem • Definition of capabilities and competencies (individual and organizational) necessary to engage systemic innovations to advance the state of CSG and evolve context • Definition of system capacity (resources, infrastructure) and compatibility for engagement of systemic innovation initiatives

(continued)

Table 3 (continued)

Governance development activity	Purpose	Objectives
Transformation	Implementation of systemic metasystem governance developmental strategy, decisions, actions, and initiatives to influence system trajectory, advancement of CSG state, and contextual development	• Holistic deployment planning and resource allocation for initiatives in support of metasystem governance development • Assignment of responsibilities and accountabilities for achievement of transformation initiatives • Exploration of the potential failure modes and mitigation actions necessary to increase probability of success of launched initiatives • Launching of selected initiatives to enhance the metasystem • Integration and assessment of ongoing initiatives in relationship to CSG development strategy and priorities • Identification and assessment of emerging 'rogue' initiatives against the deficiencies, pathologies, and priorities (blueprint) for strategic CSG development
Evaluation	Assessment of the effectiveness of metasystem initiatives, ongoing strategic performance of the metasystem, and development of the metasystem	• Identification of the minimal set of indicators (measures) that serves to show progression of the metasystem development efforts and shifting state of the metasystem • Assess effectiveness of initiatives undertaken for systemic metasystem transformation • Provide feedback for continuing relevance of transformation strategy in the light of new system knowledge, understanding, and contextual changes

(continued)

Table 3 (continued)

Governance development activity	Purpose	Objectives
Evolution	Setting and monitoring the trajectory and maturation of metasystem governance and system identity	• Monitor the long-range purposeful trajectory of the system in response to internal and external shifts • Enhance the continuing maturity (CSG state advancement) of the system of interest, taking the long view, uncorrupted by short-term aberrations • Ensure continuity, sustainability, and viability of the system in relationship to changes in the system, context, and environment • Prevent system erosion through methodical development consistent with shifting demands on the system of interest within the context and environment

conversation concerning execution of CSG development. Second, the conversation and actions invoked in CSG development are directed to enhance the overall function of CSG. This is achieved by engaging activities targeted to make improvements in the state of CSG and context for the system of interest. There is an advantage that accrues from the depth of exploration that should be achieved on the 'front end' of CSG development. In particular, engagement in CSG without a workforce commensurate to the engagement offers limited probability of success.

A third critical consideration for CSG development stems from the explorations and mapping of historical, presently existing and future initiatives in relationship to CSG development priorities. This serves as a 'litmus test' to question the relationship of initiatives to CSG development. If initiatives are truly targeted to improving the system, their utility with respect to addressing priorities, deficiencies, and identified needs should be capable of withstanding scrutiny. Thus, decision-makers are provided actionable intelligence concerning the contribution of different 'well meaning' activities currently underway or being contemplated to improve CSG. If development initiatives, either ongoing or being considered, cannot be 'justified' as to their relevance to the most pressing needs for improving the state of CSG and context, they should be called into question.

CSG development is not envisioned as an easy approach to system improvement. On the contrary, it is viewed as a difficult development path. This path is fraught with potential obstacles that should be considered by individuals or entities contemplating engaging the approach presented for CSG development.

6 Exercises

The following exercises provide an opportunity to examine the concepts presented in this chapter through several questions.

1. For a situation of your choosing, identify elements from the depiction of the complex system problem domain identified in Fig. 1. What is the significance of the nature of this domain for practitioners? Future systems?
2. Identify a 'system superhero' that you have come across. Identify why this system superhero might be detrimental to the long-range survivability of the system(s) they continually save. What can be done in the case of a system superhero unwilling to 'relinquish' their superpowers over a system?
3. What do each of the three fields supporting CSG (governance, general systems theory, management cybernetics) bring to CSG development?
4. Succinctly explain the essence of the CSG paradigm identified in Fig. 4. What difficulties might be encountered in the deployment of this paradigm?
5. What guidance and cautions might you suggest for practitioners who might be considering initiation of a CSG-based initiative?

References

1. Ackoff RL (1997) Systems, messes and interactive planning. Soc Engag Soc Sci 3(1997):417–438
2. Adams KM, Hester PT, Bradley JM, Meyers TJ, Keating CB (2014) Systems theory as the foundation for understanding systems. Sys Eng 17(1):112–123. https://doi.org/10.1002/sys.21255
3. Baranowski C, Ampleman M, Whitler J, Posner A (2012) EPA's climate ready water utilities (CRWU) initiative. Proc Water Environ Fed 2012(1):600–604
4. Beer S (1979) The heart of the enterprise. Wiley, New York
5. Beer S (1981) Brain of the firm: the managerial cybernetics of organization. Wiley, Chichester
6. Beer S (1984) The viable system model: its provenance, development, methodology
7. von Bertalanffy L (1950) An outline of general system theory. Br J Philos Sci 1(2):134–165. https://doi.org/10.1093/bjps/I.2.134
8. von Bertalanffy L (1968) General system theory: foundations, developments, applications. George Braziller, New York
9. von Bertalanffy L (1972) The history and status of general systems theory. Acad Manag J 15(4):407–426. https://doi.org/10.2307/255139
10. Calida BY (2013) System governance analysis of complex systems. Doctor of Philosophy (Ph.D.) Dissertation. Engineering Management and Systems Engineering. Old Dominion University
11. Calida BY, Keating CB (2014) System Governance: emergence of practical perspectives across the disciplines. Gheorghe AV, Masera M, Katina PF (eds) Infranomics: sustainability, engineering design, and governance, pp 269–296. New York: Springer
12. EPA (2012). *Adaptation strategies guide for water utilities.* Environmental Protection Agency. http://water.epa.gov/infrastructure/watersecurity/climate/upload/epa817k11003.pd
13. Farr J, Buede D (2003) Systems engineering and engineering management: keys to the efficient development of products and services. Eng Manag J 15(3):3–11
14. Hammond D (2002) Exploring the genealogy of systems thinking. Syst Res Behav Sci 19(5):429–439. https://doi.org/10.1002/sres.499
15. Heylighen F, Joslyn C (1992) What is systems theory? In: Heylighen F, Joslyn C, Turchin V (eds) Cambridge dictionary of philosophy. Principia Cybernetica Web, Brussels. Retrieved from http://pespmc1.vub.ac.be/SYSTHEOR.html
16. Jackson MC (2019) Critical systems thinking and the management of complexity, 1st edn. Wiley, Hoboken, NJ
17. Jaradat R, Keating C (2014) Fragility of oil as a complex critical infrastructure problem. Int J Crit Infrastruct Prot 7(2):86–99
18. Katina PF, Hester PT (2013) Systemic determination of infrastructure criticality. Int J Crit Infrastruct 9(3):211–225. https://doi.org/10.1504/IJCIS.2013.054980
19. Keating CB, Katina PF (2011) System of systems engineering: Prospects and challenges for the emerging field. Int J Syst Syst Eng 2(2/3):234–256
20. Keating CB (2014) Governance implications for meeting challenges in the system of systems engineering field. In: 2014 9th international conference on system of systems engineering (SOSE), IEEE, Adelaide, pp 154–159. https://doi.org/10.1109/SYSOSE
21. Keating CB, Kauffmann P, Dryer D (2001b) A framework for systemic analysis of complex issues
22. Keating CB (2014) Governance Implications for meeting challenges in the system of systems engineering field. In: IEEE system of systems engineering conference, Adelaide, Australia June 9–14, 2014
23. Keating C B, Katina PF (2012) Prevalence of pathologies in systems of systems. International Journal of System of Systems Eng 3(3/4): 243–267. https://doi.org/10.1504/IJSSE.2012.052688

24. Keating CB, Katina PF, Bradley JM (2015) Challenges for developing complex system governance. In: IIE Annual Conference. Institute of Industrial and Systems Engineers (IISE), Proceedings, p 2943
25. Keating CB, Bradley JM, Katina PF (2016) Systemic Analysis of Complex System Governance For Acquisition. Proceedings of the International Annual Conference of the American Society for Engineering Management. Huntsville 2016
26. Keating CB, Morin M (2001) An approach for systems analysis of patient care operations. Journal of Nursing Administration 31(7/8): 355–363
27. Klir GJ (1972) Preview: the polyphonic GST. In: Klir GJ (ed) Trends in general systems theory
28. Korzybski A (1994) Science and sanity: an introduction to non-Aristotelian systems and general semantics. (Preface by Robert P. Pula.). Fort Worth, TX: Institute of General Semantics
29. Laszlo E (1996) The systems view of the world: a holistic vision for our time. Hampton Press, Cresskill Lector Publishing Company, Iraklio, pp 65–80
30. Laszlo A, Krippner S (1998) Systems theories: their origins, foundations, and development. In: Jordan JS (ed) Systems theories and a priori aspects of perception. Elsevier Science, Amsterdam, pp 47–74
31. Naphade M, Banavar G, Harrison C, Paraszczak J, Morris R (2011) Smarter cities and their innovation challenges. Computer 44(6):32–39
32. Rittel HWJ, Webber MM (1973) Dilemmas in a general theory of planning. Policy Sci 4(2):155–169. https://doi.org/10.1007/BF01405730
33. Skyttner L (2005) General systems theory: problems, perspectives, practice, 2nd edn. World
34. Von Bertalanffy L (1956) A biologist looks at human nature. The Scientific Monthly, 82(1): 33–41
35. von Foerster H, Mead M, Teuber HL (1953) Cybernetics: circular causal and feedback mechanisms in biological and social systems. Josiah Macy, Jr Foundation, New York
36. Waddington CH (1957) The strategy of genes: a discussion of some aspects of theoretical biology
37. Warfield JN (1976) Societal systems: planning, policy and complexity. Wiley-Interscience, New York
38. Whitney K, Bradley JM, Baugh D E, Chesterman CW (2015) Systems theory as a foundation for governance of complex systems. Internation Journal of System of Systems Engineering, 6(1–2): 15–32

Complex System Governance Reference Model

Charles B. Keating and Polinpapilinho F. Katina

Abstract This chapter provides an in depth exploration of the reference model for complex system governance (CSG). The CSG reference model is explored as the set of interrelated functions and associated communication channels that must be performed for a system to remain viable (continue to exist). To provide this exploration, this chapter is focused on three primary objectives. First, the background for the CSG reference model is developed. This background places the CSG reference within the larger scope of the emerging CSG field. Following the introduction to the role of the reference for CSG field development, the conceptual foundations are examined. These foundations include management cybernetics, systems theory, and system governance. Included in the management cybernetics discussion are the 10 communication channels that are used in the CSG reference model. Second, the CSG reference model is developed. This development explores the nine metasystem functions that constitute CSG. Each function is examined for the primary role, responsibilities, and representative products from the function. The functions provide 'what' must be done to execute CSG for a complex system. Third, the future directions for further development of the CSG reference model are explored. The fit of the reference model within the larger scope and development of the CSG field is examined.

Keywords Complex system governance · Reference model · Management cybernetics

C. B. Keating (✉)
Old Dominion University, Norfolk, VA, USA
e-mail: ckeating@odu.edu

P. F. Katina
Department of Informatics and Engineering Systems, University of South Carolina Upstate, Spartanburg, South Carolina, USA

C. B. Keating et al. (eds.), *Complex System Governance*, Topics in Safety, Risk, Reliability and Quality 40, https://doi.org/10.1007/978-3-030-93852-9_7

1 Introduction

Complex system governance (CSG) is an emerging field in the earliest stages of development. The introduction of this field has been previously suggested in several different works, most recently, Keating et al. (2019). In this previous work, CSG was defined as "*Design, execution, and evolution of the metasystem functions necessary to provide control, communication, coordination, and integration of a complex system."(p. 6)*. Without repetition of earlier work, at a high level, the following elements of the definition are elaborated as essential foundations for our present purposes:

1. ***Communication***—the flow and processing of information within and external to the system that provides for consistency in decisions, actions, and interpretations made with respect to the system.
2. ***Control***—invoking the minimal constraints necessary to ensure desirable levels of performance and maintenance of system trajectory, in the midst of internally or externally generated perturbations of the system.
3. ***Coordination***—providing for interactions (relationships) between constituent entities within the system and between the system and external entities such that unnecessary fluctuations are avoided.
4. ***Integration***—continuous maintenance of system unity as a dynamic balance between autonomy of constituent entities and the interdependence of entities necessary to invoke a coherent whole. This interdependence produces the system identity (uniqueness) that exists beyond the identities of the individual constituent entities.
5. ***Design***—purposeful and deliberate arrangement of the governance system consistent with the achievement of desirable performance outputs and outcomes.
6. ***Execution***—performance of the system design within the unique system context, subject to the emergent perturbations stemming from both dynamic interaction with the environment as well as internal elaborations within the system.
7. ***Evolution***—the change of the governance system in response to internal and external shifts. These shifts may be in response to new knowledge, environmental perturbations, internal system perturbations, or emergence.
8. ***Metasystem***—the set of interrelated higher level functions that provide for governance of a complex system.

However, as compelling as this definition might appear, it creates a necessary but not sufficient set of conditions to fully articulate and prepare for practice related to CSG. In the earlier Keating et al. [19] work, the need for a 'reference model,' coupled with a corresponding development framework, was identified as one of the critical elements for CSG field development. The other two elements included: (1) setting of a comprehensive research agenda to direct purposeful development of the field, and (2) introduction of an initial set of challenges to focus research around areas with the potential to address some of our most vexing problems in dealing with complex

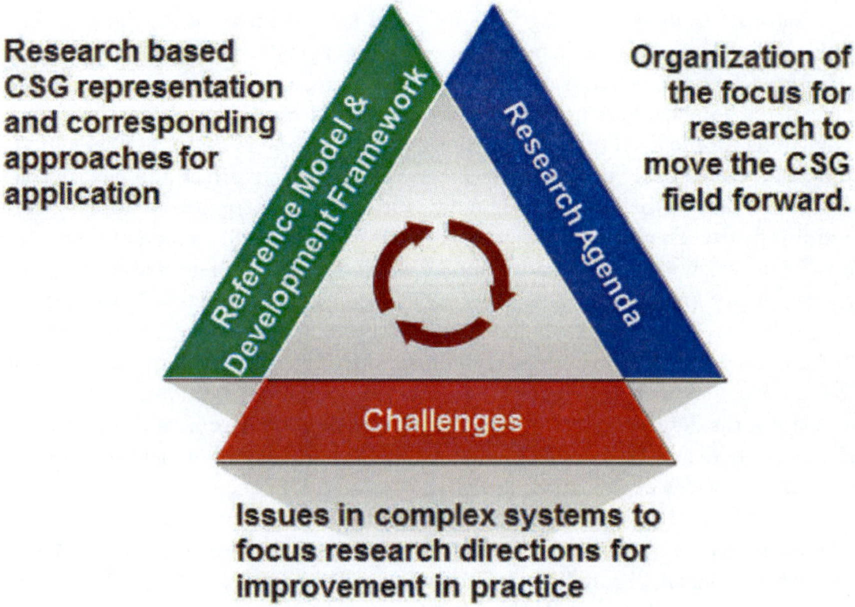

Fig. 1 CSG reference model fits within the developing CSG field

systems. Thus, the need for a CSG reference model was identified as one of three critical elements in a triad for the development of the CSG field as first expounded in Keating et al. [19] and shown in Fig. 1.

The first element of the triad for development of the CSG field includes setting of a comprehensive research agenda to guide holistic field development in an integrated fashion. The research agenda serves to position current and ongoing research within the larger context of research and entities undertaking research related to the CSG field. It offers an organizing approach to accelerate development of the field with the ultimate objective of engaging coherent and rigorous research to improve practice. The second element of the triad involves generation of a CSG reference model and corresponding CSG development framework. The reference model establishes a conceptually grounded representation of a complex governance system from a theoretically, axiomatically, and axiologically consistent frame of reference. In addition, the corresponding CSG development framework provides a corresponding guide for the methodological, method, and application of CSG to successfully bridge knowledge to practice. In effect, the development framework serves to advance the CSG field while bringing it to the world of the practitioner, offering a comprehensive approach for the analysis and methodical development of governance for a complex system. The third and final element of the triad involves the introduction of an initial set of challenges, around which research can be undertaken to advance the CSG field and begin elaboration of the research agenda. The CSG reference model serves an important role in the developing CSG field and is a precursor to the development framework and an essential element of the research agenda.

To serve the primary purpose of expounding the CSG reference model, the chapter is organized to accomplish three primary objectives. First, we set the conceptual foundations for the CSG reference model. This foundation is based on Beer's [3, 4, 6] metasysem as described in the viable system model. In effect, the metasystem is stationed 'above' or 'beyond' the entities it serves to integrate, coordinate, and control [15, 19]. Therefore, the metasystem construct is ideally suited as a starting point from which to develop and ground CSG in the conceptual underpinnings of systems theory and management cybernetics. The second objective is focused on introducing the CSG reference model. The nine functions and subfunctions of the CSG reference model are developed with respect to their primary purpose and set of requirements that must be achieved in fulfilment of the function/subfunction. Third, we examine the future directions for further elaboration and development of the CSG reference model and the fit of this model within the larger development of the CSG field. As part of this examination, the role of the CSG reference model in relation to the other CSG field development areas is explored. The chapter concludes with a set of implications that the CSG reference model holds for the emerging field of CSG, along philosophical, theoretical, axiomatic, axiological, methodological, method, and application development challenges.

2 Conceptual Foundations for the CSG Reference Model

In setting the stage for the development of the CSG reference model, we focus on management cybernetics as a foundation upon which to build. Management cybernetics, or sometimes referred to as organizational cybernetics, was developed by Beer [3, 4, 6] in the form of the viable system model (VSM) and described by Beer as the 'science of effective organization'. In addition, we rely on systems theory as a philosophical, theoretical, and axiomatic basis for our development of the CSG reference model. In this section, we identify the two primary conceptual bases for our reference model development.

Systems theory provides a strong theoretical grounding for the CSG field as well as the constituent CSG reference model. Systems theory has been previously linked to the CSG field Keating et al. [19] and identified by Adams et al. [1] as a set of axioms and associated propositions (principles, concepts, and laws) that seek to describe the behavior of both natural and manmade systems. The concepts of systems, and the emergence of systems theory, are certainly not new. In fact, the foundations of systems thinking have been traced as far back as the ancient Chinese work *The I Ching*, translated as book of change and dated to be at least 5000 years old [23]. This work noted the dynamic nature of changing relationships among elements. Additionally, the central philosophical tenet of systems thinking, *holism*, can be traced back to the writings of Aristotle, who suggested that 'the whole is more than the sum of its parts'. In one of the most cogent presentations of systems theory,

Adams et al. [1] have consolidated the body of systems theory knowledge around a set of organizing axioms and corresponding propositions (principles, concepts, and laws). For brevity, we have included Table 1 that specifies the systems theory axioms (following the work of Adams et al. [1]) and draws the implications for the current development of the CSG reference model. For an extended discussion on systems theory the reader is referred to the more complete work of Adams et al. [1] and for explication of the nature of systems theory to the CSG field the works of Keating et al.

Table 1 Systems theory axioms and governance

Systems theory axiom	Complex system governance reference model Implications
Centrality Axiom—central to all systems are emergence and hierarchy and communication and control	• Deal with emergent conditions and perturbations • Define relationships for accountability and responsibility • Information for consistent decision, action, and interpretation • Monitor and maintain performance while preserving maximum autonomy
Contextual axiom—meaning in systems is derived from the circumstances and factors that surround them	• Compatible with the context and environment within which the system exists • Flexibility based on shifting context • Articulates, monitors, interprets, and responds to context and contextual shifts
Goal axiom—systems achieve specific goals through purposeful behavior using pathways and means	• Establish, monitor, and maintain strategic direction and identity • System purpose, goals, and objectives consistency • Coherence in identity • Cohesive force that maintains integrity of the system in focus
Operational axiom—systems must be addressed in situ, where the system is exhibiting purposeful behavior	• Guide system strategic execution • Consistency in system behavior and performance • Production of outputs and outcomes consistent with expectations
Viability axiom—key parameters in a system must be controlled to ensure continued existence	• Measurement of system performance • Monitor and process internal and external fluctuations • Regulate key parameters essential to continued system existence
Design axiom—purposeful imbalance of resources and relationships	• Maintain and evaluate system model against execution • Model the present and future system • Establish exchange in system (matter, energy, information)
Information axiom—systems create, process, transfer, and modify information	• Information needs for decision, action, and interpretation support • Efficiency in exchanges • Dynamic information access, availability, or utility

[19] and Whitney et al provide a detailed development. As the CSG field becomes established, systems theory offers a strong theoretical foundation upon which to anchor the field, following the development of Adams et al. [1] and adapted from the earlier work of [18]. It is important to note that the axioms, as well as constituent propositions, do not operate independently or in mutual exclusivity of one another.

The contributions of systems theory to the emerging CSG reference model are summarized as: (1) grounding the model in a strong philosophical and theoretical basis, (2) reliance on a philosophic/theoretical foundation that has withstood the test of time, and (3) establishes a multidisciplinary foundation that supports model application across a spectrum of fields and applications.

The VSM serves as an excellent foundation for the development of the CSG reference model. The essence of the VSM [3, 4, 6] related to the development of the CSG reference model is held in two primary contributions. First, the VSM is concerned with the design for requisite variety [2], which basically states that the control in a system is a determined by the degree to which the regulator of a system is capable of matching the variety (complexity) being generated external to the system (from the environment). Hence, 'requisite' is the variety that must be generated to regulate and maintain system viability within established or desired limits. If external variety exceeds the variety (matching) capability of the regulator (providing feedback for system adjustment to maintain key parameters), then the system will not maintain viability (existence).

From this simple relationship, [3, 4, 6] expounded the VSM as a set of functions that provides for the disposition of system variety through *filtering* (attenuating variety by limiting variety beyond the capacity of the system to respond), *amplification* (generation of larger amounts of variety from the system to better match variety being externally cast upon the system) and *transduction* (translation to preserve meaning across system boundaries). This control, through the regulation of variety, is control in the cybernetic perspective, as opposed to more pejorative interpretations of control as domination of an individual or entity that limits independence. Consistent with a cybernetics perspective of control, control is neither a good nor bad, but rather an element that exists in every viable (existing) system. In this sense, management cybernetics embraces control as: (1) necessary to ensure a system continues to exist in response to environmental perturbations, (2) only provides a minimal set of constraints [regulation] on the system necessary and sufficient to maintain performance and behavior of the system, and (3) preserving autonomy [freedom and independence of decision, action, and interpretation [19, 20] of constituent entities in a system].

A second major contribution of management cybernetics is the identification of a set of interrelated metasystem functions in the VSM that provide for the continuing viability of a system. The metasystem provides the integration and coordination necessary to ensure that a system continues to produce the products or services that allow it to meet performance levels necessary to continue to operate (exist). Failure of any of the metasystem functions would jeopardize the overall system. Beer's formulation of the metasystem provides 5 essential functions for continued systemviability. For brevity, the metasystem functions of the VSM are provided

Table 2 Metasystem functions in the VSM

VSM Metasystem function	Role of the function
Coordination system 2	• Provides for system stability by preventing unnecessary fluctuations within the set of systems being integrated by the metasystem • Promotes operational system performance by ensuring sufficient integration within the system • Acts to harmonize the system such that the system acts in unison • Limits unnecessary turbulence, increasing system efficiency as well as effectiveness
Operational control system 3	• Maintains operational performance on a day to day basis • Provides for the execution of policy, distribution of resources, and accountability within the system • Focused on near term achievement and maintenance of system performance levels
Audit and accountability system 3*	• Provides monitoring of the system to identify aberrations and invoke necessary explorations to determine the source of the aberrant behavior or unexpected variance • Essential to understand the nature of variance and focus actions to resolve variance
Development system 4	• Scans and captures information from the environment and assesses that information for strategic implications and system level impacts • Models the future and strategic evolution of the system
Policy system 5	• Provides for the strategic decisions and direction that maintain the identity of the system • Monitors and maintains a balance between the inherent tension between the long-term external focus and the short-term internal focus of the system

(consistent with earlier summaries from [13, 19] to offer a high level overview of [3, 4, 6] VSM metasystem functions (Table 2).

The third major contribution of management cybernetics and the VSM is the inclusion of *communication channels*. Table 3 is provided as a summary of communication channels for the VSM based on several works articulating Beer's VSM [3–6, 9, 10, 13] as supplemented by [16].

The metasystem construct makes several important contributions to our conceptual foundations for the CSG reference model, including: (1) since the metasystem operates at a higher logical level beyond (meta) to the elements (entities) that it must integrate, we can focus on the integration, coordination, communication, and control at a level beyond the entities that are governed, (2) being that the metasystem has been conceptually grounded in the foundations of systems theory and management cybernetics, the conceptual lineage has been established and provides a more robust foundation, (3) the 'function' view of metasystem permits a focus on defining *what* must be achieved to fulfil the function, as opposed to *how* it must be fulfilled, (4) the metasystem functions are interrelated and do not operate in isolation from

Table 3 Communication channels in the VSM

Communicationchannel	Primary functions
Command	• Provides direction to operational units • Dissemination of non-negotiable direction to the system
Resource bargain/Accountability	• Provides/determines the resources (manpower, material, money, information, support) for operational units • Defines performance levels to which operational units will be held responsible • Determines how operational units will interface for performance reporting and accountability
Environmental Scanning	• Provides design for sensing of the external environment • Identifies environmental patterns, activities, or events with system implications • Provided for access throughout the metasystem as well as governed systems
Operations	• Provides for the routine interface between operational system entities and from the metasystem to operational units
Coordination	• Provides for system balance and stability by ensuring that information concerning decisions and actions necessary to prevent disturbances are shared among operational units
Audit	• Provides routine and sporadic feedback on the performance of system operations • Investigates and reports on problematic areas
Algedonic	• Provides instant alert to crisis or potentially catastrophic situations occurring in the system • Bypasses routine communications channels and structure to identify system threats
Dialog	• Provides examination and interpretation of organizational decisions, actions, and events • Seeks alignment of perspectives and shared understanding of organizational decisions and actions in light of system purpose and identity
System learning	• Provides detection and correction of system errors, testing of assumptions, and identification of system design deficiencies • Ensures that the system continually questions the adequacy of its design
Informing	• Provide routine transmission of information throughout the system • Routes information that is not appropriate for other channels for accessibility throughout the system

one another, in effect operating as a system in and of itself, (5) the performance of metasystem functions is necessary to produce continued viability, not necessarily high performance, as a system can exist at various levels of performance, and (6) by understanding the nature and role of the metasystem functions, functions can be purposefully designed, executed, and maintained.

The final element in the conceptual foundation for the CSG reference model is the CSG field. Since we have provided an essential development of the CSG field in the introduction to this chapter, we now shift to the development of the CSG reference model.

3 Complex System Governance Reference Model

The CSG reference model is proposed as a critical element in the development of the CSG field. The reference model provides a systemic representation of CSG, built upon the intellectual foundations of systems theory and management cybernetics. The purpose of the CSG reference model is to provide an organizing construct for the interrelated functions necessary to perform CSG. The CSG reference model is designed to provide the following contributions to the emerging CSG field:

- *Common Grounding Reference Point*—the model provides a common model for identification of 'what' a governing metasystem must accomplish if the system is to remain viable. Arguably, any complex system that exists is performing the functions of the CSG reference model, albeit they may be performed at a minimal level.
- *Set of Common Functions and Requirements*—the model provides a detailed explication of the functions that must be performed for governance of any complex system. This level of detail for governance, drawing back to the foundations of systems theory and management cybernetics, is essential to CSG field development.
- *Multiple Utility*—the model informs analysis, design, maintenance, and evaluation for CSG. As such, it provides both researchers and practitioners a valuable artifact for dealing with complex systems.
- *Foundation for Field Related Development*—the model provides a foundation for other developments and contributions to the CSG field, not limited to development methodologies, methods development, tools/software development, and research.

Prior to getting into details concerning the CSG reference model, we offer a high level depiction of the model. We have been careful to point out the consistencies, as well as elaborations, from the metasystem functions identified by Beer's early works [3, 4, 6]. With respect to the metasystem functions of the VSM, the CSG reference model has the following commonalities/distinctions: (1) the numbering convention has been kept consistent to avoid confusion between the VSM metasystem functions and the CSG metasystem functions, (2) the communication function has been directly incorporated into the CSG reference model to amplify the importance of communications with respect to the other functions, (3) the number of functions/subfunctions has been extended to nine to amplify additional functions we feel are necessary and sufficient for metasystem design, and (4) we have treated the productive system in Beer's VSM as a 'black box' in our CSG reference model, allowing the focus to be on the interrelationship of the metasystem to the entities being governed, not on the entities themselves. Our departure from the strict confines of Beer's VSM metasystem formulation (proposed over four decades ago) may be unsettling to purists. However, we have elaborated, modified, and extended the metasystem of the VSM to fit the purposes of CSG. This does not cast doubt or challenge the basis or formulation of the VSM. On the contrary, it simply evolves and extends the VSM to better fit our intended use for representation of CSG. There are nine metasystem functions that we have identified for our CSG reference model. A brief depiction of the nature and role of the metasystem functions is:

- *Metasystem five (M5)—Policy and identity*—corresponds to system 5 in the VSM metasystem, focused on overall steering, giving direction and identity for the system
- *Metasystem five star (M5*)—System context*—elaborates a responsibility within the VSM system 5, focused on the specific context within which the metasystem is embedded
- *Metasystem five prime (M5')—Strategic system—monitoring* elaborates a responsibility within the VSM system 5, focused on oversight of the system at a strategic level
- *Metasystem four (M4)—System development*—corresponds to system 4 in the VSM metasystem, focusing on the long range development of the system to ensure future viability
- *Metasystem four star (M4*)—Learning and transformation*—elaborates a responsibility within the VSM system 4, focused on facilitation of learning based on correction of design errors in the metasystem and planning for transformation of the metasystem.
- *Metasystem four prime (M4')—Environmental Scanning*—elaborates a responsibility within the VSM system 4, focused on sensing the environment for trends, patterns, or events with implications for both present and future system performance and development
- *Metasystem three (M3)—System operations*—corresponds to system 3 in the VSM metasystem, focused on the day to day operations of the metasystem to ensure that the system maintains performance levels.
- *Metasystem three star (M3*)—Operational performance*—corresponds to system 3* in the VSM, focused on monitoring system performance to identify and assess aberrant conditions.
- *Metasystem two (M2)—Information and communications*—elaborates the system 2 function in the VSM to focus on the design for flow of information and consistent interpretation of exchanges (communication channels).

The detailed articulation of these metasystem functions is depicted in Table 4. As shown in Fig. 2, the functions are interrelated. None of the functions operates independent of the other functions. In addition, it is important to note that none of the functions is 'more important' than the others. Consistent with the VSM, all of the CSG reference model functions are necessary to ensure the continuing viability of the entire system. Poor performance of one metasystem function will propagate through the entire metasystem. The metasystem functions are performed through associated

Table 4 Metasystem functions for the CSG Reference Model

Metasystem function	Primary role	Responsibilities	Products
Metasystem five (M5)policy and identity	Primary function is to provide direction, oversight, accountability, and evolution of the system. Focus includes policy, mission, vision, strategic direction, performance, and accountability for the system such that: (1) the system maintains viability, (2) identity is preserved, and (3) the system is effectively projected both internally and externally	• Establishes and maintains system identity in the face of changing environment and context • Defines, clarifies and propagates the system vision, strategic direction, purpose, mission, and interpretation • Active determination and balance for system focus between present and future • Disseminates strategic plan and oversees execution • Provides for capital resources necessary to support system • Sets present and future problem space for focus of product, service, and content development and deployment • Sets strategic dialog forums • Preserves autonomy and integration system • Marketing of system products, services, content, and value • Public relations planning and execution • External mentorship development (e.g., Board of directors) • Establishes system policy direction and maintains identity of the system executed through strategic direction • Represents the system interests to external constituents • Defines and integrates the expanded network for the system (strategic partnerships) • Evolves scenarios for system transformation and implements strategic transformation direction	• Forums and mechanisms to define, maintain, and evolve system identity and focus (mission, vision, strategic direction, purpose) • Strategic system plan • Public relations plan execution and performance monitoring • Marketing plan execution and performance monitoring • Integrated system mapping • Satisficing system policies • Governance architecture for the metasystem

(continued)

Table 4 (continued)

Metasystem function	Primary role	Responsibilities	Products
Metasystem five star (M5*)system context	Primary function is to monitor the system context (the circumstances, factors, conditions, or patterns that enable and constrain the system)	• Identify system context and provide for assessment of contextual impacts on system performance (constraining or enabling) • Actively manages context • Conducts boundary spanning to determine the boundary conditions, values, and judgments for the system • Conducts inquiry into contextual barriers to system execution or development • Monitors and assesses the influence of contextual aspects for the system • Informs development of the strategic plan	• Stakeholder analysis • Contextual mapping • Contextual monitoring and development strategy
Metasystem five prime (M5')strategic system monitoring	Primary function is to monitor measures for strategic system performance and identify variance requiring metasystem level response. Particular emphasis is on variability that may impact future system viability.	• Track ongoing performance of system based on measures of performance for operations • Disseminates system performance throughout system • Identification, analysis, and maintenance of system context • Conducts inquiry into performance aberrations • Monitors and assesses the continuing adequacy of strategic system performance measures • Informs development of the strategic plan	• Measures for strategic system performance • Results of inquiry and analysis of performance issues • Recommendations for continuance, modification, or deletion of performance measures
Metasystem four (M4)system development	Primary function is to provide for the analysis and interpretation of the implications and potential impacts of trends, patterns, and precipitating events in the environment. Develops future scenarios, design alternatives, and future focused planning to position the system for future viability	• Analyzes and interprets environmental scanning results for shifts, their implications, and potential impacts on system evolution • Guides development of the system strategic plan and system development map • Informs the development of the strategic plan • Guides future product, service, and content development • Guides investment priorities • Identifies future relationships critical to system development • Identifies future development opportunities and targets that can be pursued in support of mission and vision of the system	• Planning for response to environmental scanning • Models of the present, future, and environment for the System • Strategic system development plan and system development map

(continued)

Table 4 (continued)

Metasystem function	Primary role	Responsibilities	Products
Metasystem four star (M4*)learning and transformation	Primary function is to provide for identification and analysis of metasystem design errors (second order learning) and suggest design modifications and transformation planning for the system	• Processes inputs for system wide implications • Identifies mechanisms for double loop Learning • Designs objectives, measures, and accountability for second order learning in the system • Leads in future transformation analysis • Provides future focused input to strategy development • Informs the development of the strategic plan	• Design for second order system learning • System transformation strategy • Dissemination of learning results, implications, and opportunities
Metasystem four prime (M4')environmental scanning	Primary function is to provide the design and execution of scanning for the system environment. Focus is on patterns, trends, threats, events, and opportunities for the system	• Designs for environmental scanning for the entire system (includes trends, changes, patterns, critical stakeholders, collaborative entities, research, etc.) • Executes the environmental scanning design • Maintains a model of the metasystem environment • Captures emergent environmental conditions and events • Consolidates results from environmental scanning and provides synthesis • Informs the development of the strategic plan • Disseminates essential environmental information and shifts throughout the system	• Design for environmental scanning including objectives, organization, execution, and performance monitoring • Publication of environmental scanning activities enabling coordination of targets, execution, data capture and analysis • Dissemination of scanning results, and implications of patterns, trends, threats, events, and opportunities for the system

(continued)

Table 4 (continued)

Metasystem function	Primary role	Responsibilities	Products
Metasystem three (M3)system operations	Primary function is to maintain operational performance control through the implementation of policy, resource allocation, and design for accountability	• Oversight for products, services, value, and content delivery • System planning and control for ongoing day to day operational effectiveness • Develop near term system design response to evolving operational issues and monitor operational performance measures • Operationally interprets and ensures implementation of the system policies and direction • Interpretation and translation of implications of environmental shifts for operations (based on inputs from system development) • Informs the development of the strategic plan • Determines resources, expectations, and performance measurement for operational performance • Design for accountability and performance reporting for operations	• Operational plan for production of system that generates value • Execution forums for ongoing operational maintenance • Resource planning for operational requirements • Establishes operational goals in relationship to strategic performance objectives • Sets priorities and resource allocation for operational support activities and investments • Determines performance measure targets
Metasystem three star (M3*)operational performance	Primary function is to monitor measures for operational performance and identify variance in system performance requiring system level response. Particular emphasis is on variability and performance trends that may impact system viability	• Track ongoing performance of system based on measures of performance for operations • Disseminates system performance throughout system • Conducts inquiry into performance aberrations • Informs the development of the strategic plan • Monitors and assesses the continuing adequacy of operational performance measures	• Measures for operations • Results of inquiry and analysis of performance issues • Recommendations for continuance, modification, or deletion of performance measures

(continued)

Table 4 (continued)

Metasystem function	Primary role	Responsibilities	Products
Metasystem two (M2)information and communications	Enables system stability by designing and implementing the architecture for information flow, coordination, transduction and communications within the metasystem and between the metasystem, the environment and the governed system	• Designs and maintains the architecture of information flows and communications within the metasystem, between the metasystem and environment, and between the metasystem and the governed system • Ensures efficiency by coordinating information accessibility within the system • Identifies standard processes and procedures necessary to facilitate transduction and provide effective integration and coordination of the system • Informs the development of the strategic plan • Identifies and provides forums to identify and resolve emergent conflict and coordination issues within the system	• Standard processes and procedures for internal coordination of the system • Communications architecture for the metasystem • Defined external coordination vehicles necessary for support for the system (e.g., public relations, press releases)

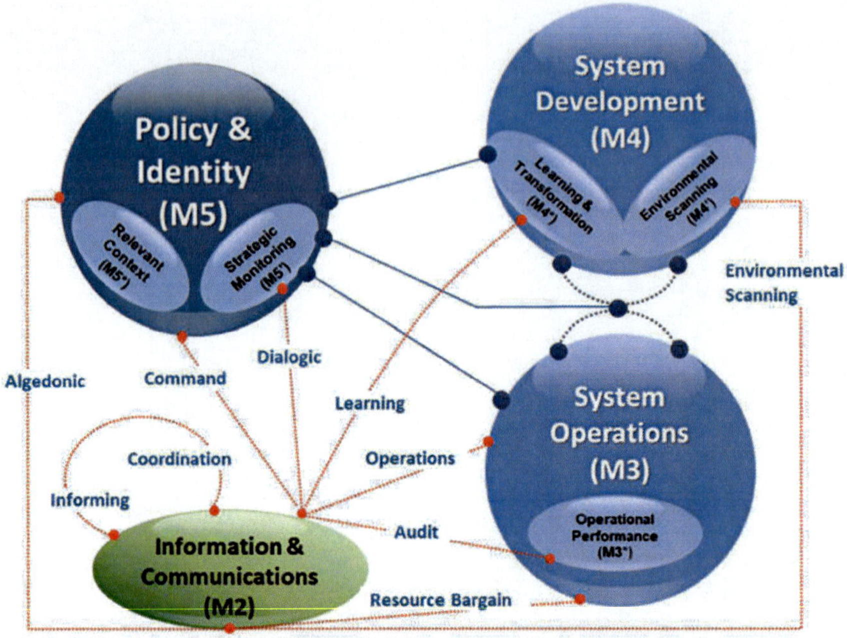

Fig. 2 CSG Reference Model interrelated functions and communications channels

mechanisms (the particular implementing devices that execute the metasystem function and exist in relationship with other mechanisms within the metasystem). The set of mechanisms and their interrelationships provide the structure that permits performance of the metasystem functions.

4 Conclusions and Implications

This chapter represents a refined development of a reference model suitable for the emerging field of CSG. We consider this model to be continually evolving, although it is thoroughly grounded in systems theory and management cybernetics. However, the complexity of the ideas and underlying theoretical foundations suggest that the model will naturally evolve as we gain experience that only time, and applications of the model can bring. The CSG reference model represents an important step forward for the CSG field. The reference model provides a foundation upon which there can be an evolution of development frameworks, corresponding methods to support application, software-based tools, and the underpinnings for applications based on deployment of the model. In addition, the model provides opportunities to make further contributions to the body of knowledge through research undertaken to further explore, test, and evaluate efficacy of the model. In effect, the building of the CSG

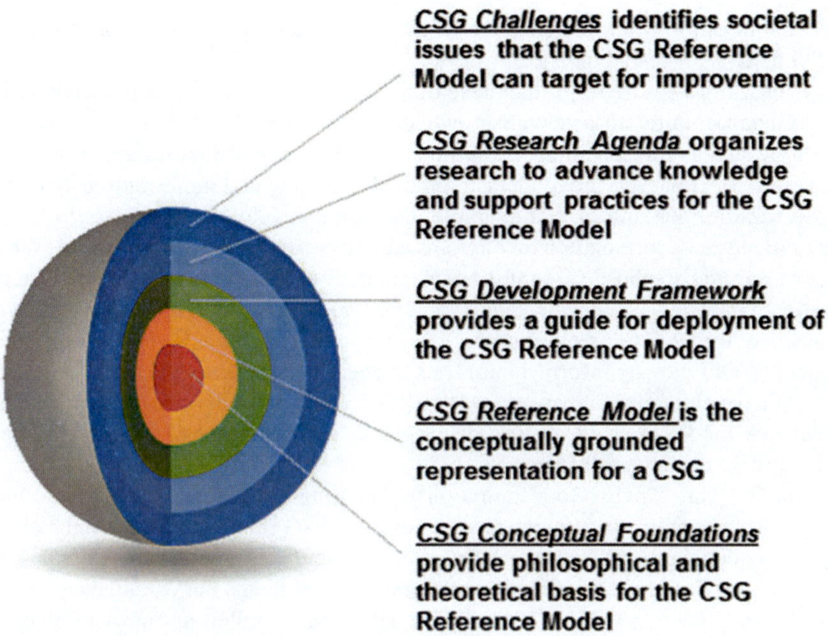

Fig. 3 CSG Reference model as an element in building the CSG field

reference model represents a necessary response to establishing a grounding frame of reference, based on the strong philosophical and theoretical linkage to systems theory and management cybernetics. Figure 3 captures the CSG reference model within the larger developing field of CSG. In effect, the CSG reference model: (1) is built upon a sound theoretical base (systems theory and management cybernetics), (2) provides an important element in the emerging CSG frameworks to guide application, (3) serves to inform critical developmental areas for research endeavors in CSG, and (4) supports meeting the application challenges that modern complex systems pose.

The current state of CSG reference model development provides two important contributions directly related to moving the CSG field forward. First, the reference model lies between the theoretical/philosophical roots of the field and the practical applications that can be built to deploy the model. As such, the model lies between the philosophical, theoretical, axiomatic, and axiological aspects of the CSG field development and the method, methodology, and application aspects of the field. The CSG reference model is a necessary development of the field to provide a bridge between the research-based and practice-based field development emphases. Second, the model supports a dialog important to the field. Continuing research articulates a sound systems theoretic grounded representation for CSG. Prior to this model, we have found little rigorously and theoretically grounded development of complex system governance. Existing models of governance Keating et al. [19] fall short on detailed development of the meaning, nature, and role that is played in CSG. The specific fit of the CSG reference model as an essential aspect of building the larger CSG field is suggested in Fig. 3.

In charting a course for further development of the CSG reference model there are several opportunities to accelerate growth of the CSG field. First, the model exists as a well-grounded representation of *what* must be achieved to fulfil the requirements for governance. However, the model should evolve as new and more rigorously developed explorations unfold. For example, the initial development of systems theory centerd around a formulation of 7 axioms and 30 corresponding propositions. While this was initially robust, care must be taken not to exclude a wider array of systems theoretic knowledge and what that formulation might bring to enhance further development of the CSG reference model and corresponding knowledge base. Second, the model should directly inform approaches to engage CSG development. The generation of a development framework(s) for CSG is essential to build the field. The CSG reference model is an ideal candidate against which rigorous CSG development frameworks can be established.

Finally, there should be concentration in preparing for application of CSG through methods and applications supportive of both the CSG reference model and a corresponding development framework. In addition, there are opportunities to widen the grasp of the CSG field by inclusion of several related fields. For example, system of systems engineering [14, 17, 18] is a field with many parallels and strong linkage to CSG. We should avoid closing off the CSG field to other related fields. The premature closing of the CSG field, while possibly temporarily pleasing, may serve to overly narrow the field early and potentially preclude insightful lines of inquiry that could broaden the utility of the CSG field.

One such framework for application of the CSG reference model is the CSG architecture framework [7, 21]. This framework describes the conventions, principles and practices for establishing complex system governance architectures. This lies in support of accomplishing the 9 governance functions, 65 related responsibilities, and 34 related outcomes suggested for CSG. The framework also integrates 30 systems theory propositions [22] as well as metasystem pathologies encapsulating 83 systems theory-based pathologies stemming from Katina [11]. The CSG architecture exist as a first attempt to operationalize the CSG reference model in a form that is 'actionable'. CSG architecture development is targeted to three primary stakeholders [7]. Among these stakeholders are included:

- ***Complex System Owner*** is a person or organizational body responsible for maintaining system viability through decision making not delegated to a metasystem governor or metasystem governance function owner.
- ***Metasystem Governor*** is a person or organizational body responsible for maintaining system viability through accomplishment of all metasystem governance functions.
- ***Metasystem Governance Function Owner*** is a person or organizational body responsible for accomplishment of one or more metasystem governance function(s).

The CSG architecture framework is considered to be a dynamic toolset for complex system owners, metasystem governors, and metasystem governance function owners. The thrust of the toolset is focused on the discovery, development, and maintenance of information necessary for development of complex system governance architecture products (model-centric outcomes/representations). These products facilitate greater understanding of a system of interest performance of complex system governance functions. The CSG architecture framework is representative of efforts to advance the CSG field through making the CSG reference model actionable. In effect, the CSG architecture framework serves as a catalyst for transition from theoretical underpinnings of CSG, and the CSG reference model, to real world application of CSG and performance of metasystem governance functions and discovery of associated pathologies. These advances are critical to the continuing maturation of the CSG field through operational applications.

In closing, we are confident that this foray into the CSG reference model has contributed to pushing the CSG field forward. CSG reference model-based applications, such as the CSG architecture framework, represent a vital step forward in depicting the functional elements of governance, their execution, and their interrelationships.

5 Exercises

1. Describe the role of the CSG reference model for the CSG field.
2. Identify a system of interest (SoI). For the SoI identify at least one mechanism (a vehicle used to implement a function) for each of the nine metasystem functions (e.g., weekly staff meeting for Operations function).
3. For a system of interest (SoI), for each of the 10 communication channels select at least one mechanism (vehicle used to fulfil the communication channel) for each of the channels (e.g., daily newsletter for Informing Channel).

References

1. Adams KM, Hester PT, Bradley JM, Meyers TJ, Keating CB (2013) Systems theory as the foundation for understanding systems. Syst Eng 17(1):112–123. https://doi.org/10.1002/sys.21255
2. Ashby WR (1958) Requisite variety and its implications for the control of complex systems. Cybernetica 1(2):83–99
3. Beer S (1979) The heart of the enterprise. Wiley, New York, NY
4. Beer S (1981) Brain of the Firm, Wiley
5. Beer S (1984) The viable system model: its provenance, development, methodology and pathology. J Oper Res Soc 7–25
6. Beer S (1985) Diagnosing the system for organizations. Oxford University Press, Oxford, England

7. Carter B (2015) A metasystem perspective and implications for governance. Int J Syst Syst Eng 6(1/2):90–100
8. Carter B, Moorthy S, Walters D (2016) Enterprise architecture view of complex system governance. Int J Syst Syst Eng 7(1/2/3):95–108
9. Espejo R, Harnden R (eds) (1989) The viable system model: interpretations and applications of Stafford Beer's VSM. Wiley, Chichester
10. Flood RL, Jackson MC (1991) Creative problem solving: total systems intervention, vol 831397533. Wiley, Chichester
11. Katina PF (2015) Emerging systems theory–based pathologies for governance of complex systems. Int J Syst Syst Eng 6(1–2):144–159
12. Katina PF, Despotou G, Calida BY, Kholodkov T, Keating CB (2014) Sustainability of systems of systems. Int J Syst Syst Eng, 5(2): 93–113
13. Keating CB (2000) A systems-based methodology for structural analysis of health care operations. J Manage Med 14(3/4):179–198
14. Keating CB, Katina PF (2011) System of systems engineering: prospects and challenges for the emerging field. Int J Syst Syst Eng 2(2/3):234–256
15. Keating CB, Katina PF (2012) Prevalence of pathologies in systems of systems. Int J Syst Syst Eng 3(3/4):243–267. https://doi.org/10.1504/IJSSE.2012.052688
16. Keating CB, Morin M (2001) An approach for systems analysis of patient care operations. J Nurs Adm 31(7/8):355–363
17. Keating CB (2005) Research foundations for system of systems engineering. In: International conference on systems, man and cybernetics, 2005, vol 3. IEEE, pp 2720–2725. https://doi.org/10.1109/ICSMC.2005.1571561
18. Keating CB (2014b) Governance implications for meeting challenges in the system of systems engineering field. In: IEEE system of systems engineering conference, Adelaide, Australia June 9–14, 2014
19. Keating CB, Katina PF, Bradley JM (2014) Complex system governance: concept, challenges, and emerging research. Int J Syst Syst Eng 5(3):263–288
20. Keating C, Rogers R, Unal R, Dryer D, Sousa-Poza A, Safford R, Rabadi G (2003). System of systems engineering. Eng Manage J 15(3):36–45
21. Walters D, Moorthy S, Carter B (2014) System of systems engineering and enterprise architecture: implications for governance of complex systems. Int J Syst Syst Eng5(3): 248–262
22. Whitney K, Bradley JM, Baugh DE, Jr CWC (2015) Systems theory as a foundation for governance of complex systems. Int J Syst Syst Eng 6(1–2):15–32
23. Wilhelm R (1968)The I Ching: or book of changes, vol 19. Taylor & Francis

Fundamentals

Context in Complex Systems Governance

Meggan Schoenberg

Abstract A complex system's identity and viability are directly related and affected by its context. It is important to identify, monitor, and manage (or mitigate risk) of system contextual elements. This chapter defines complex system context with respect to Complex System Governance (CSG) and provides a methodology to define relevant contextual elements for the practitioner to use for risk mitigation and governance. Leveraging the systems of systems engineering (SoSE) methodology as described in Keating and Adams, *Overview of the systems of systems engineering methodology* [2], and Crownover's complex system contextual framework (CSCF) [6] this article will help the practitioner identify and evaluate relative importance of contextual elements to maintain the viability and identity of a complex system.

Keywords Complex system governance (CSG) · Context · Contextual framework · SoSE or system of systems engineering · Complex system contextual framework (CSCF) · Contextual framework · Context matrix

1 Introduction

What is context? Context informs understanding and perspective and clarifies meaning. Understanding a painter's environment, how they perceived their environment, and life at the time of painting a specific masterpiece can add meaning and explanations for the painter's choices to use light colors or dark colors, happy faces or melancholy faces, stills or abstracts, and even the subject of the painting. In this way, context is often thought of in hindsight and upon reflection of a great masterpiece and therefore considered informative and benign. However, it can also have an effect. Consider the case of "New Coke." Coca-Cola felt pressure by its rival Pepsi-Cola who was winning the well-publicized "Pepsi Challenge" where Pepsi was chosen in a blind taste test more often than Coca-Cola. In response, they developed a new formula, tested it in their own blind taste test 190,000 times, and on April 23, 1985,

M. Schoenberg (✉)
National Centers for System of Systems Engineering, Old Dominion University, Norfolk, VA, USA

C. B. Keating et al. (eds.), *Complex System Governance*, Topics in Safety, Risk, Reliability and Quality 40, https://doi.org/10.1007/978-3-030-93852-9_8

the company chairman and CEO, Roberto Goizueta, announced at a press release the "New Coke." Much to the surprise of Coca-Cola, shares on the New York Stock Exchange dropped, and by June, the company was fielding 5,000 angry phone calls a day. Groups organized to protest and one Seattle consumer even filed suit against the company to force it to provide the old drink.[1]

> The problem, though, is that the company had underestimated loyal drinkers' emotional attachments to the brand. Never did its market research testers ask subjects how they would feel if the new formula replaced the old one. (Klein 2020)

Seventy-nine days later, "Coca-Cola Classic" was back with a corporate apology. Thankfully, they were able to recover from this oversight, but many companies and projects do not recover from their failure to recognize the context of their decisions. That is why it is important to not only look back and understand through context, but to look forward and identify context that may impact a system and the governance of that system. This chapter defines context for Complex System Governance and provides a methodology to identify contextual elements for the practitioner to mitigate costly mistakes and steer complex systems toward identified goals.

2 Defining Context in Complex System Governance

A Context Vignette—How context matters to system viability and identity

The manager of a database development team in a software development company had worked hard to improve moral, productivity, and quality of the team after accepting the position. The projects they worked on were deployed and actively used 24 hours a day, 7 days a week. The team rotated being on call after hours to fix bugs in the live environment that hindered sales. Prior to her efforts the person on call normally spent most nights of their weeklong, after-hours duty, fixing bugs. Some nights over a million dollars in revenue was lost due to software bugs. Within 6 months after implementation of measures to improve moral, productivity, and quality, no bugs were identified as a result of their work in production and the whole team slept well at night. For a brief period, the team's moral was high. After a few months of continued rest, the new manager started sensing moral declining again. Not understanding how this could happen when the team was getting rest, quality of the code was at its highest ever, and the team produced more code than any other division of the company; she began to look at the environment for clues. She quickly discovered that at the monthly awards and recognition meetings the other divisions were getting all the awards and her team was only briefly recognized when they came to the rescue of the other teams. The company had a system to reward heroic efforts in the middle of the night, but no system for rewarding good quality. The perception was that the other teams were working harder and deserved more credit. The new manager attempted to explain this oversight to her boss and part-owner in the company who could not understand as he had no real experience with software development. He decided to combine the successful database development team with another team who received regular heroic awards. The new manager left the company. Within 18 months, the company was out of business.

---

[1] Event and data regarding the Coca-Cola case were extracted from www.History.com (Klein 2020).

The term context is not easily defined and even harder to articulate where it starts and where it ends. "Context shifts and dances, it slips and sides. It insists on its mystery, yet it demands we come to terms with it every single day" [7]. The word "context" is like the word "love"; we tacitly understand it and try to define it with other words, but always seem to fail in capturing its full meaning. To understand what the word means, we have to ironically understand the context for which it is used. In this chapter, we discuss context with respect to complex systems and Complex System Governance (CSG).

Systems theory offers the contextual axiom: *Meaning in systems is derived from the circumstances and factors that surround them* (Keating and Bradley, Complex System Governance Reference Model, 2015). For this reason, CSG metasystem five star (M5*), system context, is a function intrinsically linked to policy and identity, metasystem five (M5). The M5 function provides direction, oversight, accountability, and evolution of the system. The M5 focus includes policy, mission, vision, strategic direction, performance, and accountability of the system such that (1) the system maintains viability, (2) identity is preserved, and (3) the system is effectively projected both internally and externally. Figure 1 is provided as a reference point for CSG metasystem functions [9]. It is no wonder that being able to articulate the circumstances and factors surrounding a complex system is an important factor to describe the system's identity. However, it is a little harder to understand the importance and relevance of context and being able to articulate system context in order to maintain viability and preserve identity of a system. The first context vignette shows how a new manager's failure to understand the context of her development team with respect to the company culture and experience level of the leadership caused the identity and viability of the team and her position to be redefined. When the leadership of the company failed to understand the contextual environment, the company was lost. This example is given to show the importance of understanding context and how it can have severe and possibly fatal results on the viability of a system when not understood, monitored, and governed.

Specifically, ***metasystem five star (M5*)—system context*** is focused on the specific context within which the metasystem is embedded where context is the set of circumstances, factors, conditions, or patterns that enable or constrain execution of the system.

Before going any further, it is important to discuss the definition of context within CSG. While it appears intuitive to say context is the set of circumstances, factors, conditions, or patterns that enable or constrain execution of the system, it is not intuitive when you attempt to differentiate what is considered context and what is not. This is true because contextual elements are part of the system, part of the environment, and part of the interactions.

Within CSG literature, context is consistently defined as "the set of circumstances, factors, conditions, patterns, or trends that enable or constrain execution of the system" [9]. CSG considers the environment as separate but related. Environment is defined as "the aggregate of all surroundings and conditions within which a system operates" [8]. If you are confused, do not worry, there is good cause to

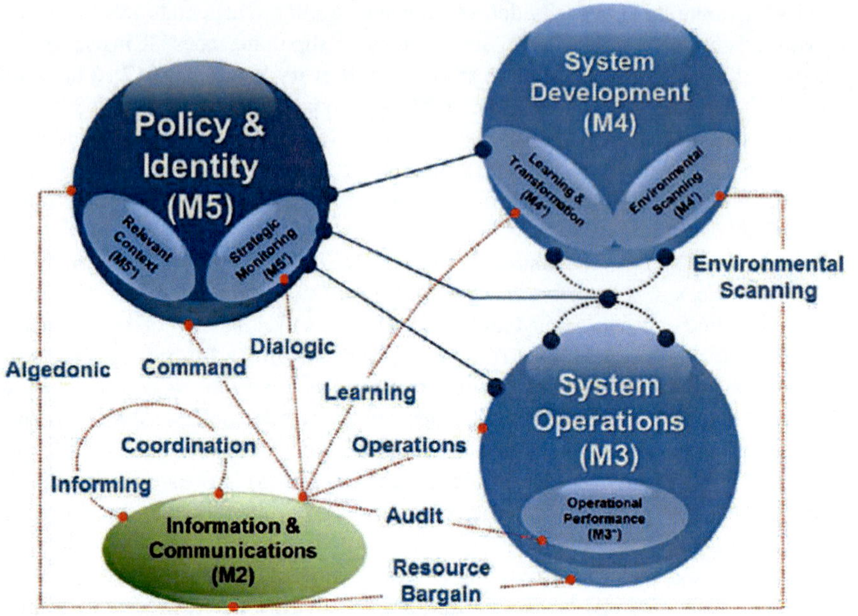

Fig. 1 Main CSG functions

be confused. In the book *Digital Ground: Architecture, Pervasive Computing, and Environmental Knowing*, Malcolm McCullough says:

> "Context" is not the setting itself, but the engagement with it, as well as the bias that setting gives to the interactions that occur within it. "Environment" is the sum of all present contexts. [10]

In Crownover's dissertation, however, the environment is explained as everything outside the boundary of the system that "touches" the system but does not necessarily act on the system. Where context includes elements within, on, and around the system that have an intimate relationship with the system rather than a mere coexistence [6]. These definitions appear to be nearly opposite. However, if McCullough is implying that an environment is the sum of perceptions of all that have interacted with a setting, then he is really not that far from the CSG complimentary axiom and principle that a complex system is described through the varying world views of those that perceive it and by reasonable correlation, so is our understanding of the context of a complex system. For clarity of purpose, this book will differentiate a complex system context from its environment as Crownover describes, acknowledging that perception is always in the eye of the beholder. Figure 2 shows a complex system embedded in its environment and context. The complex system is not cleanly defined, and similarly the context is not cleanly defined. The environment is external to the system and only becomes a part of the system context when it is engaged. A definition of a specific complex system environmental element is based on the world views of

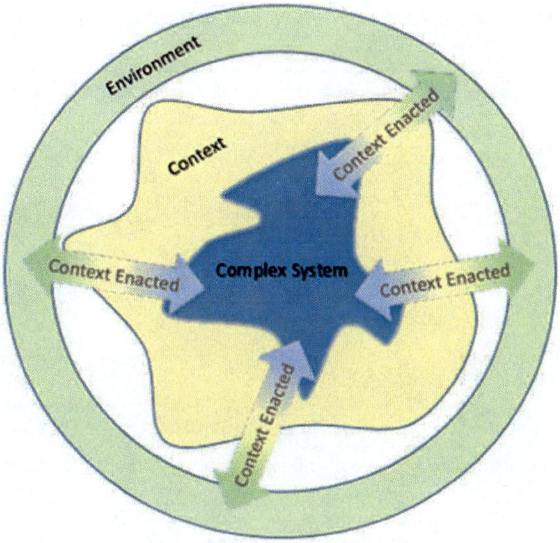

Fig. 2 Interaction of environment and context on a complex system

stakeholders and their understanding of the complex system but are only relevant when the stakeholders can articulate how the environmental element affects or is affected by the system or in some way contributes to the identity of the system. From an organizational perspective, Crownover describes the differentiation this way:

> The environment includes all of the systems outside of the organizational (system) boundary - e.g., government systems, national systems, ecological systems, transportation systems, etc. These systems are all part of the environment, but the systems themselves are not part of the context of the system of interest. Rather, the system context includes how the actions of the governance system enables or constrains the system in carrying out its purpose.

So, while we can describe the environment of a complex system, if we cannot articulate how an environmental element interacts with the complex system or provides meaning to the system identity, it is not a relevant contextual environmental element. This becomes very important when we are practicing CSG and conducting systems analysis on complex systems.

Defining and understanding system context appear easy at first since we have a tacit understanding based on our own individual perception. However, practitioners face the same realities when trying to define system context as they do with defining modern complex systems. Modern complex systems face uncertainty, interdependence, complexity, emergence, and ambiguity. A clean, perfect contextual framework of a given complex system is as impossible to articulate as a clean, perfect articulation of the given complex system. Table 1 defines these realities with respect to complex systems context.

Our understanding of the realities of complex systems and their contextual framework have evolved from systems theory and from a theoretical foundation for our understanding of context in CSG practice. Table 2 describes a core subset of relevant

Table 1 Complex system and complex system context realities

Reality	Definition	Understanding system context
Uncertainty	Incompatible knowledge casting doubt for decision/action consequences	Stakeholders have incompatible views on the contextual elements of a system, relevance of individual and aggregate contextual elements to the system governance and identity, degree of influence, and probability of influence on a complex system. The fact that complex systems themselves are uncertain exacerbated the already difficult task of finding a complimentary view of relevant context that is sufficient to ensure system viability and clarity of system identity
Interdependence	Mutual influence among systems, where the state of each system influences, and is influenced by, the state of interrelated systems	Contextual elements include enacted elements like organizational systems, political systems, environmental systems, social systems, and other systems that interact in ways that redefine how they influence and are influenced by a specific complex system. This interdependence adds complexity and variability causing drifts in the contextual framework that must be monitored and governed to minimize uncontrolled variability and negative impact on system identity
Complexity	System so intricate and dynamically interconnected that complete understanding, prediction, control, or explanation is impossible	Context is constantly changing and evolving and fundamentally defined by its interaction or relationship to the complex system. The complexity of the system, the contextual elements, and the varying perceptions and perspectives on system context make it impossible to have a complete understanding of contextual elements and how they influence and are influenced by the system and other contextual elements. This in turn makes accurate prediction, control, or explanation impossible
Emergence	Unpredictable events and system behaviors that cannot be predicted and are only known after they occur	Emergence happens in the contextual elements changing the enacted context of a complex system of interest forcing emergence in the complex system and potentially other areas of the contextual framework

(continued)

Table 1 (continued)

Reality	Definition	Understanding system context
Ambiguity	Lack of clarity in understanding/interpretation of both the system and context within which it exist	Ambiguity in the system of interest causes ambiguity in the understanding of the system's context. Additionally, context is an interpretation, which varies by individual, their ability to articulate, and their ability to have a clear understanding of context—which is nearly impossible

systems theory principles and how they apply to system context.[2] A more complete list can be found in Baugh, Bradley, Chesterman Jr., and Whitney's "Systems Theory as a Foundation for Governance of Complex Systems" [12]; Adams' "Systems Principles: Foundation for the SoSE Methodology" [1]; and Clemson's "Cybernetics: A New Management Tool" [5].

This theoretical understanding provides a foundation for Crownover's complex system contextual framework (CSCF) and a concept of complex system context as follows:

a. Complex system context includes events, incidents, factors, settings, or circumstances that in some way act on or interact with the system, perhaps as enabling or constraining factors.
b. Complex system context includes an "enacted" environment, which captures system/environment interactions and interdependencies [11]. However, system context and system environment are conceptually distinguishable.
c. Complex system context is a construct or interpretation of properties of a system that are necessary to provide meaning to the system, above and beyond what is objectively observable.
d. Complex system context is reflexive in nature, resulting in context further defining the system while the elements of the system are part of the self-same context. The meaning and significance of context have to be contextualized within a specific situation, domain, discipline, or practice.
e. Complex system context does not have a true reality, or there is no correct interpretation of context. The systems principle of complementarity applies equally to system context as to the system itself.

To help understand context in practical terms, W. B. Max Crownover used grounded theory methodologies to develop a framework [6]. The next section will review Crownover's complex system contextual framework (CSCG).

[2] Principles and description are quoted from Whitney, et al.'s *System Theory as a Foundation for Governance of Complex System* journal article [12].

Table 2 Systems theory principles and how they apply to system context

Principle	Description	Understanding system context
Boundary	The abstract, semipermeable perimeter of the system defines the components that make up the system, segregating them from environmental factors and may prevent or permit entry of matter, energy, and information	System boundaries are a necessary component to articulate a system's identity. System context helps to define boundaries by providing perspective and the ability to understand a system within its environment or setting. A CSG practitioner may set a system boundary of a software development project to include first line management, but not the CEO if the CEO is far removed from the project; or the project system boundary may include the CEO if the organization is small and the CEO is directly involved in projects. As context changes, the system boundaries should be reviewed, and conversely, when the boundaries change, the system contextual framework should be reviewed
Circular causality	An effect becomes a causative factor for future "effects," influencing them in a manner particularly subtle, variable, flexible, and with an endless number of possibilities	When something in the contextual framework of a system changes, it will affect the system. The effect of the change may in turn cause more system changes and further change the context of the system. Using the software project example of a complex system, if a project is developed in a company that is employee owned and then the company goes public, this contextual change may have significant impact on the project system. Employees may become less motivated and resistant to the change. This may cause communication problems and productivity problems. People may begin to turn over, and critical corporate knowledge may be lost resulting in restructuring the product or selling off that part of the company. Now the project system's context has changed. The effect becoming a causative factor for future effects may go on and on, affecting both system and context framework

(continued)

Table 2 (continued)

Principle	Description	Understanding system context
Complementarity	Two different perspectives or models about a system will reveal truths regarding the system that are neither entirely independent nor entirely compatible	System context is perceived by individuals and therefore interpreted and articulated in different ways. Each perspective adds information to build a contextual framework for a complex system and understand how the elements influence or are influenced by the system. For example, managers may help to establish organizational and leadership context while the financial analyst may provide fiscal and budgetary context. For the CSG practitioner, it is important to talk to stakeholders to build a complementary contextual framework that is satisfactory to all or critical stakeholders
Control	The process by means of which a whole entity retains its identity and/or performance under changing circumstances	System controls may cause contextual changes to steer a system toward a particular goal or to mitigate risk. For example, if the software team of our software project system is all over the age of 50 and a new 20-something hire joins the team to provide expertise in a new programming language, the CSG practitioner may prepare the team before the young developer joins by educating them on the generational differences and setting some new policies. In this way, the practitioner may mitigate negative consequences of a change to the system by adjusting the cultural contextual element

(continued)

Table 2 (continued)

Principle	Description	Understanding system context
Dynamic equilibrium	An entity exists as expressions of a pattern of processes of an ordered system of forces, undergoing fluxes and continuing flows of matter, energy and information in an equilibrium that is not static	Complex systems seek equilibrium. When something changes across the system and relevant system context, the forces will work for or against the change to bring the system back to what it knows as "normal." If change is desired, deliberate change management is required to create the "new normal." System context may effect unwanted change by implanting barriers or floodgates. Therefore, understanding the context of the system is important to ensure efficient system governance and change management
Holism	A system must be considered as a whole, rather than a sum of its parts	Holism cannot be achieved without the inclusion of context. To have a holistic understanding of a complex system, you have to understand the environment and how it influences the system. In our software project system example, to holistically understand the project system, you have to understand the stakeholder's role and influence on the system. The CEO may not be an internal component of the project system as he or she is far removed, however, the CEO may change the direction of the company away or toward the project product causing a reduction in force or the need to rapidly hire. While contextual elements may be external to the system, the enactment or how they influence and define the system is intrinsically part of the system and must be considered to avoid system failures

(continued)

Table 2 (continued)

Principle	Description	Understanding system context
Minimal critical specification	This principle has two aspects: negative and positive. The negative simply states that no more should be specified than is absolutely essential; the positive requires that we identify what is essential	Minimal critical specification applies to system context as well. If system context is defined with too many elements, it will take more resources to monitor and become overwhelming. If you define the system context too narrowly, you may miss a critical element and cause a type II error, missing the identification of a problem before it is too late, or a type III error, solving the wrong problem [4], because you have the problem source wrong as it was not identified in the system context definition
Purposive behavior	Purposive behavior is meant to denote that the act or behavior may be interpreted as directed to the attainment of a goal, i.e., to a final condition in which the behaving object reaches a definite correlation in time or in space with respect to another object or event	Governing a system through CSG and M5* (context) allows for system adjustments and changes that are purposeful and deliberate to meet desired goals and end states with some level of control over emerging system changes because the system and system context have been actively monitored
Requisite saliency	The factors that will be considered in a system design are seldom of equal importance. Instead, there is an underlying logic awaiting discovery in each system design that will reveal the significance of these factors	Complex system contextual elements and their impact on or by a system is understood relative to all other contextual elements. In this way, identifying elements with greater influence can focus monitoring and governing processes
Satisficing	The decision-making process whereby one chooses an option that is, while perhaps not optimal but good enough	Defining system context and identifying which elements are critical to deliberately monitor require satisficing. To achieve requisite saliency, the CSG practitioner must interview many stakeholders for a complementary perspective. Not all perceptions are the same, therefore a satisficing definition of system context must be the goal

3 Complex System Contextual Framework (CSCF)

The CSCF is the first contextual framework that can be used to guide a CSG practitioner to identify contextual elements and how they are enacted on a system. The CSCF is composed of categories, called meta-elements; sub-categories; elements; attributes; and dimensions. As research in CSG and contextual theory expands, other frameworks and methodologies may emerge. In the exercises for this chapter, you are encouraged to develop your own framework and apply it to a representative complex system.

The CSCF framework starts with four meta-elements: human, systemic, methodological, and environmental as described in Table 3. Meta-elements are a conceptual superset that logically group contextual elements.

The following hierarchies in Figs. 3, 4, 5, and 6 show the elements and attributes for each meta-element and are followed by a discussion of their applicability to complex systems and CSG.

The human meta-element recognizes the "human factor." Many people can recall an individual leader or team member that had a significant impact on a project or team. A leader can have a significant impact on a system as a decision maker. They can be good communicators, good decision makers, and morale building, or they can be confusing, poor decision makers, and morale draining. It is important to understand the stakeholders of a complex system, their roles, their level of influence, type of influence, how they are affected by other stakeholders, their relationships, the influence on and from relationships, their experience level in relevant areas, and their world views. This information is captured in the role-related sub-category and respective elements. It is also important to understand individually and at various group levels the culture, values, and relevant perspectives. Effective leaders take time to understand these factors before making major changes to an organization to ensure the most effective and least resistant path to success without collateral damage. As a CSG practitioner, these contextual elements are critical to effective governance and problem solving. The CSCF addresses these contextual elements in the perceptual sub-category of the human meta-element.

Table 3 Complex system contextual framework meta-elements

Meta-element	Description
Human	Related to the various aspects of human involvement in complex systems, specifically looking at the roles people play and the perspectives they bring
Systemic	Related to the various aspects of dealing with complex systems that stem from systemic principles and concepts and from taking a systems view
Methodological	Related to the aspects of dealing with complex systems that stem from specific approaches or methodologies being applied or considered for application
Environmental	Related to the aspects of dealing with complex systems that are related to the system's environment

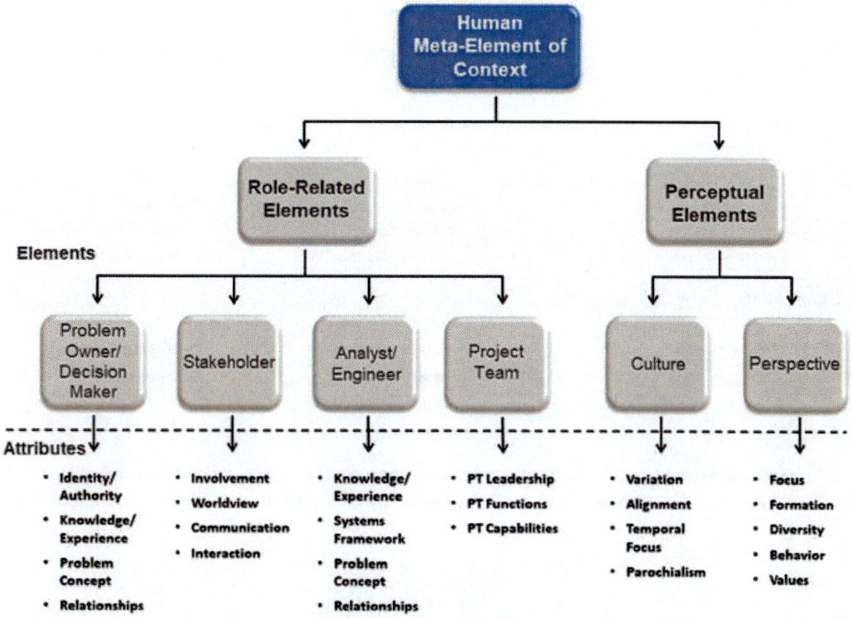

Fig. 3 **Human meta-element**—related to the various aspects of human involvement in complex systems, specifically looking at the roles people play and the perspectives they bring

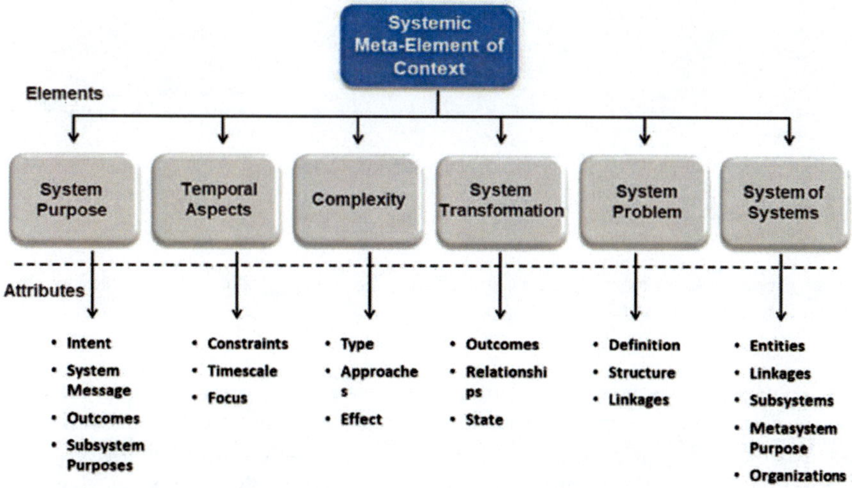

Fig. 4 **Systemic meta-element**—related to the various aspects of dealing with complex systems that stem from systemic principles and concepts and from taking a systems view

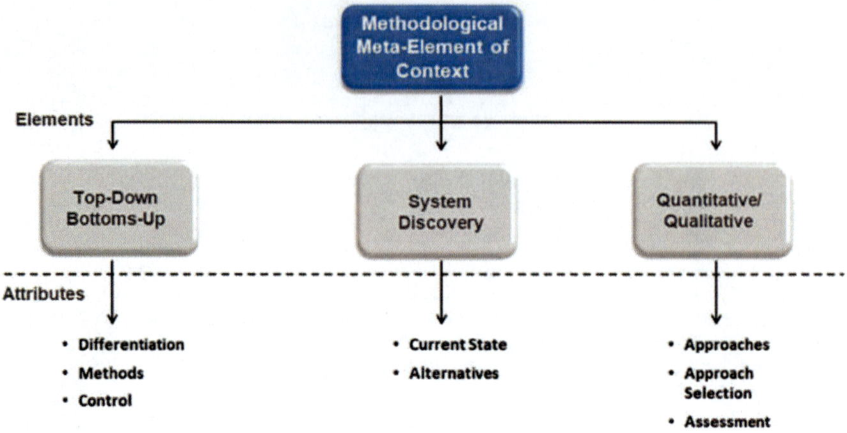

Fig. 5 **Methodological meta-element**—related to the aspects of dealing with complex systems that stem from specific approaches or methodologies being applied or considered for application

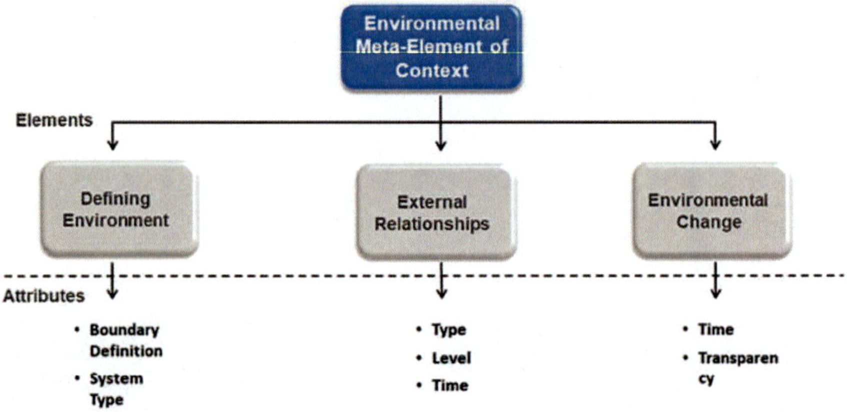

Fig. 6 **Environmental meta-element**—related to the aspects of dealing with complex systems that are related to the system's environment

The human meta-element category may be one of the most difficult for the CSG practitioner because people and relationships are complicated. If the practitioner assumes that policy and rules can be established and enforced to control people in a way that they do not have to take time to understand the contextual elements, attributes, and consequences on the system, they may incur a type III error in managing complex system problems. A type III error is when the wrong problem is solved precisely as described in part 2 of the chapter vignette.

A Context Vignette, Part 2—How Context Matters

In the chapter vignette, the organization had accepted a culture of high salaries for long hours

of work with an accepted high turnover rate of developers. When developers complained for long hours, they received bonuses making them temporally happy again. Leadership philosophy was that happy developers produce more code which produce more revenue. Unfortunately, they fail to see that the quality of the code was causing even larger losses in revenue and that very few of their developers where trained in software quality methodologies. The failure to understand and mitigate the experience level of both the owner-manager and the developers resulted in a decision to solve the wrong problem and the business began to fail.

The systemic meta-element results from a recognition that a "system is affected because it is being viewed as and conceptually constructed as a system" [6]. This is particularly true when CSG practitioners are using CSG to govern a complex system. The CSG practitioner and systems analyst have their own experiences, perspectives on the system, varying levels of influence on the system, and values. A systems analyst with a high degree of CSG or systems thinking experience and inclination will have a different perspective and effect on the system than a systems analyst who does not.

The system analyst example exposes a link to the human meta-element. It is the perception of the stakeholders that either perceive or do not perceive the system as a system and agree or disagree on the system elements. When determining elements like the system's purpose, the worldview of stakeholders is required. Hopefully, there is a high degree of commonness or complementary views that provide a satisficing systems purpose statement. This is also true for the other elements: temporal aspects, complexity, system transformation, system problem, and system of systems representation.

Similar to the systemic meta-element, the methodological meta-system results from the recognition that the methodologies used to define, analyze, and govern a system also affect the system. These methodologies affect the system by producing resulting decisions, discussions, and actions. The tools and approaches can also affect the system during execution knowingly or inadvertently much like quantum particles whereby the very act of observing the particle affects the state. Additionally, this meta-system is also linked to the human meta-system. The human perspective determines the type of methodology to use and how it is used. It also determines if a methodology and the results are accepted and actionable. For example, if the systems analyst utilized a qualitative method, but the culture of the organization only recognizes quantitative methods, the analysis result may not be accepted by the organization.

Finally, the environmental meta-element captures the contextual elements most often thought of and discussed, but perhaps not in the way expected. This meta-element category does not give us a nice list of neat, tangible environmental elements to check off like political environment, socioeconomic environment, or physical setting; rather, it sets the stage to uniquely define environmental elements with salient influence on the system. "What is required is not simply a matter of providing a textbook definition of environment, but rather the articulation of the system-specific criteria utilized to delineate or demarcate the environment. Doing so requires development of a consistent approach for determining what is and what is not part of the system" [6]. The defining environment element assists the practitioner to use a deliberate approach while recognizing that the approach itself may affect the system.

The external relationships element is where the practitioner uses the element attributes and dimensions to identify environmental elements (circumstances, factors, conditions, and patterns) that have a relationship with the system that may affect the system or be affected by the system. The element is labeled "external relationships" because the contextual element is not the building next door, but the knowledge that the building next door provides shade to my building thereby providing some level of cooling and a cheaper electric bill. A change in the external element will cause a change in my system. The focus on relationships also has the advantage of focusing CSG practitioners on relevant environmental elements.

The environmental change element of the environmental meta-element addresses "the importance of the system having awareness of and being able to respond to environmental change" [6]. To effectively analyze and govern a system, it is imperative to understand how the system responds to internal and external change. Change management requires a human, system, methodological, and environmental contextual intelligence for success.

4 Context in Practice

This section will use the systems of systems engineering (SoSE) methodology as described in Keating and Adams, *Overview of the systems of systems engineering methodology* [2] in the International Journal Systems of System Engineering, to provide a representative practical application that may be use in CSG. The SoSE methodology, as shown in Fig. 7, is built from foundational systems principles and

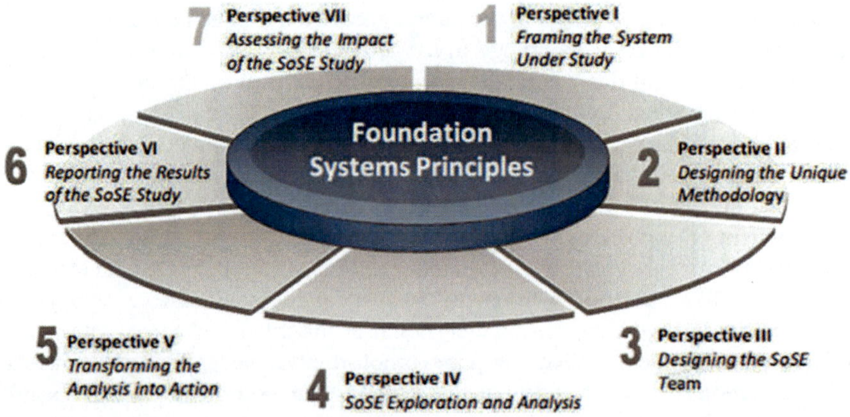

Fig. 7 System of system engineering methodology

seven perspectives. It can be used to address complex system problems or to establish a framework for persistent governance.

In Table 2 of this chapter, we discussed a core set of system principles that are foundational to the SoSE methodology. Here we will provide a short overview of the seven perspectives of the SoSE methodology followed by a more comprehensive discussion of perspective I. By framing the system under study in perspective I, we will show how the practitioner may build the contextual framework for a system. This framework can be used for system governance or problem solving.

[3]*Perspective I, Framing the system under study*

This perspective is designed to rigorously structure the system problem, the contextual setting and environment within which the problem system exists. Key execution elements in this perspective include.

1. Identify the wide context for the system under study—establish the circumstances, factors, conditions, and patterns that characterize the situation surrounding the SoS at a high level.
2. Characterize the system under study—understand the basic structure and characteristics of the SoS under study, including the SoS's objectives, functions, environment, resources, components, and management.
3. Characterize the complex nature of the system domain under study—establish the complex nature of the SoS and problem domain.
4. Present the system domain as characteristically complex—present the SoS under study as a complex systems problem.
5. Frame the SoSE problem—depict the problem situation by expressing the structure, elements of processes, and the situation.
6. Define study purpose, reformulated problem statements and objectives—clearly explain the nature, purpose, high-level approach, and objectives for the effort.
7. Conduct stakeholder analysis—explicitly account for and address the multiple interests (rational and irrational, inside and outside) which can impact achievement of system objectives.
8. Conduct contextual analysis—account for the set of circumstances, factors, conditions, values, and/or patterns that are influential in constraining and enabling the SoS engineering process, the SoS solution/recommendation design, SoS solution/ recommendation deployment considerations, and interpretation of outputs/outcomes stemming from the effort.

Perspective 2 Designing the unique methodology

This perspective designs a unique methodology based on the problem and the problem context.

9. Construct high-level design for the study—construct a unique high-level methodology that will adequately support the study objectives and the SoS context. Must be compatible with the problem and problem context.

[3] Perspective overview taken from [2].

10. Develop the analytic strategy—create the design for quantitative and qualitative exploration (data collection and analysis) necessary to understand and make decisions concerning the SoS under study.
11. Develop assessment criteria and plan—construct a set of measurable criteria to be used during and after the problem study to ensure continued fit of problem, context, methodology, and capability to meet study objectives.

Perspective 3 Designing the SoSE team

This perspective designs the team to undertake the SoSE study, taking into account the nature of the SoS problem and the team resources, skills, and knowledge that can be brought to bear for the problem.

12. Assess team knowledge, skills, and abilities (KSA)—develop an inventory of team knowledge, skills, and abilities that may be used in the study.
13. Match team KSA to the analytic strategy and unique methodology—based on the KSAs, establish assignments, roles, and expectations for the team in performing the study. Team expectations and selection of task leaders are established.
14. Establish means of team expectation and performance assessment—construct a set of measurable criteria that can be used during and after the SoS problem study to evaluate the performance of the team.

Perspective 4 SoSE exploration and analysis

This perspective is designed to explore and conduct the emergent analysis by executing the analytic strategy and SoSE management plan (SoSEMP). Its constituents include.

15. Build the SoSE management plan (SoSEMP)—the SoSEMP defines how the SoS study will be organized, the structure of the team, and how the SoSE process will be designed to provide products that directly support the study goals and objectives requirements.
16. SoSE exploration and analysis—explore and analyze each study objective by executing the analytic strategy.

Perspective 5 Transforming the analysis into action

This perspective is designed to transform the results of the emergent analysis by guiding implementation of derived recommendations. Its constituents include.

17. Define implementation goals, objectives, and activities—clearly explain the nature of the implementation, purpose, high-level approach, and objectives necessary to support the desired SoS outputs and outcomes.
18. Modify the SoSE management plan (SoSEMP)—add activities to the integrated schedule that ensures that the tasks from the implementation objectives tree are properly resourced to support the implementation goals, objectives, and activities.

19. Implementation of the exploration and analysis recommendations—change, modify, or construct processes for the SoS under study to implement recommendations.

Perspective 6 Reporting the results of the SoSE study

This perspective reports the results of the SoSE effort to capture the transformation of the analysis into action. It comprises.

20. Developing the engineering report—develop a coherent set of artifacts (data, analyses, correlations, etc.) that can provide specific findings and recommendations that directly impact the SoS problem under study.
21. Internal evaluation of the engineering report—evaluate the study report using the set of measurable performance criteria previously developed.

Perspective 7 Assessing the impact of the SoSE study.

This perspective is designed to assess the impact of the report on the real-world SoS problem under study. The final two of the 23 perspective-related elements are assigned here, and they are

22. Evaluating the initial impact of the engineering report—evaluate the impact that the SoSE study report had on the real-world system problem and its environment.
23. Plan for follow-up and follow-through—evaluate the impact analysis and develop a set of actions to follow-up and follow-through on the analysis.

In perspective 1, we identify key contextual elements and their potential impact on the system. These contextual elements should be reviewed and considered in all following perspectives and actions. They will likely evolve as the practitioner continues to advance their understanding of the system. When changes are made to the system contextual understanding, a retroactive look at completed tasks should be reviewed and modified if the advanced systems context understanding warrants.

Perspective 1 has eight elements as described above. The first element, identification of the wide context for the system under study, establishes the importance of context from the very beginning. "Engineers must understand and ultimately represent context if they are to move a system or SoS of interest from some current state to a different, desired state" [3]. "Complex systems cannot be understood independent of the context within which they exist" [3].

In the beginning, there exists a tacit knowledge of the complex system under study which is not unanimously shared by all stakeholders. Sometimes the framing effort has a problem statement if the effort is to solve a complex system problem; and sometimes a set of disparate documents exist that contain elements of a system description, like a contract vehicle, vision statement, engineering design document, or software development plan. Either way, the first step is to establish a problem statement or basic system description. A system description should be a short description of the complex system including goals for management and desired outputs and outcomes.

These statements should be reviewed and updated by all key stakeholders, producing a complimentary and satisficing statement.[4] With this very basic understanding of the system and effort, element 1 of the perspective 1, developing the wide context, may begin.

The output of this effort, and subsequent contextual analysis efforts, will be called the system's contextual framework. The practitioner may establish the system's contextual framework in many ways. A recommended approach that is encouraged to be tailored and modified to the benefit of the effort and stakeholders understanding will be used as a representative approach.

One way to capture the system's contextual framework is to develop a matrix with the columns identified in the contextual matrix, Table 4, and the contextual elements identified in the rows. Use Crownover's CSCF, [6] the established problem or system description statement, and the factors in the table to guide the first pass population of the matrix. In each subsequent perspective element, this matrix should be reviewed and updated as new information is discovered. The practitioner should interview as many stakeholders and experts as possible to ensure a complimentary matrix solution.

As the matrix is developed, the practitioner may begin to form or confirm system boundaries. Contextual elements may be internal, external, and boundary crossing, but in order to understand the impact to the system under study, the practitioner has to have an understanding of what is part of the system and what is part of the system environment (external). Generally, something is internal to the system if it can be managed and governed through adjusting system controls. For example, usually federal laws and regulations are considered external contextual elements, but company and department policies may be internal to the system. However, remember that complex systems will always have ambiguity and uncertainty, so do not expect perfection. A minimal critical specification is the goal. The practitioner should consider as much as can be identified but focus on the most important and impactful elements.

Element 2 of perspective 1, characterize the system under study, will use the problem/initial statement and the system's contextual framework developed in element 1 as an input to develop a list of system characteristics including the systems definition, components, objectives, functions, environment, resources, and governance structure. The systemic meta-element elements, system purpose, temporal aspects, complexity, system transformation, system problem, and system of systems, should be considered in the holistic characterization. The contextual framework matrix may be used for the system environment characteristic of element 2. The expression of the environment characteristic (external to the system, but a change in the element will cause a change in the system) is the meta-element section rows of the matrix. Similarly, the human meta-element rows should contain and/or be updated

[4] If the world views of stakeholders are so different that a satisficing statement cannot be written, a type four error has occurred. A type four error is engaging in a problem solution with incompatible or divergent "philosophical" perspectives. These efforts do not often end well unless the opposing philosophical perspectives are addressed satisfactorily.

Table 4 Contextual matrix

Column	Description	Factors
Element number	A number to uniquely identify each row	
Element name	A short name of the contextual element that uniquely identifies it and gives a basic understanding of what it is or how it affects the system	
Element description	Describe the element with enough detail that the reader can understand the element, how it is unique, and the data captured in the rest of the columns	
Meta-element	The primary CSCF meta-element category of the element described in this row	The practitioner should identify if this row element is a result of a specific meta-element or the interaction of meta-elements elements
Element type	The element type may be the CSCF attribute or some other designation (like circumstance, factors, conditions, values, or patterns) that allows the practitioner to logically group the elements in a useful way	This column should aid the practitioner in quickly identifying which rows need to be updated in subsequent passes of analysis. It also should facilitate the risk matrix and mitigation process
Influence/impact on the system	What effect does this row element have on the system? What is the interaction with the system and other contextual elements?	Consider how this row element may influence or impact the system should something change in the system or the row element. Understanding the influence of this row element includes both current influence and how changes may influence
Factors required for influence/impact	What are the factors of this row element that are required to happen or exist to influence the system as discussed in the previous field (influence/impact on the system)? Factors should be identified as factors for current, steady-state influence, and factors of change	The practitioner should consider other contextual elements, system changes, row element changes, or events. This field should help the practitioner identify what must be monitored to effectively govern the system

(continued)

Table 4 (continued)

Column	Description	Factors
Type of impact	How does the current, steady-state factor affect the system? If one or more of the change factors for influence of the row element happens, how is the system or other contextual items expected to be affected?	Tools like SWOT (strengths, weaknesses, opportunities, and threats) analysis can help the practitioner to more completely understand the impact. Impact may be an opportunity, or it may be a threat that is realized by the strengths and weaknesses of the contextual elements' influence on the system
Probability of occurrence	This field does not consider the current, steady-state influence of the row element on the system. It should show the likeliness of opportunity or risk identified in the previous column (type of impact) should one of the factors of change happen?	This may be quantifiable or qualitative
Severity of occurrence	This field does not consider the current, steady-state influence of the row element on the system. It should show the severity of impact on the system from the changes factors if they were to occur	Severity is not bad or good, it is a variable of degree. A change may affect every sub-component of the system such that it changes the very identity of the system or it may only change one part of a sub-system that has minimal influence The practitioner should consider compounding influence in this field. Many times a simple change causes another change which causes other changes, aggregating to a much higher overall impact on the system
Control mechanisms	What mechanism can be put in place to control changes, mitigate risks, or influence changes?	In addition to considering the risk management control mechanisms, consider that system steady state is not always desired. When change is desired, how can the practitioner use this row element to facilitate or instigate while controlling for a particular outcome?
Level of control	Based on the influence and impact of this row element, should the practitioner monitor closely or is periodic assessment sufficient?	The practitioner should consider what is required to maintain steady state, and what is required for change when populating this field

by the governance structure characteristic definition. The process to define system resources will likely identify, clarify, and/or update internal and external contextual elements in the contextual framework as well. For example, military technology development programs have to operate within the federal government appropriation cycle and rules. If an appropriation bill is not passed at the beginning of the fiscal year, many programs are directly affected. This is an external contextual element that is very important to the system.

This step will identify sub-systems and internal contextual elements as each sub-system exists within its own context. The practitioner may want to add a column to the contextual framework matrix to relate the contextual element row to the system or one of its sub-systems and an associated column to describe the relationship with other sub-systems, the super-system, or the external environment. Again, there are many contextual elements, so focus on the elements that have the most impact on the system or the most potential to impact the system if altered.

The characteristics that describe the complex nature of the system and the complex domain of the system under study are the subject of elements 3 and 4 of perspective 1. Understanding the complexity of the system is directly related to the complexity of the response required to govern or affect change in a system. The systems theory and CSG law of requisite variety states that "control can be obtained only if the variety of the controller is at least as great as the variety of the situation to be controlled" [12]. Therefore, a careful assessment of the complex system characteristics is expected to update the contextual framework control mechanism at the very least.

The characteristics of complex systems are related to the meta-elements in the contextual framework matrix as described in Table 5. This table should help the practitioner understand the characteristics and how to use the CSCF to assess the complex system.

In element 5, a rich diagram is constructed to describe the complex system. It may not directly represent the context of the system as described in the matrix; that is okay. However, the practitioner may want to add a column to the contextual framework matrix that maps elements onto the system diagram for reference.

Element 6 of perspective 1 relooks at the problem statement or system initial description system and objectives to refine them based on the analysis of previous elements. The contextual framework matrix is both an input and an output of this element as the systemic meta-element rows may evolve now that there is a deeper, holistic understanding of the system and its context.

The human meta-element rows in the system context framework matrix should make short work of element 7 of the perspective 1 and conduct stakeholder analysis. At this point in the process, the practitioner is taking another look with a focus on the stakeholders. Adams and Meyers offer several characteristics and tools for a thorough analysis of stakeholders in Adams and Meyers [3]. The practitioner should update the matrix with any revelations.

Element 8 of perspective 1, conduct contextual analysis, takes one more pass at contextual framework matrix now that the framing of the system is almost complete. This step focuses on identifying and understanding the contextual elements impact

Table 5 Complex system characteristics relation to CSCF meta-elements[5]

Characteristic	Description	Metasystem relationship
Hyper-turbulent conditions	A complex system is said to exhibit hyper-turbulent conditions when the environment of the problem is highly dynamic, uncertain, and rapidly changing	The temporal aspects of the systemic meta-element address the rate of change and the agility for change. The practitioner should evaluate the level of turbulence in the system and how contextual elements may affect the turbulence
Ill-defined problems	A complex system problem is ill-defined when circumstances and conditions surrounding the problem are potentially in dispute, not readily accessible, or lack sufficient consensus	The system problem element of the systemic meta-element attributes definition, structure, and linkages describes the contextual elements respective to the problem definition. The methodological meta-element addresses the contextual elements analysis methodology. The practitioner needs to consider how the ill-defined nature of the system and problem affects the system and the governance or problem-solving methodology
Contextual dominance	In complex systems, the technical aspects of the problem are overshadowed by the contextual or soft aspects	Practitioners often attempt to solve system problems with technical solutions that do not work because of contextual impact like human factors. The contextual matrix developed in the process outlined in this chapter should define the contextual element, their relative system impact, and the severity of the impact which can be defined as the element's dominance. The methodological meta-element assigns contextual elements to how we recognize and use this information in the analytical process

(continued)

[5] Descriptions in this table are quoted from Adams and Meyers "Perspective 1 of the SoSE methodology: framing the system under study". [3]

Table 5 (continued)

Characteristic	Description	Metasystem relationship
Uncertain approach	There is no clear or prescribed approach to proceed with the SoSE effort that guarantees success. Standard processes applied to the ill-defined problem have failed or are likely to fail to resolve the issue	The methodological meta-element reviews the contextual elements associated with the analytical methods used to define and solve complex system problems. Since the approach taken is uncertain, effects of errors on the system and the methodology processes should be considered
Ambiguous expectations and objectives	A complex system cannot be perfectly defined and understood the same way by stakeholders; therefore, ambiguous expectations and objectives may result	The elements of the systemic meta-element and the perceptual elements in the human metasystems are the predominant linkage to the CSCF; however, the effect or relationships with all other elements and metasystems describe ambiguity of expectations and objectives
Excessive complexity	A complex system often has excessive complexity such that traditional approaches are not adequate for management	This characteristic may use the complexity attributes of the systemic meta-element to assess the excessiveness and the system effects
Pluralistic perspectives	Complex systems have a high degree of variability of individual perspectives, objectives, and perceived interests. Stakeholders are not expected to identify the boundaries of a system or the system problem the same way	The perceptual elements in the human metasystems are the predominant linkage to the CSCF; however, the effect or relationships with elements of the systemic, methodological, and environmental metasystems are also critical to understanding the system context
Extended stakeholders	Complex system stakeholders include extended stakeholders like those that emerge throughout the life of the system, environment contextual stakeholders, and the CSG practitioner in addition to the traditional stakeholders	The human meta-element and the external relationships of the environmental meta-element describe the extended stakeholders and their relationship to the system or to the relevant environment, but a review of relationships with the systemic and methodological meta-elements may uncover not so obvious stakeholders

(continued)

Table 5 (continued)

Characteristic	Description	Metasystem relationship
Emergence	Complex systems experience emergence—the occurrence of events and system behaviors that result from interactions between system elements cannot be predicted and are only known after they occur [9]	Characterization of emergence in the contextual framework is difficult. A review of the systemic meta-element and its relationships with the human and environmental meta-elements with respect to elements, attributes, and dimensions that may contribute to excessive emergence or system stabilization may reveal important information for the contextual framework
Ambiguous boundaries	The inclusion and exclusion boundary criteria are arbitrary and necessarily qualitative in nature	The systemic and environmental meta-elements help to shape the understanding of system boundaries. In the methodology meta-element, the ability of the methodology to develop complimentary system boundaries and at the same time recognize the imperfection of their declaration will have contextual relevance to the system
Unstable planning foundations	The complex, ill-defined, and turbulent nature of complex systems makes efforts to plan specific engineering or analysis efforts subject to sudden, and potentially radical, shifts	This characteristic provides an important consideration for the examination of the methodological meta-element
Information saturation	There is too much data and information to understand a complex system holistically. It is challenging to sort through the information and achieve requisite saliency	This characteristic is applicable to all meta-elements. Recognizing that perfection is not possible, and that there is too much information, requisite saliency is required

(continued)

Table 5 (continued)

Characteristic	Description	Metasystem relationship
Identity coherence	Identity is the set of fundamental values, patterns, and attributes that provides a consistent reference point and baseline logic for making grounded decisions, taking consistent actions, and providing self-reinforcing interpretation of abstract events. Modern technical enterprises are confronted with an accelerating pace, blurring of ethical value systems, and complexity. Incoherent or ambiguous identity deepens the inability to achieve consistency in the face of the new realities facing organizations working on complex SoSE problems	In the methodological meta-element review, it is important to make sure that all, or the right stakeholder, have been interviewed in the system definition and framing process to provide a complimentary and satisficing statement of identity which includes elements of the system meta-element
Large number of systems elements	Complex systems have a high degree of variety in system elements. System elements are simply parts or components of the system and are unlimited in variety	While it is impossible to define every element of a system, the systemic meta-elements of a system are a starting point. However, the real importance of this characteristic is the understanding that requisite saliency is necessary and a good representation of a system is better than none
Interaction between systems elements	The interactions between system parts, referred to as relationships, are critical to framing the system as they govern the properties of the system	The system of systems and systems transformation elements of the systemic meta-element; the external relationships element of the environmental meta-element; and the role-related elements of the human meta-element directly address relationships. The practitioner should review system and sub-system interactions and contextual element interactions

(continued)

Table 5 (continued)

Characteristic	Description	Metasystem relationship
Predetermined attributes	Many system attributes are predetermined from the independent, purposefully designed, acquired, and managed sub-systems from which they are constructed	The sub-system purpose attribute of the system purpose element and the system of systems element of the systemic meta-element should provide the practitioner with an understanding of the predetermined attributes of the system
System evolution over time	Does the system change over time or is it a static system? Complex systems change over time	The temporal aspects, complexity, and system transformation elements of the systemic meta-element address how the system may change and respond to changes over time, but should be reviewed with the environmental change and external relationships element of the environmental meta-element and the human meta-element
Sub-systems pursue own goals	Sub-systems have been designed to operate independently such that each sub-system has a purposive objective or behavior and is a producer of some end result, objective, or goal. A natural tension between the connectedness required for operation within the complex system and autonomy required to meet the goals and objectives of its own operations will exist. How do the sub-systems and the resulting tensions affect the complex system?	The sub-system purposes attribute of the system purpose element, and the system of systems element of the systemic meta-element should provide the practitioner with an understanding of the sub-system goals and how they impact the larger system as well as the environment

(continued)

Table 5 (continued)

Characteristic	Description	Metasystem relationship
Predominantly open to environment	Complex systems are affected by their environment and are, in a way, a sub-system, to even larger complex systems	This characteristic speaks directly to the environmental meta-element and helps the practitioner to understand that not only do the environmental elements affect the system, but the system affects the larger systems embedded in the environment. There is an important interconnectedness which is the foundation of holistic systems thinking
Interaction organization	Most complex systems are part of other larger systems and themselves contain sub-systems. The concept of hierarchy recognizes the openness of the system and relationship to its environment from an interdependent perspective	Similar to the factors discussed in the "predominantly open to environment" row of this table, this characteristic articulates the notion that the system under study is a sub-system to larger systems and is therefore embedded in a bigger hierarchy. This means that the practitioner should consider how the system under study may be affected by the contextual elements of larger systems in its environment. However, remember the concept of requisite saliency and bound the review to areas of greatest importance
System affected by behavioral influences	Complex systems are routinely affected by the behavioral influence from the systems environment	A review of the behavior of contextual elements in the human, methodological, and environmental element is applicable. The behavior of an environmental system like a political system is as important to understand as the behavior of leaders who operate within the system and have a direct influence on the system

on the system and scoping the matrix to a minimal critical specification. The practitioner should review the influence on the system, factors for influence, impact, probability of occurrence, severity of occurrence, control mechanisms, and level of control columns in correlation to the refined system framing outputs. This review is the last step in the perspective 1 process before a brief exercise is designed to help the practitioner understand the implications of the outputs created in perspective 1 of this methodology.

In the final exercise of perspective 1, the practitioner is asked to examine the implications for the system under study, the organization conducting the analysis, the individuals conducting the analysis, and the SoSE discipline. While this step is not required, it is encouraged for CSG practitioners. As CSG is still young, reflection and feedback will help future practitioners.

Once the complex system's contextual framework matrix has been developed with respect to the system framing, stakeholders who seek to govern the system to maintain an identity and achieve a desired goal or solve system problems should monitor contextual elements with periodic assessments. As risks are realized, the matrix control mechanisms may be used to mitigate negative impacts or enhance positive impact. If a change in the system's identity, purpose, or goals is desired, the matrix can be used to identify areas to stimulate change with minimal negative consequences. The matrix should continually be updated as all complex systems evolve and change.

5 Implications of Context in Complex System Governance

In this chapter, we defined complex system context with respect to CSG, a framework to help identify system context, and a methodology to frame a system in preparation for solving system problems or to actively govern. The framework and methodology were representative and can be modified or adapted as required. It is recommended and encouraged for the practitioner to find the approach that works best for the specific effort. However, a holistic approach should not be abbreviated to focus only on the quantifiable, "low-hanging fruit" contextual assessment. Contextual influences directly affect the identity and viability of the system. Many practitioners are uncomfortable identifying human-based influences on a system, but they cannot be ignored as they often have the most influence.

Context can be very difficult to articulate, and no perfect solutions are possible. "There's no way to map every single factor in even a simple real-world environment, but it's possible to take snapshots from different perspectives, at various key moments, and bring them together into something more like a collage of snapshots that come nearer to telling the entire story" [7]. However, if the importance of understanding a system in context, and all the contextual factors that may influence a system, is recognized, the practitioner will have a higher probability of success meeting their goals.

The knowledge and tools discussed in this chapter are designed to help the practitioner achieve the greatest success possible. However, there remains plenty of room for further research in the area. It would be ideal if an easy-to-use model and method for identifying critical contextual items and a system for monitoring, assessing, and governing those critical items existed. Tools like checklists, measures of effectiveness, mitigation techniques, and rules for what is critical and what is not would make a new practitioner more effective and faster. The exercise section of this chapter will challenge you to try and develop these tools for a specific system. After the exercise, consider how the solutions can be applied generically across all complex systems.

Exercises

1. From your experience, select a complex system and use the methodology in this chapter to frame the system context. For this initial familiarization exercise, multiple stakeholder perspectives are not required. It should be completed by the practitioner only.
   a. Did you discover areas of context that were not obvious to you before you started?
   b. How did this exercise help you understand the system under study?
   c. How could this process be improved or modified to better accommodate the analysis of the complex system selected for this exercise?
2. Develop a set of categories, like a checklist, for practitioners to consider to describe the environmental complex system contextual items.
   a. How will this checklist help the practitioner?
   b. When would these categories not work for a practitioner?
   c. When you need to make decisions, do you consider all of these categories? Explain your answer.
3. Using the complex system contextual framework (CSCF) described in this chapter, explain how elements from different metasystems interact. Select three separate examples for this exercise.
   a. How might these interactions affect the system?
   b. Describe how this examination of element interactions would modify your response to exercise 1?

References

1. Adams KM (2011) Systems principals: foundation for the SoSE methodology. Int J Syst Syst Eng 2(2/3):120–155
2. Adams KM, Keating CB (2011) Overview of the systems of systems engineering methodology. Int J Syst Syst Eng 2(2/3):112–119
3. Adams KM, Meyers TJ (2011) Perspective 1 of the SoSE methodology: framing the system under study. Int J Syst Syst Eng 2(2/3):163–192

4. Adams KM, Hester PT (2012) Errors in systems approaches. Int J Syst Syst Eng 3(3/4):233–242
5. Clemson B (1984) Cybernetics: a new management tool. Gordon and Breach Science Publishers, Phillidelphia, pp 199–257
6. Crownover WM (2005) Complex system contextual framework (CSCF) a grounded-theory construction for the articulation of system context in addressing complex systems problems. Old Dominion University Submitted Dissertation. Old Dominion University, Norfolk, VA
7. Hinton A (2014) Understanding context environment, language, and information architecture. O'Reilly Media, Sebastopol
8. Keating CB (2015) Complex system governance: confronting 'System Drift.' Old Dominion University, Norfolk, VA
9. Katina PF, Keating CB (2016) Complex system governance development: a first generation methodology. Int J Syst Syst Eng 43–74
10. McCullough M (2004) Digital ground: architecture, pervasive computing, and environmental knowing. The MIT Press, Cambridge, Massachusetts
11. Weick KE (1995) Sensemaking in organizations. Sage Publications Inc., Thousand Oaks, CA
12. Whitney K, Bradley JM, Baugh DE, Chesterman CW Jr (2015) Systems theory as a foundation for governance of complex systems. Int J Syst Syst Eng 6(1/2):15–32

Metasystem Pathologies in Complex System Governance

Polinpapilinho F. Katina

Abstract This chapter articulates the utility of *system pathology* to *problem formulation* in complex systems. In *Complex System Governance*, the problem formulation is pivotal in the initialization stage involving system context and framing. Context revolves around establishing the particular circumstances within which the system of interest is embedded. Framing involves establishing the design configuration and execution of the system. This chapter suggests that pathology identification is a crucial part of problem formulation. However, there is a lack of qualitative studies suggesting a direct link between system pathology and problem formulation. The goal of this chapter is to make this link explicit. This is done through the following steps: (i) reviewing systems-based methodologies while emphasizing problem formulation, (ii) understanding the various perspectives of system pathology, (iii) development of an enhanced model (UPGR^{2}IDS) for classifying system pathology at the metasystem level, (iv) high-level application of the model in complex system governance. The chapter concludes with questions to engage the reader.

Keywords Complex system governance · Complexity · Management cybernetics · Metapathology · Problem formulation · General systems theory · System pathology · Systems theory-based pathology · Viable system model

1 Introduction

The operating landscape for systems in the 21st century has been described as ambiguous, complex, emergent, interdependent, and uncertain [53, 68, 73]. This sentiment is not new. In fact, it appears that [101] assertion that: "the frequency and magnitude of organizational failures and the subsequent impacts are increasing at an alarming rate" [101, p. ii] remains strong. Moreover, it is evident that our systems "have become hopelessly interconnected and overcomplicated, such that in many cases even those who build and maintain them on a daily basis can't fully understand

P. F. Katina (✉)
Department of Informatics and Engineering Systems, University of South Carolina Upstate, Spartanburg, SC 29303, USA
e-mail: pkatina@uscupstate.edu

C. B. Keating et al. (eds.), *Complex System Governance*, Topics in Safety, Risk, Reliability and Quality 40, https://doi.org/10.1007/978-3-030-93852-9_9

them any longer" [11, p. 2]. Arguably, issues such as these contribute to a need to shift from "traditional" science marked by mechanistic worldview to a more ecological and systemic worldview marked by understanding fundamental system interconnectedness and interdependence [57, 85]. In the latter, one can find a call for innovative methodologies (e.g., system of systems engineering), methods (e.g., fragility assessment), tools (e.g., 3-D printing tools), and technologies (e.g., blockchain technology) to address complex systems and their operating environments. This thinking is fortified in ideas of needing to see phenomena in "wholes" and their "interactions" and forms an essential tenet of systems science [6, 30, 97, 24, 121]. However, the term "systems science" does not have a single and commonly accepted definition and is often interchangeable with "systems thinking" and "systems theory." Interestingly, these terms are commonly attributed to proponents such as Anatol Rapoport, Norbert Weiner, Karl Ludwig von Bertalanffy, and Ross Ashby [82, 86]. These proponents wanted to provide an alternative approach to *reductionism* (in the scientific method), commonly referred to as *holism*.

In the traditional scientific method, complex organisms are nothing more than the sum of their parts. This suggests that a complex organism can be reduced to constituent elements [57, 23]. However, as doubts regarding this thinking emerged, researchers became more interested in notions of Holons—a totality [24]. Succinctly, the argument for general systems theory (GST) began in the 1920s when von Bertalanffy stated that "since the fundamental character of the living thing is its organization, the customary investigation of the single parts and processes cannot provide a complete explanation of the vital phenomena. This investigation gives us no information about the coordination or parts and processes" (1972, p. 410).

There are three related, but different terms associated with GST: systems science, systems technology, and systems philosophy. *Systems science* places emphasis the "scientific exploration and theory of 'systems' in various sciences (e.g., physics, biology, psychology, social sciences), and general systems theory as the doctrine of principles applying to all (or defined subclasses of) systems" [24, p. 414]. This aspect of GST deals with knowledge of the connected "wholes" as opposed to detailed and isolated systems. *Systems Technology* is the aspect of GST dealing with "problems arising in modern technology and society, including both 'hardware' (control technology, automation, computerization, etc.) and 'software' (application of systems concepts and theory in social, ecological, economical, problems)" [24, p. 420]. In this view of GST, the assumption is that solving the world's most pressing issues (e.g., pollution, economies, healthcare, and international conflicts) cannot be done in isolation from one another [108, 24, 119]. Arguably, the increasing levels of ambiguity, complexity, emergence, interdependence, and uncertainty associated with the operating landscape for systems in the 21st century confirms this dilemma. Consequently, the systems technology element of GST is concerned with the development of unique sets of methods and tools enabling discovery of problematic situations, understanding and bringing about positive change.

The third element of GST is *systems philosophy*. In this case, philosophy addresses the philosophical underpinnings (i.e., paradigm) within which GST rests. These underpinnings include (i) systems ontology, (ii) systems epistemology, and (iii) the nature of man. *Systems ontology* deals with how an observer views reality. In this case, reality can be characterized along with two opposing extremes: realism and nominalism. Realism is focused on reality being described as independent of the observer. In principle, every person is able to verify every aspect of the objective reality. Anything that cannot be verified in this way is not part of the objective reality. Nominalism is rooted in the notion that reality is a "construction" of the observer and thus does not exist independent of the observer. In principle, every person's "truths" are subjective and depend on "personal experiences." Epistemology deals with how people obtain and communicate knowledge. Again, two opposing characterizations are used: positivism and anti-positivism. Positivistic epistemology holds that the knowledge is objective, reducible, and concrete. In contrast, anti-positivistic epistemology holds that knowledge is subjective, irreducible, and fallible. The third element of systems philosophy addresses the nature of human beings. On the one hand, people are viewed as deterministic, but on the other, they are viewed as voluntaristic in nature. A deterministic view of people suggests that there is a predetermined outcome, deducible from a cause-effective relationship. The voluntarism perspective suggests that people have "free will" with respect to determining relationships and that they exist beyond predetermined outcomes or absolute cause-effect relationships. Unsurprisingly, GST tends to lean toward nominalist ontology and anti-positivistic epistemology while accounting for voluntarist nature of man.

Extended discussions regarding systems philosophy, ontology, and nature of man are discussed elsewhere [28, 49, 67]. However, the proceeding discussion is meant to highlight two points: First, our systems and their operating environment are increasingly ambiguous, complex, emergent, and uncertain. Second, ideas grounded in GST can be used to offer alternative insights into our systems and their operating environment. GST is fundamentally different along the elements of systems science and systems technology. However, much of its benefits are derived from looking at systems and their environment along systems philosophy dimensions of ontology, epistemology, and human nature.

Interestingly, although GST was initially developed in mathematical terms and emphasizing isomorphic relationships [23], it was observed that "much of his [von Bertalanffy's] writing reflects a deeper concern with the mechanistic and reductionist orientation of then current models in biology and psychology" [57, p. 436]. This is supported by the work of von Bertalanffy [24] who later stated that "classical science in its various disciplines, such as chemistry, biology, psychology, or the social sciences tried to isolate the elements of the observed universes… We have learned, however that for an understanding not only the elements but their interactions as well are required—say, the interplay of enzymes in a cell, the interactions of many conscious and unconscious processes in the personality, the structure and the dynamics of social systems, and so forth" (pp. 414–415).

A single and universally accepted GST has yet to emerge. However, the aspect of GST describing isomorphic concepts, laws, principles, and theorems applicable to different systems is evident in literature [6, 36, 49, 60–70, 65]. For example, the work of [6, 121], thirty (30) propositions—inclusive of laws, principles, and theorems—are proposed as means to investigate situations from a systems viewpoint along seven (7) axioms (i.e., centrality, context, design, goal, information, operational, and viability). Furthermore, over 80 principles can be attributed to GST [64]. In this chapter, a deliberate effort is undertaken to use this view of GST as the basis to address deep systemic issues affecting complex systems.

Specifically, the role of GST in problem formulation is examined in relation to the initialization stage of CSG involving (i) particular circumstances, conditions, patterns, or trends within which system of interest is embedded (i.e., context) and (ii) design, configuration, and execution of system functions (i.e., framing). Framing and context provide entry into systems [5, 43, 64, 94]. Yet, there remains a "lack of clarity as to what problem definition is or how to do it" [38, p. 30]. Undoubtedly, initialization is an essential element of problem formulation. As the term implies, this phase is directed to providing an initial understanding of the situation of the system of interest and its operating environment [79]. Typically, two primary facets are considered: framing and context. Framing involves establishing the current state of the system of interest, including its nature as well as structure [83]. In this case, framing assessment can provide a set of representations that serve to depict the current state of the system of interest and its environment. Similarly, context is pivotal in developing an understanding of the system and its context. However, context is concerned with circumstances, factors, conditions, and patterns in which the system is embedded [77]. Since every system is embedded within a specific context, context assessment can provide a set of circumstances enabling or constraining the system of interest. Obviously, the fidelity of the initialization phase will depend on the observer. This evokes the issues of systems ontology and epistemology in which reality associated with the current state of the system of interest, its environment, and circumstances enabling or constraining the system of interest may be a "construction" of the observer (nominalist ontology) and realizing that how the associated knowledge is communicated may follow the path of subjectivity, irreducible, and fallible (i.e., anti-positivistic epistemology). Regardless of one's placement on the scales of ontology and epistemology, it is evident that initialization should enable better understanding of the nature and structure of the system (framing) and discovery of circumstances affecting system performance (context). Arguably, context assessment in the initialization phase includes the articulation of system pathologies. In this case, a pathology is taken as circumstance, condition, factor, or pattern that acts to limit system performance or lessen system viability [and growth], such that the likelihood of the system achieving performance expectation is reduced [72]. However, there is a lack of qualitative studies suggesting a relationship between system pathology and system framing. Current research attempts to close this gap through a systemic examination of the potential relationship between system pathology and problem formulation phases of different systems-based methodologies.

To support this purpose, the rest of this chapter is organized as follows: Sect. 2 is a review of methodologies grounded in GST. The aim of this section is to establish foundations and utilities associated with different systems-based methodological approaches, including the emerging field of CSG. Emphasis is placed on the *problem formulation* phase. Section 3 elaborates on problem formulation by considering *system pathology*. This section aims to provide varying perspectives on system pathology that must be examined as part of understanding systems. In Sect. 4, a model classifying system pathology at the metasystem level is developed. A description of the model's components and characteristics are provided, along with implications for system-based methodologies, especially complex system governance. Section 5 is an abbreviated case application of the model directed at the DoD acquisition system reform and acquisition system. Implications (and future research directions) are drawn along a system of interest (e.g., acquisition system), practice, and model development. The chapter concludes with end-of-chapter questions to engage readers interested in the triplet of system pathology, problem formulation, and complex system governance.

2 Systems-based Methodological Approaches

A methodology, according to Jackson [61], is a set of "procedures for gaining knowledge about systems and structured processes involved in intervening in and changing systems" (p. 134). Interestingly, there is no shortage of methodological approaches used to explore and gain knowledge about systems. Table 1 provides a summary of systems-based methodologies. Since selection and use of a specific methodological approach depend on the nature of the problem and system at hand and the purpose of the analysis [38], it is reasonable to assume that results will vary with each approach. Moreover, issues of ontology and epistemology should not be ignored, especially when dealing with complex situations [106].

The inclusion of *complex system governance* (CSG) in Table 1 serves two purposes: First, CSG leans toward a "soft" system approach and recognizes that CSG can have a large number of relevant variables as well as myriads of interactions. Aspects of optimization, while desired, might be impossible for this approach. Moreover, a "hard" systems approach "assumes that problems are set in mechanical-unitary contexts. Hard methodologies take it as a given that it is relatively easy to establish clear objectives for the system in which the problem resides—so context must be unitary. They then try to represent that system in a quantitative model that simulates its performance under different operational conditions—something only possible if the system is simple and the context mechanical" [61, p. 30]. For CSG, a systems approach recognizes a large number of relevant variables as well as myriads of interactions. Aspects of optimization, while desired, might be impossible for this approach. Therefore, emphasis must be on structures (and mechanisms) that enable/disable system behavior of viability and performance. Second, CSG is primed as a promising conceptually grounded field, and a methodology is capable of providing insights into complex systems and their operating landscape. As a field, CSG is the "design, execution, and evolution of the metasystem functions necessary to provide control, communication, coordination, and integration of a complex system" [74, p. 264]. GST is rooted in GST's laws, principles, and theorems used to understand the structure, behavior, and performance of complex systems [6, 71, 85, 23, 121] and management cybernetics, the science of effective organization

Table 1 A summary of systems-based methodologies

Classification	Methodology	A brief description of the methodology	Primary proponents
Hard Systems Thinking	Systems Analysis	This methodology is largely dependent on feedback loops and black boxes of cybernetic management. It aims to optimize sociotechnical systems based on fixed parameters such as cost and benefits. Systems analysis includes several phases discussed elsewhere (see e.g., [56, 93]	[15, 44, 56]
	Systems Engineering	This approach places emphasis on defining technical and business customer needs with the goal of producing quality products that meet user needs. A generic life cycle model for systems engineering along with its stages are discussed elsewhere [26, 37, 105]	[26, 59]
	Operations Research	This approach is commonly associated with determining maximum (or minimum) variable (e.g., profit, performance, yield, loss, risk) inventory, allocating, waiting time, replacement, competitive, and combined processes. Operations research was developed to deal with complex organizations that are under control of management [35, 62]. A generic model associated with this approach is discussed elsewhere [35, 60, 62]	[35]
Soft Systems Thinking	Systems Dynamics	System dynamics is concerned with limits of growth and understanding of the system structure using feedback loops as the main determinants of system behavior. Mathematical in nature, system dynamics involves four major variables: system boundary, network of feedback loops, variables of "rates" or "flows" and "levels" or "stocks," and leverage points [31]	[51, 107]
	Organizational Cybernetics	Organizational cybernetics embodies the idea that organizations are black boxes characterized by complexity, self-regulation, and probabilistic behaviors. Central to this approach is the viable system model, which is based on a neurocybernetic model, consisting of five essential subsystems that are aligned with major viable organizational functions. The viable system model is a model rather than a methodology as it does not have a clear set of prescribed phases for deployment. However, two general stages of system identification and system diagnosis are discussed elsewhere [63]	[18–20]
	Strategic Assumption Surfacing and Testing	This approach is grounded on the premise that formulation of right solutions to the right problem requires uncovering critical assumptions underlying policy, plan, and strategy. The articulation of critical assumptions should enable management to compare and contrast and gain new insights on their assumptions when dealing with a "wicked" situation	[90, 95]
	Interactive Planning	Developed by Russell L. Ackoff, this methodology focuses on creating a desired future by designing desirable present conditions. It is made up of two parts: *idealization* and *realization*. These parts are divisible into six interrelated phases [4]	[1–3]
	Soft Systems Methodology	Attributed to Peter Checkland and his colleges at Lancaster University, this methodology emerged as a response to the need for methods that can be used to intervene in "ill-structured" problem situations where it is important to learn about systems while still focusing on "goal-seeking" endeavors that answer "what" should be done and "how" it should be done [63]. Checkland and Scholes [32] suggests that understanding context was largely ignored in systems engineering. His research was aimed as providing more rigorous attempt to tackle problematic situations though addressing issues such as context	[32, 122]

(continued)

Table 1 (continued)

Classification	Methodology	A brief description of the methodology	Primary proponents
	Systems of Systems Engineering Methodology	This methodology is intended to provide a high-level analytical structure to explore complex system problems [7]. Proponents of this approach suggest that developing and enhancing our understanding of complex systems requires a "rigorous engineering analysis [System of systems engineering methodology that invests heavily in the understanding and framing of the problem under study" [7], p. 113). In [41] research, a three-phase approach (i.e., defining SoS problem, abstracting the system, modeling and analyzing the system for behavioral patterns) is suggested. However, [8, 9, 7] suggest a seven (7) stage process consistent of twenty-three (23) constituent elements	[9, 7, 76]
	Critical Systems Heuristics	Developed by Werner Ulrich, this methodology is concerned with "unfairness in society" [63]. This approach promotes emancipatory systems thinking for planners and citizens alike. Synonymous with this methodology are three phases [116]	[116, 117]
	Organizational Learning	Developed by Chris Argyris and Donald Schön, this methodology is concerned with single-loop and double-loop learning where management of an organization is able to contrast "expected outcomes" with the "obtained outcomes." Contrasting these outcomes involves learning based on errors discovered during single-loop learning and provides the basis for modifying organizational norms, policies, and objectives [47]. A key premise of this methodology is that learning and adapting new knowledge must be generated at the individual as well as at organizational level [12–14]	[13, 14]
	Sociotechnical Systems	Attributed to Eric Trist, Ken Bamforth, and Fred Emery and their work at the Tavistock Institute in London, this methodology is concerned with a joint optimization of both social/soft including human and technical aspects of organizations [99]. This methodology involves several steps as postulated by [99] for redesigning sociotechnical systems [109]	[33, 34, 110]
	Total systems Intervention	Developed in the early 1990s by Robert Flood and Michael Jackson, this meta-methodology emerged out of the recognition of strengths of capabilities of individual systems approaches, the need for pluralism in systems thinking, and calls for emancipatory ideas in systems thinking—in reference to critical systems thinking [63]. This methodology is based on the premise that contemporary systems-based methodologies are not complementary. [86] thus suggested that a successful complex organizational intervention might require a "combination" of any set of systems-based approaches. This methodology is underpinned by principles of complex situations and consists of three phases [48, 50]	[48, 50, 61]
	Complex System Governance	Complex system governance (CSG) is an emerging field, representing an approach to improve system performance through purposeful design, execution, and evolution of nine essential metasystem functions that provide for the control, communication, coordination, and integration (C3I) of a complex system. CSG is anchored in GST and management cybernetics, emphasizing the effective performance of metasystem functions necessary to maintain system viability [74, 118]	[78, 71, 79]

[18–20, 36]. Subsequent research has resulted in a three-stage (i.e., initialization, readiness level assessment, and governance development) methodology for implementing CSG in complex situations [73]. The purpose of initialization is to establish the present state of CSG, including framing of the governance metasystem and the unique context of a complex system. This purpose is achieved through three elements: (i) setting the initial profile for context and framing of CSG, (ii) providing a 'baseline' against which changes in context and CSG can be examined, and (iii) re-initializing the CSG profile as development initiatives are undertaken and the CSG landscape shifts. Although these elements are concisely part of initialization for CSG methodology, "problem formulation" is a general state for systems-based methodological approaches. It aligns with the previous methodological research, especially the need to understand the system's state, including its context. This issue is clearly identifiable when one examines the purpose associated with problem formulation phases in different methodologies [69]. It is reasonable to add *"identification of system pathologies"* to this listing since pathologies may act to limit expected system performance. A system pathology can be defined as a circumstance, condition, factor, or pattern that acts to limit system performance, or lessen system viability [and growth], such that the likelihood of [the] system achieving performance expectation is reduced [72].

Moreover, there is a wide acknowledgment of the importance of problem formulation. For example, Quade [100] suggests that a major element of problem formulation relates to being "dissatisfied with current or projected state of affairs" [100, p. 23]. To enable successful succeeding steps for problem resolution, the analyst must attempt to bring as much clarity as possible to the situation under study [119]. Such efforts might as well involve identification of the nature of problems and development of implementable solutions that might include economic, political, technological, and other constraints that exist [100]. Consequently, how the analyst views the situation has a major implication on problem formulation.

Problem formulation is also related to overall systems success. Wellington [120] suggests that "the correct solution of any problem depends primarily on a true understanding of what the problem really is, and wherein its difficulty, we may profitably pause upon the threshold of our subject to consider first, in a more general way, its real nature—the causes which impede sound practice; conditions on which success or failure depends; directions in which error is most feared" [120, p. 1]. Consequently, problem formulation phase "has subsequently been considered the most critical stage in policy analysis" [43, p. 2] and is "probably the single most important routine, since it determines in large part…the subsequent course of action" [92, p. 274]. To these acknowledgments, we add the ability to reduce the probability of precisely solving the wrong problem, otherwise known as type III error [81, 94, 96, 98], reducing the possibility that cost will spiral out of control and well-maintaining credibility of involved parties [89]. Hence, system context assessment (and framing) is not limited to traditional problems but also includes the identification of pathologies. The following section elaborates on the idea of system pathology, including the various perspectives and implications for practitioners tasked with problem formulation.

3 System Pathology in Complex Systems

Webster's New Explorer Encyclopedic Dictionary suggests that the term pathology is derived from two ancient Greek terms: *pathos* (i.e., suffering, experiencing, and emotions) and *logia* (i.e., study of) [91]. The term's usage appears to have emerged in the early 17th century in association with the examination of dead bodies in hopes of uncovering the cause of death [88]. Earlier attempts to uncover causes of death were often linked to understanding structural and functional changes, and paying close attention to physical changes played a critical role in understanding structural changes (morphological) in living things [115]. During the middle ages, it was widely believed that life was sustained by humors. The medical philosophy of humoral theory held that, for example, the human body was filled with four basic well-balanced substances of black bile, yellow bile, phlegm, and blood [29]. The world then held that these basic substances were directly related to four elements of earth, fire, water, and air and that this sustained life. Imbalance (i.e., excess or deficit) in the humors was the cause of diseases and death [29, 115]. This philosophy was replaced by a more scientific cellular theory of Rudolf Virchow and bacteriological theory of Louis Pasteur where disease is understood via microscopic analysis of infected cells [29]. The discoveries of disease-causing microbes (e.g., bacteria, virus, and fungi) suggested that symptoms could be observed and treated to prevent structural and functional changes in the human body [88]. However, since a symptom is only indicative of an underlying problem, it becomes necessary to examine the underlying causes of symptoms so that proper treatment can be prescribed. Therefore, the term "pathology" can be used in relation to observing symptoms and determining the cause of disease and death. Additionally, pathology is also related to understanding structural and functional morphological changes and encompasses disease etiology, disease pathogenesis, cell morphologic changes, and consequences in living organisms [84]. However, the term pathology can be used in connection to inanimate systems. For example, Barnard's [16] work on formal organizations describes functional and scalar pathologies that affect organizational growth. The functional status of a system describes the individual conditions such as privileges, rights, immunities, duties, and obligations that can affect the performance of an organization. Scalar status pathology describes organizational conditions related to superiority in organizational hierarchy and jurisdiction and their effects on corporate growth [16]. In this instance, system pathology relates to organizational management structures that can limit an organization's growth.

In policy analysis, an area of research that tends to address alternative policies and how each achieves a given goal, given the relationship between the policies and goals, problem identification plays an essential role. During problem formulation, goals are determined; a boundary is set; context must be understood as well as the target social system and initial approaches [100]. Interestingly, if pathology is defined as "discrepancies [in social systems] between cherished goals and reality - whose existence and undesirability can be taken for granted" [43, p. 38], the complexity involved in understanding social issues, suggests that the concept of social pathology might change as noted by Becker's [17] statement: "a problem is not the same to all interested parties" (p. 7) and complemented by [43] statement: a problem may not "necessarily [be] the same to all disinterested parties, or even to the same researcher" (p. 25).

Pathology also describes deviations (or shortcomings) in subsystem functions of the viable system model in management cybernetics. The viable system model (VSM) is based on the seminal work of Stafford Beer. Using principles of communication and control, Stafford Beer's work, supplemented by additional research [46, 75], envisioned the necessary and sufficient subsystems of productive (S1), coordination (S2), operations (S3), monitoring (S3 Star [*]), system development (S4), learning and transformation (S4*), and system policy and identity (S5) as well as their functions for organizational viability (continued existence) despite turbulent environmental conditions [18–20]. Additionally, Beer also postulated that "viable systems of all kinds are subject to breakdown. Such breakdowns may be diagnosed, simply in the fact that some inadequacy in the system can be traced to malfunction in one of the five subsystems, where in turn one of the cybernetic features ...will be found not to be functioning" [21].

With this view in mind, Beer [21] postulated that management ought to give attention to the configuration of systems to avoid several pathologies that might affect organization viability. Beer's [21] work on pathology is supplemented by [102] expansive research on organizational viability. Three broad categories of organizational pathologies (i.e., structural, functional, and informational) are suggested [102]. Structural pathologies are "related to an inadequate treatment of total complexity faced by an organization" [102], p. 142). Functional pathologies are deficiencies associated with "each of the organizations that compose the total organization...The aim is to see whether the essential functions (systems) necessary for the organization's viability exists and work adequately" [102, p. 142]. The communication (and information) pathologies pertain to a lack of (or inadequacies in) mechanisms that must enable the transfer of information between subsystems and the environment [21,102]. In this instance, system pathology is related to the performance of organizational functions that can constrain system viability.

System pathology, as described above, is related to the viability of a system and a class of systems known as "system of systems" [72, 76]. Using 41 primary objectives of the six (6) VSM based subsystem functions researchers [18–20, 72] developed 41 metasystem pathologies "indicative of inadequacies in the design, execution, or interpretation of the performance of the system[s] of systems" [72, p. 253]. In this case, the term metasystem is purposely used to indicate something beyond individual system objectives; the objective is to address issues above and beyond single system functions, missions, or objectives [45, 83]. This appears to support the notion of "system of systems" where interest is at the system-of-systems level beyond individual systems. The articulated metasystem functions form part of the governing structure that integrates complex autonomous systems (i.e., system of systems) to achieve functionality beyond constituent systems. The proposed system of systems pathologies can work to constrain the functions that enable system-of-system viability.

System pathology is also used in connection with intelligence systems. In intelligence, Sheptycki suggests [104], knowledge is created from acquired data which may lead to taking specific actions. At a general level, there is a *directive activity* in which the "customer's intelligence needs" are identified and established. A *collection activity* in which information pertinent to customer needs is gathered and a processing activity in which analysis takes place to convert information into consumable "intelligence packages." A *dissemination activity* is also involved in which "intelligent packages" are given to customers and then a *final activity* which involves a joint assessment of what was done and what should take place [104]. To accomplish these activities, there is a need for collaborative effort involving different organizations at different levels of intelligence. In a crime policing environment, two pertinent issues can dictate this dialog. First, the principles of information flow for "intelligence [systems] are supposed to flow upward in the data pyramids" [104, p. 313]. However, since different agencies can operate on different pyramids of intelligence, it is not feasible to have a standard operating procedure across all intelligence systems that might not be feasible [104]. Second, given that the structure of intelligent systems is multi-agency, there is a need for movement of information between and or across information hierarchies [104]. However, in a multi-agency hierarchy of intelligence systems, different agencies are bound to operate at different levels on the intelligence landscape. These two issues, coupled with the desire to transform intelligence organizations into more effective systems, create the right conditions for several organizational process pathologies. In this instance, system pathology relates to agency structures, policies, strategies, technologies, and day-to-day processes constraining societal policing.

The above viewpoints are consistent with the work of Troncale [112, 113], who suggests the emulation of over 2000 years of history of medicine should be used to solve systems problems. Specifically, it is suggested that medical thinking and its approaches can be used to postulate key system processes including, among others, processes of adaptation, binding, emergence, feedback, input, and redundancy, describing how systems work as well as how they do not work. Specifically, Troncale [114, p. 12] states that "each of the key systems processes could be examined in case studies for not achieving the function they normally perform in making a system sustainable. That quickly would yield a 'taxonomy' or 'classification' of possible dysfunctions that is much more detailed than currently possible. Each systems process would then name an entire category of dysfunctions for SEs [system engineers] to be on the lookout for or avoid by design." In this instance, system pathology is related to dysfunction in key system processes based on the examination of biological processes.

A related view of system pathology is presented by Davidz [40] in relation to systems engineering. It is suggested that there is a lack of enhanced formal methods that can be used to uncover deficiencies in systems engineering processes. In this field, a system is defined as an integrated set of elements, subsystems, or assemblies that accomplish a defined objective. These elements include products (e.g., hardware, software, and firmware), processes, people, information, techniques, facilities, services, and other support elements. Davidz notes that systems pathology can be extended to systems that execute the systems engineering processes; such dysfunctions in the execution should be avoided [40]. Dysfunctions in the execution can be avoided using a medicine field analogy where understanding of causes, detection, and treatment of diseased states in different organisms is possible due to methodical characterization of organism processes. For systems engineering, an accumulation of understanding of causes, detection, and treatment of dysfunctional execution enables the realization of healthy systems through diagnosis, prevention, and/or treatment of systemic dysfunctions in programs [40, 39].

At this point in this exploration, the following observations are made: First, it should be clear that there is a variety of perspectives on system pathology ranging from dysfunctions and failure in medicine, organization (systems and process), as well as computer hardware systems. For example, Bobba et al. [27] discuss seven pathologies in hardware transactional memory: FriendlyFire, StarvingWriter, SerializedCommit, FutileStall, StarvingElder, RestartConvoy, and DuelingUpgrades. These pathologies degrade performance and are indicative of deviation from normal expected behavior [27]. And in these perspectives, there exist numerous categories, classifications, and characteristics of pathologies, each correct in their own right. Again, there is no one correct view of system pathology. The correctness might as well depend on the purpose of the analysis and the system of interest.

Second, a system pathology is inherently bad for any given system since it can negatively constrain expected performance. This issue is reinforced by uniformity calling for better understanding pathologies (i.e., failure and dysfunctions) in animate and inanimate-designed systems. In this chapter, inanimate-designed systems include but are not limited to any one system. For those tasked with problem formulation, there is a need for tangible methods, tools, and techniques for implementation of the present research to enable understanding of system pathologies during the phase of problem formulation. For a practitioner, such methods must provide rigorous systematic procedures to identify and assess pathologies and provide insights into possible remediation measures against pathologies.

For our present purposes, the nature of system pathologies in complex systems can be captured in the following vignette describing critical points and suggested relevance to practitioners and system development:

•

All systems are subject to the laws of systems. Just as there are laws governing the nature of matter and energy (e.g., physics law of gravity), so too are our systems/organizations subject to laws (principles, laws, and concepts) defining the behavior and performance of organizations. These system laws are always there, always-on, non-negotiable, unbiased, and explain system performance.

Violations of systems laws carry consequences. Irrespective of noble intentions, ignorance, or willful disregard, violation of system laws carry real consequences for system performance. In the best case, violations degrade performance. In the worst case, violations can escalate to cause catastrophic consequences or even eventual system collapse.

Violations of systems laws are the basis for pathologies. Pathologies are circumstances, conditions, factors, or patterns that act to limit system performance or lessen system viability, such that the likelihood of a system achieving performance expectations is reduced. When system performance fails to meet expectations, violations of systems laws are always in question.

Table 2 Aims of general systems theory

Sources	The aims of general systems theory
[6, 57, 24]	• To investigate the isomorphy of concepts, laws, and models from various fields and to help in the useful transfer from one field to another • To encourage the development of adequate theoretical models in the fields which lack them • To minimize the duplication of theoretical efforts in a different field • To promote the unity of science through improving communications among specialists

Previous research in GST produced over 80 system theory-based pathologies [55, 69, 66]. This set of pathologies emerged from the contrasting meaning of concepts of GST as they relate to problem formulation. Using a thesis that failure to adhere to principles of GST decreases the likelihood of achieving expected system performance, Katina [60–70] used grounded theory method and QSR International's NVivo®10 software package to analyze systems theory text "data" for "significant word or phrase" (Saldana 2013, p. 42) and then thinking critically about the meaning as it relates to phenomena at hand (Mason, 2002). A detailed account of these systems theory-based pathologies is found elsewhere [61–70]. These pathologies form the basis for the remainder of this chapter. Moreover, the present research aligns with the 1954 bylaws of the *Society for General Systems Research* (since renamed: *International Society for the Systems Sciences).* Table 2 states the original aims of general systems theory as indicated in society's bylaws. In this case, the laws and concepts of GST are the basis for problem formulation.

4 UPGR^{2}IDS Model: Classifying Systems Theory-based Pathologies

In science, a model can represent an idea, an object, a process, or a system and is used to describe and explain phenomena that cannot be experienced directly [103]. The assessment of individual system theory-based pathologies is labor-intensive (see a case application in Katina [69]); in this case, a model can be used to offer an explanation of pathology phenomena at the metasystem level. The objective of the model is not to describe in detail each aspect of metasystem pathology. On the contrary, the purpose is to briefly acquaint readers with the dimensions and nature of metasystem pathologies and to provide a reference for further exploration, if desired. By nature, any model is somewhat subjective, and no doubt other classifications or categorizations are possible. Each is correct from a particular vantage point.

Certainly, there is no accepted guide or one "correct" way to group pathologies. In fact, [111] research recognizes that his hierarchical tree of concepts stemming from GST was only meant as one of "many [possible] alternative hierarchies among

P.S.C.'s [principal systems concepts] that could be logically supported and empirically demonstrated for real systems" (p. 36). After using phases of the grounded theory method to create a model for discovering pathologies in principles of GST, eight categories are emerged that appear to provide an umbrella covering the entire set of systems theory-based pathologies. This set of *metasystem pathologies* is clustered along with eight themes: (1) systemic Dynamics, (2) system Goals, (3) systemic Information flow, (4) systemic Process and activities, (5) systemic Regulation, (6) systemic Resources, (7) systemic Structures, and (8) Understanding of systems forming the *UPGR²IDS* model. Table 3 describes each metapathology and pathology attributes:

- *Systemic understanding pathology*—a set of systemic pathological conditions related to the theme of human understanding of complex systems. This theme is developed from GST concepts suggesting that the human capacity for understanding plays a major role in how one deals with complex systems
- *Systemic process pathology*—a set of systemic conditions affecting processes of complex systems. This theme emerges out of concepts of GST describing several processes (internal and external) to the system that must take place to ensure system development, stability, and continued viability
- *Systemic goal pathology*—a set of systemic pathological conditions affecting system performance in terms of goals. This theme emerged from GST concepts suggesting that complex systems have goals, and those goals can be achieved through effective use of certain GST concepts
- *Systemic regulatory pathology*—a set of systemic conditions affecting a system in terms of control and regulation. This theme emerges from concepts of systems theory suggesting that a certain level of control is required to guide complex system development and enabling growth, stability, and continued viability
- *Systemic resources pathology*—a set of systemic conditions affecting a system in terms of resources and resources utilization. This theme emerges from concepts of GST suggesting a need for resources in enabling system development. In addition, the manner in which resources are utilized can have an adverse effect on system productivity
- *Systemic dynamic pathology*—a set of systemic pathological issues affecting system performance from the view of the dynamic nature of complex systems. GST suggests that complex systems continuously interact with other systems to produce performance. There is a need to consider interactive nature of complex systems, their subsystems, and the interplay with their environment
- *Systemic information pathology*—a set of systemic conditions affecting a system in terms of information and communication. GST suggests that the performance of a complex system is related to the ability to create, transmit, receive, and extract meaning from information (i.e., messages)
- *Systemic structure pathology*—a set of systemic pathological conditions pertaining to the structure of a system. GST suggests that all systems can be characteristically organized in certain patterns and relationships to enable achieving maximum performance

Table 3 UPGR2IDS metapathologies and their corresponding pathology attributes

Elements of UPGR2IDS	Attributes of the metapathology	A brief description of attributes
Systemic <u>*U*</u>nderstanding pathology	Basins of stability pathology	A condition in which a system's stability is reduced because of the inability to recognize different system, configurations, and their periods of transitions. It has been suggested that complex systems have three configurations: *order*, *chaos*, and *transition phase*. Each configuration requires different resources and produces different consequences
	Circular causality pathology	A situation in which a traditional (linear) causality model of thinking is applied without recognizing the intricate interactions in subsystems of a complex system. Under the traditional model of thinking, event A is directly related to B (i.e., causes), and in turn, B causes C. Emphasis is placed on finding single causes while ignoring a multitude of other factors
	Complementarity pathology	A situation in which an organization ignores other perspectives/models that are not entirely compatible with the established-predominate perspectives including missions, goals, and objectives. An organization in this case mistakenly assumes that there is only one "right" perspective
	Darkness pathology	A situation in which a system is operated upon under the assumption that all its relevant aspects including behaviors are known

(continued)

This section provided a detailed breakdown of the eight metapathologies supporting the UPGR^{2}IDS model for classifying GST-based pathologies. A comprehensive description of each metapathology, including individual attributes (i.e., systems theory-based pathologies), detailed accounts of dimensions of pathologies, and relation to systems theory in terms of problem formulation, is provided. The

Table 3 (continued)

Elements of UPGR2IDS	Attributes of the metapathology	A brief description of attributes
	Eudemony anthology	A situation in which precedence is placed on the financial profitability of a system above any other measures. This situation involves ignoring import measures that are desirable in describing overall well-being since they are not easily quantifiable. Specifically, the literature suggests that the overall well-being of a system, including people and the society at large, is related to having the right balance in material, technical, physical, social, nutritional, cognitive, spiritual, and environment
	Pathology of holism	A situation in which the management assumes a mode of operation suggesting that behaviors of an integrated system are possessed in its subsystem parts. This pathology is different from the pathology of emergence in that it suggests that understanding of a system cannot be maintained past a particular point of reduction. Under the pathological condition of holism, there are system properties (i.e., behaviors) that cannot be deduced from parts; likewise, there are subsystem behaviors that cannot be deduced from the system
	Incompleteness pathology	Operating a system upon the assumption that the traditional terms of discourse/frame of reference are both consistent and complete. Any given frame of reference/framework is always incomplete

(continued)

Table 3 (continued)

Elements of UPGR2IDS	Attributes of the metapathology	A brief description of attributes
	Reification pathology	A situation in which reality is distorted because of confusing abstract ideas to concrete entities. Young's (1964) words make it more apparent: this pathology occurs when "an analytic or abstract relationship [is treated] as though it were a concrete entity" (p. 109)
	Requisite parsimony pathology	A condition in which a system fails because the human element of the organization has assumed more activities than what can reasonably be handled. The number is limited to seven plus or minus two
	Requite saliency pathology	A condition in which organization productivity is reduced due having undifferentiated importance of organizational missions and objectives. This pathology is related to having *spurious saliency* (i.e., the organization is emphasizing the wrong elements, out of proportion to what they deserve), *unproductive emulation* (i.e., members of the organization might be behaving as those who help create rather than resolve problems), and *having a cultural lag* (i.e., not operating using a common established knowledge base)
	Synchronicity pathology	A situation in which phenomena about a system appear to be meaningfully related but are ignored since it is impossible to be explained in terms of causality-language
	Transcendence pathology	The assumption that stability and viability in complex systems can only be achieved within the confines of reality as defined and understood within the objective realm of "scientific" approach

(continued)

Table 3 (continued)

Elements of UPGR2IDS	Attributes of the metapathology	A brief description of attributes
	Ultra-stability pathology	A condition in which a system can fend off known and anticipated disturbances, but it is not sufficiently designed to fend off unknown disturbances without changing its internal structures; stability at a logically higher level
	Undifferentiated coding pathology	This pathology deals with the issue of "objectivity" and "subjectivity" in understanding issues affecting systems. More specifically, this pathology is a situation in which reality and knowledge are directly attributed to observable results such that anything that does not involve human sensors such as eyes, ears, and touch are not valued
Systemic ***p***rocess pathology	Consequent production pathology	A condition in which there is failure to focus on the underlying structure of the system causing the outcomes/outputs, desired, or otherwise. The focus should be on attempting to (re)calibrate the structures of the system in order to produce an improved product or service
	Diminishing returns pathology	A condition in which management mistakenly assumes that increasing the workforce increases the productivity of the organization as a whole without expanding the landscape of operations
	Events of low probability pathology	A situation in which a complex system is expected to accommodate all scenarios including those of low probability. More specifically this pathological condition indicates that it is an error to attempt to be all things to all people at all times

(continued)

Table 3 (continued)

Elements of UPGR2IDS	Attributes of the metapathology	A brief description of attributes
	Maximum power pathology	A situation in which a system lacks ability to maximize its production through increased capacity for intake and transformation rate. Failure to be able to keep up with demand
	Sociotechnicality pathology	A condition in which an organization has a preference for either the social (i.e., soft/human) aspects or the technical (i.e., technology in the workplace) aspect of an organization but not both
	Sub-optimization pathology	This condition elaborates on several other pathologies including emergence, holism, and satisficing. It suggests that independent improvement of subsystems does not always improve the performance of the integrated system as a whole
Systemic goal pathology	Equifinality pathology	A situation in which a system is operated with a belief that there exists only one approach/method to achieve a final desired state—including goals and missions. There might indeed be one approach/solution; however, the issue at hand is whether other alternative approaches can be examined and taken into consideration
	Multifinality pathology	Involves the notion of experience and the fact that humans have a tendency to draw premature conclusions regarding complex situations that they have previously experienced. Consequently, it is an error for one to anticipate the same results using the same approach; outcomes might vary widely based on subtle situational differences

(continued)

Table 3 (continued)

Elements of UPGR2IDS	Attributes of the metapathology	A brief description of attributes
	Purposive behaviorism pathology	A situation in which the purpose of the system is unguided and primarily based on intended results as opposed to what the system produces
	Satisficing pathology	A condition in which the management team of a system searches for the best possible solution (i.e., optimization) instead of searching for a good-enough solution (i.e., satisficing)
	Unity pathology	A situation in which a system lacks an integrated system purpose or having an identity that is not easily distinguishable from other systems
	Viability pathology	Concerned with failure to balance two related elements: subsystem autonomy and integration of the whole and system stability and system adaptation
Systemic regulatory pathology	Autonomy pathology	A situation in which a subsystem does not have the ability to act as an independent agent without the constraints of a higher system. Autonomy in this case might include being able to make decisions and taking actions
	Balance of tensions pathology	A situation in which the system lacks a governing structure that can relieve tension among different subsystems/elements. The governing structure (i.e., the metasystem structure) can be used to balance tensions along the dimensions of (1) independence of subsystems and missions of the whole, (2) structured design and self-organization, and (3) maintaining stability and allowing for change commensurate with unpredictability in the system/environment

(continued)

Table 3 (continued)

Elements of UPGR2IDS	Attributes of the metapathology	A brief description of attributes
	Control pathology	A condition that emerges out of having ineffective control mechanisms. It has been suggested that control is what "permits the system to adapt and remain viable."
	Cybernetic stability pathology	A condition in which a system lacks a sufficient number of external connections. This is like a freestanding structure. It has been suggested that an increased number of connections makes a system more stable and easily adaptive
	Dialecticism pathology	A condition in which a system lacks the ability to detect errors and learn. More specifically, this condition involves the lack of means to correct errors through single-loop learning where reflection is made on what is good/bad about operations
	Feedback pathology	A situation in which a system lacks the means to improve its behaviors because of insufficient scanning processes. Scanning processes provide the basis for bringing the system close to a desired state
	Frame of reference pathology	A situation in which a system lacks standards by which it can be judged. In this case, a standard is not a sufficient measure for the truth of the judgment, but it is a reliable indication of how the system and its elements are
	Homeorhesis pathology	A situation in which a system lacks mechanisms to guide and enable it to return it to a pre-set path or trajectory following an environmental disturbance

(continued)

Table 3 (continued)

Elements of UPGR2IDS	Attributes of the metapathology	A brief description of attributes
	Homeostasis pathology	A situation in which a system lacks monitoring mechanisms that can be used to alert of any external changes affecting system's essential internal variables. Systems can use negative feedback to reduce fluctuations in the output caused by the environment
	Iteration pathology	A situation in which a system lacks means to account for continuous comparison of the first iteration to the norm to discover errors. Similar to a continuous process that keeps comparing actual state and the desired state of the system, the iteration process provides the means to measure errors in a timely manner
	Least effort pathology	A situation in which a system attempts to move forward by selection of a path of high resistance. Started differently, this is a situation in which a system pursues its goals using methods and tools that are deemed inefficient
	Minimal critical specification pathology	A situation in which a system is managed by prescribing a detailed account of what must be done and how it must be done. In managing complex systems, it is recommended to minimal specifications
	Pareto pathology	A condition in which significant efforts are undertaken to alter the "80/20 production curve." This pathology stems from assuming the existence of a "causal-interrelationships" are evident in simple systems
	Redundancy of potential command pathology	A condition where subsystems lack the "freedom" to decide and act on behalf of the system

(continued)

Table 3 (continued)

Elements of UPGR2IDS	Attributes of the metapathology	A brief description of attributes
	Requisite hierarchy pathology	A situation in which the regulatory body of an organization is not well-designed to match the variety of the organization. This pathology is evident in situations where the variety of the system is higher than what the regulatory body can handle
	Requisite knowledge pathology	A situation in which an organization simply has a bad regulator. A bad regular for an organization is simply a regulator that is not well-informed of the relevant facts that enable organizational viability
	Requisite variety pathology	Specifically addressing channels linking the regulator and system, this is a situation in which the regulatory entity of an organization has insufficient capacity to address the variety of the system
	Subsidiarity pathology	A situation in which local issues need to always be solved by a higher authority. A local issue is a subsystem issue, and a local authority must solve it
	First cybernetic control pathology	A situation in which a system lacks ability to compare system behavior against a set standard. When the comparison is done, the system might lack a mechanism to enable corrective measures and actions to be undertaken

(continued)

research results presented in this section articulate systems conditions affecting system performance (i.e., pathologies); these conditions are described in terms of failure to adhere to and/or violating concepts of GST. The following vignette draws on the principle of eudemony to illustrate how the principle can be violated as a basis for a corresponding pathology.

How do we violate the Principle of Eudemony?

At the basic level, this principle suggests that the well-being in complex systems involves more than financial profitability. Beyond money, a sense of well-being and happiness involves

Table 3 (continued)

Elements of UPGR2IDS	Attributes of the metapathology	A brief description of attributes
	Second cybernetic control pathology	Similar to the *first cybernetic control pathology* and addressing control in terms of communication, this pathology suggests that a system might go out of control if its communication elements are incapable of providing sufficient regulations to address variety. In this case, communication provides regulations that enable the system to address any disturbances that might impede the system
	Third cybernetic control pathology	This is a grave warning against tinkering with an unbroken system. It states that a system can only be brought into control (i.e., a more preferred state), if and only if it has gone out of control
Systemic resources pathology	Buffering pathology	A condition in which a system lacks surplus resources. In essence, the system is being operated without slack. In this case, slack is reserved and might be defined as "capacity in excess of immediate needs."
	Pareto optimality pathology	A situation in which a measure, for instance, allocation of resources, is undertaken to improve one part of a system and is believed to have no adverse effects on other parts of systems. In welfare economics, it has been shown that it is not possible to make one part of the system better without making another part worse-off

(continued)

the right balance in material, technical, physical, social, nutritional, cognitive, and spiritual. Unfortunately, the past 500 years of Western Civilization have devoted most of its conscious awareness to the possession of things and money. People are often overworked, and while buying things, they can't afford. For example, some data suggest that 50% of the homeowners in the US is underwater. Hence, a pathology of eudemony is a situation in which precedence is placed on financial profitability above any other measures. Financial profitability (i.e., money)

Table 3 (continued)

Elements of UPGR2IDS	Attributes of the metapathology	A brief description of attributes
	Patchiness pathology	A situation in which a system lacks ability to increase diversity in terms of consumption of resources from the environment. This pathology does not apply to situations where the environment has only one resource. Counter to the *pathology of omnivory* which primarily addresses diversification of internal structures, patchiness pathology addresses complex system failure to "acquire" a taste for different resources such that "if one set of resources declines, there will not be any other to take their place."
	Redundancy of resources pathology	A condition in which a system is designed and operated under the assumption of optimum efficiency. Under this condition, the allocated resources, for example, might be exactly what is needed—no more no less. In other words, critical redundant resources are not provided
Systemic *d*ynamic pathology	Adaptation pathology	A situation in which neither the internal structures of a system is able to change in response to external disturbances, nor the system being able to lessen environmental changes affecting it
	Dynamic equilibrium pathology	A situation in which system expected performance is reduced due to imbalance in interactions with external systems

(continued)

should only be taken as an enabler—a means to an end). Violating the principle of eudemony suggests that one lacks a sense of self is unbalanced with social surroundings, and thus not in tune with the universe [87]. Money should be seen as a "constraint" to happiness [22] An organization suffering from this pathology will tend to place more emphasis on money, which, as suggested by [22], does not necessarily improve the quality of life.

Models serve a variety of uses—from providing a way of explaining complex phenomena to presenting hypotheses. Similarly, the present model serves to (i) explain categories of dysfunctions that affect system performance, (ii) offer ways

Table 3 (continued)

Elements of UPGR2IDS	Attributes of the metapathology	A brief description of attributes
	Emergence pathology	A condition in which management assumes behaviors of the system whole can be directly inferred from properties of subsystems, independent of subsystem interaction. In this case, management fails to recognize that complex systems exhibit behaviors beyond those of the individual subsystems
	Environmental-modification pathology	A condition in which a system fails to negotiate its environment. As indicated by the pathology of adaptation, systems can either change themselves or change the environment. The pathology of environmental-modification places more emphasis on the efforts undertaken to influence the environment of the system
	High-flux pathology	A situation in which the rate of arrival of resources to systems is less than failures. Related to recovery time, the pathology of high-flux suggests the need to have resources arrive as soon as failure occurs. The lag in arrival of resources has implications on system stability
	Morphostasis pathology	A condition in which stability of an organization is reduced by resisting change; preferring the *status quo*
	Over-specialization pathology	A situation in which a system is so specialized that it cannot afford to change
	Polystability pathology	A circumstance in which a system is managed as if system-level equilibrium is similar to its subsystems. Subsystems have their own equilibriums which are different from that of the system

(continued)

Table 3 (continued)

Elements of UPGR2IDS	Attributes of the metapathology	A brief description of attributes
	Punctuated equilibrium pathology	A situation in which the long periods of stasis (i.e., relative calmness) become the basis for a potentially catastrophic event
	Relaxation time pathology	A situation in which a system experiences too many changes at the same time. When a system is continuously bombarded with many changes, it becomes incapable of processing or assimilating any of the changes and becomes chaotic
	Safe environment pathology	A situation in which a system fails to create a permanently stable environment
	Self-organization pathology	A condition in which management fails to work with the self-organizing tendencies of complex systems. This condition happens when an organizing structure limits autonomy of its subsystems by using global patterns to influence local interactions
	Steady state pathology	A condition in which one focuses on the steady state of a system whole while ignoring steady states of subsystems. This is an error since a system cannot be in a steady state if any of its subsystems are not in steady states
	System environment pathology	Concerned with understanding the relationship between the system and its environment. A complement to *pathology of boundary*, this involves a failure to understand a line of demarcation distinguishing environment from system

(continued)

Table 3 (continued)

Elements of UPGR2IDS	Attributes of the metapathology	A brief description of attributes
	Red Queen pathology	A condition in which a system fails to survive because of its inability to compete with other systems in the same environment. This goes beyond the idea of adapting, evolving, and proliferation inasmuch as they relate to gaining a competitive advantage. It relates to the idea of simply surviving inasmuch as surviving means that an organization takes all the running it can do, just to stay in the same place
Systemic information pathology	Channel capacity pathology	Shannon-Hartley's channel capacity pathology has to do with the lacking ability of a communication channel to transmit different messages without channel modification. A well-designed communication channel accounts for noise (i.e., any factor in the process that works against the predictability of the outcome of the communication process) in transmission
	Communication pathology	The receiver of information is unable to receive information as intended by the sender. Communication is broadly defined as "all of the procedures by which one mind may affect another."
	Equivocation pathology	A situation in which communication channels of a system are inefficient in delivering intended signal (i.e., messages) from one point to the next. In delivering messages (i.e., information), the sender may wish to conceal the meaning so that only the intended receiver can decipher and understand its meaning. In a secret system, the receiver is able to understand the meaning

(continued)

Table 3 (continued)

Elements of UPGR2IDS	Attributes of the metapathology	A brief description of attributes
	Information redundancy pathology	A situation in which little and perhaps insufficient effort is dedicated to reducing error in information transmission. More specifically, it suggests that transmission of information (i.e., communication) can be enhanced through making redundancy of transmitted messages
Systemic structure pathology	Flatness pathology	A situation in which the structure of governance is an inverted pyramid. This is a situation in which there is a "larger the number of administrators relative to that of producers."
	Hierarchy pathology	A situation in which a system lacks a basic structure of a hierarchy. A hierarchy provides a regulatory structure that enables "organization" of the system to generate desired system performance/behavior
	Internal elaboration pathology	A condition in which the management style creates silos due to overemphasis on development of policies and procedures of subsystems and people management
	Morphogenesis pathology	A situation in which a system fails to remain stable after creating a new and radically different structure (system) elaborating on the existing structures as conditioned by morphocatalyst influencing the system
	Omnivory pathology	A situation in which system's internal structures (i.e., pathways) cannot be modified to increase their ability to intake a diverse number of resources. Systems that can take in a diverse number of resources are more stable since a decline on one of the resources will not affect the system

(continued)

Table 3 (continued)

Elements of UPGR2IDS	Attributes of the metapathology	A brief description of attributes
	Organizational closure pathology	A situation in which a system lacks a critical part in the structure that provides closure
	Recursiveness pathology	A violation of the *theorem of system recursion* defined as a condition in which the system in question is incapable of defining itself as a viable system containing viable systems and being contained in a viable system
	Resilience pathology	A situation in which a system, when it experiences a disturbance, has no ability to quickly return to its previous configuration
	Robustness pathology	A situation in which a system lacks the ability to use simple and/or complex mechanisms to withstand environmental changes without modifying the system
	Separability pathology	A situation in which subsystems are tightly coupled together such that a small disturbance is reflected throughout the entire system. In other words, the tight coupling in a large number of subsystems along with positive feedback creates the right conditions for a single breakdown in one of the subsystems to have a major effect on other subsystems and the system as a whole
	Genesis of structure pathology	Addresses the need to initiate and maintain communications among forming structures in a system
	System boundary pathology	A situation in which a boundary (i.e., line of demarcation) of a system is fuzzily defined. A line of demarcation provides a minimum description distinguishing a system from its environment

(continued)

Table 3 (continued)

Elements of UPGR2IDS	Attributes of the metapathology	A brief description of attributes
	System context pathology	An attempt to address systemic issues (or systems) independent of the context in which they are embedded. It is impossible to understand and draw the meaning of a system independent of its context

to categorize pathologies in complex systems, and (iii) explain the sources of each category of dysfunction in terms of attributes of the metapathology. Second, the suggested model can be used as a basis for refined and enhanced development of "better" categories of metapathologies. Again, any model is somewhat subjective, and no doubt other classifications or categorizations are possible. Therefore, other categories are possible, especially through case study applications in different systems. These can only enhance the suggested metapathologies as well as their attributes. Each is correct from a particular vantage point. Third, these pathologies should not be seen as existing in isolation. In GST, emphasis is placed on the whole. In fact, the examination of parts or systems in isolation is said to be incapable of yielding a complete picture of a phenomenon [24]. The same view is taken in the research where pathologies affecting systems do not exist in isolation; they are always related. This phenomenon is not new and is exhibited elsewhere in relation to the nine essential metasystem functions in the CSG model. Finally, it is suggested that the development of a "cure" is dependent on understanding the "disease." In this case, this analogy suggests that it is now possible to suggest and develop "cures" for these categories.

Additionally, research has implications on GST-grounded methodological approaches, especially their problem formulation phase. For example, CSG development methodology calls for initialization (problem formulation phases), readiness level assessment, and governance development. Initialization is the first stage of CSG development, and as the term implies, it is directed to provide an initial understanding of the situation. This understanding encompasses two primary facets: framing and context. Framing establishes the nature and structure of the system of interest and serves to articulate the current state of the system under exploration. Context assessment addresses the setting within which the system of interest is embedded. Therefore, a rigorous application of the UPGR^{2}IDS model enables the establishment of frame and context of the system of interest, provides a foundation for the second stage of CSG development (i.e., readiness level assessment, subsequently reformulated as *Development Mapping*), and is therefore instrumental to the third stage (i.e., governance development) which establishes, executes, and evaluates the continuous development of the governance through activities that enhance system governance and improve readiness level and thus enhance the state of CSG. Figure 1 is modified from [79] to depict the three stages of CSG development.

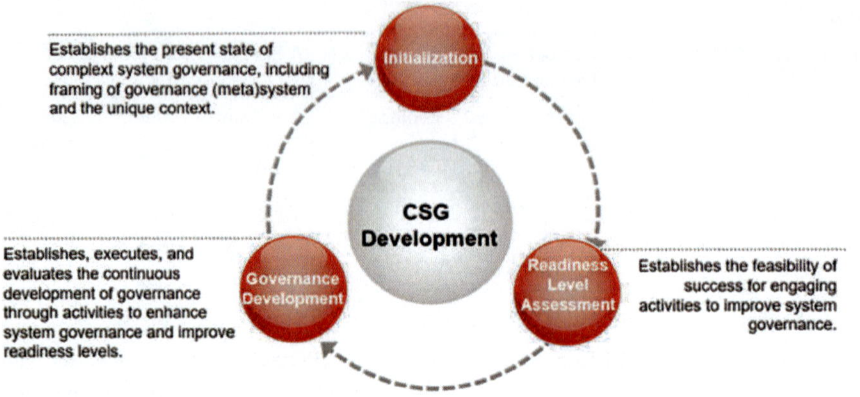

Fig. 1 The three stages of CSG development methodology

5 UPGR^{2}IDS Model Application: The Case for Acquisition System

In this section, a limited case application of the UPGR^{2}IDS model is explored in the acquisition system. The state of the acquisition system is generally not considered to be strong. This might be attributed, similar to many other government topics, to a lack of straightforwardness associated with extremely complicated questions. However, the importance of the acquisition of systems and other materials and supplies to equip a nation cannot be overstated. For example, if the armed forces do not have proper equipment for battle, their existence would be threatened, and so too would the existence of the nation itself. If one takes the acquisition process as a function of government (and not a function of industry), then the acquisition process is subject to the rules and regulations of the governing government. And if it sees the acquisition process as involving many different systems and subsystems, many of which are complex in their own right, shifting depending on the system involved, then the most complex version of the federal acquisition system is the *Department of Defense (DoD) acquisition system*. The defense acquisition system is actually a composition of three different systems that are linked together: (i) the acquisition system, which creates the systems and delivers them to the warfighter, (ii) the requirements system, which generates the requirements from which the acquisition systems develops products and (iii) the planning, programing, budgeting, and execution process, which is the way the department of defense asks for and gets the money it needs from Congress. Figure 2 captures the most well-known attempt to document the complexity of the acquisition system as a wall chart illustration of the DoD acquisition system indicating principal phases and decision points. The associated complexity stems from several major sources, including:

- The complexity of the programs (e.g., a navy aircraft carrier is considered the most complex system ever designed)

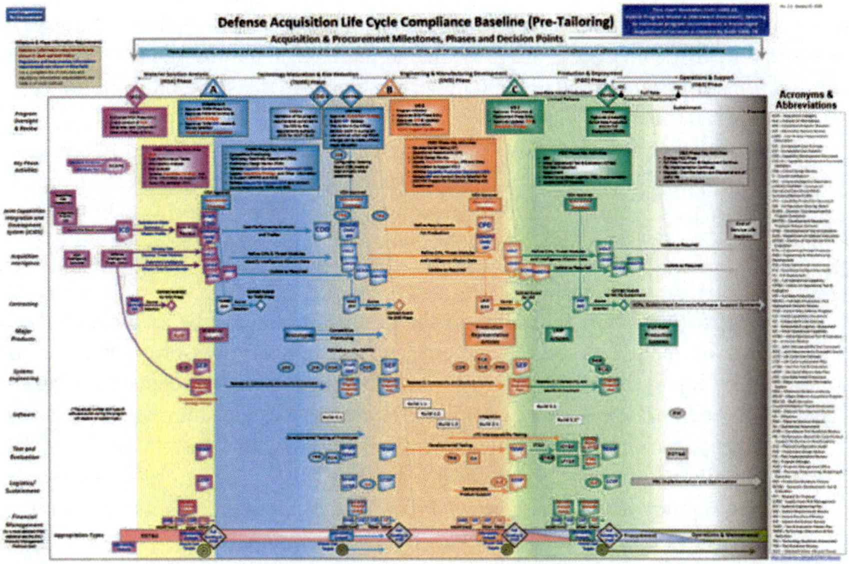

Fig. 2 Department of defense acquisition chart, Adapted from [10]

- The need to integrate defense systems into a very complex existing system with many interphases and relationships already in place
- The very harsh environment that defense systems must operate in (i.e., complex and lengthy testing protocols)
- Complexity is also driven by the need to follow government rules and regulations. This must be done by all parties involved in the acquisition processes
- Complexity is also needed for different stockholders, including many profit organizations, trying to influence the processes in their favor
- Complexity is driven by adding complexity to the system. For example, the buildup of rules, processes, and reviews built into the system from each new leader and from generation after generation of congress

When the system acquisition process is invoked, it is intended to establish an agreement between two organizations under which one party acquires products or services from the other. The acquirer experiences a need for an operational system, for services in support of an operational system, for elements of a system being developed by a project, or for services in support of project activities. The start of an acquisition process begins with the determination of, and agreement on, user needs. The goal is to find a supplier that can meet those needs [58].

However, failure in the system acquisition process is immense. For example, a 2008 United States Government Accountability Office report [52] indicated that there was (i) an average schedule delay of 21 months and (ii) an average budget overrun of 26 percent in the acquisition process [42]. In dollar terms, the combined cost overrun of all studied programs was over $295 billion, up from $42 billion for a similar study conducted just seven years earlier [42]. Interestingly, failures in the system acquisition process are projected to increase in the future [42]. Numerous investigations have attempted to elicit underlying factors that have prevented success in acquiring systems [25, 53, 54].

The developed system pathology approach can be used to provide a different level of analysis and insight into the acquisition system. For present purposes, the nature of system pathologies in complex systems (including DoD acquisition) can be captured in the following critical points and their suggested relevance to acquisition system development:

- ***The acquisition system is subject to the laws of systems*** Just as there are laws governing the nature of matter and energy (e.g., physics law of gravity), so too are acquisition systems subject to laws (principles, laws, concepts defining the behavior, and performance of complex systems). These system laws are always there, always-on, non-negotiable, unbiased, and explain system performance. *For example, the **theory of system boundary** [70] suggests that every system (including acquisition) has a set of boundaries that indicates some degree of differentiation between what is included and excluded in the system.*
- ***Violations of systems laws carry consequences***—Irrespective of noble intentions, ignorance, or willful disregard, violation of system laws carries real consequences for acquisition system performance. In the best case, violations degrade performance. In the worst case, violations can escalate to cause catastrophic consequences or even eventual system collapse. *In the case of the violation of system boundary for acquisitions, it might involve failure to understand the logical separation between the acquisition system and its environment. Moreover, having a wrong boundary (i.e., too narrow/too broad) ensures that the acquisition system cannot be divided into any meaningful portions to enable rigorous examinations of the system, including goals, interactions, and input/outputs.*
- ***Violations of systems laws generate associated pathologies***—pathologies are circumstances, conditions, factors, or patterns that act to limit system performance or lessen system viability, such that the likelihood of a system achieving performance expectations is reduced. When system performance fails to meet expectations, violations of systems laws are always in question. *In the case of acquisitions, the violation of system boundary is the basis for system boundary pathology, which is a situation in which a boundary (i.e., line of demarcation) of a system is fuzzily defined. A line of demarcation provides a minimum description distinguishing a system from its environment. Without a line of demarcation, Troncale* [111] *argues that it is not possible to attain system goals. In this case, the difference between a system and its environment cannot be unknown. This creates the right conditions for pursuing too narrow or too wide of a scope of system objectives. Having a wrong boundary ensures that the acquisition system cannot be divided into meaningful portions to enable rigorous examination, including goals, interactions, and input/outputs. A wrong view of boundaries gives a false impression of a system of interest, resulting in the pursuit of wrong problems and the development of wrong solutions.*

Previous research articulates an individual set of GST laws and principles, pathologies, and applicability for acquisition system Reform [80]. Present research offers linkage to metapathologies using the proposed UPGR^{2}IDS model. Lacking more rigorous case application, Fig. 3 provides a broadly defined application of the model suggesting the inclusion of GST in acquisition system reform and acquisition system development with eight metapathologies that might be used to better inform future acquisition system design, execution, development, and reform. Immediately, two issues emerge: First, the suggested metapathologies have not been addressed in DoD acquisition system reform and acquisition system. Second, the application of UPGR^{2}IDS enhances framing and context assessment (i.e., Initialization) which then forms the basis for subsequent stages (i.e., *Readiness Level Assessment/Development Mapping* and *Governance Development*) for acquisition system development using the CSG development methodology.

In consideration of the present work in relationship to acquisition system development, four primary implications are offered:

1. *Acquisition system reform has proceeded without the inclusion of GST*—this is not totally unexpected. Acquisition has developed as a practice-based field. Notwithstanding the absence of GST, there is also a recognizable absence of consistent grounding on any theoretical basis. Therefore, the conclusion is offered that suggests an emphasis on a stronger theoretical linkage, which may include GST, might be beneficial for acquisition system reform.
2. *GST offers a different perspective and inquiry framework for the examination of acquisition system reform*—GST emphasizes understanding system design, execution, and development from the standpoint of a well-grounded mature

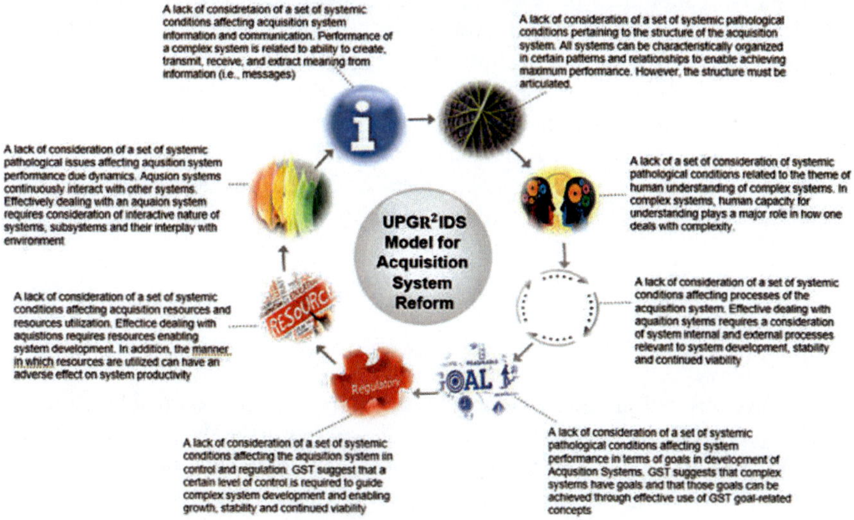

Fig. 3 Application of UPGR2IDS model in system acquisition as a basis for Initialization

body of knowledge. It offers a language, given as the set of propositions, which serves to explain the behavior/performance of complex systems while providing some predictive power.

3. *Acquisitionsystem development breakthrough might be supported by focusing on the underdeveloped "conceptual" emphasis*—the scarcity of literature targeted to the conceptual (philosophical, theoretical, axiomatic) aspects of the acquisition system, suggests that this might be an area with substantial promise for enhancing acquisition system reform. As the preponderance of work has eluded this area, there might be significant breakthroughs to reform dilemmas.
4. *Focus on GST (laws, principles, and concepts) violation might provide new and novel insights for acquisition system reform*—since GST-based research application is non-existence in the acquisition system, there is potential for new and insightful thinking. This might offer a shift in the trajectory of acquisition system reform that has not yet been achieved.

Clearly, GST-based research can offer insights into complex systems. A large utility associated with the UPGR^{2}IDS model remains in the execution in different domains. Undoubtedly, this calls for the model's adaptation to identify, represent, and develop countermeasures to metapathologies. However, it must be stated that complexity dictates that if a metapathology countermeasure works in a given situation, it may not work in subsequent situations.

A key feature of current research efforts is the identification of metapathologies during problem formulation phases of any systems-based methodology. Since problem formulation is relevant to subsequent phases in system approaches, there remains a need to see how the inclusion of pathologies in framing and context enhances current practices. This can only be achieved via case study applications. Additionally, it should be obvious the model (UPGR^{2}IDS) is cyclic and iterative, rather than linear, to be repeated over time to facilitate learning and continuous improvement. Finally, the application of the developed model can serve to refine the model itself. Some aspects of the model are expected to improve over time with increased applications. Moreover, and through case applications, it may become evident that some metapathologies are more prevalent than others. This might serve, for example, as the basis for the deployment of similar countermeasures and the development of new countermeasures for specific system problems.

6 Exercises

The following questions are meant to engage readers (researchers and practitioners alike) dealing with problem formulation regales of industry. The hope is that these questions will enable one to think differently and (hopefully) take actions they would otherwise not have taken. As one goes through these questions, one must have working knowledge their system of interest:

1. What methodological approach does your organization use to intervene in complex situations?
2. Explain the philosophical underpinnings associated with your organizational methodological approaches.
3. What method(s) do you use to address problem identification in your organization?
4. Do these methods account for system pathologies? Explain.
5. Differentiate symptom, pathogen, pathogenesis, and pathology.
6. Explain why there are varying perspectives on system pathologies.
7. Systems theory-based pathologies are grounded in the violation of laws, principles, and theorems of GST: General Systems Theory. What is GST? How can GST be violated; given three examples.
8. What is a *systems theory-based pathology?*
9. How can a systems theory-based pathology be used to explain why systems fail?
10. Metapathologies *are derived from* systems theory-based pathologies. Explain the ontological and epistemological underpinnings of metapathologies.
11. Each metapathology contains a number of specific (attribute) pathologies. Select one specific attribute pathology and explain how it relates to GST.
12. The language associated with GST is not common. Explain why?
13. A system pathology is inherently bad for any given organization/*system* since it can negatively constrain expected system performance. Think of the meaning of the term 'inoculation.' How can pathology to make a system/organization more robust.
14. Eight metapathologies are suggested. Identify these as a selected organization and suggest potential solutions.
15. Literature contains both successes and failures of programs. Use your newly acquired knowledge to explain both successes and failures of a selected program.

References

1. Ackoff RL (1974) Redesigning the future: a systems approach to societal problems. John Wiley & Sons, New York, NY
2. Ackoff RL (1981) Creating the corporate future. Wiley, New York, NY
3. Ackoff RL (1981) The art and science of mess management. Interfaces 11(1):20–26
4. Ackoff RL (1999) Re-creating the corporation: a design of organizations for the 21st century. Oxford University Press
5. Adams KM, Hester PT (2012) Errors in systems approaches. Int J Syst Syst Eng 3(3/4):233–242. https://doi.org/10.1504/IJSSE.2012.052683
6. Adams KM, Hester PT, Bradley JM, Meyers TJ, Keating CB (2014) Systems theory as the foundation for understanding systems. Syst Eng 17(1):112–123
7. Adams KM, Keating CB (2011) Overview of the system of systems engineering methodology. Int J Syst Syst Eng 2(2/3):112–119. https://doi.org/10.1504/IJSSE.2011.040549

8. Adams KG, Meyers TJ (2011) Perspective 1 of the SoSE methodology: Framing the system under study. Int J Syst Syst Eng 2(2/3):163–192
9. Adams KM, Keating CB (2009) SoSE methodology rev 0.2 (No. NCSoSE Technical Report 009-2009). National Centers for System of Systems Engineering, Norfolk, VA
10. Ambrose M, Flores J (2019) Defense acquisition life cycle compliance baseline [Defense Acquisition University]. Defense Acquisition University
11. Arbesman S (2016) Overcomplicated: technology at the limits of comprehension. Current, New York
12. Argyris C (1985) Strategy, change, and defensive routines. Pitman, Boston, MA
13. Argyris C, Schön D (1978) Organizational learning: a theory of action perspective. Addison-Wesley, Reading, MA
14. Argyris C, Schön D (1996) Organizational learning II: theory, method, and practice. Addison-Wesley, New York, NY
15. Atthill C (1975) Decisions: west oil distribution. P.B. Educational Services, London, UK
16. Barnard CI (1946) Functions and pathology of status systems in formal organizations. In: Whyte WF (ed) Industry and society. McGraw-Hill, New York, NY, pp 46–83
17. Becker HS (ed) (1966) Social problems: a modern approach. Wiley, New York, NY
18. Beer S (1979) The heart of the enterprise. John Wiley & Sons, New York, NY
19. Beer S (1981) The brain of the firm: the managerial cybernetics of organization. Wiley, Chichester, UK
20. Beer S (1985) Diagnosing the system for organizations. Oxford University Press, Oxford, UK
21. Beer S (1984) The viable system model: its provenance, development, methodology and pathology. The Journal of the Operational Research Society, 35(1): 7–25. https://doi.org/10.2307/2581927
22. Beer S (1978) Platform for change. John Wiley
23. von Bertalanffy L (1968) General system theory: foundations, developments, applications. George Braziller, New York, NY
24. von Bertalanffy L (1972) The history and status of general systems theory. Acad Manag J 15(4):407–426. https://doi.org/10.2307/255139
25. Berteau DJ, Levy R, Ben-Ari G, Moore C (2011) Defense Industry access to capital markets: Wall Street and the Pentagon. Center for Strategic and International Studies, Washington, DC, pp 1–27. https://www.csis.org/analysis/defense-industry-access-capital-markets
26. Blanchard BS, Fabrycky WJ (2006) Systems engineering and analysis, 4th edn. Pearson, Prentice Hall, Upper Saddle River, NJ
27. Bobba J, Moore KE, Volos H, Yen L, Hill MD, Swift MM, Wood DA (2007). Performance pathologies in hardware transactional memory. In: Scheideler C, Gibbons P (eds) Proceedings of the 34th annual international symposium on Computer architecture. ACM, San Diego, CA, pp 81–91. http://dx.doi.org/https://doi.org/10.1145/1250662.1250674
28. Burrell G, Morgan G (1979) Sociological paradigms and organizational analysis. Ashgate Publishing, Burlington, VT
29. Bynum WF, Porter R (1997) Companion Encyclopedia of the history of medicine, 1st edn. Routledge, New York, NY
30. Cabrera D, Cabrera L (2015) Systems thinking made simple: new hope for solving wicked problems, 1st edn. Odyssean, Ithaca, NY
31. Calida BY, Katina PF (2015) Modelling the 2008 financial economic crisis: triggers, perspectives and implications from systems dynamics. Int J Syst Syst Eng 6(4):273–301. https://doi.org/10.1504/IJSSE.2015.075487
32. Checkland PB, Scholes J (1990) Soft systems methodology in action. John Wiley & Sons Ltd, Chichester, UK
33. Cherns A (1976) The principles of sociotechnical design. Hum Relat 29(8):783–792. https://doi.org/10.1177/001872677602900806
34. Cherns A (1987) Principles of sociotechnical design revisited. Hum Relat 40(3):153–161
35. Churchman CW, Ackoff RL, Arnoff EL (1957) Introduction to operations research. Wiley, New York, NY

36. Clemson B (1984) Cybernetics: a new management tool. Abacus Press, Tunbridge Wells, Kent, UK
37. Componation P, Collopy P (2012) Systems engineering theory: addressing complexity and uncertainty in space system architecting. In: AIAA SPACE 2012 conference & exposition. American Institute of Aeronautics and Astronautics, Pasadena, California. http://dx.doi.org/https://doi.org/10.2514/6.2012-5278
38. Crownover MWB (2005) Complex system contextual framework (CSCF): a grounded-theory construction for the articulation of system context in addressing complex systems problems (Dissertation). Old Dominion University, Norfolk, VA, United States
39. Davidz HL, Nightingale DJ (2008) Enabling systems thinking to accelerate the development of senior systems engineers. Syst Eng 11(1):1–14. https://doi.org/10.1002/sys.20081
40. Davidz HL (2018) Systems engineering pathology: leveraging science to characterize dysfunction. In: Disciplinary convergence in systems engineering research. Springer, Cham, Switzerland, pp 683–696. https://doi.org/10.1007/978-3-319-62217-0_47
41. DeLaurentis DA, Sindiy OV, Stein WB (2006) Developing sustainable space exploration via a system-of-systems approach. American Institute of Aeronautics and Astronautics, San Jose, CA. http://arc.aiaa.org/doi/pdfplus/https://doi.org/10.2514/6.2006-7248
42. Deloitte Development LLC (2009) Can we afford our own future? Why A&D programs are late and over-budget — and what can be done to fix the problem. Deloitte Touche Tohmatsu Limited, London: UK. https://www2.deloitte.com/content/dam/Deloitte/us/Documents/manufacturing/us-ad-canweaffordourownfuture-0127.pdf
43. Dery D (1984) Problem definition in policy analysis. University Press of Kansas, Lawrence, KS
44. Digby J (1989) Operations research and systems analysis at RAND, 1948–1967 (No. RAND/N-2936-RC). RAND, Santa Monica, CA, 4p
45. Djavanshir GR, Khorramshahgol R, Novitzki J (2009) Critical characteristics of metasystems: toward defining metasystems' governance mechanism. IT Prof 11(3):46–49. https://doi.org/10.1109/MITP.2009.45
46. Espejo R, Harnden R (1989) Viable system model: interpretations and applications of stafford Beer's VSM. Wiley & Sons Inc., New York, NY
47. Fiol CM, Lyles MA (1985) Organizational learning. Acad Manag Rev 10(4):803–813. https://doi.org/10.2307/258048
48. Flood RL (1995) Total systems intervention (TSI): a reconstitution. J Oper Res Soc 46(2):174–191
49. Flood RL, Carson ER (1993) Dealing with complexity: an introduction to the theory and application of systems science. Plenum Press, New York
50. Flood RL, Jackson MC (1991) Critical systems thinking: directed readings. John Wiley, New York.
51. Forrester JW (1961) Industrial dynamics. MIT press Cambridge, MA, Cambridge, MA
52. GAO (2008a) Defense acquisitions: assessments of selected weapon programs (No. GAO-08-467SP). United States Government Accountability Office, Washington D.C.. https://www.gao.gov/new.items/d08467sp.pdf
53. GAO (2008b) Defense acquisitions: Zumwalt-class destroyer program emblematic of challenges facing Navy Shipbuilding (No. GAO-08–1061T). U. S. General Accountability Office, Washington, DC, pp 1–17. https://www.gao.gov/products/GAO-08-1061T
54. GAO (2008c) DOD business transformation: Air Force's current approach increases risk that asset visibility goals and transformation priorities will not be achieved (No. GAO-08–866). U. S. General Accountability Office, Washington, DC. https://www.gpo.gov/fdsys/pkg/GAOREPORTS-GAO-08-866/html/GAOREPORTS-GAO-08-866.htm
55. Gheorghe AV, Vamanu DV, Katina PF, Pulfer R (2018) Critical infrastructures, key resources, and key assets. Springer International Publishing, Cham, Switzerland. https://doi.org/10.1007/978-3-319-69224-1_1
56. Gibson JE, Scherer WT, Gibson WF (2007) How to do systems analysis. Wiley-Interscience, Hoboken, NJ

57. Hammond D (2002) Exploring the genealogy of systems thinking. Syst Res Behav Sci 19(5):429–439. https://doi.org/10.1002/sres.499
58. INCOSE (2015) Systems engineering handbook: a guide for system life cycle processes and activities. INCOSE
59. INCOSE (2011) Systems engineering handbook: A guide for system life cycle processes and activities In: Cecilia H (ed) INCOSE, San Diego, CA
60. Jackson MC (1990) Beyond a system of systems methodologies. J Oper Res Soc 41(8):657–668
61. Jackson MC (1991) Systems methodology for the management sciences. Plenum Press, New York, NY
62. Jackson MC (2000) Systems approaches to management. Springer, New Jersey, NY
63. Jackson MC (2003) Systems thinking: Creative holism for managers. John Wiley & Sons Ltd, Chichester, UK
64. Katina PF (2015) Emerging systems theory–based pathologies for governance of complex systems. Int J Syst Syst Eng 6(1/2):144–159. https://doi.org/10.1504/IJSSE.2015.068806
65. Katina PF (2016) Metasystem pathologies (M-Path) method: phases and procedures. J Manag Develop 35(10):1287–1301. https://doi.org/10.1108/JMD-02-2016-0024
66. Katina PF (2017) Problem formulation research: from systems theory to discovery of system pathology. Mod Manage Forum 1(2):1–11
67. Katina PF, Keating CB, Jaradat RM (2014) System requirements engineering in complex situations. Requirements Eng 19(1):45–62. https://doi.org/10.1007/s00766-012-0157-0
68. Katina PF, Pinto CA, Bradley JM, Hester PT (2014) Interdependency-induced risk with applications to healthcare. Int J Crit Infrastruct Prot 7(1):12–26. https://doi.org/10.1016/j.ijcip.2014.01.005
69. Katina PF (2015b) Systems theory-based construct for identifying metasystem pathologies for complex system governance. Ph.D. Old Dominion University, United States--Virginia. http://search.proquest.com.proxy.lib.odu.edu/docview/1717329758/abstract/29A520C8C0A744A2PQ/2
70. Katina PF (2016b) Systems theory as a foundation for discovery of pathologies for complex system problem formulation. In: Masys AJ (ed), Applications of systems thinking and soft operations research in managing complexity. Springer International Publishing, Geneva, Switzerland, pp 227–267. http://link.springer.com/chapter/https://doi.org/10.1007/978-3-319-21106-0_11
71. Keating CB, Bradley JM (2015) Complex system governance reference model. Int J Syst Syst Eng 6(1–2):33–52. https://doi.org/10.1504/IJSSE.2015.068811
72. Keating CB, Katina PF (2012) Prevalence of pathologies in systems of systems. Int J Syst Syst Eng 3(3/4):243–267. https://doi.org/10.1504/IJSSE.2012.052688
73. Keating CB, Katina PF (2015) Editorial: Foundational perspectives for the emerging complex system governance field. Int J Syst Syst Eng 6(1/2):1–14
74. Keating CB, Katina PF, Bradley JM (2014) Complex system governance: concept, challenges, and emerging research. Int J Syst Syst Eng 5(3):263–288. https://doi.org/10.1504/IJSSE.2014.065756
75. Keating CB, Morin M (2001) An approach for systems analysis of patient care operations. J Nurs Adm 31(7–8):355–363
76. Keating CB, Rogers R, Unal R, Dryer D, Sousa-Poza AA, Safford R, Rabadi G (2003) System of systems engineering. Eng Manag J 15(3):35–44
77. Keating CB (2010) Balancing structural tensions in complex systems. Presented at the Proceedings of the american society for engineering management. ASEM Press, Fayetteville, Arkansas, pp 1–8
78. Keating CB (2015) Complex system governance: theory to practice challenges for system of systems engineering. In: 2015 10th System of systems engineering conference (SoSE) (pp 226–231). https://doi.org/10.1109/SYSOSE.2015.7151955
79. Keating CB, Katina PF (2016) Complex system governance development: a first generation methodology. Int J Syst Syst Eng 7(1/2/3):43–74. https://doi.org/10.1504/IJSSE.2016.076127

80. Keating CB, Bradley JM, Katina PF, Arndt C (2018) A systems theory-based examination of failure in acquisition system reform. In: Proceedings of the 14th NPS annual acquisition research symposium. Monterey, California, 9–10 May 2018 [SYM-AM-18-033]
81. Kimball AW (1957) Errors of the third kind in statistical consulting. J Am Stat Assoc 52(278):133–142. https://doi.org/10.2307/2280840
82. Klir GJ (ed) (1972) Trends in general systems theory, 1st edn. John Wiley & Sons Inc., New York, NY
83. Krippendorff K (1986) A dictionary of cybernetics. The American Society for Cybernetics, Norfolk, VA. http://repository.upenn.edu/asc_papers/224
84. Kumar V, Abbas AK, Fausto N, Aster J (2010) Robbins and cotran: pathologic basis of disease, 8th edn. Saunders, Philadelphia, PA
85. Laszlo E (1996) The systems view of the world: a holistic vision for our time. Hampton Press, Cresskill, NJ
86. Laszlo A, Krippner S (1998) Systems theories: their origins, foundations, and development. In: Jordan JS (ed) Systems theories and a priori aspects of perception. Amsterdam, The Netherlands, Elsevier Science, pp. 47–74. http://www.academia.edu/713345/Systems_Theories_Their_origins_foundations_and_development
87. Li J (2013) The visible hand: from struggling survival to viable sustainability. In: Proceedings of the 56th annual meeting of the ISSS, San Jose, CA, pp 1–19. http://journals.isss.org/index.php/proceedings56th/article/view/1959
88. Long E (1965) A history of pathology. Ann Intern Med 63(4):741–741. https://doi.org/10.7326/0003-4819-63-4-741_4
89. Lynn LE (1980) The user's perspective. In: Majone G, Quade ES (eds) Pitfalls of analysis, vol 8. John Wiley & Sons, New York, NY, pp 89–115
90. Mason RO, Mitroff II (1981) Challenging strategic planning assumptions: theory, cases, and techniques. Wiley-Interscience, New York, NY
91. Merriam-Webster (2006) Webster's new explorer encyclopedic dictionary. Federal Street Press, Springfield, MA
92. Mintzberg H, Raisinghani D, Théorêt A (1976) The structure of the "unstructured" decision processes. Adm Sci Q 21(2):246–275
93. Miser HJ, Quade ES (1988) Handbook of systems analysis: craft issues and procedural choices, vol 2. North-Holland, New York, NY
94. Mitroff II (1998) Smart thinking for crazy times: the art of solving the right problems. Berrett-Koehler Publishers, San Francisco, CA
95. Mitroff II, Emshoff JR (1979) On strategic assumption-making: A dialectical approach to policy and planning. Acad Manag Rev 4(1):1–12. https://doi.org/10.2307/257398
96. Mitroff II, Featheringham TR (1974) On systemic problem solving and the error of the third kind. Behav Sci 19(6):383–393. https://doi.org/10.1002/bs.3830190605
97. Mobus GE, Kalton MC (2015) Principles of systems science. Springer-Verlag, New York. www.springer.com/us/book/9781493919192
98. Mosteller F (1948) A k-sample slippage test for an extreme population. Ann Math Stat 19(1):58–65
99. Pasmore WA (1988) Designing effective organizations: the sociotechnical systems perspective. John Wiley & Sons Inc., New York, NY
100. Quade ES (1980) Pitfalls in formulation and modeling. In: Majone G, Quade ES (eds) Pitfalls of analysis, vol 8. Wiley-Interscience, Chichester, England, pp 23–43
101. Rasmussen J, Batstone R (1989) Why do complex organizational systems fail? World Bank Environmental Working Paper, No. 20
102. Ríos JP (2012) Design and diagnosis for sustainable organizations: The viable system method. Springer Berlin Heidelberg, New York, NY
103. Science Learning Hub (2018) Scientific modelling. Science Learning Hub. https://www.sciencelearn.org.nz/resources/575-scientific-modelling
104. Sheptycki J (2004) Organizational pathologies in police intelligence systems: some contributions to the lexicon of intelligence-led policing. Eur J Criminol 1(3):307–332

105. Smartt C, Ferreira S (2012) Constructing a general framework for systems engineering strategy. Syst Eng 15(2):140–152. https://doi.org/10.1002/sys.20199
106. Sousa-Poza AA (2015) Mission engineering. Int J Syst Syst Eng 6(3):161–185. https://doi.org/10.1504/IJSSE.2015.071453
107. Sterman JD (2000) Business dynamics: systems thinking and modeling for a complex world. McGraw-Hill, New York, NY
108. Strijbos S (2010) Systems thinking. In: Frodeman R, Klein JT, Mitcham C (eds) The Oxford handbook of Interdisciplinarity. Oxford University Press, USA, New York, NY, pp 453–470
109. Taylor JC, Felten DF (1993) Performance by design: sociotechnical systems in North America. Prentice Hall, Englewood Cliffs, NJ
110. Trist EL, Bamforth KW (1951) Some social and psychological consequences of the Longwall Method of coal-getting: An examination of the psychological situation and defences of a work group in relation to the social structure and technological content of the work system. Hum Relat 4(1):3–38. https://doi.org/10.1177/001872675100400101
111. Troncale L (1977) Linkage propositions between fifty principal systems concepts. In: Klir GJ (ed) Applied general systems research: recent development and trends. Plenum Press, New York, NY, pp 29–52
112. Troncale L (2011a) Can a theory that integrates the natural systems sciences help systems engineering of defense against security threats? In: 2011 Eighth International Conference on Information Technology: New Generations (ITNG). Las Vegas, NV, pp 947–952. https://doi.org/10.1109/ITNG.2011.213
113. Troncale L (2011b) Would a rigorous knowledge base in systems pathology add significantly to the systems engineering portfolio? In: CSER'11 Proceedings, Conference on Systems Engineering Research, April 14–16 2011. Redondo Beach, CA, pp 1–17. http://lentroncale.com/wp-content/uploads/2011/12/SysPath-CSER11-Troncale-article.pdf
114. Troncale L (2013) Systems processes and pathologies: creating an integrated framework for systems science. In: INCOSE International Symposium vol 23(1), pp 1330–1353
115. van den Tweel JG, Taylor CR (2010) A brief history of pathology. Virchows Arch 457(1):3–10. https://doi.org/10.1007/s00428-010-0934-4
116. Ulrich W (1983) Critical heuristics of social planning: a new approach to practical philosophy. Paul Haupt, Bern/Stuttgart
117. Ulrich W (1987) Critical heuristics of social systems design. Eur J Oper Res 31(3):276–283. https://doi.org/10.1016/0377-2217(87)90036-1
118. Walters D, Moorthy S, Carter B (2014) System of systems engineering and enterprise architecture: implications for governance of complex systems. Int J Syst Syst Eng 5(3):248–262. https://doi.org/10.1504/IJSSE.2014.065755
119. Warfield JN (1976) Societal systems: planning, policy and complexity. Wiley-Interscience, New York, NY
120. Wellington AM (1887) The economic theory of the location of railways. Wiley, New York, NY
121. Whitney K, Bradley JM, Baugh DE, Chesterman CW (2015) Systems theory as a foundation for governance of complex systems. Int J Syst Syst Eng 6(1–2):15–32. https://doi.org/10.1504/IJSSE.2015.068805
122. Wilson B (1984) Systems: concepts, methodologies, and applications. John Wiley & Sons Inc., New York, NY

Environmental Scanning for Complex System Governance

Dale Baugh

Abstract Complex systems are comprised of subsystems that operate together to achieve a common goal that is not achievable by any of the individual subsystems. Complex system governance (CSG) is defined as design, execution, and evolution of the metasystem functions necessary to provide control, communication, coordination, and integration of a complex system. A key governance function of CSG is the ability to scan the environment for information that helps guide and influence the system's present operations as well as future development. While experiencing a significant increase in environmental complexity, the governance function must be capable of increasing its ability to achieve a future state. This chapter provides a perspective, grounded in systems theory, on the environmental scanning function for CSG. This perspective suggests an approach to environmental scanning as a critical function to improve prospects for continuing complex system viability.

Keywords Environmental scanning · Viability · Complex system governance · Systems theory · Management cybernetics

1 Introduction

Environments create both problems and opportunities for organizations [1]. Few would argue that organizations today face unprecedented challenges in maintaining survival and success [2]. Houghton [3] found that some of the contributors to organizational changes come from the effects of Moore's law (doubling of computing capacity every two years),Toffler's law (culture is changing more rapidly today than in the past),the Bandwidth scaling law (network bandwidth is increasing at a similar rate to computing capacity),Reed's law (indicates that the value of group forming options grows even faster than the growing value of the network itself); the law of accelerating returns (the rate of technological information exponential growth is increasing at an independent exponential rate); and the trend of unpredictability (unpredictability emerges in different ways in different aspects of our culture).

D. Baugh (✉)
National Centers for System of Systems Engineering, Old Dominion University, Norfolk, US

C. B. Keating et al. (eds.), *Complex System Governance*, Topics in Safety, Risk, Reliability and Quality 40, https://doi.org/10.1007/978-3-030-93852-9_10

Political, regulatory, educational, governmental, financial, and other environmental parameters are also changing at a seemingly increasing rate. Given these trends, change likely will remain one of the substantial ongoing conditions of the twenty-first century [3]. Are these change trends contributory to the short life span of large organizations? What aspects of organizational design: structures, cultures, communications or behaviors, are vulnerable to this pace of change? Additional research into the relationship of environmental changes and how large companies deal with these changes could provide insight into how these organizations can improve their ability to survive and remain viable (independent existence as a producing system) with changing environment [4] even across centuries such as the Sumitomo Group and Stora Enso.

The rate of environmental change experienced by large organizations can quickly reach that of an overwhelming nature and in fact, has overwhelmed many. See Figs. 1 and 2 [5]. The relatively short life span of large organizations is at least in part attributable to the impact of environmental changes on our complex organizations [6].

Systems theory and management cybernetics have provided a framework for assessing and improving performance of complex organizations. The classic work in management cybernetics is The Heart of the Enterprise [71]. Beer defined management cybernetics as the "science of effective organization" (p. 4). From systems theory [7], we get a focus on the complexity of systems and the implications for control of systems. Contemporary work in system of systems engineering provides for a system of systems as a set of systems integrated and coordinated through a higher order "metasystem." This metasystem is a construct of convenience, existing at a different logical level than that of the systems being integrated, allowing us to intellectually "break the unity" of the system of systems for purposes of analysis [8]. The metasystem performs governance functions for the governed systems.

By considering the principles grounded in management cybernetics and systems theory to assess the role and practice of how a system engages with its rapidly changing environment, perhaps, new insights into organizational design and performance can be developed. Ultimately, this could improve a large organization's viability. Bringing stability to large companies' futures could lead to better organizational performance and, as a result, bring the associated benefits of improved performance to many involved in the global economy. There is economic value in improving organizational performance. There are also opportunities for benefits to governmental organizations that are equally valuable to a population's economy when environmental change tends to demand organizational growth and hierarchical spread to adjust to the change. Seizing these opportunities requires improving the performance of governmental organizations through new and informed insights into environmental changes.

It is somewhat anticlimactic to state that being able to effectively understand the environment within which a system is embedded is critical. As a starting point for understanding environmental scanning, we default to [2] its description as "internal communication of external information about issues that may potentially influence an organization's decision-making process." From this starting point, environmental

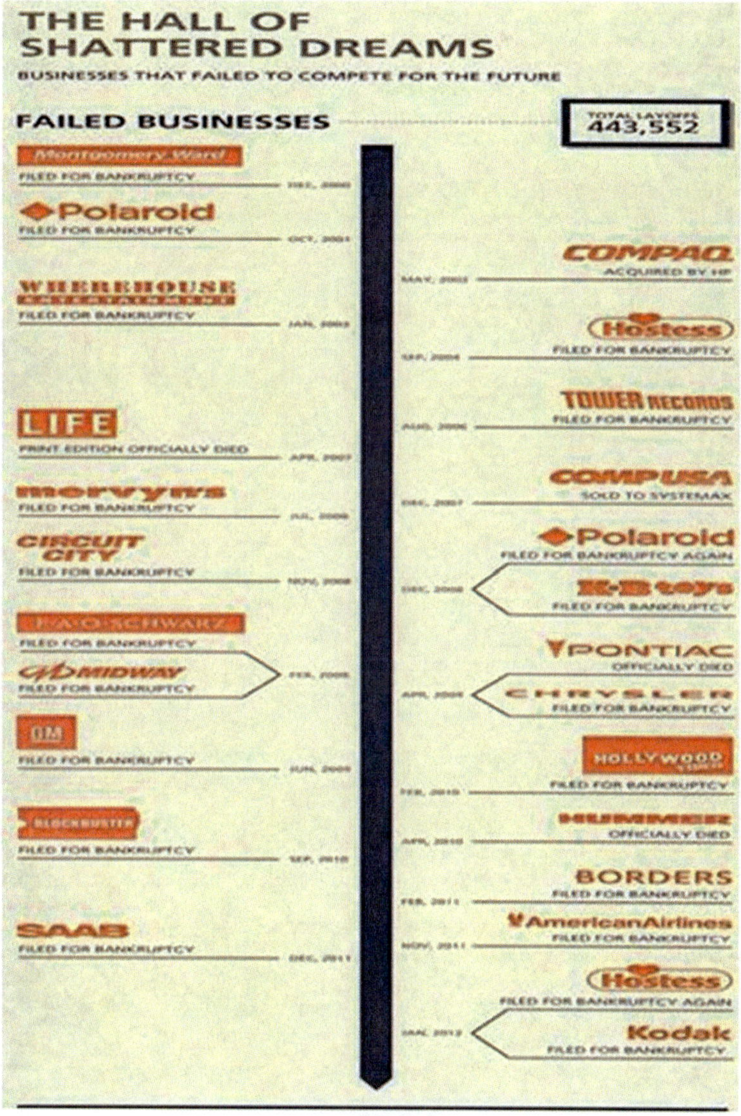

Fig. 1 No organization is too big to fail or too small to succeed—sobering stats on organizational failures [5]

scanning for CSG is examined at a deeper level. Figure 3 provides a capsule summary of the development for this exploration of environmental scanning for CSG. Six different developmental areas are explored, including complexity, systems theory, cybernetics, management cybernetics, current state, and future directions. Each of these areas are explored in this chapter.

Fig. 2 Over 70% of organizations were expected to fail within 40 years of starting [5]

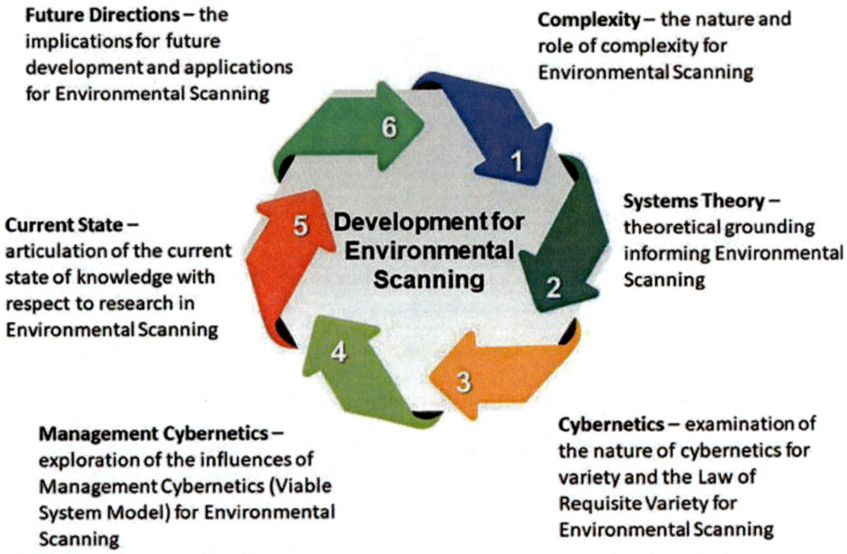

Fig. 3 Overview of the chapter organization

First, complexity is explored as a driving force for understanding the need for environmental scanning. This section is focused on the nature and implications that complexity holds for complex systems and environmental scanning. Second, the influences of systems theory for environmental scanning are discussed. Systems theory provides an essential basis for the systemic worldview as an underpinning for environmental scanning for complex systems. Third, the cybernetic roots informing environmental scanning are examined. Specifically, the role of environmental scanning in relationship to variety, and Ashby's Law of Requisite Variety is explored. Fourth, management cybernetics roots for environmental scanning are examined from Stafford Beer's viable system Model. The role and function of environmental scanning from the perspective of management cybernetics is developed. Fifth, having established the conceptual underpinnings for environmental scanning, a perspective and current state for environmental scanning in CSG are developed. This current state examines the most recent research-based perspective that is currently emerging. Additionally, the future implications for research and practice related to environmental scanning for CSG are suggested. The chapter concludes with several exercises related to environmental scanning for complex systems.

2 Complexity as a Driving Force for Environmental Scanning

It is helpful at this point to insert a definition of complex, as it is used frequently in this chapter in describing a system (organization) of interest. To focus on the applicability of complexity to this chapter, we can apply the framework for complexity sense-making from the work done by Snowden and Boone [9]. They posited that in a framework of domains, complexity falls into five identifiable domains: simple, complicated, complex, chaotic, and disorder. "Complicated" is a domain where cause and effect can be determined over time and from various viewpoints but are not immediately known as they are in the "simple" domain. In the "complex" domain, cause and effect are not initially known but emerge as patterns over time involving many agents of the system. In the "chaotic" domain, no relationship between cause and effect is discernible, as many unknowables exist. For this paper, the "complex" domain attributes closely match what could be considered similar common attributes of large companies [10]. Thus, for purposes of this examination of environmental scanning, "complex system" will refer to large companies (systems) that have the attributes found in the complex domain. At a fundamental level complexity is denoted by a large number of variables/entities, rich interconnections/relationships among the variables/entities, dynamically shifting/evolving over time, and subject to emergence. Beyond complexity, the domain for complex systems is marked by a host of attributes [11] that include ambiguity (lack of clarity in system definition and context), uncertainty (inability to reduce a system to cause effect relationships to define action), holistic influences (the range of system attributes across technology, human, social, organizational, managerial, policy, and political dimensions), and high contextual impacts (circumstances, factors, and conditions that act to enable or constrain a system). In light of complexity and the domain for complex systems, environmental scanning is instrumental in helping to navigate such questions as: What is the lifespan of a system and how does the environment affect that lifespan? How long should a system exist given a shifting environment within which it operates? What attributes of a system's design are contributory to a system's lifespan and effectively addressing, adapting, and navigating changing environmental circumstances? Can a system's design be improved for extending its lifespan given the environmental shifts being faced? Are systems theory and environmental scanning relevant to addressing the structures and/or functions of a system that have an influence on the lifespan of a system? Can systems theory be applied to large organizations and their fitness to exist in their environment? If it can, perhaps there could be an expectation of improvement in organizational viability through application of systems theory axioms and propositions to enhance prospects for continuing viability. Although environmental scanning may not offer a total response to these vexing questions confronting complex systems, it can offer some degree of preparedness to better navigate difficult circumstances.

Complexity for modern systems and the environments within which they exist is not diminishing. On the contrary, complexity in both systems and their environments is expanding and accelerating. Thus, being cognizant of the shifting complexity in the environment is a critical aspect that supports enhanced prospects for present and future viability. Environmental scanning plays a crucial role as a major function in CSG. Environmental scanning provides advanced intelligence on the trends, patterns, and events in a system environment. This vital CSG function assumes responsibility for the structure, design, the attendant processes, and strategies for environmental scanning. In the face of environmental turbulence, this permits the early identification, processing, and guidance for response development to ensure continuing viability of a complex system. The outward and future focus of environmental scanning is essential to assure future system viability. Environmental scanning provides advanced warning of potentially catastrophic events stemming from emergence in the environment. In this respect, emergence are those events that cannot be known or predicted in advance. To the degree that a system is able to quickly identify, assess, and respond to emergent environmental events, which will ultimately determine system performance and help to maintain viability. Additionally, positive aspects of the environment are identified as potential opportunities with implications for system development. In essence, environmental scanning facilitates taking a proactive role in rigorous design, execution, and development of the scanning function for a complex system. Thus, environmental scanning is key to effective functioning of CSG.

3 The Influence and Implications of Systems Theory for Environmental Scanning

General systems theory (GST) grew out of dissatisfaction with a reductionist view of organizations created during the machine age. Reductionism, from the mechanistic perspective, held that all relationships between parts and between parts and the whole were explained by cause and effect relationships. This yielded a mechanistic understanding of organizations [12]. Von Bertalanffy [13] applied his knowledge from the study of living organisms to organizations, leading to GST or systemic thinking. Systemic thinking is about understanding how the parts relate to each other and create larger wholes. These larger wholes lead to self-organization, understanding language, emotion, interactive processes, and learning how to handle situational complexity [14]. Thus, from a systems theory perspective, for an organization to be capable of adapting to environmental changes and to have a profitable future, it needs to be considered as a living organism instead of a passive machinery system [15]. In recent years, systems theory has developed from multiple viewpoints so that there is no longer a widely agreed upon definition of systems theory [16]. Adams et al. "proposed systems theory as a unified group of specific propositions which are brought together by way of an axiom set to form the construct of a system". Systems theory intends to provide explanations for real-world systems. These explanations can lead to increased understanding and provide improved levels of explanatory power and predictive capability for the real-world systems, such as large organizations, where there is a need to improve [17]. Considerable literature has been developed since the 1960s that examines the relationship between organizational theory and systems theory [18]. Kahalas [18] posits that in open systems where an organization is influenced by its environment that the organization is not predictable and that its survival is not guaranteed. He further posits that organizational barriers are erected in response to the environmental changes since the environmental input is not entirely controllable. He concludes that systems theory may provide "an essential perspective for studying organizations" by focusing on the complex interrelationships among organizational variables and by providing a "set of concepts useful in describing and analyzing the relationships" [18].

Adams et al. [16] proposed systems theory as a unified group of propositions formed from a set of axioms that establish the construct of any system. They suggest that this set of axioms can be used to understand a system's operations. Utilizing their formulation as a general approach to aid in understanding a system's behavior could provide a unique approach to developing a framework for assessing the function of environmental scanning. Their work was refined by Whitney et al. using discoverer's induction and furthered to describe the axioms provided and their role in systems of systems [19]. These axioms are listed in Table 1 along with the implications for environmental scanning. Matching the axiom characteristics to the corresponding properties of the governance function from Table 1, results in Table 3. Table 3 shows how systems theory axioms are capable of providing explanations for how environmental scanning performs its role in CSG. Exploring environmental scanning from

Table 1 System axioms and their implications for environmental scanning

Axiom	Implications for environment scanning
The Contextual Axiom states that *system meaning is informed by the circumstances and factors that surround the system*. The contextual axiom's propositions are those which bound the system by providing guidance that enable an investigator to understand the set of external circumstances or factors that enable or constrain a particular system	System context is influenced by externally driven events, trends, and patterns. Environmental scanning is critical to provide intelligence of aspects of the environment that can have both negative as well as positive implications for a system and the context within which a system is embedded
The Goal Axiom states that *systems achieve specific goals through purposeful behavior using pathways and means*. The goal axiom's propositions address the pathways and means for implementing systems that are capable of achieving a specific purpose	System purpose can be constrained or enabled by what transpires in the environment. Therefore, effective environmental scanning is critical to support advanced identification, assessment of impacts, and determination of appropriate responses for maintenance of system viability
The Design Axiom states that *system design is a purposeful imbalance of resources and relationships*. Resources and relationships are never in balance because there are never sufficient resources to satisfy all of the relationships in a system's design. The design axiom provides guidance on how a system is planned, instantiated, and evolved in a purposive manner	System resources are derived from the environment. Environmental scanning provides for monitoring the sources for resources essential to provide inputs for a system. The continuous monitoring of environmental activities with the potential to impact the flow of resources for a system is a primary responsibility for environmental scanning
The Operational Axiom states that *systems must be addressed in situ, where the system is exhibiting purposeful behavior*. The Operational Axiom's propositions provide guidance to those that must address the system in situ, where the system is functioning to produce behavior and performance	Every system is embedded in an environment. Environmental scanning establishes and maintains the model of the system environment. The 'in situ' nature of operation of a system requires knowledge of the environment that impacts, and is impacted by, a system
The Centrality Axiom states that *all systems are central to two propositions: emergence and hierarchy, and communication and control*. The centrality axiom's propositions describe the system by focusing on (1) a system's hierarchy and its demarcation of levels based on emergence arising from sub-levels; and (2) systems control which requires feedback of operational properties through the communication of information	Environmental scanning provides information to all system functions to identify emergence occurring in the environment with implications for system viability. Additionally, constraints emanating from the environment are captured and assessed for system implications. Environmental scanning provides for the designed flow and interpretation of information from the environment to the system and from the system to the environment
The Information Axiom states that *systems create, possess, transfer, and modify information*. The information axiom provides understanding of how information affects systems	Environmental scanning provides the design for information flows and interpretations between the system and its environment. The purposeful design for this information exchange with the environment is a responsibility of the environmental scanning function

(continued)

Table 1 (continued)

Axiom	Implications for environment scanning
The Viability Axiom states that *key parameters in a system must be controlled to ensure continued existence.* The viability axiom addresses how to design a system so that changes in the operational environment may be detected and affected to ensure continued existence	Environmental scanning is essential to determine external impacts on key system parameters being monitored for system viability. Environmental scanning provides early identification, processing, and response coordination for potential threats, as well as support, for system viability

the systems theory perspective provides new insights to complex system viability and shows how system viability can be enhanced through improvements in the system's governance functions.

The complex system governance reference model [20] presents the function of environmental scanning in relationship to other metasystem functions of CSG. The reference model conceptual foundations reside in systems theory and management cybernetics and establish the model's philosophical and theoretical underpinnings [20]. The model identifies the environmental scanning functions as assuming responsibility for modeling the environment and design the mechanisms for identification, processing, and response to environmental events, trends, and patterns that impact system viability. The environmental scanning function works in conjunction with other metasystem functions. Thus, environmental scanning must be integrated with other functions to provide a seamless performance of CSG. It is noteworthy that scanning the environment is not limited to the external environment. It also includes looking internally at the system to assess the current state of functions and to assess system responses to external changes before transformation and development functions can begin. This internal scanning is performed by a variety of the CSG functions. However, it is noteworthy that the relevant context function is targeted to understanding the inner workings of circumstances, factors, and conditions that enable or constrain a complex system. In contrast to the environmental scanning function (focused on the external environment), the relevant context function (focused on internal system impacts) still shares the emphasis on "scanning." Thus, there is much to be gained in the scanning for complex systems in the joint integration between the internally and externally directed scanning efforts.

Environmental scanning functions as identified in the CSG reference model are grounded in systems theory. This grounding helps relate the functions of environmental scanning into the overall governance framework for a complex system. The purpose of such governance is stated to be "focused on purposeful development of control, communication, coordination, and integration functions necessary to produce and sustain desirable levels of system performance" [20], (p. 1). Thus, the environmental scanning function as it operates in CSG is supportive of system viability. This occurs through the role of environmental scanning to identify, process, assess implications, and inform system response to mitigate negative consequences or capitalize on opportunities emerging from the environment.

In sum, systems theory provides a set of axioms (taken for granted assumptions that require no additional validation) and propositions (the set of laws, principles, and concepts that define, explain, and predict the behavior/performance of complex systems). This "language" of systems theory informs the systemic worldview that is essential to understanding environmental scanning for CSG. In effect, systems theory provides the conceptual/theoretical foundations that inform environmental scanning design, execution, and development for CSG. These foundations include the language and worldview upon which the development, deployment, and practices related to environmental scanning rest.

4 Variety and the Cybernetic Roots of Environmental Scanning

W. Ross Ashby developed the Law of Requisite Variety from his biological observations [21]. Subsequently, he found applications of his law in the control of complex systems [69]. The basics of his law are rooted in the concept of variety, where variety can be defined as the possible number of distinguishable states that a system of interest may take on [21]. For a complex organization, composed of many people, the possible organizational state at any time, considering the people, could be approximated by 2^n where n is the number of people involved. In this situation, each person has two possible states, e.g., agree or disagree (Beer [71], pp. 31–39). In actuality, with expanding the number of states for each system element, the equation becomes:

$$V = Z^n, \quad \text{where} \tag{1}$$

V represents the variety in the system as the number of different possible states
Z is the number of different states that an element in the system can occupy, and
n is the number of entities in a system

For a complex system, the calculated variety (as a measure of complexity) quickly and exponentially approaches infinity. A straightforward demonstration (Fig. 4) easily makes this point. Suppose that beyond Beer's example of each person having only an "agree" or "disagree" option, we expand the number of individuals (system entities) and the possible number of states that might be taken by each individual in that system. In this example, take 17 members on a system team participating in a conference call. Also, permit each member of the 17 entity system to be capable of occupying one of nine different states of engagement (highly engaged, engaged, somewhat engaged, neither engaged nor disengaged, somewhat disengaged, disengaged, highly disengaged, checked out, or interested). Thus, the calculated variety (measured complexity) for the conference call is calculated as $V = 9^{17}$, or approximately 16.7 quadrillion.

This example demonstrates the exponential rise in variety (measure of complexity) for a very limited number of system elements and states that they might occupy. Now,

Variety – a measure of complexity for a simple non-trivial system, calculated as the number of different states that the system is capable of occupying

Variety Equation $V = z^n$

V is calculated Variety

z is the number of states that can be occupied by the elements in the system

n is the number of entities in the system

Variety Example – determine the variety for a conference call that has 17 members with each member being able to occupy one of 9 different states (highly engaged, engaged, somewhat engaged, neither engaged nor disengaged, disengaged, highly disengaged, checked out, or uninterested)

Calculated Variety **$V = 9^{17}$ = 16,667,181,699,666,569 or Approximately 16.7 Quadrillion different states that can be occupied by the system**

Variety of a complex system exponentially increases in a system to infinity for all practical purposes. The variety that can be generated from the environment for a complex system also exponentially increases to infinity in practicality.

Fig. 4 Demonstration example of the relationship between variety and complexity

if this complex organization of "n" people is exposed to an array of environmental conditions determined by changes in technology, law, and global markets among others, it is easy to see that adding different "states" will escalate the variety for all intents and purposes to infinity. Thus, variety generated inside a complex system and the variety generated by the environment can provide for immeasurable amounts of variety. Since variety is also a measure of complexity, both environments and organizations can become complex without much effort. Why this is important is best understood in Ashby's term "requisite" when it comes to understanding variety. Requisite variety as applied by Ashby is related to the concept of regulation. Ashby stated that "any regulator (if it conforms to the qualifications given) must model what it regulates" [22]. Stated in terms of variety, Ashby's Law of Requisite Variety says that a regulator's capacity to regulate cannot exceed its capacity as a channel for variety. For good regulation to occur, the regulator must be able to handle the variety in the system it is attempting to regulate [23]. If it is not capable of that, the system could become overwhelmed by the changes that do not fall under the regulator's control. For the purposes of this chapter, regulation is treated as a function of governance. Tying these concepts together helps us understand that to govern a complex system, the governance function must be capable of regulating the system of interest in the face of significant environmental variety. The governance function must be able to generate (manage) variety commensurate with the variety generated by its environment if it is to be a successful governor. If the governance function cannot match the variety of the environment or obtain requisite variety, the environment could have undue influence over the system it is affecting. This could lead to poor system performance, system failure, or some other unmanageable system outcome. Applying the concept of requisite variety to large companies or governmental organizations implies that a robust governance function is fundamental in a highly variable organizational environment for the success (viability) of the organization. In addition, environmental scanning is key to sensing external variety that must be regulated if a system is to remain viable.

A Vignette for Environmental Scanning

A large healthcare company existed in a highly turbulent environment. Their industry was heavily regulated. There were sudden shifts in regulations, standards, and requirements that had to be met to maintain compliance. The risk was to be "out of compliance" with required regulations. Lack of compliance would be met with consequences ranging in different forms of admonishment, increased oversight, or monetary fines. In the exploration of the environment and how they were tracking emergence in the environment, it was recognized that the emergence of "crises" were often due to the lack of advance identification and processing of regulatory changes in the environment (different regulatory and oversight agencies at the federal, state, and local levels). In further introspection, there was a realization that the environmental "scanning" vehicles were not identifying, processing, disseminating, and assuring responses that would keep the organization in compliance with all of the "new" regulations. Instead, in one instance, the organization identified a difficult noncompliance issue to a regulation that had shifted a full six months prior to the recognition of an "out of compliance" state. In effect, the organization had lost six months of reaction time to make system modifications to accommodate the new regulatory requirement. Instead, the noncompliance became a full blown crisis—forcing hasty analysis, urgency for response, and shifting scarce resources to address the noncompliance immediately. Additionally, the organization faced a severe narrowing of the possibilities of what could be done immediately to address the noncompliance "crisis." This unfortunate set of circumstances was routinely repeated for the organization. The recognition of deficiencies in the environmental scanning function was a timely revelation of the source of much of their crisis mode of operation—that only became a crisis due the lack of purposeful design and execution of an environmental scanning system.

5 The Management Cybernetics Perspective for Environmental Scanning

The early works in cybernetics were focused on information transmission and controlling a situation. This "mechanical" focus fell short of dealing with the trends in environmental changes discussed above. W. Ross Ashby took the mathematical aspect of cybernetics and simplified this technical approach to traditional language through the field of biology and living organisms [24]. He expanded the scope of cybernetics into network communications with his "Law of Requisite Variety" that says: "R's capacity as a regulator cannot exceed R's capacity as a channel of communication" [24]. Ashby's work was predicated on the communications being measured and being well defined so that their variety could be measured in a straightforward way. Beer [25, 71] took the work of Ashby and related it to the design and diagnosis of organizations. Beer's works were the source of the field of management cybernetics. His crowning work was the definition of the viable system model (VSM). The VSM made it possible to see complex organizations (those involving many people) as a recursive set of closed Webs of communications systems. Beer developed his work by studying living organisms at first, and then organizations composed of living beings. Beer's work in management cybernetics resulted in a model for viable systems that can be used to assess and evaluate complex system performance. This assessment model can be used for the understanding and the redesigning of complex systems. His work on the VSM has stood without much debate for several decades. The recursive aspect of complex systems was helpful in the diagnostics of, and the strategic design for, viable complex systems.

One of the recursive systems Beer identified in the VSM was system four (S4), whose purpose was to interface with the complex system's environment for helping to predict a future state for the system [4]. See Fig. 5 (adapted from Keating and Katina [73]) for a presentation of the VSM. Beer identified system four (S4) most closely with system intelligence, making sense of external changes, and proposing innovative directions for the organization based on its current condition [27]. S4 provides strategies that are future-oriented, helping the organization in study adapt

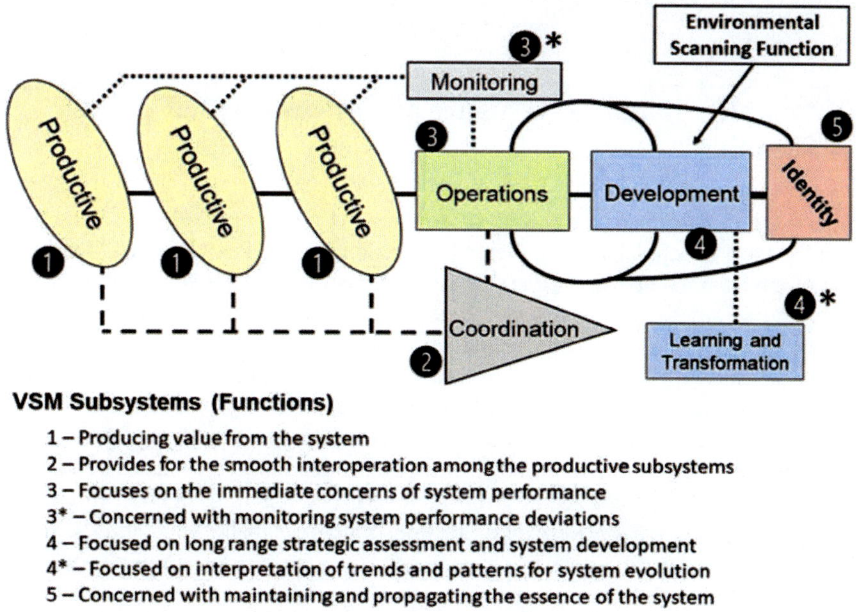

Fig. 5 VSM Showing five subsystems

to environmental changes. This is done so that the organization "can invent its own future (as opposed to being controlled by the environment)" [28]. Understanding the functionality of S4, in its role of interfacing a system with its environment, could be helpful in understanding how a system's governance function could act to support that system's viability in a changing environment. However, there is little research available as to the "how" and "why" this intelligence function is to operate in support of making sense of exponentially changing environments. The VSM provides a framework for assessing organizational systemic strengths and weaknesses but has traditionally been underutilized for organizational analysis (Brocklesby and Cummings [29]) and this appears to still be the case. Perhaps, this is due to a lack of emphasis in consistent employment of environmental scanning as an information provider for organizational analysis.

Management cybernetics has identified the role and value that environmental scanning provides for complex systems. In short, the role of environmental scanning in viable systems is to: (1) provide environmental sensing to identify patterns, trends, or events with the potential to disrupt the system or present opportunities for the system, (2) process environmental shifts such that the system can be in a position to respond to negative events or capitalize on positive events, and (3) act as a vehicle for increasing the regulatory capacity of the system to provide "requisite variety" that matches variety generated from the environment. However, the specific detailed development and execution of environment scanning are not rigorously developed in management cybernetics. While we may speculate as to why this is the case, the reality is that environmental scanning remains underdeveloped in management cybernetics and elsewhere.

6 Complex System Governance and Environmental Scanning

"When the rate of change outside exceeds the rate of change inside, the end is in sight"—Jack Welch, Chairman, General Electric [30], (p. 419). CSG and the function of environmental scanning are directed to ensure that "the end is not in sight" by providing advanced cues of environmental shifts and the impacts on longer term system viability. CSG can be defined as: performance of the (metasystem) functions necessary to provide direction, communication, control, and change essential to ensure continuing system viability [8]. The governance functions direct the system of interest to achieve its goals, while that system is experiencing change. From these two concepts, one can deduce that the amount of change today's large companies are experiencing creates a need for robust governance functions in order to survive. Thus, an organization's governance functions must be capable of dealing with the impact of a rapidly changing environment. All system functions are necessary to maintain viability. However, if the environmental scanning governance function is not sufficient to deal with the system environment, viability might be challenged.

In fact, with poorly performing environmental scanning, perhaps it is inevitable that the system of interest will decay until such time that it can no longer independently exist (remain viable).

Digging deeper into the role of the environmental scanning function for CSG, there are several essential aspects that serve to define the function. The primary function of environmental scanning for CSG is to provide the design and execution of scanning of the system environment. Environmental scanning represents a critical function if a system is going to remain viable. The specific role of environmental scanning has been articulated as [11]: Designing for environmental scanning for the entire system such that trends, changes, patterns, critical stakeholders, collaborative entities, research, etc., are routinely taken into consideration. Thus, this invokes Beer's original intent to provide "intelligence" for the system.

Environmental scanning is achieved by a system of integrated mechanisms that must execute the scanning function. The degree to which these environmental scanning mechanisms are effectively executed influences: (1) the early identification and processing of events that may be detrimental to current system performance or impede future system development (e.g., emergent disruptive technologies), (2) identification of potential opportunities to bolster system trajectory for the future (e.g., incorporation of AI into product lines), and (3) proactive generation of influence to move the environment in particular directions (e.g., establish a demand signal for an emerging market niche).

The environmental scanning function also has responsibility for the maintenance of a model of the environment, in conjunction with the metasystem M4 development function, as well as the scanning interface to that environment. This model is dynamic and must specify the scanning mechanisms, interrelationships, processing approach, and definition of relevant aspects of the environment. Thus, with respect to conducting environmental scanning, several precise actions are invoked. Among these are the explicit mapping of the environment, design of mechanisms to facilitate emergent event capture, definition of the approach (process, procedures, system) for processing of events for system implications, consolidation of scanning results, and dissemination to the relevant system actors that can make the scanning results actionable.

Environmental scanning is one of the nine metasystem functions for CSG. It plays a vital role, in conjunction with other functions, for understanding the impact and implications for emergent events occurring in the relevant environment for a system. Developmental research for environmental scanning in CSG is currently underway. In the following section of this chapter, the state of research in environmental scanning is explored. This "current state" of research continues to evolve. However, it is sufficiently mature such that it would be remiss not to include this current state of knowledge for environmental scanning.

7 Finding Common Ground in the Literature for Environmental Scanning in CSG

The literature discussed above demonstrates the contribution of each field of study-management cybernetics, environmental scanning, CSG, and systems theory, to the viability of complex organizations. As these concepts have been developed in mostly independent fields of study, little research has been done on how the fundamental principles from each field-systems theory, CSG, and management cybernetics could be applied to the environmental scanning process to establish an approach to enhance complex system viability. In addition, environmental scanning has been recognized across a host of different fields. Table 2 provides an overview of the treatment of environmental scanning in the literature. It was developed from the cited literature to show where common attributes exist across the fields of study. From Table 2, it can be observed that environmental scanning is recognized as important in the development and propagation of organizations. This observation suggests that further study to identify the details of these relationships could be productive in advancing the design of complex systems in support of their viability.

In summary, the existing ES literature indicates that writings on this topic are relatively few, are diverse from their respective field of study, are not grounded in any one perspective, and have multiple definitions and applications of this function. The application of the ES function in CSG is certainly a necessary endeavor to advance the field.

Specifically, since the governance function is responsible for system change (design) essential to ensure continuing system viability [8], it is useful to identify the function of governance that is most influential in dealing with the system of interest's changing environment. Keating [8] provided a summary for primary metasystem functions as: coordination, operational control, development, and policy. The metasystem function most closely related to dealing with the system of interest's changing environment is the development function M4. The purpose of this function is to scan and capture information from the environment and assess that information for system impacts. Therefore, it follows that the process of environmental scanning is at the center of the development governance function. What is a framework for performing the environmental scanning function that best supports the governance function of development? Given that systems theory is fundamental in cybernetics and governance as well as system design could systems theory be fundamental to a framework for environmental scanning that is closely related to system viability? Fig. 6 shows the overlapping relationships among management cybernetics and CSG. It is noteworthy that systems theory provides a common frame of reference for both fields. These overlapping relationships imply that systems theory propositions can inform CSG and be influential in supporting system viability.

Table 2 A survey of environmental scanning treatment across different fields of study

Field of study	ES elements	References
Business management	Collection of information about events; interpret information for strategic planning	D'aveni [10], Beal [31], Mayer et al. [32], Ojo and Abdusalam [33], Wambua [34], Kelly [75]
Cybernetics	Adapting to environmental change; predicting the future; adaption for viability	Ashby [24], Beer [26, 71], Heylighen [35], Heylighen & Joslyn, [36], Nechansky [37], Pickering [38]
CSG	Design, deploy, monitor for sensing of environment for both present and future system viability	Keating and Bradley [20], Baugh [39], Calida [40], Carter [41], Keating et al. [74], Walters et al. [82]
Information science	Assess strength and weaknesses in support of future plans; influence decision making processes	Abels [42], Choo [43], Maier et al. [44]
Management	Perceive environment and respond; spotters; keep abreast of happenings; gather relevant information; gather data about events	Ackoff [45], Elenkov [46], Leonard [47], Milliken [48], Samsami et al. [49], Saviano and Di Nauta [50]
Marketing	Learning about events to cope with environment; adapt to changing conditions at leadership level; stay competitive	Frazier [51], Saxby et al. [76] Stanwick et al. [52, 79], Spitz and Ludlow [53]
Planning	Criteria based screening of information; futurism; foresight; discern information from signals; create understanding for decision making; systemic collection of external information in order to reduce randomness; detecting trends for strategic planning;	Bryson [54], Clemens [55], Fahey and King [56], Fahey et al. [57], Kahalas [18], Silverblatt and Korgaonkar [58]
Political science	Detect trends and developments that could shape the future	Daft et al. [1], D'aveni [10], Clemens [55], Bouhnik and Gan [59], Zheng and Carter [60], Tversky and Kahneman [81]
Systems theory	Sensing of environment for implications of both present and future system viability	Von Bertalanffy [13], Espejo [14], Whitney et al. [19], Ackoff [61], Ireland [62], Keating and Katina [63], Richardson [64], Thomas [65], Waelchlif [66], Skyttner (1996), Thompson [80]
Futurism	Anticipate an organizations knowledge of the future. Define status quo, define preferred sate, commit logical and predictable actions	Fahey et al. [57], Voros [67], Conway [68], Slaughter [77], Beer [70], Dunagan [72]

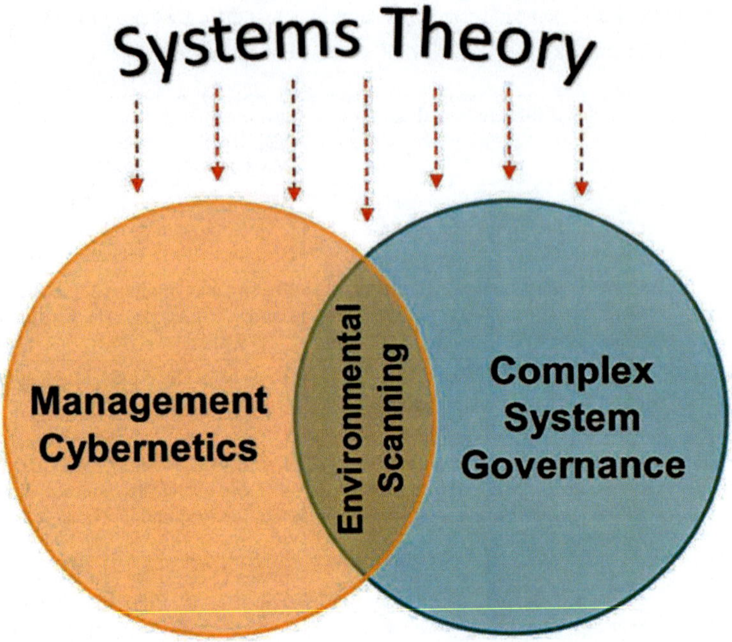

Fig. 6 Environmental scanning at the intersection of management cybernetics and CSG

8 Environmental Scanning—The Current State of Understanding

Environmental scanning has been described as the "internal communication of external information about issues that may potentially influence an organization's decision-making process" [2], (p. 40). The process of environmental scanning has been shown to help focus an organization's planning on external forces that could threaten the organization's stability and future existence (viability). Studies have shown that there is a relationship between scanning processes and organizational performance as measured by profitability and growth [31]. Environmental scanning has been recognized as playing an important role in helping organizations reduce their chances of being blind-sided into poor or reactive performance [2]. Though many organizations recognize the importance of environmental scanning, "past studies indicate that very few organizations have adopted a systematic and structured approach to this task" [31], (p. 272). Fahey et al. [32] go further to explain that their findings support the conclusion that organizations recognizing the need for environmental scanning do not yet have sophisticated systems and have not integrated their outputs into the strategic planning process. Subramanian et al. [31] found support for a relationship between scanning systems and performance in Fortune 500 companies, but they also concluded a better perspective is needed in scanning

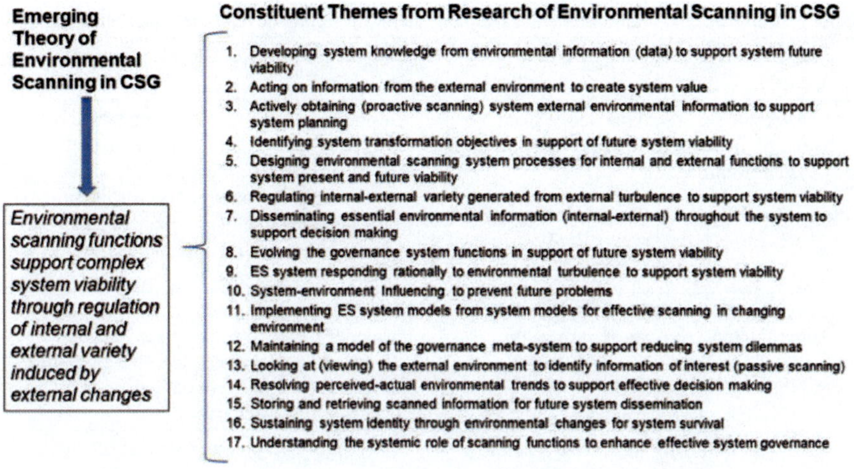

Fig. 7 Emerging theory and supporting themes of environmental scanning in CSG

practices and the relationship that exists between environmental scanning and organizational performance. The past depictions of the lack of effective environmental scanning, as well as the recognized importance of this function, appear to still be the case. However, there is ongoing research that is seeking to provide an advancement to our understanding of environmental scanning. This advancement will be instrumental in establishing the underlying theoretical/conceptual foundations as to the nature of environmental scanning. Better understanding of environmental scanning-related phenomena will provide a stepping stone to advance the supporting methods, tools, and technologies that will enable practitioners to better perform the design, execution, and development of environmental scanning.

The current state of research for environmental scanning in CSG is captured in Fig. 7. Following initial work from Baugh [33], this research has been evolving over several years and is now coming to fruition. The research is sufficiently mature to present as the current state of knowledge for environmental scanning in CSG.

The implications of this research for environmental scanning in CSG are significant. Table 3 is provided to relate each of the emerging themes of environmental scanning in CSG to a set of implications for CSG development and applications for environmental scanning.

There are three major implications of the current state of research for environmental scanning in CSG. First, the rigorous development of the themes (Table 3), following a grounded theory method approach [34–36], has provided a research-based articulation of environmental scanning role in performance of CSG and beyond. Previously, while environmental scanning has been acknowledged, and partially articulated, the comprehensive and holistic depiction of environmental scanning represents a major theoretical/conceptual step forward. Thus, the emerging CSG

Table 3 Implications of environmental scanning themes for CSG development and practice

	Environmental scanning theme	Implications for CSG development and practice
1	Developing system knowledge from environmental information (data) to support system future viability	Information from the environment is critical to CSG. However, more critical is the processing of that information in a way that supports maintenance of future viability. This provides intelligence stemming from the environment and implications for the system
2	Acting on information from the external environment to create system value	Information from environmental scanning provides the basis for responsive action. These actions may seek to either minimize threats to the system or capitalize on opportunities discovered
3	Actively obtaining (proactive scanning) system external environmental information to support system planning	CSG requires a constant state of planning for current operations as well as future trajectory maintenance. Accurate information, and processing of the information for system meaning, is critical to support both operational (near term) decisions as well as strategic (evolutionary) decisions
4	Identifying system transformation objectives in support of future system viability	Environmental scanning plays a critical role in helping to determine necessary transformation to maintain viability. The discovery and processing of the patterns, trends, and emergent events stemming from the environment are critical inputs for transformation directions
5	Designing environmental scanning system processes for internal and external functions to support system present and future viability	Environmental scanning effectiveness is determined by the design for the system and processes that will be used to execute the scanning function. The utility of environmental scanning is found in the degree to which the design is effective in supporting decisions, actions, and interpretations to support present and future viability
6	Regulating internal–external variety generated from external turbulence to support system viability	Environmental scanning plays a vital role in matching (absorbing) variety generated from the environment. This variety absorbing capacity for environmental scanning is critical to system viability

(continued)

Table 3 (continued)

	Environmental scanning theme	Implications for CSG development and practice
7	Disseminating essential environmental information (internal–external) throughout the system to support decision-making	A primary role for environmental scanning is the dissemination of the information to appropriate system entities that can take action based on the information. The design for environmental scanning must include the dissemination approach for information to the point where it can support decisions and become actionable
8	Evolving the governance system functions in support of future system viability	Environmental scanning provides insights in the continuing adequacy of the metasystem design based on shifts in the environment. This early warning of environmental shifting patterns and trends can provide additional time and potential directions to better position the system for future viability
9	ES system responding rationally to environmental turbulence to support system viability	The processing and deliberation of meaning from environmental scanning provides a "metering" of information. This metering can limit irrational decisions that may be based on limited, incomplete, inadequately processed, or incorrect information from the environment
10	System-environment influencing to prevent future problems	The environmental scanning function can take a "proactive' stance to influence the environment in ways beneficial to the system. In this sense, potential escalations in problems might be precluded for both present and future system development
11	Implementing ES system models from system models for effective scanning in changing environment	As the environment for a complex system changes, so too must the design for environmental scanning of that environment. Thus, the design for environmental scanning must be in continuous evolution
12	Maintaining a model of the governance metasystem to support reducing system dilemmas	The explicit model of the metasystem is essential to understand how "variety" stemming from the environment is matched. Variety introduces the possibility for dilemmas in the system. Effectively identifying and matching variety precludes or mitigates dilemmas
13	Looking at (viewing) the external environment to identify information of interest (passive scanning)	A primary role of environmental scanning is to identify and absorb important information accrued through passive examination of the environment for emerging trends, patterns, or events

(continued)

Table 3 (continued)

	Environmental scanning theme	Implications for CSG development and practice
14	Resolving perceived-actual environmental trends to support effective decision-making	Environmental scanning takes the role of ensuring that perceptions of conditions are in actually correct. This "validation" of environmental conditions supports more informed decision-making
15	Storing and retrieving scanned information for future system dissemination	Information from the environment has relevance at different points in time. Environmental scanning must identify information that should be reserved for future dissemination and store that information for future access and dissemination
16	Sustaining system identity through environmental changes for system survival	Shifts in the environment offer different challenges that test the identity for a system. Environmental scanning functions to process environmental shifts and their implications for identity and continued system viability
17	Understanding the systemic role of scanning functions to enhance effective system governance	Environmental scanning must be designed from a systemic worldview, with the role of enhancing governance functions. Ultimately, effectiveness in governance is, in part, dependent on the performance of scanning functions

field body of knowledge is extended for the critical environmental scanning function. Second, the foundation is set to begin establishing the corresponding array of methods, tools, and techniques to permit the rigorous examination, assessment, and insights for development of environmental scanning to improve prospects for CSG performance. Thus, stemming from the conceptual/theoretical base, more rigorous and comprehensive environmental scanning can be supported. Third, ultimately, improvements in environmental scanning are targeted to improve practices related to performance of CSG functions. The practitioner emphasis, moving from theory to method to application, is essential if CSG is to become accessible to practitioners.

9 Evolving Directions for Environmental Scanning in CSG

Research into the viability of complex systems is certainly not confined to investigating the governance functions, the assessment framework of the VSM/CSG, or environmental scanning. Any field or fields of study that can provide insight into how complex systems (organizations) can remain viable in a rapidly changing environment offers potential benefit. This chapter does not intend to suggest that a sole reliance on systems theory, or systems theory-based derivative methods/methodologies, is the only fruitful approach for enhancement of system viability. However, consistent with systems theory is the concept of holism. A holistic approach for investigation of complex system viability, from the environmental scanning perspective, should be given strong consideration as this chapter suggests. Keating [37] established critical foundations for research in complex systems. To assist in the further development of research in complex systems, he suggested a framework consisting of five logical levels (philosophic, axiomatic, methodological, application, and method). This holistic framework provides a level of research organization, grounded in systems theory, to ensure more robust consideration of complex systems research. We can now apply this robust research framework to this chapter's topical information to provide an informative set of questions that can guide the investigation of phenomena related to environmental scanning. This set is neither complete nor static in its development. However, it is fundamental to encourage more research dialog in this specific topical area. Table 4 presents a set of topical research questions, adapted to environmental scanning in CSG that are mapped against the critical foundations of complex system research taken from Keating [37]. These research questions can be used to stimulate future development of the environmental scanning function from the systems theory perspective. This future research could contribute to the developing field of research in CSG. When applied to large private companies and organizations, it could prove valuable in minimizing unintended organizational losses and negative impacts related to premature failures (loss of viability). When applied to governmental organizations, it could prove valuable in minimizing unintended largess, costly regulation, process and procedure ineffectiveness, and reduce loss of viability due to upset constituencies or upsets concerning errant policies.

One of the frontier development directions for environmental scanning is producing a set of instruments to measure, monitor, and represent scanning. The ability to scan the environment must also enable practitioners to make more informed decisions with respect to maintaining present and future viability.

10 Implications for Practice and Practitioners

Why are some organizations successful over an extended period, while others fall by the wayside? One-third of the organizations listed in the 1970 Fortune 500 had been acquired, dismantled or merged with other organizations by 1983 [75]. Research into the viability of larger organizations (Fortune 500 size) indicates that their lifespan, on the average, is about 50 years, significantly less than a human being [83], and perhaps, they are experiencing even shorter lifespans in contemporary times. A contemporary study by Ellen De Rooij of the Stratix Group in Amsterdam shows that the average life expectancy of all firms, regardless of size, measured in Japan and much of Europe, is only 12.5 years [38]. While we see these shorter life spans, a handful of large organizations have been in existence for more than a hundred years (e.g., Sumitomo Group, Stora) [83]. So, why do so many organizations have short lifespans? The impact of

Table 4 Topical research questions mapped against the critical foundations of research for environmental scanning for CSG

Dimension	Description	Research questions for environmental scanning
Philosophical	The philosophical underpinnings used to inform the perspective of purposeful decision, action, and interpretation	• What are the philosophical underpinnings of the governance functions of complex systems and their implications for environmental scanning? • What philosophies are evolving in the area of environmental scanning? • What paradigms are emerging in understanding complex system viability and the contributions of environmental scanning to that viability?
Axiomatic	That which is accepted as source knowledge from the fields of interest	• What areas of CSG and environmental scanning do new propositions need to emerge? • What existing propositions of systems theory need further exploitation in relationship to environmental scanning?
Methodological	The guiding frameworks that inform engagement of complex systems problems	• What frameworks can be constructed to direct the development or analysis of environmental scanning for CSG? • What are the diagnostics for environmental scanning as part of CSG development?
Method	Identity of the specific techniques, tools, or processes that are appropriate to a specific application	• What tools or techniques can be developed to assess environmental scanning effectiveness? • What methods would be employed to select the proper tools for environmental scanning?
Application	The attempts to deploy the chosen methodology	• How can environmental scanning methodologies be deployed in different system settings and contexts? • What are the best practices for environmental scanning that lead to greater system viability?

these short life spans in terms of human and financial capital is difficult to measure, but it must be a staggering number at the global level. The Enron failure alone cost individual employees hundreds of thousands of dollars in retirement fund losses. The Florida Retirement System that had invested in Enron lost over $280 million, and over 28,000 employees at Andersen's U.S. operations, Enron's auditing firm, were at risk of losing their jobs [78]. These companies, or organizations, represent large, complex systems. Can systems theory axioms and propositions provide insights to enhance viability of organizations such that the huge losses we have seen can be reduced? Can an understanding of CSG functions bring greater clarity on organizational survival in a complex, changing world? Can the application of environmental scanning functions within the complex system governance paradigm lead to enhanced strategic planning, executive decision-making, long-range planning, and variety management improvement in such a way as to enhance the viability of a system that is purposely trying to improve in a changing environment?

One of the essential functions in executive decision-making is the ability to scan the organization horizon, identify and understand the changes that are occurring or may occur, and to build and implement strategies to meet the organization environment's new demands [39]. Implementing the environmental scanning function causes the decision-maker to move out of the view of the near-term issues into a macro-view of the organization's internal and external environment. The environmental scanning function will move the decision-making process from a reactive mode to a proactive mode as adjustments are made to the scanned intelligence. Although many organizational leaders have never purposely conducted an environmental scan [39], there are purposeful ways to conduct environmental scans of the internal and external factors that could impact an organization. There is complacency with existing, stable, and comfortable organizational models and frameworks of thinking that have worked well in the past. However, for many decision-makers, either out of necessity or out of a lack of understanding, the tacit choice is to ignore external changes. As presented above in the introduction, many organizations that were once successful have closed their doors because they did not correctly identify and respond to changes in their external environments. Perhaps, some of those Fortune 500 organizations that have failed from changes in their external environments could have remained viable if their executive decision-makers had paid attention to their changing external environment. This would entail identification and implementation of changes to their organizational model (governance system) to meet the new demands stemming from a rapidly shifting environment.

11 Summary

A systems theory-based development of a framework for the environmental scanning function, within the governance functions for complex systems, could reveal new understanding about system viability. As all complex systems intrinsically perform governance functions, understanding that improvement in those functions can lead to improvements in governance performance. The ideal outcome would be to learn how the governance function improvements can contribute to enhanced system performance. When applied to practice, such revelations could improve our ability to perform the governance functions in ways that enhance viability. Enhanced system viability, when applied to our large companies, has the potential to reduce economic losses, sustain workplace jobs, support a robust economy, and even support organizational growth in a highly competitive and changing organizational environment. Enhanced system viability when applied to governmental organizations, has the potential to reduce operating costs and maximize performance in supporting the interests of served populations in a highly competitive and changing global economic environment.

Environmental scanning is one of the functions of complex system governance. It is the system's window into the outside world. Its functionality is relatively new in today's organizations, but its role in governance has been recognized for some time. Significant development of methods and applications for environmental scanning is promising in support of more purposeful governance functionality. Applying systems theory axioms and propositions to the developmental framework for environmental scanning have the potential to bring the benefits of systems theory to future developmental activities to enhance environmental scanning. Three of the potential benefits are:

- First, a holistic approach to expand research in environmental scanning. A holistic approach treats the environmental scanning function as part of the larger set of governance functions. Thus, this considers how environmental scanning interfaces with other important governance functions. Researching environmental scanning in isolation could lead to suboptimization of the environmental scanning framework development. Not only must the environmental scanning function inform other CSG functions but it also must also be informed by those other functions.
- Second, considering management cybernetics and Ashby's law of requisite variety. Variety in the environment continues the trend of escalation and acceleration. Thus, the need for a matching variety in the governance function capabilities is addressed as a critical emphasis through the environmental scanning function. Developing a framework for this function that considers the command, control, coordination, and communication aspects of the governance functions in an accelerating change environment is fundamental to successful governance, and by extension system viability.
- Third, considering complex system viability. The literature on this topic readily demonstrates the significant negative impact of short-lived systems (organizations). Stafford Beer's VSM was developed with the objective of sustaining a viable system through variety engineering. Thus, the VSM provided a fundamental framework for analysis and assessment that has at its core the design of a viable system. In effect, this represents "variety engineering," a primary objective of CSG. Environmental scanning is critical to enable viability through variety engineering

Failure to address the complex system functions that lead to system viability leaves us where we are today, with short-lived systems in the world where the environment is changing at an ever-increasing rate. Short-lived organizations take a significant toll on our economy and our workforce. Progress toward greater viability in our organizations has the potential to improve our lives and utilization of resources.

The corresponding role and importance of environmental scanning to assure that the future cannot be understated.

12 Exercises

1. The systems theory-based section of this chapter describes three overarching functions of environmental scanning as it functions in complex systems governance scanning, development, and transformation. Think of a complex system in your experience and describe the issues faced by that complex system using those three functions.
2. What are some of the ways that a system agent (member) can identify opportunities for system transformation by combining internal and external information?
3. For the environmental scanning function in complex system governance, identify reasons why this function may not be successful and strategies that might increase the probability of that function's success.
4. How can the environmental scanning function be applied in response to Ashby's law of requisite variety? What are some of the challenges that might be encountered when establishing requisite variety in a CSG scenario?
5. How can the CSG reference model be used to address systemic deficiencies, not just readily recognizable symptoms that may appear on the surface? What role would environmental scanning play in this analysis?

References

1. Daft RL, Sormunen J, Parks D (1988) Chief executive scanning, environmental characteristics, and company performance: an empirical study. Strateg Manag J 9(2):123–139
2. Albright KS (2004) Environmental scanning: radar for success. Inf Manag J 38(3):38–45
3. Houghton (2013) Thinking & teaching tools for digital thought, 14 October 2013 [online]. http://www.wcu.edu/ceap/houghton/readings/technology_trends.html. Accessed 11 November 2014
4. Beer S (1984) The viable system model: its provenance, development, methodology and pathology. J Oper Res Soc 35(1):7–25
5. Solis B (2013) No business is too big to fail or too small to succeed—sobering stats on business failures, pp 1–18 [online]. http://www.briansolis.com/2013/02/no-business-is-too-big-to-fail-or-too-small-to-succeed-sobering-stats-on-business-failures/. Accessed 6 July 2014
6. Duncan R (1972) Characteristics of organizational environments and perceived environmental uncertainty. Adm Sci Q 17(3):313–327
7. Flach JM (2012) Complexity: learning to muddle through. Cogn Technol Work 14(3):187–197
8. Keating Charles B (2014) Governance implications for meeting challenges in the system of systems engineering field. In Proceedings of the 2014 9th International Conference on System of Systems Engineering (SoSE), Adelaide, Australia June 9–13, 2014 (pp 154–159). Adelaide, Australia
9. Snowden DJ, Boone ME (2007) A leader's framework for decision making. Harvard Busin Rev 85(11):68–76, 149

10. D'aveni RA (1989) The aftermath of organizational decline: a longitudinal study of the strategic and managerial characteristics of declining firms. Acad Manag J 32(3):577–605
11. Keating CB, Katina PF (2019) Complex system governance: concept, utility, and challenges. Syst Res Behav Sci 36(5):687–705. https://doi.org/10.1002/sres.2621
12. Ackoff RL (1973) Science in the systems age: beyond IE, OR, and MS. Oper Res 21(3):661–671
13. Von Bertalanffy L (1972) The history and status of general systems theory. Acad Manag J 15(4):407–426
14. Espejo R (1994) What is systemic thinking? Syst Dyn Rev 10(2–3):199–212
15. de Geus A (1998) The living company. Wash Q 21(1):197–205
16. Adams K, Hester PT, Bradley JM, Meyers TJ, Keating CB (2014) Systems theory as the foundation for understanding systems. Syst Eng 17(1):12. https://doi.org/10.1002/sys.21255
17. Adams KM (2011) Systems principles: foundation for the SoSE methodology. Intern J Syst Syst Eng 2(2/3):120. https://doi.org/10.1504/IJSSE.2011.040550
18. Kahalas H (1977) Long range planning—an open systems view. Long Range Plan 10(5):78–82
19. Whitney K, Bradley JM, Baugh DE, Chesterman CW Jr (2015) Systems theory as a foundation for governance of complex systems. Intern J Syst Syst Eng 6(1/2):15–32. https://doi.org/10.1504/IJSSE.2015.068805
20. Keating CB, Bradley JM (2015) Complex system governance reference model. Intern J Syst Syst Eng 6(1/2):16. https://doi.org/10.1504/IJSSE.2015.068811
21. de Raadt JDR (1987) Ashby's law of requisite variety: an empirical study. Cybern Syst Intern J 18(6):517–536
22. Conant RC, Ross Ashby W (1970) Every good regulator of a system must be a model of that system. Int J Syst Sci 1(2):89–97
23. Ashby WR (2011) Variety, constraint, and the law of requisite variety. Emerg Compl Organ 13(Nos. ½):190–207
24. Ashby WR (1956) An introduction to cybernetics. Chapman & Hall LTD, London. https://doi.org/10.2307/3006723
25. Beer S (1985) Diagnosing the system for organizations. Wiley, Chichester, West Sussex, New York
26. Beer S (1981) Brain of the firm: the managerial cybernetics of organization, 2nd ed., J. Wiley, Chichester, Eng., New York
27. Sung D, You Y, Song J (2008) The conceptual mechanism for viable organizational learning based on complex system theory and the viable system model. Online submission, ERIC, EBSCOhost. Accessed 11 November 2014
28. Espejo R, Gill A (1997) The viable system model as a framework for understanding organizations. Phrontis Limited & SYNCHO Limited [online]. http://www.phrontis.com. Accessed 11 November 2014
29. Brocklesby J, Cummings S (1996) Designing a viable organization structure. Long Range Plan 29(1):49–57
30. Tomkovick C, Miller C (2000) Perspective—riding the wind: managing new product development in an age of change. J Prod Innov Manag 17:413–423
31. Subramanian R, Fernandes N, Harper E (1993) Environmental scanning in US companies: their nature and their relationship to performance. MIR: Manag Intern Rev 33(3):271–286
32. Fahey L, King WR, Narayanan VK (1981) Environmental scanning and forecasting in strategic planning—the state of the art. Long Range Plan 14(1):32–39
33. Baugh D (2015) Environmental scanning implications in the governance of complex systems. Intern J Syst Syst Eng 6(1–2). https://doi.org/10.1504/IJSSE.2015.068812
34. Charmaz K (2006) Constructing grounded theory: a practical guide through qualitative analysis. SAGE Publications Ltd, Thousand Oaks, Calif
35. Charmaz K (2014) Constructing grounded theory (2nd ed.). Thousand Oaks, Calif, SAGE Publications, Incorporated
36. Charmaz K (2017) Constructivist grounded theory. J Posit Psychol 12(3):299–300
37. Keating CB (2005) Research foundations for system of systems engineering. 2005 IEEE International Conference on Systems, Man and Cybernetics, vol 3, pp 2720–2725

38. De Rooij E (1996) A brief desk research study into the average life expectancy of companies in a number of countries. Stratix Vonsulting Group, The Netherlands
39. Gosnell K (2020) Environmental scanning: how CEOs can stay ahead of the curve and beat the competition. Business.com, Jan. 24, 2020. https://www.business.com/articles/what-is-environmental-scanning
40. Beal RM (2000) Competing effectively: environmental scanning, competitive strategy and organisational performance in small manufacturing firms. J Small Bus Manage 38(January):27–47
41. Mayer JH, Steinecke N, Quick R, Weitzel T (2013) More applicable environmental scanning systems leveraging "modern" information systems. IseB 11(4):507–540. https://doi.org/10.1007/s10257-012-0207-7
42. Ojo O, Abdusalam DO (2011) Scanning the business environment and its implications on organisational performance: a case study. Intern J Inform Busin Manag 3(2):326
43. Wambua J, Omondi M (2016) Factors influencing the environmental scanning of organizations in manufacturing sector: a case study of Kenya Breweries limited. Strat J Busin Change Manag 3(2, Article 11):233–253
44. Heylighen F (1992) Principles of systems and cybernetics: an evolutionary perspective. Cybern Syst 92(1992):3–10
45. Heylighen F, Joslyn C (2001) Cybernetics and second-order cybernetics. In Meyers RA (ed) Encyclopedia of physical science & technology (3rd ed., pp. 1–24). Academic Press
46. Nechansky H (2013) Issues of organizational cybernetics and viability beyond Beer's viable systems model. Int J Gen Syst 42(8):838–859. https://doi.org/10.1080/03081079.2013.777719
47. Pickering A (2002) Cybernetics and the Mangle: Ashby, Beer and Pask. Soc Stud Sci 32(3):413–437
48. Calida BY (2013) System governance analysis of complex systems. Unpublished Doctoral Dissertation, Old Dominion University
49. Carter B (2016) Systems theory based architecture framework for complex system governance. Unpublished Doctoral Dissertation, Old Dominion University
50. Abels E (2002) Hot topics: environmental scanning. Bull Am Soc Inf Sci Technol 28(3):16–17
51. Choo CW (1999) The art of scanning the environment. Bull Am Soc Inf Sci Technol. https://doi.org/10.1002/bult.117
52. Maier JL, Rainer RK Jr, Snyder CA (1997) Environmental scanning for information technology: an empirical investigation. J Manag Inf Syst 14(2):177–200
53. Ackoff RL (1967) Management misinformation systems. Manag Sci 14(4):B-147-B-156. https://doi.org/10.1287/mnsc.14.4.B147
54. Elenkov D (1997) Strategic uncertainty and environmental scanning: the case for institutional influences on scanning behavior. Strateg Manag J 18(4):287–302. https://doi.org/10.1002/(SICI)1097-0266(199704)18:4%3c287::AID-SMJ865%3e3.0.CO;2-B
55. Leonard A (2000) The viable system model and knowledge management. Kybernetes 29(5/6):710–715
56. Milliken FJ (1987) Three types of perceived uncertainty about the environment: state, effect, and response uncertainty. Acad Manag Rev 12(1):133–143. https://doi.org/10.5465/AMR.1987.4306502
57. Samsami F, Khodadad Hosseini SH, Kordnaeij A, Azar A (2015) Managing environmental uncertainty: from conceptual review to strategic management point of view. Intern J Busin Manag 10(7):215–229. https://doi.org/10.5539/ijbm.v10n7p215
58. Saviano M, Di Nauta P (2011) Project management as a compass in complex decision making contexts: a viable systems approach. In Proceedings of the 12th International Conference on Product Focused Software Development and Process Improvement (pp. 112–119). ACM
59. Frazier GL (1983) On the measurement of interfirm power in channels of distribution. J Mark Res 20(2):158. https://doi.org/10.2307/3151682
60. Stanwick P, Cronin JJ, Taylor SA (2015) Environmental scanning, environmental uncertainty, and the capabilities of the firm: a proposed framework. In King RL (ed) Proceedings of the 1991 academy of marketing science (AMS) annual conference (vol. 14, pp. 106–110). Springer International Publishing. https://doi.org/10.1007/978-3-319-17049-7_22

61. Spitz AE, Ludlow RR (2015) Futuristics and innovation in marketing. In Rogers III JC, Lamb CW Jr (eds) Proceedings of the 1983 academy of marketing science (AMS) annual conference part of the series Developments in Marketing Science: Proceedings of the Academy of Marketing Science (pp 153–157). Springer International Publishing. https://doi.org/10.1007/978-3-319-16937-8_36
62. Bryson JM (1988) A strategic planning process for public and non-profit organizations. Long Range Plan 21(1):73–81
63. Clemens R (2009) Environmental scanning and scenario planning: A 12 month perspective on applying the viable systems model to developing public sector foresight. Syst Pract Action Res 22(4):249–274. https://doi.org/10.1007/s11213-009-9127-y
64. Fahey L, King WR (1977) Environmental scanning for corporate planning. Bus Horiz 20(4):61–71. https://doi.org/10.1016/0007-6813(77)90010-6
65. Silverblatt R, Korgaonkar P (1987) Strategic market planning in a turbulent business environment. J Bus Res 15(4):339–358. https://doi.org/10.1016/0148-2963(87)90005-1
66. Bouhnik D, Giat Y (2015) Information gatekeepers—aren't we all ? Inf Sci Intern J Emerg Transd 18:127–144
67. Zheng DE, Carter WA (2015) Leveraging the internet of things for a more efficient and effective military. Rowman & Littlefield
68. Ackoff RL (1974) The systems revolution. Long Range Planning, (December), pp 1–20
69. Ashby WR (1958) Requisite variety and its implications for the control of complex systems. Cybernetica 1(2):1–13. http://pcp.vub.ac.be/Books/AshbyReqVar.pdf
70. Beer S (1979) The Heart of Enterprise. John Wiley & Sons
71. Beer S (1994) Decision and Control: The Meaning of Operational Research and Management Cybernetics. John Wiley & Sons Ltd (UK)
72. Dunagan, J (2012) Designer governance. Futures, 44(9):832–844. https://doi.org/10.1016/j.futures.2012.07.002
73. Keating CB, Katina PF (2012) Prevalence of pathologies in systems of systems. Int J Syst Syst Eng 3(3):243–267
74. Keating CB, Katina PF, Bradley JM (2014) Complex system governance : concept, challenges, and emerging research. Int J Syst Syst Eng 5(3):263–288
75. Kelly EP (1997) The Living Company: Habits for Survival in a Turbulent Business Environment. (Review). Academy of Management Executive Executive, 11(3)95–97. http://proxy.lib.odu.edu/login?url=//search.ebscohost.com/login.aspx?direct=true&db=edsgao&AN=edsgcl.57445340&site=eds-live&scope=site
76. Saxby CL, Parker KR, Nitse PS, Dishman PL (2002) Environmental scanning and organizational culture. Mark Intell & Plann 20(1):28–34. http://www.emeraldinsight.com/doi/full/10.1108/02634500210414747
77. Slaughter RA (2000) A new framework for environmental scanning. Foresight 1(5):387–397
78. Sridharan UV, Dickes L, Caines WR (2002) The Social Impact of Business Failure: Enron. Am J Bus 17:11–22. https://doi.org/10.1108/19355181200200006
79. Stanwick P, Cronin JJ, Taylor SA (2015) Environmental Scanning, Environmental Uncertainty, and the Capabilities of the Firm: A Proposed Framework. In: King RL (Ed), Proceedings of the 1991 Academy of Marketing Science (AMS) Annual Conference. Springer International Publishing, 14:106–110. https://doi.org/10.1007/978-3-319-17049-7_22
80. Thompson KR (2005) Axiomatic theories of intentional systems: Methodology of theory construction. Sci Inquiry J 7(1):13–24
81. Tversky A, Kahneman D (1986) Rational Choice and the Framing of Decisions. J Bus 59(4–2):251–278

82. Walters D, Moorthy S, Carter B (2014) System of systems engineering and enterprise architecture: implications for governance of complex systems. Int J Syst Syst Eng 5(3):248. [online] https://doi.org/10.1504/IJSSE.2014.065755 (accessed 2 November 2016)
83. De Geus A (1997) The Living Company. Harvard Business Press

Communications for Complex Systems Governance

Charles W. Chesterman Jr.

Abstract This chapter will expose the role, nature, and definitions associated with the Communication Mechanisms related to Complex System Governance (CSG). The need and utilization of CSG have been identified and explained in the previous chapters; however, the purpose and role of the Communication Mechanisms will be addressed here. The exploration of the Communication Mechanisms will start with the conceptual foundation of Communications and move on to the nature, structure, and interactions of Communication Mechanisms with influences. This understanding of the conceptual and practical basis will then be followed for the practitioner with vignettes and scenarios to demonstrate the importance of Communications Mechanisms, how to interpret limitations on understanding of their function and the complexities that they attempt to model. Finally, there will be several exercises that can be used to enable the practitioner to identify missing mechanisms or make improvements to current practices in their organization.

Keywords Communication channels · Viable system model (VSM) Communication channels · Complex systems governance · Communication design · Management cybernetics

1 Introduction

At the same time, as "systems engineering grew out of engineering in the 1940s and 1950s" [1, p. 48], work was being accomplished in communications. C. E. Shannon in his journal article "A Mathematical Theory of Communication" points out that the "fundamental problem of communication is that of reproducing at one point either exactly or approximately a message selected at another point" [2, p. 623]. Shannon described communication as a system containing five parts: *Information source*—produces a message or sequence of messages to be communicated to the receiving terminal, *transmitter*—which operates on the message in some way to

C. W. Chesterman Jr. (✉)
Old Dominion University, Engineering Management & Systems Engineering, Norfolk, USA
e-mail: cchester@odu.edu

C. B. Keating et al. (eds.), *Complex System Governance*, Topics in Safety, Risk, Reliability and Quality 40, https://doi.org/10.1007/978-3-030-93852-9_11

produce a signal suitable for transmission over the channel, *channel*—the medium used to transmit the signal from the transmitter to receiver, *receiver*—performs the inverse operation of that done by the transmitter reconstructing the message from the signal, and *destination*—is the person or thing for whom the message is intended.

Subsequent work on communications such as the works of [3] The Two-Step Flow of Communications: An up-to-date report on a hypothesis; [4] toward a behavioral theory of Communication, [5] organization structure and Communications, [6] A Transactional Systems Model of Communications: Implications for Transactional Analysis' and [7] Verbing Communication: Mandate for Disciplinary Invention has expanded Communication of Shannon into Communications Theory and Information Theory. Importantly, Losee states "when communication is defined in terms of informative processes, one can study both the information that is conveyed and the processes that carry it. Definitions of communication often involve terms such as knowledge, belief, meaning, or intention" [8, p. 2]. Losee makes the case for a process model that is comprehensive, in that it deals with both the process of communicating, the content of the communication and what can occur when the communication is received. Craig in his work on Communication Theory, summarizes that with respect to cybernetics "in contrast to other traditions of communication theory, cultivates a practical attitude that appreciates the complexity of communication problems and questions many of our usual assumptions about differences between human and nonhuman information-processing systems" [9, p. 142]. From the perspective of Communication theory, information theory, and cybernetic theory, a synthesis of what the communication process accomplishes can be simply stated as; communication is taken as the process by which meaning is assigned and conveyed to create a shared understanding.

Complex System Governance (CSG) as described by [10] in their paper *Complex system governance: concept, challenges, and emerging research* are built on System theory and Management Cybernetics and incorporate as one of their cornerstones the metasystem as described in [11] Viable System Model (VSM). Communication as one portion of Management Cybernetics (communication and control for effective system organization) provides for "the flow and processing of information within and external to the system that provides for consistency in decisions, actions, and interpretations made with respect to the system" [12, p. 265]. Nyström points out that the VSM "has been used for diagnosing different kinds of organizations at different levels where its use highlights existing or missing communication patterns and information flows in different communication channels and relates findings to a viable system" [13, p. 523].

The development of Complex System Governance (CSG) works on addressing "three primary shortcomings in addressing modern complex system problems" [12, p. 226]. Firstly, there are complex systems with "constituent problems, requiring inquiry and solutions that lie beyond the limited grasp of technology-centric approaches" [12, p. 226]. This is especially true for complex systems with problems for which "solutions must cross the entire spectrum of organizational, managerial, human, social, policy, and political dimensions" [12, p. 226]. Secondly, current system-based solution sets "have not yet managed to bridge the divide between the hard, technical, objective-based aspects of complex systems and the soft, non-technical, subjective aspects" [12, p. 226]. Finally, the "landscape for modern systems has changed appreciably into a much more complex problem space" [12, p. 226]. As many readers can appreciate, the landscape includes "difficulties encountered across the holistic range of technical, organizational, managerial, human, social, information, political, and policy issues" [14, p. 2944].

The described metasystem functions account for system performance by purposeful development of control (constraints necessary to ensure consistent performance and future system trajectory), communications (flow and processing of information necessary to support consistent decision, action, and interpretation throughout the system), coordination (providing for effective interaction to prevent unnecessary oscillations within and external to the system), and integration (maintaining system unity through common goals, designed accountability, and balancing system and constituent interests [12, p. 265]. This purposeful development of control, communications, coordination and integration will find that the core attributes of Control will be Identity and Communications. The Complex System Governance (CSG) Reference Model also known as the Metasystem Governance Reference Model has nine metasystem functions included in the metasystem. From the CSG Reference Model, the information and Communications (M2) function, primary responsibilities, and product descriptions are listed in Table 1.

The development of information theory by Shannon when dealing with physical systems has progressed and works very well as new technologies have been introduced as communication systems. The work on CSG has been described as the

Table 1 Information and communications (M2)

Function	• Designs, establishes, and maintains the flow of information and consistent interpretation of exchanges (through communication channels) necessary to execute metasystem functions
Primary responsibilities	• Designs and maintains the architecture of information flows and communications within the metasystem, between the metasystem and environment, and between the metasystem and the governed system • Ensures efficiency by coordinating information accessibility within the system • Identifies standard processes and procedures necessary to facilitate transduction (Note: See Sect. 3 for Transduction write up) and provide effective integration and coordination of the system • Identifies and provides forums to identify and resolve emergent conflict and coordination issues within the system
Products	• Standard processes and procedures for internal coordination of the system • Communications architecture for the metasystem • Defined external coordination vehicles necessary for support for the system (*e.g.*, public relations, press releases)

Drawn from: Metasystem Governance Reference Model, National Centers for System of Systems Engineering, Old Dominion University, C. Keating, 11/19/2014 "design, execution, and evolution of the metasystem functions necessary to provide control, communication, coordination, and integration of a complex system" [10, p. 264]. The evolution of CSG and the contents of this book the reader will find is a holistic approach focused on the metasystem. As described in Complex System Governance Metasystem functions, the metasystem construct only defines *"what"* must be performed to maintain system viability (existence). It does not specify *"how"* a system is configured, or what devices (mechanisms) the system implements to achieve the metasystem functions [12, p. 228].

Drawing from Management Cybernetics and the VSM are a set of communication channels that provide specific functionality for the metasystem functions. Table 2 below provides a summary of communication channels from several works articulating Beer's VSM [11, 15–20] and supplemented by [21]. The "how" relative to communication will be discussed in the next section.

2 Defining Communication for Complex Systems

As earlier indicated, communication is taken as the process by which meaning is assigned and conveyed to create a shared understanding. Communications and the channels of communication, whether it is between two individuals or the members of an organization or a society in general, have a single functionality. The number of individuals that are senders or receivers does not change that singular functionality. Likewise, the content or the package that was developed by the sender and intended for the receiver still is in support of this singular functionality. With respect to communications, there are a set of Communication Mechanisms or concepts (direction, mode, product, and technology (conveyance)), as drawn from the research, that operate in the process by which meaning is assigned and conveyed to create a shared understanding. This shared understanding meets the purposeful design of the Metasystem needs of Variety Attenuation, Variety Amplification, and Transduction. The concepts are not independent of each other and individually or independently do not answer the who, what, when, and how questions of communication. Secondly, if one were to start with one of the concepts the reviewer would need to go to the other concepts to achieve any degree of understanding that they interactively bring to communications. While this interrelationship of the concepts would appear to be like a software program with a "continuous do loop" and never achieving a result, a better understanding of cybernetics will lay a richer foundation so that the reviewer may successfully migrate among the concepts and achieve a foundational understanding.

Jackson [1] in discussing systems points out that the concept of *variety* is a depiction which indicates the number of states that a system can exhibit and "according to Ashby's law of requisite variety, systems can only be controlled if the would-be controller can command the same degree of variety as the system" [1, p. 9]. It was Norbert Wiener in 1949 that coined the term "cybernetics" formed from the Greek for *steersman* "for the entire field of control and communication theory, whether in the machine or in the animal" (Wiener [22], p. 19). In the twenty-first century,

Table 2 Communication channels

Communicationchannel	Primary functions
Command	• Provides direction to operational units • Dissemination of non-negotiable direction to the system
Resource bargain/accountability	• Provides/determines the resources (manpower, material, money, information, support) for operational units • Defines performance levels to which operational units will be held responsible Determines how operational units will interface for performance reporting and accountability
Operations	Provides for the routine interface between operational system entities and from the metasystem to operational units
Coordination	• Provides for system balance and stability by ensuring that information concerning decisions and actions necessary to prevent disturbances are shared among operational units
Audit	• Provides routine and sporadic feedback on the performance of system operations • Investigates and reports on areas on problematic areas
Algedonic	Provides instant alert to crisis or potentially catastrophic situations occurring in the system bypasses routine communications channels and structure to identify system threats
Dialog	• Provides examination and interpretation of organizational decisions, actions, and events • Seeks alignment of perspectives and shared understanding of organizational decisions and actions in light of system purpose and identity
System learning	• Provides detection and correction of system errors, testing of assumptions, and identification of system design deficiencies • Ensures that the system continually questions the adequacy of its design
Informing	• Provides routine transmission of information throughout the system • Routes information that is not appropriate for other channels for accessibility throughout the system
Environmental Scanning	• Provides for designs, deployment, and monitoring of the environment • Enables capture, processing, and accomplishing analysis to form patterns, trends, and determine increase/decrease in relationships between an organization and the external environment • Supports determining potential implications for both present and future system viability

the term is often used in a rather loose way to imply "control of any system using technology." In other words, it is the scientific study of how humans, animals, and machines control and communicate with each other.

The work of Stafford Beer moved Cybernetics from the control and communications of Wiener to organizational cybernetics or variety engineering. Beer presents the Viable System Model (VSM) with the book *Decision and Control* [23], followed by *The Heart of Enterprise* [11] and finally *Brain of the Firm* [15]. The VSM reflects Beer's neuro-cybernetic model that, with its five subsystems, imitates the human brain and body and their functional requirements. Like the body, "viable systems maintain equilibria behavior only by multiple contact with whatever lies outside themselves" [23, p. 257]. This ability to contact is a principal function of a channel of communication. But not only is there contact, and while it is complex,

It is characteristic of a viable system that all its parts may interact; not indeed to the extent that all possible permutations of all possible parts with all other possible parts must manifest themselves, but to the extent that subtle kinds of interaction drawn from all these permutations can and do take place [23, p. 257]

This does not overwhelm the system because there is control as Variety Engineering provides. With continuous interaction between the five subsystems, Beer draws upon biology and the process of homeostasis, through which control and equilibrium are achieved. With respect to variety, Beer indicates that "ONLY variety absorbs variety" [11, p. 89] and that this law of requisite variety is accomplished because it is required by nature. Accordingly, it also means that when systems are designed, there needs to be mechanisms of amplification (projection) and attenuation (filtering) of variety included.

Figure 1 below graphically shows Beer's First Principal of Organization [11, p. 96], displaying the relationship between the management unit, the operational unit that is regulated by the management unit and the environment for the operational unit. As there will be transmission of variety between all three elements via a channel

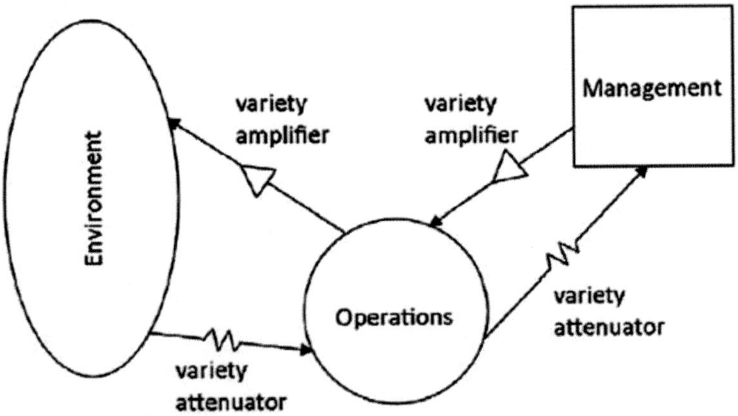

Fig. 1 Relationship of units and channels of Communication (adapted from [11, p. 96])

of communication, with the proper design of amplifiers and attenuators, there will be diffusion and equivalency of variety over time. Unlike natural systems, "it is management's job to DESIGN the necessary amplifiers and attenuators" [11, p. 97] as the engineering of variety in a complex system. The next section will describe the communication mechanisms.

3 Communication Mechanisms

With respect to communications, there are a set of Communication Mechanisms or concepts (direction, mode, product, and technology (conveyance)), as drawn from the research, that operate in the process by which meaning is assigned and conveyed to create a shared understanding. The rest of this section will discuss each of these mechanisms as well as other cybernetic terms that influence communication.

- **Direction**

Communication is framed as a minimum of two participants and associated with the participants is a channel of communication or some type of conveyance. The works of: [2] A Mathematical Theory of Communication, [11]. The Heart of the Enterprise, and [24] Metasystem Governance Reference Model (MGRM) for Complex Systems and many others articulate this communications framework. Within this framework, the participants are identified with the roles source and receiver. When the process of feedback is added between the same set of participants, the roles of the participants are exchanged.

The source will always be associated with the origination/creation/designing or establishment/mandating a packet of information. The receiver will always be the intended receiver of the packet of information. In this framework, the packet of information, to be fully described in the section on product, will always be originated/created/designed or established/mandated for the use of the receiver. While it is possible that the packet of information may be received by others, the design by the source is always intended for the receiver. The receiver having acquired the packet of information takes the action intended because of the design and use of the channel of Communication concepts by the source. Accordingly, a direction convention that the packet of information always is created by the source and then conveyed to the receiver (direction is ***from*** **-** ***to***).

Contained in the mechanism of direction is an element of time. Wiener in discussing the reception of sense organs (receivers in the nervous system), states that there is a "input–output is a consecutive one in time and involves a definite past-future order" (Wiener [22], p. 55). As will be discussed in following sections, development of direction for the source and receiver has a specific structure and all models of communication including Shannon's have a direction. While it is true that two individuals may meet and start an interaction or conversation, they do so because one of the two individuals initiates an interaction. The motive behind the initiation is always associated with the source.

- **Mode**

The technology associated with Communication has significantly progressed from what could be recognized as original communication capabilities to the current spectrum of communication capabilities. While Shannon was dealing with telecommunications (telegraph, telephone, television, telephony, teletype, and telegraphy) which was concerned with electromagnetic signals, there are many communication capabilities associated with a channel of Communication that are not electromagnetic. Such terms as auditory (hearing), balance, biochemical, electromagnetic, haptic, kinesthetic, olfactory (smell), pain, tactile (touch), taste, temperature, or visual (sight) reflect the senses that humans have.

The mechanisms of mode addresses questions, such as how does the receiver receives the packet of information, can the same packet of information be constructed as different packages such as a written report or could it be a verbal report or both? Additionally, does the direction of the packet of information affect how it was received? Lastly, within the context of Variety Engineering, would the design associated with Variety Engineering have an impact on the optimum construct for a packet of information?

These questions are centered on how the packet of information was packaged for the receiver. Examples are that the packet of information such as a document was provided to the intended party (receiver) or that the document was part of an agenda for a face to face meeting. Looking at this scenario, one can realize that the packet contained text or something written (nonverbal) and written and presented (nonverbal and verbal). Looking back on the development of Communications theory by Shannon, the choice of mode by the source addresses concerns voiced by Weaver with respect to Shannon's Model of Communication,

> "LEVEL B. How precisely do the transmitted symbols convey the desired meaning? (The semantic problem.)
> LEVEL C. How effectively does the received meaning affect conduct in the desired manner? (The effectiveness problem.)" [25, p. 2].

where the mode chosen enables the receiver to receive the desired meaning of the packet of information and will take the appropriate action. While it is recognized that there can be a breakdown in the channel of Communication, to be addressed in the technology (conveyance) section, mode helps reveal part of the purposeful design associated with Communications and channels of Communication supporting CSG. The choice with which mode to use depends on the motive and intent of the source and on when to initiate communication. Secondly, with respect to increased technological advances instead of a single option for technology (conveyance), there are many modes that can be used. Finally, if there was a "blind" person as the expected receiver, the source chose not to use a visual mode but rather both tactile and verbal modes. Again, there is an interrelationship between the Communication Mechanisms.

- **Product**

Communication as described in this chapter is the process by which meaning is assigned and conveyed to create a shared understanding. Previously, the term message or package of information was presented as the result of developing meaning by the source. Shannon identified the generation of the associated package of information with the source. Shannon's model of Communications did not include feedback, but Communication theory has advanced to the transmission model or standard view of Communication, and finally, the transaction model correcting the issue of feedback. The transaction model has a basic premise of individuals (source/receiver) simultaneously engaging in the sending and receiving of messages so that the "message" may not be the sole creation of the source but can change depending on other contributors/individuals or other influences. An example of many influences or contributions is the advertising industry where the intended receiver is "targeted" in order to ensure that receiver, having acquired the packet of information, takes the action intended because of the design and use of the channel of Communication concepts. Literature on the advertising industry speaks to the term of "messaging" and where one finds the model of message creation being likened to the lifecycle of a physical product (e.g., car). Rather than thinking solely in terms of a package of information the term product moves the concept of "meaning" (formally called package of information) to the contextual level of a Communication Mechanism.

Unlike direction or mode, products have a specific structure in their development, and while it may directly be part of a life cycle, there are many cases where the meaning that is assigned and conveyed can have a life cycle in its development. This perception may not be obvious immediately, but to consider the effects of variety, there is a correlation. The following quote from Beer's *Designing Freedom* provides an example of communication interrelated with variety but that the communication is also a process.

But not for nothing is that store called departmental. There is a shoe salesman and a cake salesman that is what organizational structure is for to carve up the total system variety into subsystems of more reasonably sized variety. ... But if the store is careful, it will have an information bureau—which exists precisely to absorb this excess variety. [26, p. 8]

The above example lays out the effect of a business purposely organizing itself and communicating to the environment (public) and the business organization structure (internal) by departments. This conveyance to the public and its shared understanding of where to find goods and services (business departments) also demonstrates that communication is not necessarily an instantaneous event and this is especially true with respect to variety attenuation and variety amplification.

Most products can be related with one or more modes. As an example, a written document can be read (not vocal) by the receiver, a written document can be delivered as a speech (vocal) by the source to a group (receiver). Same product with but two different mode associations. For this example, the direction in both cases was from a source to the receiver, but the product could have been developed by staff (source) and given to the program manager (receiver), and then, the program manager (source) delivers at a professional gathering of peers (receiver). The creation of the product has associated with it motives and intent like mode. As the source was responsible for both the choice of product and the mode, it can be assumed that motive and intent are solely with the source and the product and mode reflect the development from a design. As always, the product is required to create a shared understanding by the receiver.

The creation of a shared understanding has the implication that the language or culture is likewise shared between the source and the receiver. Most of the readers of this chapter understand that their industry, their organization or their city contain associations with language or culture that in some cases are unique or are shared. For those that served in the military, there are words or expressions that are unique to a specific branch. There is significant research dealing with speech communities that are the subject of Ethnography of Communication.

The idea of intent or motive by the source in formulating the design of the product has been mentioned, and it is the intent or motive that resides with the source that is the driver or the pretext for communication and the context of what was being accomplished. The creation of a shared understanding is to achieve a goal, where for the most part the formulation of the product was accomplished through bargaining and incorporated as a dialectic process to achieve a completed entity. This is consistent with speech communities that continually discover and exchange new speech where Sennett says that as a dialectic process "with the explicit meaning of statements and tends to lead to closure and resolution," (Sennett [27], video: see 18:30–30:00) the speech community achieves closure and a new formulized speech.

There are writings that indicate that the receiver may be active or passive, where receiver being active when the product (packet of information) is consumed. Receiver passivity could imply that the product is received, and no action is taken or that there is no receiver. Weaver wrote on two problems, "how precisely do the transmitted symbols convey the desired meaning" and "how effectively does the received meaning affect conduct in the desired way" [25] p. 2). With respect to both problems, the actions of the source are heavily invested in forming a product that the receiver expects and knows what to accomplish or that the receiver receives and takes action without a prior expectation.

A Communications for Complex System Governance Vignette—Message on a Beach

The message on the beach provides an excellent example of the issue of active or passivity of the receiver. The scenario is that the survivors of a shipwreck have organized themselves to obtain and move brush or debris to an area above the high water mark. Here, the group has used the material to form piles on the beach in the construct of large size letters that read SOS. Rather than creating lines or arrows or some other set of symbols, the group has created a product with a single meaning. In this vignette, the design of the communication is purposeful, where the direction (from-to) is the survivors to air crews. While the air crews may vary from location to location and day of the week, they are the intended recipients of the product. This is similar to the example of the department store in two specifics; firstly, the communication may not be instantaneous. Secondly, the receiver is trained to have a shared understanding of the symbology of SOS. The mode is nonverbal for this vignette and while the delivery of the product may appear to be passive the receiver will be active when the product is consumed.

- **Technology (Conveyance)**

There is a large body of literature that speaks to channel of Communications, and there are many examples of specific type of channels such as Algedonic, Coordination, Anti-Oscillation or Coordination, Command, Audit, Resource Bargaining, Accountability, Homeostat, Policy Intervention, Resource, and Provision. This could lead one to the question what are the number of channels required, or would a channel by its characteristic determine what could be conveyed or what is the composition or technology used for a channel of Communication. These questions directly affect the functionality of Communications with respect to CSG.

The use of the Beer VSM terms is indicative of the source's intent, to be discussed later, and underlies the design involved with creating the product more so than the specific mechanism of how a channel accomplishes the conveyance. An excellent example is the Algedonic Channel where the intention "to quickly convey information in the event of emergency or failure in the (M2-M3- M3*-M4) management system (an organizational "override" channel)" (O'Gradey [28], p. 5). The necessity of override is easily met with current electronic technology and could enable the M1 (productive function) to communicate to the M5 (policy and identity function); however, as a practice this need shows a failure in mode design or pathologies associated with direction, mode, product, technology (conveyance). Failure of mode design would be to use non-electronic technology such as face to face meetings, library of documents and personnel changing location as their failure suggests no accommodation of an "emergency." This lack of accommodation has everything to do with poor design rather than the abilities of a special Algedonic Channel by itself to achieve the shared understanding.

The technology (conveyance) of the channels of Communication is not all an electronic conveyance. As most readers are aware, there is a wide spectrum of conveyance such as face to face meetings, library of documents and even personnel changing location to achieve communication. The choice of technology is left to the source using capabilities. Within the context of current technology, the diagram of Fig. 2 could be reshaped into a network model of one function to many functions. A good example is a Web service and email where an email is generated by one function and sent to one or more different functions at the same time. This example complies with the communication direction of "from-to" and by design in support of the product, the email is to select function/s. A slightly different example is broadcast, which conforms to direction, is one to many but there is the desire for less control on the receivers. The process of advertising though would indicate that there is considerable work in product generation so that the receivers are targeted, a form of selection of receivers. The channels of Communication and support of Complex System Governance will now be discussed.

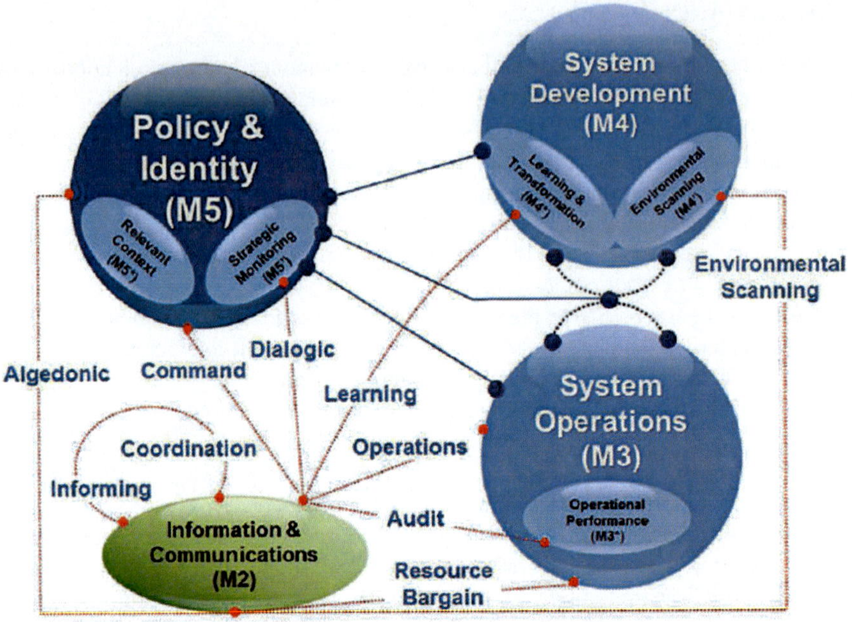

Fig. 2 Relationship of units and channels of Communication (used by permission of C. Keating)

4 Channels of Communication and Complex Systems Governance

The nine interrelated CSG functions, that form the metasystem, are shown in Fig. 2. For this representation, there are single red dashed lines for channels of communication all labeled with the terms of Algedonic, Coordination, Anti-Oscillation or Coordination, Command, Audit, Resource Bargaining, Accountability, Policy Intervention, Resource, Environmental Scanning, and Provision similar to the Beer VSM as well as some that are not labeled in blue solid lines (M5 to M4).

The effort associated with the design of the channels of communication is not treated consistently in the literature on systems theory. It is only with the emergence of CSG and the interrelationship of the CSG functions that a focus on the criticality of communication design has been recognized. The work by [29] into identifying what are identifiable pathologies for the metasystem for complex system governance has helped highlight the importance of communication design. Part of the lack of awareness is that most individuals are not involved with the creation of a new complex system, rather they enter into an existing complex system and are not sufficiently aware of systemic issues. Table 3 below elaborates the critical role of communication for each of the Meta System Functions.

The issue of building channels of communication as original design or what to do if the lack of a channel is determined and is significantly important. Most literature

Table 3 Meta system functions and communication channels

Meta system	Nature and role	The role of communications be elaborated for each of the functions to demonstrate the criticality of communications for the function
Metasystem Five (M5) Policy and Identity	Corresponds to System 5 in the VSM metasystem. Focused on overall steering, giving direction and identity for the system	Strong channels to insure: • accurate distribution of complete Product • timely distribution of product • maintenance of linkage to all receivers • maintain organizational standards and Identity
Metasystem Five Star (M5*) System Context	Elaborates a responsibility within the VSM System 5 Focused on the specific context within which the metasystem is embedded	Strong channel to insure: • proper storage and processing for distribution of Product
Metasystem Five Prime (M5') Strategic System Monitoring	Elaborates a responsibility within the VSM System 5 Focused on oversight of the system at a strategic level	Strong channel to insure: • proper handling of product exchanges • proper storage of product
Metasystem Four (M4) System Development	Corresponds to System 4 in the VSM metasystem Focusing on the long-range development of the system to ensure future viability	Strong channel to insure: • accurate distribution of complete Product • proper capture of product • Proper storage and distribution of product
Metasystem Four Star (M4*) Learning and Transformation	Elaborates a responsibility within the VSM System 4 Focused on facilitation of learning based on correction of design errors in the metasystem and planning for transformation of the metasystem	Strong channel to insure: • proper handling of product exchanges • proper storage of product
Metasystem Four Prime (M4') Environmental Scanning	Elaborates a responsibility within the VSM System 4 Focused on sensing the environment for trends, patterns, or events with implications for both present and future system performance	Strong channel to insure: • proper handling of product exchanges • proper storage of product • proper interface with external environment

(continued)

Table 3 (continued)

Meta system	Nature and role	The role of communications be elaborated for each of the functions to demonstrate the criticality of communications for the function
Metasystem Three (M3) System Operations	Corresponds to System 3 in the VSM metasystem Focused on the day to day operations of the metasystem to ensure that the system maintains performance levels	Strong channel to insure: • accurate distribution of complete product • capable of handling large volume of product exchanges
Metasystem Three Star (M3*) Operational Performance	Corresponds to System 3* in the VSM Focused on monitoring system performance to identify and assess aberrant conditions	Strong channel to insure: • proper handling of product exchanges • proper storage of product • capable of handling large volumes of product exchanges
Metasystem Two (M2) Information and Communications	Elaborates the System 2 function in the VSM Focus on the design for flow of information and consistent interpretation of exchanges (communication channels)	Designed for large product exchanges and clarity of product

Adopted from [24]

treats communications from the "as built" state and does not address original design and building the channels of communication as well as what would be the future condition and how to bring about change. This area will be discussed as part of the use of an assessment tool. The good news is that with respect to a life cycle view of channels of communication, the advances in technology have increased the options of conveyance and decreased concerns and issues relative to the receiver not adequately developing an understanding of the product. Secondly, as in advertisement, the model of communication has moved to participatory communications where the product is interactively developed. Another example of participatory communications can be seen in software development using the Agile Methodology (*Manifesto for Agile Software Development, 2001)* where an application is developed in short bursts, released prior to final testing (beta format) for the user to use and comment on. These comments on the beta product are absorbed with internal directions for the next iteration.

The conveyance of the product may be limited by the composition of the channel of communication or it may be the construct of the channel of communication that may enhance the spectrum of products that are conveyed. Channel capacity was at the heart of Shannon's work and as Reissberg points out, "it is important to note that the provision of channel capacity depends to a high degree on technology" [30,

p. 42]. With a perspective that the channel of communications can be disrupted or can have saturation in product, the consideration of channel capacity in a complex system makes the issue part of the design sequence or a sub-element of the design methodology and will be discussed as part of the use of an assessment tool.

Channel composition or construct has a direct relationship to the mode, direction, and product. The intelligent design by the source consists of integrating direction, mode, and product dependent upon the intended meaning and equally the intended receiver. When the desired result is not achieved, the source makes modifications to the communication mechanisms used, or if there is change in the environment associated with the complex system; likewise, the source makes modifications. An extreme example is personnel changing location, either to affect Variety Attenuation, Variety Amplification, or bring new leadership or management. It is clear that, either because of environmental, technological, or personnel changes, communication channels are not static, but always evolving dependent upon the requirements of the complex system (organization). Besides the aforementioned Mechanisms of Communications, there are other impacts on Communication which will now be discussed.

- **Variety Attenuation and Variety Amplification**

Beer's first principle of organization highlights the need for regulation. The intent of Complex System Governance is to provide to the observer of a complex system of interest the lenses to understand this regulation and the mechanisms of Variety Attenuation and Variety Amplification. The intent of purposeful design is that the transmission of variety between all meta functions, as well as the interface with the environment, that with proper design of amplifiers and attenuators, there will be diffusion and equivalency of variety over time. Unlike natural systems, "it is management's job to DESIGN the necessary amplifiers and attenuators" [11, p. 97] as the engineering of variety in a complex system.

Associated with each channel of communication, there are the mechanisms of Variety Attenuation and Variety Amplification. The Communication Mechanisms that are associated with Variety Attenuation and Variety Amplification are mode, product, and technology (conveyance) and it is the use of these mechanism/s that enable Variety Attenuation or Variety Amplification. Variety Amplification is not restricted to a specific direction between CSG functions nor for any identified channels of Communication the mechanisms of Variety Attenuation and Variety Amplification are always present. This clearly provides great flexibility for the source when going about communication design. When discussing channel capacity, the technology used has a direct effect. If there were several different channels connecting functions and not knowing which channel had a higher capacity or rate, then a general communication design would favor greater amounts of Variety Attenuation.

- **Transduction**

Transduction is the translation of information across the boundaries of systems where [11] described this important function "Transduction" of bringing stimulus into a system. Beer captured this in his third principle of organization indicating the capacity of transduction with respect to variety, "Wherever the information carried on a channel capable of distinguishing a given Variety crosses a boundary, it undergoes Transduction; the Variety of the transducer must be at least equivalent to the Variety of the channel" [11, p. 101]. When considering control "autonomic control must correct imbalances to the internal environment; the first necessity is to detect the change; receptors then alter their state, transducing the change into efferent impulses which then go to the control center" [15, p. 103]. Beer noted that "System four is the innovation generator that uses existing channels and transducers through which to stimulate and interrogate the problematic environment" [11, p. 238]. Beer does not indicate that transduction is any less important for the S1 than the S4, nor is there an indication that the makeup of the mechanism would be different. The role of Transduction is equally as important in CSG. A good example of the description of Transduction is provided by Espejo,

> That communications between agents and actors need transducers. Transducers are media that transform signals from one expression into another expression that is more appropriate to the receiver. They are necessary every time that signals cross a boundary; they change an ontology into another making signals more meaningful to receivers. A decoder alters the input code into internally meaningful code, and an encoder alters the output code into externally meaningful code [31, 32, p. 1023]

When Beer was discussing Variety Amplifiers and Variety Attenuation, "when they are not designed, they simply occur because Ashby's law asserts itself" [11, p. 92]. The work of [33] points out several cases where "facing failed design, requisite variety asserts itself in other ways so that Ashby's law always holds and varieties are balanced" [33, p. 565]. Unlike Variety Attenuation and Variety Amplification, none of the literature on the VSM? implies that Transduction creation would be part of emergence or the balancing of variety. From a cybernetic perspective, the absence or non-operation of Transduction is a pathology that is part of a failure in design of the channel of communication.

Understanding the requirement for Transduction, one might question if a mechanism of Transduction is relevant with respect to channels of communication design, particularly, with respect to current technology, or the possibility that the design of channels of communication having reached stability and maturity with respect to Variety. Transduction is most important when there is an interface with the environment as well as use of two technologies. Thus, Transduction can always be considered as an integral part of the communication design process.

- **Identity**

The previous sections have described the Communication Mechanisms as well as Variety Attenuation and Variety Amplification. The Complex System Governance has a foundation in Systems theory and as identified by [34] an [35] where there are a set of axioms and associated propositions (principles) that seek to describe the behavior of systems. This set of axioms and propositions speak directly to Communications as well as to Identity. As all systems are unique, Identity should be seen as having a direct impact on the purposeful design of the system and its channels of Communication. As the Communication Mechanisms provide the service for communication and creating a shared understanding, Identity should be considered as impacting upon the selection, design, and implementation of Communication Mechanisms.

Collected in Table 4 below are a set of statements from literature with respect to Identity.

What the reader will notice is that Identity does not have a single definition, rather there are a set of terms such as primary activities, persistent structure, purposes of a system, relative to system boundaries and environment, accommodates attributes (beliefs, values, motives, and experience), and Identity is communicated internally for operation and externally additionally for messaging. The nature of Identity is dynamic and evolves interactively due to external and internal changes. The reading of the source material with respect to Identity finds that Identity and Communications are the core attributes of Control.

With respect to channels of communication, the Identity of the Complex System needs the channels of communication to support the primary activities for which the entities in the system respond to the system inputs as well as convey the output. The

Table 4 Identity

Identity is the collection of primary activities of a viable system [36, p. 110]
Sustaining a coherent identity supports consistent decision, action, interpretation, and strategic priorities [10, p. 269]
Identity is the persistent structure of the organization (measure of identity) [37, p. 60]
Identity of the organization can be expressed in terms of the purposes it is to pursue [1, p. 89]
Identity must express and represent the purposes, but, obviously, should not be the sole repository of identity [1, p. 89]
The identity derived from purposes need to be derived considering the state of the organization's environment and the opportunities and threats that exist [1, p. 89]
Professional identity accommodates attributes, beliefs, values, motives, and experiences in terms of which people define themselves in a professional role (Schein 1978, from [38, p. 229])
Organizationally professional identity is seen to evolve interactively with role change (Ashforth and Saks 1995). (from Khuong et al. [38, p. 229]
The collective message conveys an organization identity through every form, manner, and medium of communication to the respective stakeholders (Mohamad et al. [39], p. 117)
A business has relationships with stakeholders in its environment. These relationships are necessary for the business to maintain its identity as distinct from other businesses. Maintaining a separate identity defines a business' success and survival (Regev et al. [40], pp. 696–697)
The number of norms that a business maintains is very large. Examples of such norms are the stability of a business' name, its reputation, its revenues, its profits, its number of employees, etc. The norms maintained by the business define its identity. A norm is stable but not necessarily static. It may change over time as the business adapts to its environment, for example, when the revenues grow as the business adapts to a growing market (Regev et al. [40, p. 697]
Once the boundaries of the organization, along with its identity and purpose, have been clarified, the next step is to identify the relevant environment where our organization carries on its activities (Rios [41], p. 1535)

achieved or designed structure of the channels of communication provides a persistent structure that actively supports the selective purposes of a system. The channels of communication have the interface with the system boundaries and the external environment. The system is dependent upon information from the environment as it is "in the now" and information that can impact and shape the environment "in the tomorrow." These influences of information, while they will evolutionarily modify the Identity, exist on a time scale that is subject to the nature of the system, with the internal changes occurring at a different rate due to beliefs, values, motives, and experience of the individuals associated with the metasystem functions. Finally, the information generated internally as well as from external sources is conveyed externally to reflect a messaging of the systems Identity.

This background on Identity brings forth a similar conundrum as Transduction. The relationship of Identity with the metasystem function of M5: Policy and as such Identity (intent/motive) is similar to Transduction as part of the underlying influences on the channel of Communication Design concepts.

A Communication in CSG Vignette—Message in a Bottle or Message on a Record

Some friends of mine found a bottle in their house that contained a piece of paper. Upon retrieving the piece of paper and examining it, they found that with time the writing was now undiscernible and there was an absence of any symbology. From what you have read was this communication? If you thought yes, then you may want to reconsider the write up on product and technology (conveyance). So how do you produce a product, what mode and technology (conveyance) does one use if the product is not intended for earth and is to pass more than 1.6 light-years into the future. The story of the Voyager Golden Records describes the purposeful design to carry sounds and images selected to portray the diversity of life and culture on Earth. The mission flight path was designed to travel beyond the solar system well into interstellar space. The Golden Record was designed to use the basic underlying principles of physics and mathematics to explain how to construct a device for playing the record, to identify the origin of the spacecraft and record relative to 14 pulsars and again with instructions to help establish when the record was created by using a ultra-pure source of uranium-238 that has a steady decay rate to enable the finder the ability to calculate the time elapsed since a spot of uranium was placed aboard the spacecraft.

5 Developing Individual Communication Relationships and Communication Mechanisms

The generation of a clear picture of the operations of the communications mechanisms for a complex system of interest can be accomplished through a variety of techniques. Borrowing from the Design thinking process the lifecycle of the communication channels can be designed and or discovered using a process that is highly iterative and involves moving back and forth—framing and reframing problems as shown in Fig. 3 below within the following categories:

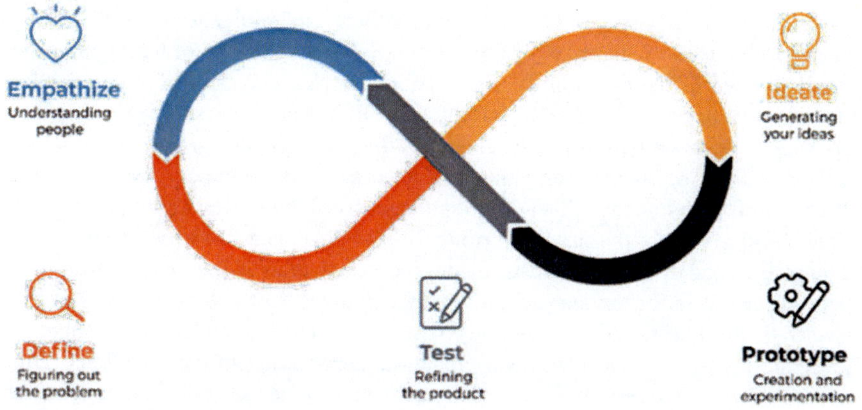

Fig. 3 Design thinking process

- Empathy - ask questions of the people in the problem space to understand their needs
- Define—use the information to develop insights, draft use cases, and establish a point of view.
- Ideate—brainstorm a myriad of ideas and suspend judgment. Quantity is encouraged.
- Prototype—build a rough but tangible sketch, model, or functioning apparatus
- Test—get feedback from real users

An area that is difficult for many practitioners is the empathy stage and a useful technique is the use of a survey instrument that is based upon Communication Mechanisms. The use of the survey instrument enables practitioners to construct and identify communication mechanisms with respect to a channel of communications. While the system of interest may identify with one or more channels of communication, the use of the survey instrument will support the identification and organization of the channels of communication. This understanding provides the basis for a more informed design, an assessment as well as the means to be able to center the system of interest in a framework such that changes may be identifiable.

Survey participation is recommended not to be confined to only a selected area of a Complex System. It is recommended to be given to a wide spectrum of individuals that participate in the system of interest, as well as the stakeholders either interior or exterior to the system of interest. Likewise if possible, it is recommended that participation be extended to individuals external to the system of interest that has a relationship to the system of interest.

It is recommended that the survey instrument design be organized so that the survey respondents will provide descriptive answers vice simple Yes or No or responses to multiple choice options. While the practitioner, after reading this book will have acquired a better grounding in systemic terms, as the survey instrument is for general use, it is strongly recommended to not use terms found in the above

Communications mechanism construct unless there is significant pre-instructional material to be used by the survey participants. Secondly, the use of terms common in the language of the community that participates in the system of interest or by stakeholders will remove undue and conflicting influences on the survey participants responses. Additionally, the practitioner may find that greater granularity may be required as the Communication Channels are described, so the survey instrument can be modified to leverage this additional detail from the survey participants.

The development of the survey wording follows the concept of enticement, in that the language used in the survey instrument is chosen to draw in the survey participant where the result is for the survey participants to invest in the thoroughness of the answers to the questions thus ensuring that the answers provided are thoughtful and deliberate to the best of the participants understanding of the questions. This personalization of the responses is partially achieved by the use of the language of the community and secondly by the perceived orientation between the survey participant (Individual) relative to other identified individual/s or group. The wording of the questions follows the relationship as shown in Fig. 4 below, where the participant is relative to identified individuals or groups and the questions do not imply separation or detachment from a group.

The flow of the survey Instrument is to take the survey participant on a path through a set of questions where the answers of one participant are grouped with responses from other participants that will serve to frame the channels of Communication. Secondly, the content of each question is unique and does not depend on the previous or follow-on questions' content to provide support. The construct of the question content is organized to lead the participant to formulate an answer that is unique to the question and not a continuation of the previous question(s). Hence, the terms used in the question form a simple graphic image that allows the participant to provide

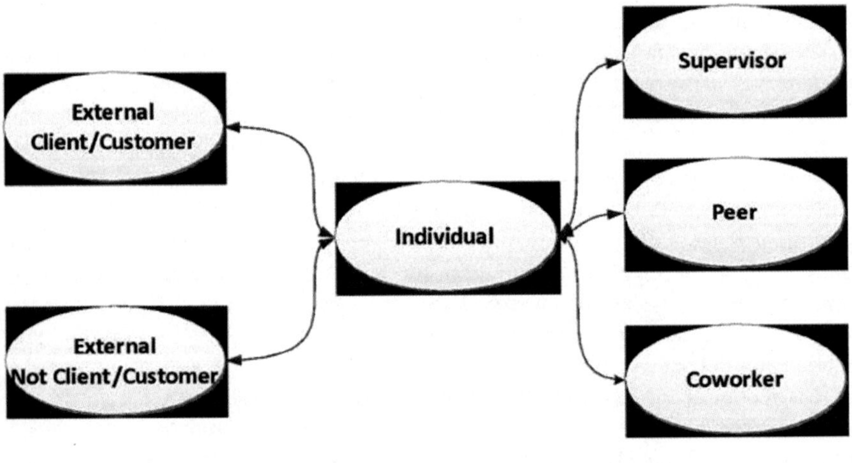

Fig. 4 Survey instrument participant relationship with individual/s or groups

their own shape in terms that he/she is familiar with, not grasping to the previous questions text to be provided as a "copy and paste" response.

The optimum flow of the survey starts with the survey participant. As it is his/her perspective as well as the perspectives of his/her fellow participants that will be building the framework, the enticement starts with their description and elaboration on who they are, where they practice within the system of interest. With care, the questions enable the participant to describe their Identity and that of the group they are part of as well as building to include the Identity of the system of interest. As the participant moves further from his/her close associations, their responses can become more stilted and less descriptive. With attention to the content of the questions, this tendency of distillation can be avoided. Secondly, shifting from the "whom" to the "what" begins the shaping of what is the purpose either of the individual or group associations to communications for the system of interest.

Previously, the concept of what is a simple and complex system was described. Within this description and as shown in Fig. 1, there is the environment as well as the system of interest with the parts of operations and management. Individuals relative to the functions associated with Governance of Complex Systems may associate with these different areas depending upon the contextual shape they formulate in response to a question. Therefore, the questions to develop the survey participants' understanding of the areas of environment, operations, and management need to be formulated for specific functions or grouping of types of functions. This is best facilitated in a survey by leading the survey participant to choose one of several paths, such as member of a group/project(s), supervisor, supervisor, and not directly involved in project(s) and supervisor and a member of project(s). These paths support development/description of the channels of Communication associated with the several functions associated with Governance of Complex Systems, the differentiation of peer from coworker and relationships based upon supervision.

Within the selected paths outlined above, the purpose of the questions will be similar across each path. The content of the questions will differ from path to path. The survey participant has been brought to the point where the questions should allow for full description of the specific channels of communication graphically shown in Fig. 4. Here, the wording of the questions is relative to the ten channels, allowing for an acknowledgment of a channel or the absence of a channel. Where there is a channel of Communications, the development of what is the product that is developed is to be exposed as well as the direction relative to the product, the specific technology (conveyance) of the channel of Communication and the mode. The use of the channel of Communication Mechanism terms should not be used. For example; direction can be asked by the term "from-to." Additionally, for a particular product, is there more than a single option for direction, technology (conveyance) as well as mode or any variation. Though one would expect that the survey participants may only deal with other individuals (supervisor, peer, coworker) that are internal to the system of observation, the questions developed for all paths ought to include the external individuals as shown in Fig. 4. Even for survey participants that are internal to a system of interest, their responses will add richness relative to Identity, purpose as well as better define where the environment is fixed.

The underlying aspect of CSG is that there is a design, execution, and evolution of the metasystem functions that are necessary to provide control, communication, coordination, and integration of a complex system. [42] suggests that governance suggests continuous achievement of (1) *Direction:* sustaining a coherent identity and vision that supports consistent decision, action, interpretation, and strategic priorities, (2) *Oversight Design:* providing control and integration of the system and corresponding initiatives, and (3) *Accountability:* ensuring efficient resource utilization, performance monitoring, and exploration of aberrant conditions. The "communication" as a metasystem function is achieved by channels of Communication. They exist, whether they were created as part of a purposeful design may not be known by the practitioner with respect to the system of interest. But what is known is that channels of Communication are critical function enablers that are either internal or external to the other metasystem functions so that metasystem products can be created, distributed continuously, without restrictions, in a manner consistent with the design, mode, and technology (conveyance). This chapter has described the Communication Mechanisms and the relationship between the metasystem and the channels as well as a methodical process to generate a clear picture of the operations of the communications mechanisms. By concentrating on the areas of direction, oversight Design and accountability, the practitioner will form a clear picture of the channels of Communication for the system of interest. From this picture, a shared perspective can be realized and the many challenges of communication will be exposed.

Exercises:

The following exercises provide an opportunity to examine the concepts presented in this chapter through several questions.

1. The identification of Identity, purpose, and environment is significant questions for a survey participant to answer. From readings of this chapter and understanding of Complex System Governance, for a system of interest, formulate for Identity and purpose, three or four questions that lead to a development of a framework of the system of interest. Did they include descriptions for individuals that may not consider themselves as part of the system of interest? Did the questions consider change in the work of the individuals?
2. Direction originates with a function and is conveyed to one or more functions (from- to). For a system of interest, develop three or four questions that would allow for the association of Identity and or purpose with functions and the respective direction. Is the direction limited by Identity?
3. The product is the designed mechanism that accomplishes Variety Attenuation or Variety Amplification. The Meta System has numerous channels of communication and the mechanism is located at the source (originator) of an exchange in information. For a system of interest, develop three or four questions that would lead to the articulation of products associated with functions. How does one measure the effectiveness of Variety Attenuation or Variety Amplification? Is the product used for different channels of Communication? Is the product

different for the direction? As mode is the Communication form (verbal, text, etc.), does this affect Variety Attenuation or Variety Amplification?

4. Technology (conveyance) is the various methods/agencies/instruments/systems (computer, Internet, assembly, meeting, etc.) to pass information from one function to one or more functions in the system. For a system of interest, develop three or four questions that would lead to the articulation of technology (conveyance) associated with products or functions. Is there a universal technology (conveyance) for all functions? Does one expect the product to always be compatible with any technology (conveyance)?

References

1. Jackson MC (2003) Systems thinking: creative holism for managers. Wiley, Chichester, p 378
2. Shannon CE (1948) A mathematical theory of communication, part 2. Bell Syst Tech J 27(4):623–656
3. Katz E (1957) The two-step flow of communication: an up-to-date report on an hypothesis. Public Opinion Q 21(1):61–78
4. Ackoff RL (1958) Towards a behavioral theory of communication. Manage Sci 4(3):218–234
5. Hage J, Aiken M, Marrett CB (1971) Organization structure and communications. Am Sociol Rev 860–871
6. Baskin O, Bruno SJ (1977) A transactional systems model of communication: implications for transactional analysis. J Bus Commun 15(1):65–73
7. Dervin (1993) Verbing communication: mandate for disciplinary invention
8. Losee R (1999) Communication defined as complementary informative processes. J Inf, Commun Library Sci 5(3):1–15
9. Craig RT (1999) Communication theory as a field. Commun Theory 9(2):119–161
10. Keating CB, Katina PF, Bradley JM (2014) Complex system governance: concept, challenges, and emerging research. Int J Syst Syst Eng 5(3):263–288
11. Beer S (1979) The heart of enterprise, vol 2. Wiley, Chichester
12. Keating CB (2015) Complex system governance: theory to practice challenges for system of systems engineering. In: 2015 10th system of systems engineering conference (SoSE). IEEE, pp 226–231
13. Nyström CA (2006) Design rules for intranets according to the viable system model. Systemic Pract Action Res 19(6):523–535
14. Keating CB, Katina PF, Bradley JM (2015) Challenges for developing complex system governance. In: IIE annual conference. proceedings. Institute of Industrial and Systems Engineers (IISE), p 2943
15. Beer S (1981) Brain of the firm
16. Beer S (1984) The viable system model: its provenance, development, methodology and pathology. J Oper Res Soc 7–25
17. Espejo R, Harnden R (eds) (1989) The viable system model: interpretations and applications of Stafford Beer's VSM. Wiley, Chichester
18. Flood RL, Jackson MC (1991) Creative problem solving: total systems intervention, vol 831397533. Wiley, Chichester
19. Keating CB (2000) A systems-based methodology for structural analysis of health care operations. J Manage Med 14(3/4):179–198
20. Beer S (1985) Diagnosing the System for Organizations. Oxford: England: Oxford University Press

21. Keating CB, Morin M (2001) An approach for systems analysis of patient care operations. J Nurs Adm 31(7/8):355–363
22. Wiener N (1949) Cybernetics. Philos Sci 16(2)
23. Beer S (1966) Decision and control: the meaning of operational research and management cybernetics
24. Keating CB (2014) Metasystem governance reference model (MGRM) for complex systems, National Centers for System of Systems Engineering
25. Weaver W (1953) Recent contributions to the mathematical theory of communication. ETC: A Rev General Semantics 261–281
26. Beer S (1974) Designing freedom. Wiley, London
27. Sennett R (2010) The architecture of cooperation (video: see 18:30 – 30:00). Harvard Graduate School of Design. Retrieved 7 December 2016
28. O'Grady W (2009) An alternate management control framework: the viable system model (VSM). In: Submitted to Auckland regional accounting conference
29. Katina PF (2015) Systems theory-based construct for identifying metasystem pathologies for complex system governance (Ph.D.). Old Dominion University, United States—Virginia. http://search.proquest.com.proxy.lib.odu.edu/docview/1717329758/abstract/29A520C8C0A744A2PQ/2
30. Reissberg A (2010) A cybernetic approach to hurricane hazard management on O'ahu, Hawai'i. In: Proceedings of the 54th annual meeting of the ISSS-2010, vol 54, no 1. Waterloo, Canada
31. Beer S (1985) Diagnosing the system for organizations. Oxford University Press, Oxford, England
32. Espejo R (2015) The cybernetics of self-organisation. Kybernetes, vol 44, no 6/7
33. Holten R, Rosenkranz C (2011) Designing viable social systems: the role of linguistic communication for self-organization. Kybernetes 40(3/4):559–580
34. Adams KM, Hester PT, Bradley JM, Meyers TJ, Keating CB (2014) Systems theory as the foundation for understanding systems. Syst Eng 17(1):112–123
35. Whitney K, Bradley JM, Baugh DE, Chesterman CW (2015) Systems theory as a foundation for governance of complex systems. Int J Syst Syst Eng 6(1–2):15–32
36. Espejo R, Schuhmann W, Schwaninger M, Bilello U (1996) Organizational transformation and learning: a cybernetic approach to management. Wiley, p 350
37. Herring CE (2002) Viable software: the Intelligent control paradigm for adaptable and adaptive architecture
38. Khuong LN, Harindranath G, Dyerson R (2014) Understanding knowledge management software-organisation misalignments from an institutional perspective: a case study of a global IT-management consultancy firm. Int J Inf Manage 34(2):226–247
39. Mohamad B, Bakar HA, Halim H, Ismail AR (2014) Corporate communication management (CCM) and organisational performance: a review of the current literature, conceptual model and research propositions. Procedia-Soc Behav Sci 155:115–122
40. Regev G, Alexander IF, Wegmann A (2005) Modelling the regulative role of business processes with use and misuse cases. Bus Process Manage J 11(6):695–708
41. Pérez Ríos J (2010) Models of organizational cybernetics for diagnosis and design. Kybernetes 39(9/10):1529–1550
42. Calida BY (2013) System governance analysis of complex systems

Perspectives on Complex System Governance Performance

Hart Rutherford

Abstract Although it seems like an obvious idea, the literature contains vastly inconsistent uses of the word "performance" because it is difficult to agree on its attributes. Both researchers and practitioners alike continue to struggle to define "good" system performance. This chapter offers several perspectives on the performance of complex systems drawn from an extensive review of the literature of systems theory, performance measurement systems, and complex system governance. While each body of literature is vitally important, they are also incomplete by themselves to fully articulate the drivers of holistic system performance. In this chapter, we argue that the most impactful driver of complex system performance is the purposeful design and development of the metasystem governance functions of CSG. Aided by the vast body performance measurement literature and the deep rooting of systems theory, we offer an initial conceptual framework for metasystem performance that distills the literature into 3 key concepts: purpose, competence, and Learning.

Keywords Complex system performance · Metasystem · Performance measurement · System governance

1 Introduction

The world is complex, and it is growing in complexity. The modern world has delivered incredible technology, processes, and information that have the power to dramatically improve or impair the human experience. Many positive advancements have come about as the result of the engineering and implementation of new complex systems. The pace of technological advancement is rapid and accelerating and we should expect this trend to continue. Therefore, we believe that the chief challenge, we face in future is not overcoming technological hurdles. Even the most superficial review of scientific progress shows clearly that every barrier to human progress will be overcome. Rather, our biggest challenge is developing new ways to deal with complexity.

H. Rutherford (✉)
National Centers for System of Systems Engineering, Old Dominion University, Norfolk, US

C. B. Keating et al. (eds.), *Complex System Governance*, Topics in Safety, Risk, Reliability and Quality 40, https://doi.org/10.1007/978-3-030-93852-9_12

Keating et al. [1] offers an additional yet challenging perspective.

> … we suggest that complex systems operate under conditions of uncertainty (incomplete knowledge casting doubt for decision/action consequences), ambiguity (lack of clarity in understanding/interpretation), emergence (unpredictable events and system behaviors), complexity (systems so intricate that complete understanding or explanation is impossible), and interdependence (mutual influence among systems through which the state of each system influences and is influenced by the state of interrelated systems). (p. 264)

Since ever-growing complexity is a future certainty, this chapter argues that the purposeful design and development of complex systems guided by meaningful measures of their performance are essential to dealing with the "wicked" problems of our current age. More succinctly, Yuchnovicz [2] states, "The need to design, build, operate, maintain, and govern complex systems is a fact of modern life" (p. 111).

Understanding the world in terms of "systems" is a compelling way to understand the complexity of the world. Skyttner [3] states, "most often the word [system] does not refer to existing things in the world but rather a way of organizing our thoughts about the real world." This chapter begins with a discussion of systems theory which is the platform upon which we can begin to understand system complexity and produce desired performance from them.

Systems theory provides descriptive power to help us understand the world. The propositions of systems theory provide a light to peer into the structure and behaviors of systems von Bertalanffy [4]. "Thus, there exist models, principles, and laws that apply to generalized systems or their subclasses, irrespective of their particular kind, the nature of their component elements, and the relations or "forces" between them … universal principles applying to systems in general" (von Bertalanffy [4], p. 32).

Since this chapter is focused on performance perspectives, we next turn our attention to a brief review of key concepts from the traditional literature of performance measurement systems (PMS). The literature of PMS is robust and virtually endless. Ideas from this body of literature are extremely valuable in shaping our perspective of complex system performance. But we will also see there are shortcomings in the literature. Chiefly, there appears to be no significant emphasis of the metasystem functions that operate at a logical level above the operational systems and subsystems that produce the outputs of the entire system. The richness of PMS literature will be discussed as a predicate to developing innovative ideas within the framework of Complex System Governance (CSG).

We will next review CSG as an emerging approach formulated as the "design, execution, and evolution of the critical metasystem functions necessary to maintain system viability [existence]" [1], (p. 274). CSG is the first approach that acknowledges the criticality of the metasystem to produce desired system performance beyond mere viability. Keating, Katina, and Bradley [1] have advanced CSG as an improved lens through which to view the world and make sense of system complexity. It is novel because it unites the most significant ideas from systems theory, management cybernetics, and governance to form a new framework for addressing and resolving "messy" problems that are so prominent in complex systems.

To conclude the chapter, we discuss key concepts of performance from the metasystem perspective and suggest a conceptual framework to organize our thoughts on performance with CSG. We assert that the most important levers of system performance are the metasystem governance functions from which all other production is governed. This discussion of a conceptual framework on metasystem performance is nascent and the product of ongoing research. While there is room to extend these ideas with deeper theoretical grounding and practical application, we also see these emerging concepts as useful and essential to build new theories and applications.

As a note to the reader, it is important to clarify the "type" of systems that will be the focus of our discussion. Ackoff and Gharajedaghi [5] developed four general types to classify systems. This is a categorization of systems having a basis in systems theory [6] which are deterministic, ecological, animate, and social. Members of the deterministic class of systems are simple mechanisms whose essential elements have a function, but do not have a purpose. They are governed by causal laws. Ecological systems are living systems that are self-organizing and self-maintaining but their responses to the environment are deterministic in nature.

Animate systems are of a "higher" order beyond an ecological system in which "reactions [to the environment] are determined; responses involve choice" (Ackoff and Gharajedaghi [5], p. 4). They posit that an individual human is an example of an Animate system. Finally, social systems are composed of animate systems and display collective choice. Examples include organizations, governments, and other social systems [6].

This chapter uses language and examples based in the context of the social type of system as a device to help the reader engage with the following perspectives of performance using a domain of experience that should be readily accessible. While there are nuanced differences among these types of systems, it is the goal of this discussion on performance that the underlying concepts and propositions of systems theory and complex system governance are applicable to every type of system.

Finally, there are terms used throughout this chapter that need to be recognized as having precision in use. To bring clarity to the discussion that follows, the following is a list of key terms.

Complex System Governance: "The design, execution, and evolution of the critical metasystem functions necessary to maintain system viability [existence]" [1].

Conceptual Framework: The organization of concepts, beliefs, and theories that combined to form new theories of the phenomena under investigation Grant and Osanloo [7].

Governance: The action or manner of governing Merriam-Webster dictionary [8]; coordinating actions and communications to achieve goals through collaboration Calida [9].

Metasystem: An integrating structure to produce behavior and structure beyond any of the constituent subsystems Keating et al. [10].

Performance measure: "A metric used to quantify the efficiency and/or effectiveness of an action" [11].

Performance measurement: "The process of quantifying the efficiency and effectiveness of action" [11].

Performance measurement system: "The set of metrics used to quantify both the efficiency and effectiveness of actions" [11].

System: "A system is a set of interrelated components working together toward some common objective or purpose" [12], "A system is a set of interacting units or elements that form an integrated whole intended to perform some function." [3].

Notably, the definition of performance was not listed here. As this is the purpose of this chapter of perspectives, namely for the reader to develop an enhanced concept of complex system performance from a higher level of abstraction within a larger framework of ideas.

2 Perspectives from Systems Theory

The word "system" has been used to describe many different ideas from diverse domains and with varying degrees of precision. As a result, its definition is highly dependent on its context. Every day, we freely use it in the most widely divergent ways in which its meaning is typically implicit and rarely explicit: "the digestive system," "the Dewey Decimal System" "you cannot beat the system" or "the newest weight-loss system." Skyttner [3] attempts to clarify this confusion by emphasizing that "most often the word does not refer to existing things in the real world but rather to a way of organizing our thoughts about the real world."

Defining a System from a Performance Perspective

In the early twentieth century, the reductionist mindset of strict control and predictable outcomes was overcome by various complexities emerging from modern scientific disciplines. The appeal of General Systems Theory (GST), with its emphasis on holism, provided an interdisciplinary vocabulary to describe systems and a set of principles that describe their behavior. Interestingly, there is no succinct definition of systems theory. Skyttner [3] states, "each body of theory has its implied assumptions or axioms which in reality are impossible to prove and hence must be accepted as value judgments." Whitney, Bradley, Baugh, and Chesterman [13] echo this idea that, "at the present time, a universally agreed upon definition for systems theory does not exist, though the term is ubiquitous in systems literature." However, through their inductive study of 30 system propositions and subsequent colligation into axioms, they increase our confidence that these principles are "well grounded in systems theory literature and justified by practitioner experience" [13].

Systems theory provides descriptive, interdisciplinary understanding of the design, structure, and behavior of systems. Since there is no commonly held definition of systems theory, there is, not surprisingly, no single definition of a system. Blanchard and Fabrycky [12] state "a system is a set of interrelated components working together toward some common objective or purpose." Skyttner [3] provides a similar perspective with, "a system is a set of interacting units or elements that form an integrated whole intended to perform some function." There are consistencies in these definitions that suggest there are 6 essential elements of a system: input, boundary, interrelated entities, intrinsic feedback, extrinsic feedback, and output.

System propositions help us understand the behavior of systems and the forces between them von Bertalanffy [4]. They are universal in that they are applicable to all systems. They reveal insights into the behavior, structure, and interactions of systems. Having a clear view of the wisdom in the propositions is the first step in understanding system performance in a much deeper way.

Selected System Propositions Related to Performance

The selected system propositions below are briefly described along with the implications to system performance. Adapted from Whitney et al. [13], they are highlighted in Table 1.

An initial perspective that emerges from the system propositions is the idea that systems are goal-seeking toward an intended final state or some equilibrium point. For animate and social system types, goal-seeking systems change their behavior in response to the environment and feedback to fulfill a predetermined goal Ackoff [20].

Whitney et al. [13] compiled several system propositions related to goal-seeking behavior into the "Goal Axiom" in which "systems demonstrate purposive behavior in order to achieve specific goals." While it is true we can achieve greater understanding of the drivers of system performance by studying its individual elements, this strictly reductionist approach would miss the mark. Whitney et al. [13] state, "by studying the goals, goal-oriented behavior functions, and the purposes of the system, greater ability is rendered to evaluate and potentially improve system performance."

The Sub-optimization proposition is an especially relevant idea to this discussion on performance because it explicitly links the idea of goal-seeking behavior and the concept of holism. This proposition affirms the study of system elements and their interactions but also demands the intentional sub-optimization of some for the purpose of attaining greater holistic efficiency. System performance in this sense is the achievement of holistic outcomes, not merely the efficiency of each element.

Holism: The Focus of Performance

It is a commonly held tenet that systems, which could be composed of parts that themselves are subsystems, exhibit properties that their parts do not exhibit by themselves (Aristotle; Boulding [21]; Bowler [22],[23]. This is the pedigree of the aphorism that the whole is more than the sum of the parts. Systems may be part of other systems in a hierarchy of systems, which systems theory regards as a universal principle [3].

The most important perspective to have when evaluating system performance is the principle of holism. According to Whitney et al. [13], "the basic doctrine of systems theory rests in Aristotle's assertion that the whole is greater than the sum of the parts: the parts, in their structural arrangement, and engaged in respective operations and interactions, constitute the whole." Holism demands, we consider the system as a whole, rather than merely the sum of its parts. For problem situations that

Table 1 Selected Systems Theory Propositions and implications to the concept of performance

Proposition and primary proponent(s)	Description of proposition	Implications to the concept of performance
Boundary (von Bertalanffy [4]; Skyttner 2005)	The abstract, semi-permeable perimeter of the system defines the components that make up the system, segregating them from environmental factors and may prevent or permit entry of matter, energy, and information	The evaluation of system performance must account for and appreciate the full range and scope of the system of interest
Control Checkland [14]	The process by means of which a whole entity retains its identity and/or performance under changing circumstances	Systems are recursively organized in hierarchical levels. System performance is directly related to the effectiveness of control at each level of recursion, which is dependent on communication of feedback
Equifinality von Bertalanffy [15]	If a steady state is reached in an open system, it is independent of the initial conditions and determined by the system parameters, e.g., rates of reaction and transport	There are innumerable alternatives with respect to system design choices and possible interventions. Therefore, there are multiple pathways to desired system performance
Minimal critical specification (Cherns 1976, 1987)	This principle has two aspects, such as negative and positive. The negative simply states that no more should be specified than is absolutely essential; the positive requires that we identify what is essential	The selection and monitoring of the most critical drivers of system performance should be as few as possible (cf. Law of Requisite Parsimony). The larger the number of drivers, the more difficult it will be to steer system performance
Purposive behavior Rosenblueth et al. [16]	Purposeful behavior is meant to denote that the act or behavior may be interpreted as directed to the attainment of a goal—i.e., to a final condition in which the behaving object reaches a definite correlation in time or in space with respect to another object or event	Performance is evaluated relative to the system's purpose and attainment of its goal
Satisficing (Simon [17, 18]	The decision-making process whereby one chooses an option, that is, not the best, but good enough	Variety can never be fully regulated, and as a result, there will always be some level of entropy present requiring design choices and interventions that are "good enough."

(continued)

Table 1 (continued)

Proposition and primary proponent(s)	Description of proposition	Implications to the concept of performance
Sub-optimization [19]	If each subsystem, regarded separately, is made to operate with maximum efficiency, the system as a whole will not operate with utmost efficiency	While attention is paid to a system's components, the holistic outcomes of the system are the true measure of system performance

are highly complex, ambiguously bounded, nonlinear, and having a sociotechnical emphasis, we adopt a holistic viewpoint [24].

This is more than a platitude. Reductionist thinking, in contrast to holism, is rooted in the implicit assumption that every element of the system, and their relationships, can be understood and explained as a matter of discernible cause and effect. This strictly analytical approach consists of taking something apart to understand its properties, explaining the individual parts, and then aggregating these independent explanations of the parts into an explanation of the whole. This commitment to a mechanistic viewpoint of the world and its systems was appropriate for many of the needs of the industrial revolution which centered on the substitution of machines for men as sources of physical work.

In the modern era, reductionism is an incomplete explanation for complex systems which have properties that cannot be explained by merely examining its parts. For example, the human body can write a paragraph but none of its constituent parts can. To write, the human brain, nervous system, muscles, hands, and many other organs must work effectively together to produce the action of writing. In fact, it is doubtful that the behavior of "writing" could have been foreseen just by studying the elements of the human body.

Ackoff [25] emphasizes this idea by making the distinction between analytic and synthetic thinking. "In the analytic mode, it will be recalled, an explanation of a whole is derived from explanations of its parts. In synthetic thinking, something to be explained is viewed as part of a larger system and is explained in terms of its role in that larger whole." A systems thinker is more interested in putting things together than in taking them apart. A system, viewed *structurally*, is a divisible whole. But when viewed *functionally*, it is an indivisible whole in the sense that some of its essential properties are lost in taking it apart Ackoff [25].

The real danger of reductionism and a highly analytical approach is that it limits our ability to truly understand the performance of a system. The actual effectiveness of the system is more than the sum of the efficiencies of each part. Our systems principle of sub-optimization dictates: "If each subsystem, regarded separately, is made to operate with maximum efficiency, the system as a whole will not operate with utmost efficiency" [19].

Ackoff [25] uses a very helpful story to illustrate this idea. Suppose we can collect one each of every model of automobile in existence. Suppose further that we assemble a panel of automotive engineers to determine the best parts of all these automobiles.

We might begin with asking the panel to survey all the available cars to select the best carburetor. We remove that carburetor and set it aside. Then, we might ask the panel to select the best transmission, distributor, braking system, and so on. With all these "best" parts, we then ask the engineers to assemble them into a new car. Would not that be the best car? No, because it is highly unlikely the parts would fit together. Even if they could be assembled, it is doubtful they would work well together. "System performance depends critically on how the parts fit and work together, not merely how each performs independently" Ackoff [25]. It is the *interaction* of the parts that is more important than the *action* of each individual part.

So, from a systems perspective, the essence of system performance is the effectiveness of the whole, where the effectiveness of the whole is driven by the interaction and "fitness" of the parts. The propositions of systems theory are essential tools that provide an understanding of the quality of a system's design and its holistic performance determines whether a system will achieve its goal. What is unclear is how to characterize and measure system performance, grounded in an underlying systems theoretic foundation, to appropriately intervene to produce greater performance. A discussion of performance measurement follows with a goal of drawing insights from the traditional performance measurement literature to support our quest for deeper perspectives of complex system performance.

Systems Holism Vignette—Government Performance

Over the last two decades, there has been a consistent effort by the U.S. government to improve the measurement of the performance of its agencies and the efficiency of their service delivery. Seeking to promote improved government performance and greater public confidence in government through better planning and reporting of the results of federal programs, the Congress enacted the Government Performance and Results Act of 1993 (GPRA), referred to as "the Results Act" and "GPRA." The Act established a government wide requirement for agencies to identify agency and program goals and to report on their results in achieving those goals. Recognizing that few programs at the time were prepared to track progress toward their goals, the Act specified a 7-year implementation time period and required the office of management and budget (OMB) to select pilot tests to help agencies develop experience with the Act's processes and concepts.

A report published by the Government Accountability Office (GAO) entitled, "managing for results—analytic challenges in measuring performance" highlighted the following observations.

** 93% of government officials surveyed reported that implementing a performance measurement program was a "great" or "very great" challenge*

** Government officials found it difficult to translate long-term strategic goals into annual performance goals*

** Agencies found it difficult to articulate their strategic level desired outcomes and instead were only able to list their program's tactical-level activities*

** Selection of an outcome measure was impeded by conflicting stakeholder views of the program's intended results or by anticipated data collection problems.*

There is no "messier" domain in systems thinking than the government. These issues will persist well into the future until there is a significant urgency for a change in thinking and approach. Real improvement in government performance measurement and accountability to its citizens will come only with intense focus on taking a holistic systems approach to deliver services.

3 Perspectives from Traditional Performance Measurement Literature

Measuring performance is such an intuitive idea that people may feel it does not require any explanation. The clearest and most succinct support of measurement as the foundation for science is from Lord Kelvin:

> I often say that when you can measure what you are speaking about, and express it in numbers, you know something about it; but when you cannot measure it, when you cannot express it in numbers, your knowledge is of a meager and unsatisfactory kind; it may be the beginning of knowledge, but you have scarcely, in your thoughts, advanced to the stage of science, whatever the matter may be [26].

Measurement is about gaining knowledge and better understanding the behavior and functioning of systems. The traditional literature of performance measurement systems (PMS) is vast and provides important insights into the performance of complex systems.

Among other purposes, measurement provides data to evaluate the state of a system, assess the maturity of a team, motivate employees, and inform strategic plans. These are important outputs of measurement. Behn [27] and Bourne et al. [28] suggest that the primary purpose of performance measurement is to improve. So, connecting this idea of improvement with the discussion of system propositions above, we may posit that performance measurement evaluates the efficiency and effectiveness of action to improve along a goal-seeking trajectory. As we turn our attention to the definitions of performance measurement from the literature, we will consider whether these initial conceptions of system performance are complete.

Defining Performance Measurement

According to a review by Franco-Santos et al. [29] of business performance measurement (BPM) definitions, they conclude that a robust definition of performance measurement would contain a description of its *features*, *roles*, and *processes*. The features of a performance measurement system are the properties or elements that comprise a BPM. The roles are the purposes or functions performed by the BPM system. The processes are the series of actions that combine together to constitute the BPM system. However, Franco-Santos et al. [29] also express concern that scholars in the field of performance measurement tend to use terms without explaining exactly what is meant. The lack of agreement on terms and definitions creates confusion and limits the generalizability and comparability of research in this area.

According to Neely, Gregory, and Platts [11], performance measurement is a topic which is often discussed but rarely defined. Based on their exhaustive literature review of performance measurement systems, they offer the following definitions:

- **Performance measurement:** The process of quantifying the efficiency and effectiveness of action.
- A **performance measure:** A metric used to quantify the efficiency and/or effectiveness of an action.
- A **performance measurement system:** The set of metrics used to quantify both the efficiency and effectiveness of actions.

These are sound definitions upon which to begin a deeper inquiry into the performance of complex systems. A striking feature of these definitions is the consistent use of the word "action." Earlier, we stipulated that measurement is about gaining knowledge. Performance measurement, however, has the additional nuance of evaluating "actions" that, in turn, provide us with increased knowledge of a system.

Neely et al. [11] define productivity as a measure of how well resources are combined and used to accomplish specific, desirable results. They note that the management concepts related to performance measurement are rooted in manufacturing and industrial contexts. The range of cost control and quality management techniques that emerged in the twentieth century strongly emphasize operational efficiencies in product development, production, and delivery. This idea is still prominent today and influences current thinking on this subject.

Aligning a performance measurement system with an organization's strategies to develop a clear picture of overall effectiveness is difficult. One particular challenge is that an organization's strategy changes over time. As a result of feedback from operations or signals from the environment, the framework of performance measurement must also change from time to time. During those transitions, the performance measurement system must remain coherent with the high-level objectives of the organization even though the actual measures themselves are likely to change [30].

The best-known performance measurement framework is Kaplan and Norton's [31] "balanced scorecard," which is based on the principle that a performance measurement system should provide managers with sufficient information to address the following questions:

- How do we look to our shareholders (financial perspective)?
- What must we excel at (internal business perspective)?
- How do our customers see us (customer perspective)?
- How can we continue to improve and create value (innovation and learning perspective)?

While the balanced scorecard provides an interesting framework, it does not provide a strong theoretical basis to support the process of performance measurement nor does it inform system design. The practitioners of balanced scorecard gain an appreciation of diverse perspectives affecting their business or organization. Its chief innovation was to depart from the highly quantitative methods and approaches that were rooted in control systems paradigms. The information and perspectives gained from this methodology are useful. However, the balanced scorecard approach does not provide a holistic systemic perspective that acknowledges the innate behaviors of complex systems and the functioning of their elements.

Current Challenges in Performance Measurement

As organizations instrument and collect data from every facet of operations, the amount and availability of performance data have never been greater. This presents a challenge to practitioners. Selecting and accessing appropriate data, conducting analysis, deriving understanding from their meaning, and then compiling them into relevant, timely tools and methods are a highly difficult undertaking. With the rise of machine learning techniques, artificial intelligence, and advances in computer hardware speed, there has never been more capability to leverage algorithms to provide a near real-time picture of the organization. Integrating the measurement systems across an organization must be accomplished with great care to ensure the resulting picture is coherent. Great care is taken to ensure there is appropriate understanding of the organization as a system from a holistic perspective.

Another practical challenge for practitioners is to ensure that any performance measurement system is fully linked to an organization's strategy. Measuring and analyzing performance data that does not directly support the attainment of the organization's goals is a waste of resources and results in confusion for the employees. It is counterproductive to have a performance measurement system without first having a clear set of strategic goals and objectives. Here, too, is the encouragement to connect these ideas of strategic goals to the systems theory propositions related to control and goal-seeking behavior. The practitioner should have a higher-level frame of reference grounded in systems theory to purposefully design and develop the organization in such a way that it retains its systemic identity.

Finally, sustaining a performance measurement system will require continual awareness of changes in the organization and the contextual issues affecting it. The strategies of an organization will change from time to time because either they have achieved their goals and have developed new ones, or because there is a need to respond to significant changes in the environment. Any material change in the focus of the organization will require a reevaluation of the performance measurement system. This is a continual process of sensing feedback from the external factors and circumstances of the environment and making the appropriate adjustments to the performance measurement system.

Critique of Traditional Performance Measurement Paradigms

Bourne et al. [28] state "while performance measurement and management theory and practice have developed and evolved significantly over the years, various authors have started to question whether the dominant performance measurement paradigm, rooted in control systems literature, is suitable in increasingly volatile and uncertain contexts" (p. 2788). Present day performance measurement systems are a collection of tools, techniques, and processes implemented as scorecards, targets, strategy maps, performance reviews, and incentives [29]. The advantage of these approaches is the acknowledgment of the organization as a complex system with interrelated elements arranged in a way to achieve a particular purpose not attainable by each part on its own. This is a beneficial perspective when problems are clearly defined and addressed by tools and processes that map directly to the organization and its contextual issues.

Unfortunately, according to Bourne et al. [28], "current evidence shows that the reality of most organizations is rather different and performance measurement and management scholars have highlighted the potential inadequacy of existing approaches" (p. 2788). As we will discuss further on in this chapter, this potential inadequacy may be rooted in not having a full appreciation for the metasystem governance functions described within the CSG framework.

Tangen [30] states that a performance measurement system should be derived from the organization's strategic objectives. This is a nod to higher-level planning and strategies with longer time horizons. This is a common idea in popular business journals and is beneficial to organizations struggling to navigate a complex business environment with focus and discipline. But, while it may be a nuanced point, the focus of traditional performance measurement systems is strongly on the operational outputs and production of the organization and not the quality or rigor of the strategic planning process at higher-level of conceptual abstraction. These higher-level governance functions are examined further when we discuss the metasystem governance functions of CSG later in this chapter.

To illustrate this point further, Fig. 1 is adapted from Tangen's [30] work, *"performance measurement: from philosophy to practice."* The research examined the performance measurement systems and organized them into categories according to their primary area of emphasis. The result of Tangen's work showed that performance measurement approaches fall within the five categories below: financial, operational, learning, customer, and strategy. His work also showed that most performance

Selected Prominent Performance Measurement Systems Within the Traditional Literature	Financial	Operational	Learning	Customer	Strategy	Metasystem
Balanced Scorecard (Kaplan & Norton, 1992)	X	X	X	X		
Performance Prism (Neely et al., 2002)			X	X	X	
Medori and Steeple's Framework (Medori & Steeple, 2000)		X			X	
Performance Pyramid (Cross & Lynch, 1992)		X		X	X	
Activity Based Costing (Johnson & Kaplan, 1987)	X	X			X	
Sink and Tuttle Model (Sink & Tuttle, 1989)	X	X				
Theory of Constraints (Goldratt, 1990)	X	X	X	X	X	

Fig. 1 Prominent performance measurement approaches by area of emphasis

measurement approaches are applicable to more than one category. What is important for our discussion in this chapter is the absence of any prominent performance measurement schemes that explicitly emphasize the metasystem as significant drivers of overall system performance.

Connecting Performance Measurement to Systems Theory

A potential approach that may provide appropriate grounding in this area is to develop a system theory-based foundation upon which to build a more coherent framework of performance measurement. We discussed previously the relevance of the Sub-optimization proposition above. Two other key system propositions that quickly emerge to have relevance to this inquiry are Boundary and Minimal Critical Specification.

The Boundary proposition is vital in any discussion of performance measurement because the scope and definition of the system of interest must be understood and must be matched with a performance measurement system that recognizes the same scope and definition. If there are gaps or overlaps at the system boundary when the theoretical scope of the performance measurement system overlays the system of interest, then the results of performance measurement will not provide an accurate representation of the system's performance. The evaluation of system performance must account for and appreciate the full range and scope of the system of interest.

Similarly, the Minimal Critical Specification proposition requires a theoretical understanding of the system of interest to define only the most essential performance measures and no more. The greater the number of features, roles, and processes included in the performance measurement system, the more difficult it will be to manage and steer the performance of the system [29]. The system proposition of Requisite Parsimony echoes this idea which suggests limiting the number of performance measures to 7 ± 2 in order to manage them effectively. The selection and monitoring of the most critical drivers of system performance should be as few as possible. The larger the number of drivers, the more difficult it will be to steer system performance.

Each of the system propositions in Table 1 above speak not only to the nature and behavior of complex systems, but also have further implications for system performance. A useful exercise for the reader would be to evaluate each system proposition and develop potential practical applications of each.

Adoption of a system of systems approach Ackoff [20] grounded in systems theory offers a promising alternative to the challenges posed by the dominant control systems paradigm to deal more effectively with complex and uncertain environments. In defining the nature of system of systems, Keating, Padilla, and Adams [32] describe "a metasystem [as being] comprised of multiple embedded and interrelated autonomous complex subsystems that can be diverse in technology, context, operation, geography, and conceptual frame. The complex subsystems must function as an integrated metasystem to produce desirable results in performance to achieve a higher-level mission subject to constraints" (p. 24).

Another critique is the inability to define consequential terms like "good" and "improve." Traditionally, performance measures are seen as a means of quantifying the efficiency and effectiveness of action which are matched to organizational context such as goals, strategy, culture, etc. [11, 33]. However, the research of Neely et al. [33] reveals that there has been little exploration of what constitutes a "good" performance measure. It is significant to note that Neely et al. and other authors use quotation marks around the word "good" when it is used in the context of producing insight into the effectiveness of the system.

For example, an excerpt from Neely et al. [33] states the authors developed recommendations "to construct a framework which seeks to encapsulate the elements which together constitute a "good" performance measure" (p. 1131). Later in the same paper, "as discussed in this paper, few authors have explored the issue of what constitutes a "good" performance measure" (p. 1141). While Neely et al. did develop a recommended framework for performance measures, it is important to recognize that it was based on an extensive review of the traditional performance measurement literature. In other words, a broad survey of the performance measurement literature produced a synthesis of concepts that, in their judgment, constituted a "good" performance measure.

The value of what Neely et al. produced is inarguably positive from the perspective of deepening our understanding of operational and productive measures of performance. We do not argue that this work has no utility. But it appears that the propositions of systems theory, which are essential to understanding the performance of any system, were not considered. We contend that our understanding of holistic system performance can be advanced through a systems theory-based approach. CSG offers a compelling argument that its approach, grounded in systems theory, management cybernetics, and governance, provides a framework for improving system performance through the purposeful design, execution, and development of metasystem governance functions. A full discussion of CSG follows in the next section of this chapter.

The major themes of performance measurement system literature emphasize the operational aspects of system performance. The high emphasis on efficiency and effectiveness of actions and processes reflect an approach that does not appropriately address the metasystem functions that operate at a logical level beyond the operational systems they must integrate [34]. The present literature in this area tends to focus on "how" actions and processes are executed and not "what" must be done, which are expressed in these metasystem functions.

Performance Measurement Vignette—High-Performing Product or High-Performing Relationship?

It is important to remember that performance is measured from the vantage point of the observer. The approach of Rolls Royce to sell "power-by-the-hour" is the subject of a classic business case study on the alignment of the firm with the customer. In its early days, Rolls Royce produced airplane engines and sold them to customers usually along with additional contracts to repair them when they failed. The repair of engines was a significant source of revenue for the firm but also a great irritation for the customer. To put it crudely, the worse the engines were, the more maintenance they required, the more revenue Rolls Royce would make.

Of course, customers do not want unreliable engines that need frequent repair. They want a reliable product that allows planes to fly safely. Instead of selling airplane engines, Rolls Royce now contracts with its customers for "power-by-the-hour." The customer buys the power the airplane engine delivers, and Rolls Royce provides all of the support (including maintenance) to ensure that those engines can continue to deliver power. This shift in business model is important because it means the interests of clients and providers are closely aligned.

From the perspective of the customer, performance was enhanced because the nature of the relationship changed even though initially the engines were the same. The change in business model made a material difference to the holistic performance of the system.

A key insight relevant to our discussion in this chapter is the idea that developing a performance measurement system is not "passive" evaluation of operational performance. The development of performance measures changes the behavior of the organization. When leaders design and communicate measures of "success," they are also fostering the behaviors, processes, and methods to fulfill that conception of success. And when the current conception of success changes, as in the case of Rolls Royce in this vignette, then the performance measurement system must also change, which in turn drives new behaviors and processes in the organization.

4 Perspectives from Complex System Governance

Complex System Governance (CSG) is novel because it unites the most significant ideas from system theory, management cybernetics, and governance to form a new framework for addressing and resolving "messy" problems that are so prominent in complex systems. CSG was formulated as the "design, execution, and evolution of the critical metasystem functions necessary to maintain system viability [existence]" [1].

CSG is an emerging field developed as a response to the problems arising from complex systems [35]. CSG focuses on developing capabilities to better understand the contextual issues surrounding tangled problem situations and diagnose the deeper level system issues that impede system performance. Problems manifest in two dimensions: the superficial, obvious issues and the deep, systemic issues at their root. While there is an immediate benefit to addressing superficial issues, problems are never truly resolved unless the systemic issues are rooted out and addressed. CSG provides an approach that appreciates these dimensions and is quickly developing perspectives, tools, and methods to address them.

The innovation of CSG is the explicit idea that system performance is regulated by 9 metasystem governance functions, which provide control, communication, coordination, and integration. Effectiveness in purposeful design, execution, and evolution of these metasystem functions determines the level of effectiveness of the system. The collection of metasystem governance functions was organized by Keating and Bradley [36] into a reference model (Fig. 2) that built upon Stafford Beer's Viable System Model [37].

Excerpted from Keating and Bradley [36], a brief description of the nature and role of each CSG metasystem function follows:

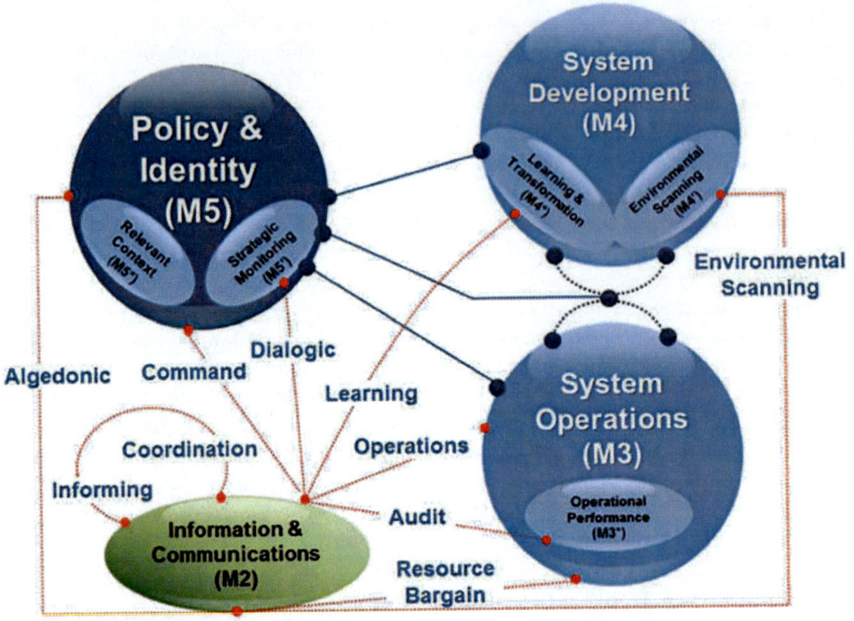

Fig. 2 Complex System Governance Metasystem Functions from Keating and Bradley [36]

- **Metasystem Five (M5)—Policy and Identity**—focused on overall steering of the system, giving policy level direction, representation of the system to external constituents, and maintaining identity for system coherence.
- **Metasystem Five Star (M5*)—System Context**—focused on the specific context within which the metasystem is embedded.
- **Metasystem Five Prime (M5')—Strategic System Monitoring**—focused on oversight of the system at a strategic level.
- **Metasystem Four (M4)—System Development**—focusing on the long-range development of the system to ensure future viability.
- **Metasystem Four Star (M4*)—Learning and Transformation** – focused on facilitation of learning based on detection and correction of design errors in the metasystem and guiding planning to support transformation of the metasystem.
- **Metasystem Four Prime (M4')—Environmental Scanning** – focused on sensing the environment for circumstances, trends, patterns, or events with implications for both present and future system performance.
- **Metasystem Three (M3)—System Operations**—focused on the day to day operations of the metasystem to ensure that the system maintains performance levels.
- **Metasystem Three Star (M3*)—Operational Performance**—focused on monitoring system performance to identify and assess aberrant or emergent conditions in the system.

- **Metasystem Two (M2)—Information and Communications**—focused on the design for flow of information and consistent interpretation of exchanges (communication channels).

The performance of these functions, required by all existing systems, supports achievement of **Control** (constraints necessary to ensure consistent performance and future system trajectory, **Communications** (flow and processing of information necessary to support consistent decision, action, and interpretation throughout the system), **Coordination** (providing for effective interaction to prevent unnecessary instabilities within and in relationship to entities external to the system), and **Integration** (maintaining system unity through common goals, designed accountability, and maintaining balance between system and constituent interests).

Metasystem governance function performance is embedded in the language describing the CSG reference model [36]. Examples of metasystem governance functions that explicitly contain performance-oriented language are listed below.

- M5 policy and identity: *"Focus includes policy, mission, vision, strategic direction, performance, and accountability for the system such that ... the system maintains viability..."*
- M5' strategic system monitoring: *"Primary function is to monitor measures for strategic system performance ..."*
- M3 system operations: *"Primary function is to maintain operational performance control ..."*
- M3' operational performance: *"Primary function is to monitor measures for operational performance."*

The descriptions of other metasystem governance functions also include the concept of performance. Keating and Bradley [36] further state that while the metasystem governance functions are unique, they have equal value. "None of the functions operates independent of the other functions. In addition, it is important to note that none of the functions is "more important" than the others. Consistent with the VSM, all of the CSG reference model functions are necessary to ensure the continuing viability of the system in focus. Deficient performance of one metasystem function will propagate through the entire metasystem." (p. 49).

The strength of CSG is its pedigree from systems theory, governance, and management cybernetics which focuses on the metasystem functions that sustain system viability [1, 37]. But the CSG metasystem governance functions depicted in Fig. 2 and described in the narrative above appear to presume a concept of performance that is not fully articulated. Indeed, Keating [38] states, "there are degrees of viability, the minimal of which is existence (p. 19.)." Further, he states, "however, viability is not "guarantee" of performance excellence."

As the reader has hopefully noticed, a central theme of this chapter is to enrich and extend the concept of viability from that of mere existence toward a richer articulation of holistic system performance. The following critique offers additional discussion of areas where concepts of performance in CSG would benefit from explication, especially in its conceptual relationship to management cybernetics.

This discussion above suggests we need to not only establish an articulation of performance for each metasystem governance function, but also develop a structure that is intrinsically holistic encompassing the entirety of the metasystem and the relationships between the governance functions. This passage also suggests performance is a matter of degree in that "better" functioning of the metasystem governance functions should produce "better" overall performance of the system.

But the language of the ideas of management cybernetics and the viable system model centers on the binary value of existence. This creates ambiguity within the CSG framework about the true nature of performance and makes it challenging to use precise definitional language. Management cybernetics offers CSG design cues for control through the model of a "metasystem." The metasystem is the set of functions that stand above the particular systems and entities that it seeks to "steer"—in the cybernetic sense of providing control—along with a set of communication channels to support this steering. The metasystem functions are arranged, as argued by Beer [39], to not merely describe the structure of a complex system, but the arrangement of essential elements to provide for viability.

Beer's [39, 40] celebrated dictum, "the purpose of a system is what it does," is a reminder there are often differences between the actual productive results of a system and the intentions of its designers. A system will always express its purpose by what it produces. Although the literature is replete with methodologies, methods, and tools to measure the outputs of a system, there appears to be little philosophical grounding to assess and measure the performance of the metasystem governance functions.

The viable system model is highly useful in describing the functions inherent in all viable systems. Viability is a much discussed term in systems literature. For example, the Viability axiom [13] "addresses how to design a system so that changes in the operational environment may be detected and affected to ensure continued existence." Viability is defined to mean the mere existence of the system. Missing is a language to describe degrees of metasystem performance beyond the state of existence.

Indeed, according to Keating [38], "all systems must perform the metasystem functions at a minimal level to maintain viability. However, viability is not a "guarantee" of performance excellence. On the contrary, viability simply assures us that the system continues to exist. There are degrees of viability, the minimal of which is existence" (p. 19). Also from Keating [38], "CSG is focused on … produc[ing] higher performing systems …" (p. 3). These are clear statements of need for a richer language to describe the degrees of effective metasystem performance.

Finally, if we were to examine the converse of "positive" performance, Keating and Katina [41] have described this same need with respect to system pathologies, which are aberrant conditions affecting system performance. These "negative" pathologies are in contrast to our present discussion of "positive" performance. "The pathologies may exist to varying degrees within a system of systems and are not presented as a binary all or nothing existence. (p. 253)." In the same way, CSG would benefit from a richer description of positive performance that provides a sense of degree of performance.

A concluding thought on CSG is that it is aptly named complex system *governance*, not complex system *engineering*. Too often, reductionist views of engineering implement basic, first-order learning to detect and correct system execution errors. On the other hand, CSG includes both first- and second-order learning to detect and carefully consider correcting fundamental errors in system structure through robust metasystem design. Keating et al. [1] note that a systemic perspective of governance requires continuous achievement of:

- "Direction—sustaining a coherent identity and vision that supports consistent decision, action, interpretation, and strategic priorities;
- Oversight design—providing control and integration of the system and corresponding initiatives; and
- Accountability—ensuring efficient resource utilization, performance monitoring, and exploration of aberrant conditions" (p. 267).

A truism from Yuchnovicz [2] is, "System performance results from system structure." Therefore, we argue that the purposeful design and continual development of the metasystem is the surest path to holistic desirable system performance rather than leaving the complex system to self-organize over time.

In the next section, we will begin to synthesize the key ideas of this chapter across the preceding perspectives of systems theory, performance measurement systems, and CSG. From these discussions, we will begin to develop new perspectives for complex system performance.

5 Synthesis of Performance Perspectives

The chief perspective offered in this chapter is that the holistic performance of a complex system need be directly tied to the performance of the metasystem governance functions present in every viable system. These metasystem functions and the quality of their operation are the most impactful drivers of overall system performance. A cursory review of the marketplace reveals hundreds of methodologies, methods, and tools to choose from to help an organization evaluate and improve their performance. A defense against adopting a management fad or an ineffective tool is the degree to which performance measurement is philosophically grounded in systems theory.

Reflecting on the discussion of systems theory above, Keating and Katina [41] echo this mandate in their discussion of systemic pathologies. "System theory (principles) first—The development and subsequent application of the system of systems pathologies must remain grounded in the conceptual foundations of systems theory from which they emerged. There will always be new tools and techniques developed to support [system of systems engineering] SoSE, as well as those that disappear. However, for any approach, technology, or method to be sustainable, it must be rooted in the underlying systems theory upon which it rests. (p. 263)".

Although performance is a value judgment of the observer, systems theory offers "… a unified group of specific propositions which are brought together to aid in understanding systems, thereby invoking improved explanatory power and interpretation with major implications for systems practitioners" [42], p. 113).

The literature on performance measurement systems is robust in describing the structure and application of operational measures. Performance measures are entirely described in terms of the productive capacity of the organization. A danger of relying solely on quantitative measurements like revenue or productive efficiency as key indicators of desirable system performance is that they obscure the holistic nature of a system.

For example, to produce revenue, a company must not only acquire materials and transform them into salable goods, but it must also actively engage critical CSG metasystem functions such as learning and transformation and scanning the environment in order to improve. Even if an organization may be unfamiliar with the term "metasystem governance function," they will readily accept that organizational learning and environmental scanning are important. Effective managers and executives have always known and appreciated the value that these dimensions bring. What is missing is an effective means to articulate what performance is in the context of metasystem governance.

Relationship Between Metasystem Governance and Performance Measurement

Figure 3 is an attempt to illustrate the relationship between the traditional measures of business performance and the metasystem governance functions of a complex system similar to an organization. Based on a review of the literature of performance measurement systems, we observe that the operational measures of productivity are "lagging" indicators in that the indicators report the outcomes and outputs of an

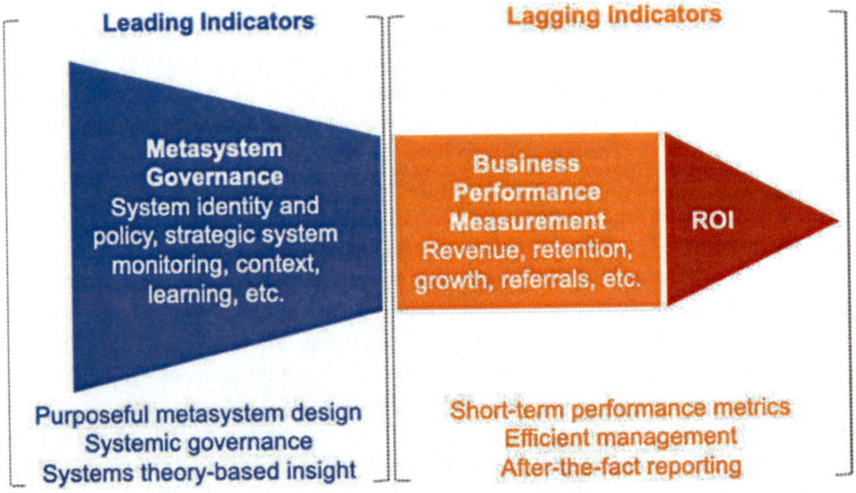

Fig. 3 Illustration of "leading" and "lagging" performance indicators

organization over a period of time. Common measures of business performance such as revenue, retention, and growth all follow the execution of plans and productive work.

Lagging indicators are signals of performance that become apparent only after a significant shift has taken place. A danger of relying solely on lagging indicators of performance is that they may foster the belief by practitioners that the current trend of organizational performance should be maintained for some time after the state of the system or its environment should have adapted an intervention to a new trajectory. Lagging indicators confirm long-term trends, but they do not predict them.

In the simplistic "rocket" graphic above (Fig. 3), we are illustrating the idea that the most impactful drivers of system performance are the metasystem governance functions described in CSG. The metasystem governance functions are "leading" indicators of holistic system performance. The purposeful design and continual development of the metasystem, rooted in the insights drawn from systems theory, are the most essential and impactful approach to produce desirable performance from a complex system. We argue that the degree to which the metasystem functions perform is directly related to the overall performance of the system. They are predictors of a change in the state or the trajectory of a system. In this way, the metasystem functions operate as a "leading" indicator of system performance.

Conception of Purposeful Design for System Performance

Throughout this chapter, we have made the case that "purposeful design of the metasystem" is a critical element to producing desirable performance from a complex system. Research has focused specifically on what "purposeful design" means from a performance perspective. Initial results of that inquiry are presented in Table 2 and reveal several concepts that appear frequently across bodies of literature and shed light on the most significant ideas related to system performance.

These core categories are words we, of course, are familiar with. But when we think of them in the context of purposeful design for metasystem performance, they take on slightly more nuance. **Competence** is more than the efficient practice of a skill. In the sense of metasystem performance, it is the ability of a system to conform to and fulfill the system's purpose. Constant assessment of system performance has the sole aim of examining the degree of correlation to system purpose. Competence is also strongly related to systemic governance described above that seeks to sustain the system's identity by providing cybernetic control and integration.

Learning places a spotlight on CSG's Metasystem Four Star (M4*) which focuses on facilitation of learning based on detection and correction of design errors in the metasystem and guiding planning to support transformation of the metasystem [36].

Table 2 Concepts related to Metasystem performance

Core category	Related concepts
Competence	Controlling, interconnecting, governing
Learning	Adapting, predicting, trajectory
Purpose	Aligning, mapping, steering, strategizing

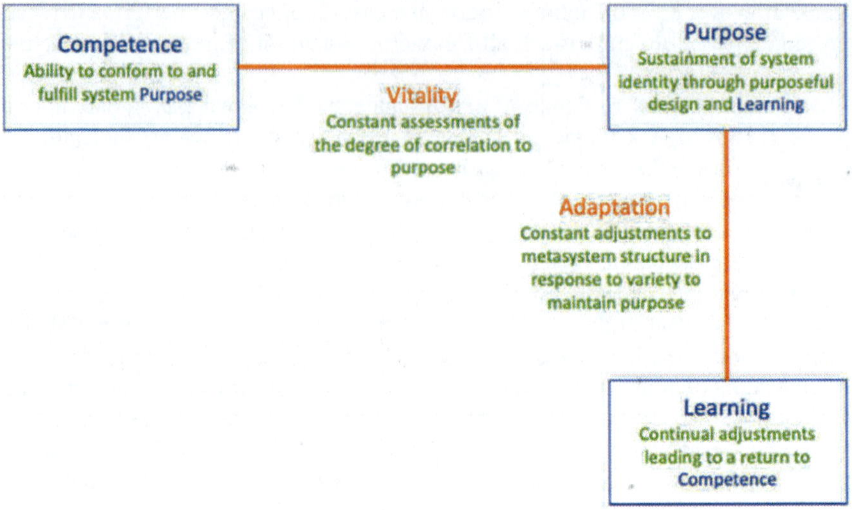

Fig. 4 Initial conceptual framework for Metasystem performance

Error in this sense is any mismatch between the goal-seeking purpose of the system and its actual execution. The objective of **Learning** is to foster continual adjustments of the metasystem leading to a return to **Competence**.

Finally, **Purpose** has a strong relationship to CSG's Metasystem Five (M5) where the focuses are on the overall steering of the system, giving policy level direction, representation of the system to external constituents, and maintaining identity for system coherence. The essence of **Purpose** is the sustainment of systemic identity.

The research has found that these high-level concepts are related. Figure 4 is an attempt to depict these relationships with descriptive labels.

These concepts are linked by mechanisms all related to performance. The first is "Vitality" used to describe the degree to which actual system execution is correlated to system identity. Beer [39] posits that all systems are viable. But the degree to which they perform at the metasystem level requires new language. The second linkage is "Adaptation" to describe our explicit focus on the metasystem as the target of our systemic intervention.

6 Summary

This chapter is an exploration of perspectives of performance at the metasystem level. These perspectives are based on insights drawn from an extensive review of three streams of literature: systems theory, performance measurement systems, and CSG.

The propositions of systems theory allow us to peer deeply into the nature and behavior of complex systems. Close examination of certain system propositions has implications to our conception of performance measurement and serves as an anchor to ground our thinking about metasystem design and development. The strong caution of systems theory is to think holistically about system structure and performance. Knowledge and appreciation of systems theory can aid the understanding of system performance by providing a framework for interpreting system behavior.

The literature of performance measurement systems is rich from the perspective of operational systems but appears to lack sufficient focus on the metasystems that operate at a higher level of abstraction. While we strongly acknowledge the vital importance of measurement systems focused on efficient and effective operations, it is an incomplete platform from which to purposefully design the metasystem.

Finally, CSG provides strong emphasis on metasystem structure and its purposeful design. The practitioner is "never done" in holistically evaluating and appropriately intervening in the governance of complex systems. One must continually account for contextual issues and the environment in which it is situated. Within the literature of CSG, the concept of performance appears frequently but used in multiple contexts and has multiple meanings.

The conceptual framework offered in this chapter is an organization of concepts that attempts to unify the high-level ideas from all three streams of literature to clarify the definition of performance from a metasystem perspective. Within the current literature, performance could mean productive output, ongoing system execution, overall mission accomplishment, or the effectiveness of people. As we have shown in this chapter, it is vital to have an understandable language of performance deeply rooted in systems theory if we are to advance further into the realm of metasystem performance.

Finally, systems of all types, not just technological ones, have embedded themselves in our society such that we cannot function well without them. Humans have evolved from primitive origins by the application of tools and systems that, in the modern age, have become exceedingly complex. As a result, the performance of a system has great bearing on our well-being. Therefore, it is essential to better understand the nature of system performance for two reasons. First, resources are scarce. Being effective consumers and stewards of resources are an imperative that we all must embrace. Second, systems have the potential to deliver great good in our society, but also have the potential for great ill. Poorly designed and poorly performing systems can cause great harm to society and to individuals. As news stories have highlighted, poor design and operation of a complex system can result in great human tragedy. The development of an improved conceptual framework of system performance would aid our understanding of systems in operation and provide understanding that might deter a great failure.

7 Exercises

- For each of the systems theory propositions above, what are the practical implications for the performance in your organization?

- Focusing on CSG's Metasystem Five (M5)—Policy and Identity describe in your own words what this means to you. Can you describe the systemic purpose of your school / business?
- In thinking about the concept of competence in Sect. 5 above, do you agree that "high competence" means conformance to purpose?
- Will an organization with a strong, clear articulation of purpose always be successful? Why or why not?

References

1. Keating CB, Katina PF, Bradley JM (2014) Complex system governance: concept, challenges, and emerging research. Intern J Syst Syst Eng 5(3):263–288
2. Yuchnovicz DE (2016) Understanding system structural tensions to support complex system governance development methodology. Int J Syst Syst Eng 7(1–3):109–129. https://doi.org/10.1504/IJSSE.2016.076124
3. Skyttner L (1996) General Systems Theory. Philadelphia: Trans-Atlantic
4. von Bertalanffy L (1968) General System Theory: Foundations, Development, Applications, rev. ed., George Braziller, New York
5. Ackoff R, Gharajedaghi J (1996) Reflections on systems and their models. Syst Res 13(1, Mar. 1996):13–23
6. Akers W (2015) An approach for the development of complex systems archetypes. Doctor of Philosophy (PhD), dissertation, Engineering Management, Old Dominion University. https://doi.org/10.25777/6xmx-r674
7. Grant C, Osanloo A (2014) Understanding, Selecting, and Integrating a Theoretical Framework in Dissertation Research: Creating the Blueprint for Your "House". Admin Issues J: Educ Pract Res 4(2):12–26
8. Governance (2022) In Merriam-Webster.com. Retrieved January 21, 2022, from https://www.merriamwebster.com/dictionary/governance
9. Calida B (2013) System governance analysis of complex systems (doctoral dissertation). Old Dominion University,Norfolk, Virginia
10. Keating C, Sousa-Poza A, Kovacic S (2005) Complex Transformation: A System of Systems Engineering (SoSE) Perspective. In: Proceedings of the American Society for Engineering Management, National Conference
11. Neely A, Gregory M, Platts K (1995) Performance measurement design—a literature review and research agenda. Int J Oper Prod Manag 15(4):80–116
12. Blanchard B, Fabrycky W (1998) Systems engineering and analysis, 3rd edn. Prentice-Hall, Upper Saddle River, New Jersey
13. Whitney K, Bradley J, Baugh D, Chesterman C Jr (2015) Systems theory as a foundation for governance of complex systems. Intern J Syst Syst Eng 6(1/2):15–32
14. Checkland P (1993) Systems Thinking, Systems Practice. New York John Wiley & Sons
15. von Bertalanffy L (1950) An outline of general systems theory. Br J Philos Sci 1(2):134–165
16. Rosenblueth A, Wiener N, Bigelow J (1943) Behavior, Purpose and Teleology. Philos Sci 10(1):18-24
17. Simon HA (1955) Abehavioral model of rational choice. Q J Econ 69(1):99–118
18. Simon HA (1956) Rational choice and the structure of the environment. Psychol Rev 63(2):129–138
19. Hitch CJ (1953) Sub-optimization in operations problems. J Oper Res Soc Am 1(3):87–99
20. Ackoff R (1971) Toward a system of systems concepts. Manag Sci 17(11):661–671

21. Boulding KE (1966) The Impact of Social Sciences, Rutgers University Press, New Brunswick, NJ
22. Bowler D (1981) General systems thinking. North Holland, New York
23. Checkland P (1999) Systems thinking, systems practice, 2nd edn. Wiley, New York, NY
24. Keating CB, Gheorghe AV (2016) Systems thinking: foundations for enhancing system of systems engineering. In System of Systems Engineering Conference (SoSE), June 2016. 11th (pp. 1–6). IEEE
25. Ackoff R (1974) The systems revolution. Long Range Plan 7(6):2–20
26. Kelvin, William Thompson, Baron (1883) Lecture on "Electrical Units of Measurement" (May 3, 1883). Popular Lect I:73
27. Behn R (2003) Why measure performance? different purposes require different measures. Public Adm Rev 63(5):586–606
28. Bourne M, Franco-Santos M, Micheli P, Pavlov A (2018) Performance measurement and management: a system of systems perspective. Int J Prod Res 56(8):2788–2799
29. Franco-Santos M, Kennerley M, Micheli P, Martinez V, Mason S, Marr B, Gray D, Neely A (2007) Towards a definition of a business performance measurement system. Int J Oper Prod Manag 27(8):784–801
30. Tangen S (2004) Performance measurement: from philosophy to practice. Intern J Product Perform Meas 53(8):726–737
31. Kaplan RS, Norton DP (1992) The balanced scorecard—measures that drive performance. Harvard Business Review, January-February 1992, pp 71−79
32. Keating C (2008) Emergence in system of systems. In Jamshidi M (ed) System of systems engineering: innovations for the 21st century. NY, USA, Wiley
33. Neely A, Richards H, Mills J, Platts K, Bourne M (1997) Designing performance measures: a structured approach. Int J Oper Prod Manag 17(11):1131–1152
34. Keating CB, Katina PF (2016) Complex system governance development: a first generation methodology. Int J Syst Syst Eng 7(1/2/3):43–74
35. Katina PF (2015) Emerging systems theory-based pathologies for governance of complex systems. Intern J Syst Syst Eng 6(1/2):144–159
36. Keating CB, Bradley JM (2015) Complex system governance reference model. Int J Syst Syst Eng 6(1/2):33–52
37. Beer S (1981) Brain of the firm. Wiley, Hoboken, New Jersey
38. Keating CB (2015b) Complex system governance: confronting 'System Drift.' Engineering management & systems engineering & old dominion university, pp 1–27
39. Beer S (1979) Heart of enterprise. Wiley, Hoboken, New Jersey
40. Beer S (1985) Diagnosing the system for organizations. Oxford University Press, New York, NY
41. Keating CB, Katina PF (2012) Prevalence of pathologies in systems of systems. Nat Cent Syst Syst Eng 3(3/4):243–267
42. Adams K, Hester P, Bradley J, Meyers T, Keating C (2014) Systems theory as a foundation for understanding systems. Syst Eng 17(1):112–123
43. Ackoff RL (1973) Science in the systems age: beyond IE, OR, and MS. Oper Res 21(3):661–671

Development

Complex System Governance Development Methodology

Charles B. Keating and Polinpapilinho F. Katina

Abstract This chapter explores a development methodology for Complex System Governance (CSG). The conceptual foundations and reference model for CSG have been articulated elsewhere as well as in this book. While these foundations provide answers to 'what CSG is' and 'why it is important', our present focus is targeted to examine 'how it is done'. This chapter aims to achieve three primary objectives. First, an introduction lays the foundation for elaboration of the CSG Development Methodology. This includes aspects of CSG and methodology essential to the following exploration of the CSG Development Methodology. Second, a 3-staged methodology for CSG development is examined. This 3-staged methodology includes Initialization (setting the current state of CSG and the supporting context), Development Mapping (analysis to determine priorities, feasibility, and capacity for CSG development), and Development (selection, planning, and execution of priority development initiatives to enhance performance of CSG functions). Third, several critical issues concerning the deployment of the CSG Development Methodology are examined. A spectrum of possible CSG methodology deployment concerns and associated issues are examined. The influence of these issues on design, execution, and the ultimate evolution of CSG for an organization (system) is examined. The chapter concludes with application-based insights for advancing CSG Development Methodology.

Keywords CSG development methodology · System development · CSG deployment

C. B. Keating (✉)
Department of Engineering Management and Systems Engineering, Old Dominion University, Norfolk, VA, USA
e-mail: ckeating@odu.edu

P. F. Katina
Department of Informatics and Engineering Systems, University of South Carolina Upstate, Spartanburg, SC, USA
e-mail: pkatina@uscupstate.edu

C. B. Keating et al. (eds.), *Complex System Governance*, Topics in Safety, Risk, Reliability and Quality 40, https://doi.org/10.1007/978-3-030-93852-9_13

1 Introduction

There are a multitude of increasing pressures stemming from complex systems and their inevitable problems. The cadence of this reality for practitioners might be captured in four themes that permeate the modern landscape of complex systems. These themes have been extolled in various forms in numerous prior Complex System Governance works ([1–4], Keating and Katina [5]). Following these works, we might summarize the themes and their implications for CSG Development in Table 1. Interestingly, although these conditions are not 'new' in the sense that they just arrived, it seems as though success in dealing with them has had minimal effectiveness. There are a myriad of approaches, both past and present, with good intentions to address our deteriorating conditions. However, we seem to be continually confounded with abysmal results, high human costs, and an increasing array of approaches and advice that falls short of expectations.

There is no definitive explanation as to why the present conditions exist or are permitted to continue to exist. Nevertheless, we offer an explanation based on several insights as to why the 'status quo' response strategies continue to exist. This continues to be the case even though system difficulties are not being adequately addressed. Although these strategies might have 'worked' in the past, their continued appropriateness and success as a response to complex systems/problems are challenged. Among these 'status quo' strategies are:

(1) *Sacrifice of Holistic Management*—Technology has been, and will continue to be, an important contributor to increasingly complex systems/problems. However, overreliance on technology to provide solutions is shortsighted. Complex systems/problems have a range of dimensions beyond technology. These additional dimensions may diminish the effectiveness of technology as the exclusive centerpiece in addressing complex systems. Other dimensions, including human, social, organizational, managerial, political, and policy can be decisive in providing resolution breakthroughs for complex systems/problems. In essence, a holistic approach is necessary and should not be sacrificed for superficial technology only treatments.

(2) *Focus on Short-Term Expedience*—Complex systems do not appear or propagate overnight. Instead, they take time to evolve to their current state. The expectation that they can be changed in a 'revolutionary' manner is shortsighted. On the contrary, complex systems require taking the 'long view'. This shifts the focus to their evolutionary development that unfolds over time, not instantaneous gratification from superficial treatment of surface issues. This is not to diminish the need to take immediate action to correct system deficiencies. However, pursuit of only short-term strategies that sacrifice system long view development invites continual operation in 'crisis' mode.

(3) *Piecemeal Development*—All systems evolve through a process of development. Unfortunately, for many this process evolves without purposeful direction. Instead, the system structure is modified in a *self-organizing* (unconstrained modification) or *ad hoc* (piecemeal modification) fashion. In contrast,

Table 1 Conditions facing system practitioners

Condition	Explanation	Implications for governance development
Uncertainty escalation	At a most basic level, uncertainty suggests that precise cause-effect relationships cannot be known for complex systems. Thus, normal approaches that assume complete knowledge and deterministic analysis (mathematical formulations) to address complex systems are incompatible with systems marked by uncertainty. As complex systems become more 'complex', uncertainty will rise—as will the inability of traditional reductionist-based approaches to successfully resolve issues	Governance development must take into consideration that uncertainty will pose several challenges, including: (1) inevitable fallibility of any approach, which requires constant questioning and adjustment, (2) appreciation of the uniqueness of each complex system, thus requiring an equally unique approach and subsequent journey for governance development, and (3) expectations that must be tempered for development outcomes, as the precise results cannot be known or predicted in advance
Ambiguity propagation	Complete knowledge and understanding for complex systems are illusionary propositions. Instead, there will always be a lack of clarity concerning the nature of each unique complex system, the unique domain within which it exists, and the unique context within which the system is embedded. Knowledge and understanding are continually refocused as they emerge over time and new knowledge of a complex system continually unfolds. The result is high levels of ambiguity, or a lack of clarity. This lack of clarity is not necessarily due to carelessness, omission, or intentional ignorance. Instead, it is natural and should be expected for complex systems	Ambiguity is simply a product of incomplete understanding of complex systems and their context. Irrespective of desire or intent, complex systems will always exist in conditions of incomplete understanding. Therefore, their development will also be mired in conditions of incomplete understanding. The challenge for governance development is found in the necessity for continuous accounting for ambiguity. This entails reduction where possible, acceptance where necessary, and accounting for development influences where feasible

(continued)

purposeful development concentrates on system modification based on appreciation of the whole system needs, priority (greatest need), and feasibility (what can be taken on within resource constraints with a reasonable possibility of success).

Table 1 (continued)

Condition	Explanation	Implications for governance development
Complexity acceleration	Central characteristics of complexity include a large number of richly interrelated elements, dynamic shifting of the system and our knowledge of that system over time, and emergence of unpredictable behavior, structure, and performance over time and operation of the system. For complex systems, complexity is not a temporary condition that will de-escalate over time. On the contrary, complexity, and its inevitable impacts, will continue to escalate as systems evolve and become increasingly interconnected, unknown, and unpredictable	CSG development must accept that complexity is not going to diminish and will most likely be exacerbated by prevailing trends. Implications for CSG development include finding better ways of dealing with elaboration (increasing interconnectedness), emergence (unpredictable patterns), and dynamics (rapid changes). Purposeful CSG development must offer continual modification of system design and execution to compensate for accelerating complexity
Holistic dominance	The landscape of complex systems is dominated by the dynamically shifting impacts of technology, human, social, organizational, managerial, political, and policy dimensions. While it would be easier to deal with singular aspects of complex systems (e.g., technology), the realities suggest that the holistic range of factors must be taken into account. In addition, these impacts can, and will be, subject to changes over time in terms of their importance and influence	For CSG development there must be an appreciation and accounting for the holistic spectrum of influences on system performance. The entire range of technology, human, social, organizational, managerial, political, and policy factors must be considered. These factors, and their interconnections, must be included in development to improve the design and execution of CSG functions
Information challenge	It is an understatement to suggest that complex systems are beset with exploding data and information. This is not new. However, the traditionally held relationships of data, information, knowledge, and wisdom must be questioned for continuing relevance and applicability for complex systems. The structuring and ordering of expanding data confound complex systems. Beyond increasing volumes, data challenges for complex systems also include veracity issues, misinformation, and accessibility issues	CSG development must be mindful of the flows and interpretation of information within and external to the system. Design and execution of CSG functions rely on information to perform. Thus, CSG development must include a focus on two aspects: (1) ensuring that information is trustworthy and (2) design for the right information is available at the right place and the right time to support consistency in decision, action, and interpretation

(continued)

Table 1 (continued)

Condition	Explanation	Implications for governance development
Contextual influences	All complex systems have a unique context within which they are embedded. This context includes the circumstances, factors, conditions, or patterns unique to the system. Context both enables and constrains a system. Similarly, the system constrains and enables the context. Separation of a system from its context is a false separation and only serves as a convenience for purposes of analysis. It is noteworthy that context is dynamic and will change over time and as system knowledge evolves and the context experiences shifts	CSG development that does not account for the context within which the system is embedded is deficient. CSG development must consider contextual influences on the system of interest, development execution, and expectations. Contextual considerations are not a one time effort. Instead, context must be continually monitored, assessed, and accounted for during CSG development. Without accounting for context, CSG development is incomplete

(4) *Treatment of Complex Systems as Simple Systems*—While the complexity of systems has continued to rise exponentially, so too has the desire to 'reduce' them. This is pursued under the false assumption that this will permit them to be more effectively addressed. The treatment as simple systems when they are actually complex systems are fraught with problems (see Kurtz and Snowden [6]. Unfortunately, this often results in oversimplification of complex systems and their problems. As Mitroff [7] suggests, the result of incorrect treatment results in solving a problem. Unfortunately, it is likely the wrong problem irrespective of how efficiently it might be solved.

(5) *Process and Event Centric Focus*—One strategy frequently deployed to engage, and 'tame', complex systems is through the establishment of standardized and repeatable processes and events. Emphasis on processes and events strategies falls short in the treatment of complex systems. There simply is no degree of process or corresponding events that can substitute for understanding the purpose, function, organization, and operation of a complex system. Processes will always fall short in achieving systemic integration. In this sense, a process-based strategy is a reductionist treatment of complexity. The reliance on events is sure to suffer the same fate for complex systems.

(6) *Complication as an Approach to Deal with Complexity*—The original intent of many systems is to provide a streamlined approach to support the effective resolution of a problem or fulfilment of a need. However, in practice, this resolution is often conveyed as adding more processes, procedures, requirements, and regulations. All of these well-intentioned complication efforts attempt to achieve mastery over complexity. Unfortunately, addressing complexity with

overcomplication is ineffective, introducing unnecessary constraints that can diminish system performance.

(7) *Emphasizing Output Over Outcome*—System outputs are identified as tangible, verifiable, and objective artifacts (products/services) consumed by external entities in the system environment which find value in the outputs. The 'output' mindset is grounded in an underlying set of values and beliefs (worldview through which all that is sensed is processed). This output worldview informs trade-offs and decisions concerning the formulation of system design, execution of that design, and activities to develop and evolve the design/execution over time. For example, many systems focus on tracking performance of cost, schedule, and technical achievement. After all, these attributes are objectively measurable. However, these indicators are 'systemically' limited in their measuring the value of a system. While these indicators (cost, schedule, performance) are necessary aspects of system performance, they alone do not provide sufficiency as a set of system judgments. Instead, more appropriate for consideration is the addition of 'outcome', which is concerned with the utility provided by the system. In effect, outcomes measure to what degree a system fulfils a need or effectively resolves a problem—from the perspective of individuals/entities that have the need or problem.

(8) *Global Control as a Goal*—Ultimately, a systems perspective of control involves establishment of a minimal level of constraints that can assure continued performance [8]. The excess constraint in a system (control) wastes resources and limits local autonomy (experiencing freedom and independence related to decisions, actions, and interpretations). The common perspective of excessive global control is what has been described as overregulation, bureaucracy, and excessive constraint without evidence of commensurate value added to the system.

Figure 1 is a summary of the relationships between the nature of the CSG landscape and the current state of common coping strategies targeted to address those issues. This does not suggest that there have not been exceptional approaches for dealing with complex systems (see Jackson [9]) or that the listings provide a definitive articulation of response strategies or the landscape. However, there is much room for improvement in developing new modes of thinking, which can in turn produce alternative methods to effectively address increasingly complex systems and their constituent problems.

There is not a universally accepted theory, methodology, method, or set of standards to assure success in dealing with the pressures of our current circumstances with complex systems. We expect this is not only the present case but will continue in the future. In fact, Jackson [9] identifies multiple systems-based approaches to deal with complex systems problems—ranging from emphasis on dealing with complex systems across technical, process, structure, organizational, and coercion emphases. Similarly, Keating [8] has identified 15 different systems-based approaches to deal with complex systems/problems. The different approaches demonstrate the difficult nature of selecting an appropriate methodology(ies) for dealing with complex

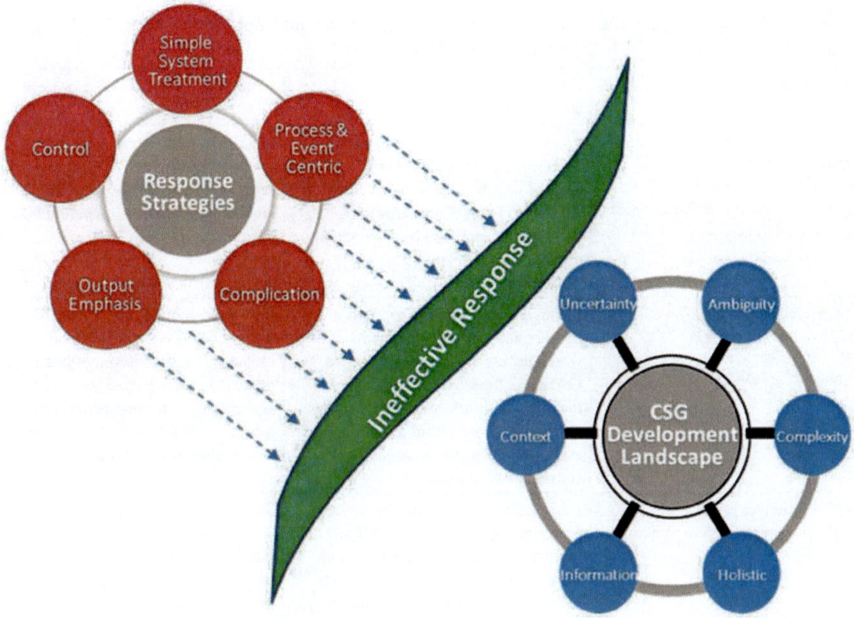

Fig. 1 Ineffective response strategies to CSG landscape

systems/problems. Nevertheless, as opposed to 'do nothing', something must be done if we are to enhance our prospects for making improvements to complex systems and more effectively addressing their constituent problems.

CSG development methodology offers a unique and distinguishable methodology for complex systems. Methodology is consistent with Checkland's [10] perspective of a methodology, which suggests that a methodology provides a framework, more specific than philosophy, but more general than a detailed method or tool. Therefore, a systems-based methodology must provide a framework that can be elaborated to effectively guide action. We have previously established our attributes for a systems-based methodology (Keating et al. [11]). Among these attributes are *transportability* (capable of application across a broad spectrum of applications), *theoretical and philosophical grounding* (linkage to theoretical body of knowldege), *actionable* (capable of leading to specific actions), *significance* (holistic capacity to address multiple problem system domains), *consistency* (approach capable of replication), *adaptable* (able to be modified to different circumstances for application), *neutrality* (sufficiently transparent to preclude biases), *multiple utilities* (capable of application for a range of system development initiatives),*and rigorous*(capableof withstanding external scrutiny). Given this set of distinguishing attributes of methodology, which will be subsequently expanded, the implications for CSG development, baseon functions and communication channels, are summarized in Table 2. Although the listing is certainly not intended to be exhaustive, it does offer insights for our thinking with

Table 2 CSG functions, associated communication channels, and implications for CSG development

CSG function	Description and associated communications channels	Implications for CSG development
Metasystem five (M5)—policy and identity	Provides for overall steering and trajectory of the system. Maintains identity and balance between current and future focus. Communication channels—Command (nonnegotiable directives) and Algedonic (system warning)	Identity provides a reference point to ensure consistency in decisions, actions, and interpretations. This serves to guide system development priorities and execution
Metasystem five star (M5*)—system context	Focused on the systems of interest specific and unique context within which the metasystem is embedded. Context is the set of circumstances, factors, conditions, trends, or patterns that enable or constrain execution of the system. No directly associated communication channels	CSG development is tempered by the context within which it must be executed. Feasible development is impacted by contextual constraints and enablers. Development must include contextual development as well as the system of interest
Metasystem five prime (M5')—strategic system monitoring	Provides oversight of strategic system performance indicators, identifying performance that meets, exceeds, or fails to meet established expectations. Informs the other functions, primarily M4 and M5 as to strategic trajectory performance. Communication channel–Dialog (examination of purpose and essence of system)	Strategic system performance takes a long view perspective for measuring CSG evolutionand trajectory over time. This is opposed to 'operational performance' which is focused more near term and local 'system centric'
Metasystem four (M4)—system development	Maintains the models of the current and future system. Focused on long range development of the system to ensure future viability(existence). No directly associated communication channels	Purposeful system development requires that the current state and the desirable future state be articulated and managed for shifts over time
Metasystem four star (M4*)—learningand transformation	Facilitates system learning based on correction of design errors in the metasystem functions and planning for transformation of the metasystem. Communication channel—Learning (identification of system adjustments to variabilities)	CSG development requires that learning (detection and correction of errors) be focused on system redesign (modification of design) as well as efficiency (execution within current design)
Metasystem four prime (M4')—environmental scanning	Designs, deploys, and monitors sensing of the environment for trends, patterns, or events with implications for both present and future system viability. Communication channel—Environmental scanning (provides intelligence of external conditions)	Understanding constraints and enablers stemming from the environment is critical to CSG development. The environment, and its shifts, will both enable and constrain CSG performance

(continued)

Table 2 (continued)

CSG function	Description and associated communications channels	Implications for CSG development
Metasystem three (M3)—systemoperations	Sees to the day-to-day execution of the to ensure that the overall system maintains established performance levels. Communication channels—Resource bargain/accountability (resource distribution and output expectations) and operations (providing directions for system operations)	CSG development must focus on the long range. However, there must be a balance established between long and short range focus and resource distribution
Metasystem Three star (M3*)—operational performance	Monitors operational system performance. Concentrates on identifying and assessing aberrant conditions, exceeded performance thresholds, or anomalies. Communication channel—Audit (provides monitoring of routine as well as emergent anomalies in system performance for variabilities)	Operational performance must be considered for CSG development. The impact of development decisions on present operations must be considered
Metasystem two (M2)—information and communications	Designs, establishes, monitors, and maintains the flow of information and consistent interpretation of exchanges (communication channels) necessary to execute metasystem functions. Communication channels—Coordination (provides for harmonizing elements within the system) and Informing (providing for routine information in the system)	Design and execution of communications is critical to both present as well as future system development. Communications must respect both present and future considerations as CSG development is undertaken

respect to the establishment of the CSG development methodology. In this chapter, we examine CSG methodology as a systems theory-based, conceptually grounded, and action-oriented approach to dealing with complex systems. These systems are subject to the problems consistent with Ackoff's [12] notion of 'messes' (interrelated sets of problems that are not well formulated, understood, or easily resolved) as well as Rittel and Webber's [13] depiction of 'wicked problems' (problems that are intractable with current levels of thinking, decision, action, and interpretation).

To serve our primary purpose of exploring CSG development methodology, we have organized the chapter to accomplish four primary objectives. First, we present an overview of the CSG methodology. This methodology is consistent with, and expands, a previous first generation methodology for CSG developed by Keating and Katina [8, 14]. This examination is focused on elaboration of a three phased approach to achievement of CSG development. Second, each of the three stages of the CSG development methodology are examined. Third, we examine critical issues in the deployment of CSG to develop a complex system. Critical issues, as well as suggested mitigation strategies, are explored. Fourth, a set of application-based insights is explored. The chapter closes with a summary that capsules the critical points of the chapter.

2 CSG Development Methodology Background

As a response to the difficulties in dealing with complex systems, CSG has emerged as an evolution of system of systems engineering (Keating and Katina [3, 5, 8]). CSG is defined as the 'design, execution, and evolution of the [nine] metasystem functions necessary to provide control, communication, coordination, and integration of a complex system.' (Keating et al. [15]). CSG is a theoretically grounded (systems theory, management cybernetics, system governance), model-driven (CSG reference model), action-oriented (definition of strategies and actions to improve a situation) approach to capture and understand complex systems. Two primary drivers of CSG, and foundations for a CSG development methodology, are the theoretical underpinnings and the derivative reference model. Although the complete coverage of CSG is beyond the scope of this chapter, we provide the essence necessary for systems theory and the CSG reference model for the CSG development methodology.

2.1 *The Essence of System Theory for CSG Development Methodology*

There is not a singular or widely accepted definition of systems theory. The most basic tenet of systems theory, *holism*, can be traced to the writings of Aristotle, who proclaimed that the whole is more than the sum of parts. Since the initial development of systems theory in the 1940's, there are a host of scholars and practitioners who have been recognized as instrumental in systems theory development, including such notable individuals as Anatol Rapoport, Norbert Weiner, Karl Ludwig von Bertalanffy, and Ross Ashby (Klir [16]; Laszlo and Krippner [17]). Systems theory emerged as an alternative to 'reductionism,' which is based on the concept that a system can be understood by successively 'breaking it down' to the level of parts. In effect, from the parts a complete understanding and objective knowledge of a system is possible from a reductionist perspective. In contrast, *'holism'*, as the most fundamental attribute of systems theory, holds that a system must be understood in terms of the emergent properties that result from interactions and relationships between elements in a system. Thus, complete system knowledge is not possible and certainly not deducible from the parts independent of their interactions. Following earlier works from Adams et al. (2014) and Whitney et al. (2015) we suggest the following points that systems theory holds for CSG:

1. Offers a set of axioms (taken for granted knowledge) and propositions (collection of principles, laws, and concepts that explain the behavior, structure, and performance of systems).
2. Suggests that violation of system propositions carry consequences and contribute to diminished system performance or outright failure.
3. Provides a theoretical and conceptual grounding that anchors CSG in a stable and enduring body of knowledge.
4. Serves to inform understanding, explanation, and 'plausible' prediction for system behavior and performance.
5. Provides insights and cues into more effective design, execution, and development of governance for complex systems.
6. Enhances capacity for more effective thinking, decision, action, and interpretation with respect to complex systems and their problems.
7. Offers a worldview, rooted in *holism*, that defines how CSG embraces complex systems, situations, and problems that are encountered.

Systems theory offers the doctrine which provides a bridge between systems science and CSG.

2.2 The CSG Reference Model

The CSG reference model is a representation that describes the specific functions and communication channels that must be performed to govern any complex system. The reference model includes nine metasystem functions. The functions (Keating and Bradley [18], Keating and Katina [5]) and their implications for CSG are identified in Table 2.

For the metasystem functions identified in Table 2 there are four important points of emphasis. First, the functions do not operate independent or mutually exclusive of one another. Instead, they are interrelated and affect, and are affected by, the other functions. Second, the functions are performed by mechanisms. Mechanisms are the governance artifacts that permit achievement of the specific functions. For example, a quarterly strategy meeting might be a mechanism to support the M4 development function. The total set of mechanisms for the metasystem functions determines the 'set adequacy' given the unique system and context. Third, the execution of the metasystem functions determines the level of governance effectiveness and ultimately system performance. Fourth, governance effectiveness will also be affected by the degree of 'purposeful' design of the metasystem functions. Without engagement in purposeful design (construction of the set of mechanisms to perform metasystem functions) it is doubtful that CSG development will achieve intended performance improvements. Thus, purposeful design of metasystem functions serves to enhance governance in a holistic fashion, avoiding a piecemeal or ad hoc approach to governance and ultimately system performance.

2.3 *Making CSG Actionable Through Methodology*

The underlying theoretical grounding for CSG is anchored in three fields, including: *systems theory* (the set of axioms and corresponding propositions that explain and predict the behavior and performance of complex systems), *management cybernetics* (described as the science of effective system structural organization), and *system governance* (the provision of system direction, oversight, and accountability). However, development must focus on a different aspect of CSG. The development focus is on examination as to 'how' CSG can be engaged to improve a complex system or address its problems. In essence, how we can make the CSG theoretical foundations and derivative CSG reference model actionable for improvement of governance for complex systems. This becomes the role of a CSG development methodology. CSG development is the 'purposeful exploration and development of governance functions for a system of interest'.

The CSG development presented in this chapter continues to evolve. This evolution progresses as new applications are engaged and our knowledge of CSG, its underlying theoretical/conceptual foundations, and learning from real world applications continue to advance the methodology. The concept of methodology is certainly not new. In fact, notwithstanding the newness of the CSG field, the current state of research in CSG development methodology is sufficient to suggest examination beyond the first generation high level articulation first posed by Keating and Katina [8, 14]. That first generation approach has evolved as new discoveries continue to emerge from intensifying research exploration and applications of the developing CSG field. This chapter articulates the current state of knowledge for CSG development methodology.

Methodology can be an imprecise term. Our current examination of CSG development methodology follows Checkland's [10] perspective which suggests that a methodology provides a framework, more specific than philosophy, but more general than a detailed method or tool. Therefore, a systems-based methodology must provide a framework that can be elaborated to effectively guide responsive action. Based on prior works on methodology development [8, 14, 19], Table 3 expands the set of attributes mentioned earlier. These attributes should be considered as essential to an effective methodology, as well as the implications specifically targeted to CSG development methodology.

Having established the conceptual foundations for CSG, we now shift attention to articulating the current state of the CSG development methodology.

3 CSG Development Methodology

CSG development is the 'purposeful exploration and development of governance for a system of interest'. CSG development methodology is the high level approach that identifies 'what' must be done for development. The specific details of 'how'

Table 3 Attributes of a methodology and CSG implications

Attribute	Description	CSG development methodology implications
Transportable	Capable of application across a spectrum of complex systems, problems, and contexts associated with the discipline. The appropriateness, or applicability, of a methodology for a range of circumstances and system problem types will confirm (or not) any claim of transportability	The CSG development methodology must remain applicable across an extensive range of systems, problems, and contexts. However, there must be a tempering for application, as no methodology has universal applicability
Theoretically and philosophically grounded	Grounded in an explicit theoretical body of knowledge and philosophical underpinnings. This guides the appropriate application, utility, and expectations of the methodology	Systems theory is the theoretical body of knowledge to which the CSG development methodology is grounded. This includes the principles, laws, and concepts that delineate the behavior, structure, and performance of complex systems. Also, philosophical (ontological, epistemological, and methodological) underpinnings are found in 'systems', instantiating conceptual foundations stemming from holism
Actionable	Provide detail sufficient to frame and guide actions appropriate for methodology application. They must not prescriptively define implementation details. However, they must define at a high level 'what' must be done to proceed and become actionable	The CSG development methodology is constructed to define 'what' must be done for application. However, the methodology stops short of providing constraining details of 'how' it must be deployed. This is left to the detailed application planning for deployment
Significance	Exhibit a capacity to drive significant understanding, action, and improvement across a holistic range of technology, organizational, managerial, human, social, political, and policy dimensions of a complex system	CSG development methodology is focused to turn advanced understanding into strategies, actions, and activities to make system improvements. These improvements cross the holistic spectrum of a complex system

(continued)

Table 3 (continued)

Attribute	Description	CSG development methodology implications
Consistency	Provide replicability of approach and results interpretation based on deployment within similar contexts. They must be transparent, with clearly delineated details regarding the design, analysis, and transformation of systems	The generalized framework for CSG development methodology provides a consistent approach. However, it also provides for tailoring to the unique nature of a system of interest and its context
Adaptable	Capable of responding to changing conditions or circumstances by allowing modifications of approach, configuration, execution, and expectations while retaining their fundamental frameworks. This is methodological latitude	The CSG development methodology retains flexibility to adjust to shifts in system knowledge, context, or problem framing. This permits adjustments prior to and during execution
Neutrality	Account for and limit external influences on design, execution, and interpretation of results. Their use must explicitly identify biases, assumptions, and limitations integral to application	Biases will exist in application of the CSG development methodology. However, care must be taken to acknowledge and minimize their potentially negative influence on design, execution, and interpretation of results
Multiple utilities	Support a range of applications, ranging from limited to comprehensive. Targeting desirable results ranging from identification of feasible system improvements (e.g., workforce development) to comprehensive system transformation (e.g., major system redesign effort)	CSG development methodology is capable of providing a range of results. These results can range from limited feasible improvement initiatives to comprehensive design for continuous system transformation
Rigorous	Sufficiently detailed to permit consistency in design for execution, irrespective of the uniqueness of the system of interest and the context with which it is embedded	A CSG development methodology must provide sufficient detail to define precisely what will be done and permit tailoring to define how it will be accomplished

the aspects of the methodology are accomplished are left to the detailed design that must be tailored to the unique circumstances, system, and context of the system. CSG development has previously been presented as a first generation methodology [8, 14]. Since this first generation, there have been significant strides forward in our understanding of deployment of CSG in operational settings. In this section, we discuss the current state of CSG development methodology based on shifts in our knowledge.

The CSG development methodology consists of three primary stages that define CSG development. Recall that the governance functions must be performed by any system that maintains viability (existence). However, just as each system is unique and exists in a unique context, the specific approach to CSG development must be tailored to appreciate that uniqueness. For succinctness, we have identified and elaborated the three primary stages of CSG development from the initial work [8, 14] describing the first generation CSG development methodology (Fig. 2).

The first stage of CSG development is *initialization.* This stage provides an initial understanding of the situation, answering the question *'What is the state of the system of interest and its context?'*. Initialization consists of two primary objectives. First, the nature and structure of the system of interest is established. This serves to articulate the current state of the system under exploration. Second, the context within which the system of interest is embedded is explored. Initialization provides a rigorous systems-based understanding of the system and its context. Completion of the *Initialization* stage provides a foundation for the second stage of CSG development, *development mapping.*

The second stage of CSG development, *development mapping*, is focused on establishing the analysis and implications for what was learned from *initialization.* This stage seeks an answer to the question, *'What do the different products from the initialization stage suggest for development?'*. This requires a deep introspection

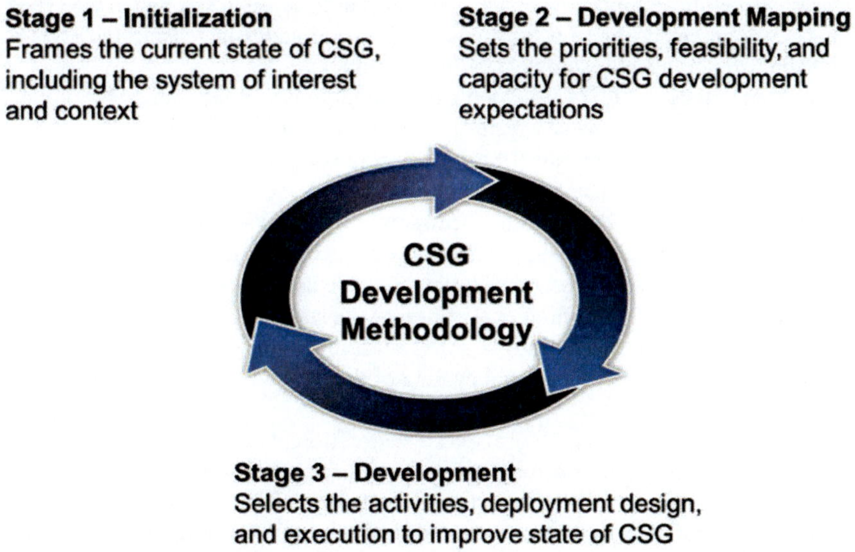

Fig. 2 Three stages of CSG development methodology

into the results of the *initialization stage*. Ultimately, the results of the *development mapping* stage specifies: (1) a profile that establishes the current state of CSG performance/maturity/sophistication, and (2) the system and context to determine the nature, types, and priority consideration for CSG development with the potential for the greatest impact.

Development is the third stage of the CSG development methodology. The two prior stages were passive and not directed to initiation of action/activities to enhance the state of CSG. This third stage identifies and engages the feasible activities (based on priorities, capability, capacity, and resource constraints) that can be undertaken in development of CSG. Three important aspects to this stage include: (1) determining the feasibility as to the different types of activities that might be successfully engaged based on the current state of CSG (established in the initialization stage), (2) the CSG development stage also includes the prioritization of activities—based on consideration of the overall state of governance, context, and resources—directed to making more informed CSG development investment decisions with greatest potential impact, and (3) selection and execution of suitable activities that are targeted to make either 'contextual' improvements or metasystem governance function improvements. Therefore, the success of initiatives can be evaluated against the shifting profile of *context* and *governance state* to which they are targeted. While not absolute in their prioritization and selection, a much more rigorous and holistic selection process can be engaged. Hence, the continuous development of CSG through an evolutionary approach is engaged. Selection of 'appropriate' initiatives results in increasing GSG capabilities and advancing the context. The result is that the 'feasible development initiative space' continues to increase in size and depth. Thus, CSG development becomes a 'virtuous circle' continually increasing the level of CSG sophistication and resulting state of performance.

The three stages of CSG development methodology are presented with a clear degree of separation. However, it must be noted that this is purely for the convenience of presentation. In reality, their separation is not clear cut. The three stages do not operate independent or mutually exclusive of one another. Instead, there is a constant comparative nature to their progression. For example, it is possible that the mapping or development stages may suggest a 'recalibration' of the initialization stage outputs/outcomes. This is particularly the case as new knowledge is generated by the continuing and deepening exploration of the state of governance/context and execution of development initiatives. After an initial 'first pass' through the three stages of development, the continual cycling of the methodology increases the state of governance.

In the following sections, we provide an elaboration of each of the three stages of the CSG development methodology. This perspective of CSG development continues to mature, as our sophistication, knowledge, and insights evolve with each successive application of CSG development. The initial works on the CSG field and methodology have provided a strong theoretical and conceptual grounding for our current state of development, including the CSG reference model [20], the emerging CSG field (Keating and Katina [5, 21]), and engaging CSG (Keating and Katina [5]).

To set a frame of reference for the detailed examination of CSG development, six overarching themes are provided for a supporting context. This context is essential to understanding the limitations and development expectations for CSG.

1. *Continuous and Deepening Development Cycle*—Engagement in CSG development is not intended to be a singular event with a clear 'stopping point'. Instead, it is cyclic in nature in that it is intended to operate on a continuing basis. Also, over successive development 'cycles' the expectation is that the state of CSG continually matures.
2. *Systemic Worldview Limits Pace and Comprehensiveness of Development*—The state of systemic thinking (worldview) held by the individuals and the system are taken into account in the initialization stage. However, the pace and depth of CSG development will be limited by the systemic worldview.
3. *Contextual Development Accrues*—Development initiatives for CSG also include development of the context in parallel with the evolving state of CSG. Context is both enabling and constraining to CSG and must be considered in development. Not taking context into consideration is short sighted and will not generate the potential advances that can result from including context.
4. *Development Targets Multiple Levels*—Development is not restricted to the complex system of interest. Instead, in addition to the system of interest, CSG development must also focus on individual, organizational, support infrastructure, and context. Thus, CSG development is holistic in the areas subject to development activities, which may span multiple areas. It is constraining to only consider the state of CSG in selection and execution of development initiatives.
5. *Emphasis on Front End Framing*—The application of CSG development is heavily weighted to the 'initialization stage' which exist at the front end. This sets the stage for all that follows. Although all stages are important, the focus on the initialization stage is critical to get correct. If the framing is insufficient, poorly performed, or incomplete there is little chance that the effort will provide the intended utility. Additionally, through repetitive 'cycling' the initialization becomes *re-initialization* which is a continual recalibration of the state of CSG and the context for the system of interest.
6. *The Metasystem is a Unifying Concept*—The metasystem is the set of functions that must be performed for systems to remain viable (continue to exist). These functions identify *'what'* must be achieved, not *'how'* they must be achieved (by specific mechanisms). Ultimately, the performance of the metasystem functions and their associated communication channels determines the level of system performance capable of being achieved. The metasystem provides governance (*communications, control, integration, and coordination*) for the system entities to operate as a unity to produce value which is consumed external to the system. CSG maintains system coherence (identity) and cohesion (unity). In effect, at the most basic level, the metasystem keeps the system from either collapsing from external pressures or flying apart from internal pressures. It is the 'glue' that allows the system to continue in the face of increasing complexity.

Thus, development has the focus of both maintaining as well as evolving the metasystem functions, their communication channels, and the context.

With this basis for the CSG development methodology we move forward to explore in depth the three stages of the methodology.

Vignette
What about the metasystem?

The importance of the metasystem is vital to system effectiveness. However, in this situation we describe a scenario where the metasystem functions of an organization were identified as underdeveloped. This organization recognized that there were difficulties in being able to adequately respond to customer issues. The university-based organization was comprised of several departments, each with its specific set of responsibilities and roles to perform in the operation. For instance, there were separate departments for finance, admissions, registration, housing, food services, and student engagement. However, participants acknowledged that there were several apparent issues, including: (1) customers rarely had an issue that was confined to one department for resolution, (2) there were coordination problems between the different departments in dealing with issues that 'fell between the cracks,' (3) while the different departments operated efficiently [by their individual measures], the overall system was deficient in performance without any true measures established beyond individual departments, and (4) communications between departments, particularly where issues were not 'owned' by an individual department was ineffective. Through a participatory CSG exploration of the system, there were several conclusions that became apparent. First, the metasystem functions that were responsible for system communication, control, coordination, and integration were largely left to be assembled and executed absent of purposeful design (largely self-organizing). Thus, the orchestration of the departments in a coherent and cohesive way was absent. Second, lacking effective mechanisms to perform metasystem functions, issues that required interactive coordination between multiple departments was difficult and inconsistent at best. The result was that each customer issue was treated as a unique case, to be managed by whomever elected to accept the challenge to resolve it by bouncing back and forth between different 'involved' departments. Third, the individual departments were managed and operated very effectively. However, absent the purposeful design of the metasystem, the system struggled to perform when problems spanned multiple departments. This system demonstrated the necessity of the metasystem to effectively integrate, coordinate, control, and communicate by design as opposed to being left to self-organization.

3.1 CSG Development Methodology Stage 1—Initialization

The first and arguably the most important stage of CSG development is initialization. This stage accomplishes two primary objectives. First, the context for the system of interest is examined. Recall that the context is the set of circumstances, factors, conditions, trends, and patterns that influence, and are influenced by, the design, execution, and evolution of the system of interest (Keating et al. [22]; Keating and Katina [5]). The establishment of the context provides a critical set of insights into what might be influential in constraining/enabling the execution of CSG development. Table 4

Table 4 Framing the context for CSG development

Context activity	Purpose	Contributions/implications
Contextual attributes identification	Identifies the nature of forces that constrain or enable the design, execution, or evolution of the metasystem	• Forces may be internally/externally generated, formal/informal, tacit/explicit, or real/perceived • Forces may include circumstances, factors, trends, patterns, or conditions that influence the metasystem • Forces may range across the spectrum of technical/technology, organizational/managerial, human/social, cultural, information, and political/policy • The worldview(s) in play for the system (including values, beliefs, and logic) must be made explicit for examination • Context and its understanding are not static. Context can and will change over a development effort. This can be by design or simply by emergence of new knowledge. Context must be kept up to date throughout a CSG development effort
Individual capacity for systemic thinking	Establishes the level of systemic thinking that exist among those (owners, operators, designers, or performers) with responsibilities for design, execution, and development of the metasystem	• The level of systems thinking for individuals is instrumental in setting context • Ultimately the development level and activities capable of being undertaken are constrained, and enabled, by the systemic thinking capacity of those who will engage CSG development • Determination of feasible activities and development expectations must be metered by the level of systemic thinking present in participating individuals • A diversity in systems thinking capacity is desirable to avoid an overly homogeneous representation

(continued)

Table 4 (continued)

Context activity	Purpose	Contributions/implications
Entity competence for systemic thinking	Provides the level of knowledge, skills, and abilities related to systemic thinking for organizations (systems) contemplating engagement in CSG development	• While individual capacity for systemic thinking is necessary for determination of CSG expectations, it alone is not sufficient for this determination • The aggregate level of systems thinking capacity will be a major determinant in CSG development and how much scarce resources need to be allocated to enhancing the aggregate level of systems thinking • At the organization (system) level the degree of joint proficiency in collective systemic thinking knowledge, skills, and abilities will meter the feasible activities and expectations for CSG development
Supporting infrastructure compatibility	Establishes the degree to which the basic physical and system support infrastructure (e.g., support systems, processes, procedures, facilities, and resources) are enabling or constraining for CSG execution and development	• Irrespective of good intentions or redesign of the metasystem for CSG, the supporting infrastructure must not be in conflict with the execution or development activities for a CSG initiative • Compatibility issues, be they conceptual or physical, must be taken into account in initialization for CSG development • Support infrastructure can be both enabling as well as disabling and subject to change over time and with new knowledge
System leadership Assessment	Identifies the degree to which the existing state of leadership in CSG is consistent with that required for development	• System leadership is a critical aspect of context for CSG • The nature and role of leadership existing in a system offers both systemic constraints and enablers for CSG development • System leadership is 'different' than traditional notions of leadership in both what is influenced, and the specific role played in CSG development

(continued)

Table 4 (continued)

Context activity	Purpose	Contributions/implications
Communication channel identification	Communications is focused on the flow and interpretation of information in CSG metasystem functions	• The different mechanisms, and their effectiveness, for performing the different communication channels required for CSG are established • Both formal and informal communications are considered • The degree to which communication is effective will limit or enable CSG development

is elaborated from earlier work in CSG development [8, 14] and provides a description of the activities, their purpose, and the contributions/implications for framing of context during the initialization stage of CSG development. Second, the current state of CSG for the system of interest is established. This constitutes the second aspect of framing for the initialization stage. The current state articulation involves mapping of the system of interest, the environment, governance architecture/requirements fulfilment, pathologies (system governance deficiencies), and balance. In sum, the initialization stage sets a 'baseline' from which further CSG development progress will be informed.

Establishing the current state of the system of interest for CSG development is the second aspect of framing conducted in the initialization stage. This requires *'framing'* for the system of interest. This operates in conjunction with the establishment of the context for the system of interest. Framing of the system of interest provides a set of representations that serve to depict the metasystem in relationship to the system(s) that it governs. The system of interest framing establishes the design configuration and execution of the metasystem, articulating the technical design details of the metasystem as well as the effectiveness in execution of that design. The major elements of the *metasystem framing* activity in the initialization stage of CSG development are described in Table 5 (adapted and elaborated from [8, 14]).

The primary product from the initialization stage is a CSG profile. This profile represents the current state of CSG and the context for the system of interest. In addition, the initialization stage is the most intense of the three stages of CSG development. However, if the appropriate level of energy and resources are not invested in the execution of this stage, the remainder of the CSG development stages are sure to be suspect at best. The initialization stage provides a baseline against which: (1) further analysis can be conducted to identify and prioritize developmental areas, (2) shifts in the 'governance landscape' can be captured, (3) the specific fit of future CSG development initiatives can be determined, and (4) holistic development of CSG can be supported based on the comprehensive picture of the state of CSG provided from the profile generated during initialization. It should be noted that what is discovered

Table 5 Framing the metasystem for CSG development

Framing activity	Purpose	Contributions/implications
System of interest identification	Identifies the system for which CSG will be examined, noting the boundary conditions as well as the metasystem governance configuration	• Definition of the system of interest must be made explicit (included entities/systems/subsystems, relationships, transformation, and boundaries) to focus the development effort • Definition of the specific criteria for inclusion/exclusion to establish the boundary conditions and separate the system from the environment is essential • The system of interest identification will be incomplete, fallible, and change over time. It should be continually modified with new knowledge and development initiative improvement results
Environment definition, mapping, and assessment	Define the environment within which the system is embedded. Focus on definition of relevant aspects outside the boundary of the system of interest. Map critical aspects of the environment and assess their implications	• Definition of the environmental forces (enabling/constraining) must be made explicit (mapped) and assessed for CSG implications • All aspects of the environment are not of equal relevance for CSG development • The most relevant (influential) aspects of the environment must be taken into consideration • The environment, its relevant aspects, and assessment will change over time and with new knowledge throughout the development effort

(continued)

in the initialization stage will evolve as new knowledge and understanding of the system of interest and context emerge throughout the effort. *In essence, the initialization stage provides the current state of CSG and captures the context within which it is embedded.*

Table 5 (continued)

Framing activity	Purpose	Contributions/implications
Governance architecture definition	Establishes and represents the particular 'architectural' views in a CSG architecture framework following Carter (2016)	• The different architectural views for CSG provide the relationships between the different mechanisms performing the metasystem governance functions • Supports the discovery, development, and maintenance of information necessary for evolution of the governance architecture • Produces (model centric outcomes/representations) of the structure, behavior, and performance of CSG • Facilitates greater understanding of the system of interest and facilitates identification and prioritization of deficiencies targeted for system performance improvement
Reference model requirements assessment	Provides an examination of the function of CSG against the requirements specified for the CSG reference model	• Assessment allows determination of 'how' the metasystem functions are performed against the 'what' must be performed (established by the requirements) • Gaps in the adequacy of coverage for metasystem functions can be identified as well as the effectiveness of individual mechanisms being used to support achievement of those functions
Metasystem pathologies identification and assessment	Establishes the degree to which variations from systems theory propositions (principles, laws, and concepts) are perceived to impact performance of CSG	• Identification of pathologies can be established against the backdrop of systems theory and the functions of the CSG reference model • Assessment of identified pathologies provides insights into the conditions that limit CSG performance

(continued)

Table 5 (continued)

Framing activity	Purpose	Contributions/implications
State of system balance	Identifies the classification for a system of interest and positions the system balance along tensions in design, change, and control dimensions. The past, present, and future state balance is identified	• Classification of system balance provides insights into the nature of the system and the corresponding implications for CSG based on the balance assessment • The specific positioning of a system with respect to the degree to which tensions are perceived to be appropriately balanced provides insights into effectiveness of CSG performance • Perceived gaps between past, present and future state balance are identified and interpreted

3.2 *Stage 2—Development Mapping*

The initialization stage for the CSG development methodology provides a significant data set that serves to *'frame'* the context and state of CSG for the system of interest. The second stage of CSG development methodology, *development mapping*, is focused on understanding the nature, meaning, and implications of the data and representations produced in the initialization stage. The original CSG development methodology [8, 14], identified this stage as *governance readiness level (GRL)* assessment. Our applications and further explorations of CSG suggested that GRL was too narrow in its formulation. Thus, *development mapping* was constructed to widen the nature and scope of this stage. The *development mapping* stage is directed to achieve three primary objectives:

1. *Process the results from the Initialization Stage.* The initialization stage produces a plethora of outputs (framing) that serve as inputs to the development mapping stage. Among these outputs are the depiction of the system of interest, the system of interest metasystem functions, and the context. Processing includes identification of the critical themes, insights, and implications stemming from the initialization stage outputs. A critical aspect of CSG development mapping is examination of the system context to identify enabling and constraining forces that influence the performance of CSG functions and will limit/enable CSG development.
2. *Identify, catalog, and rank order pathologies that exist in CSG functions with respect to their existence, impact, and resolution feasibility.* Pathologies are aberrations from normal or healthy conditions in the metasystem functions for CSG [3]. This provides a state of CSG functions performance for the system of interest. In effect, the *"CSG Landscape"* is established.

3. *Define the governance readiness level (GRL).* The GRL is a classification of the 'maturity' that exists in CSG for the system of interest. The GRL will determine the types of CSG development activities that can be undertaken with a reasonable feasibility of success. There are currently nine levels of CSG readiness (Table 6) for classification. This classification serves to position CSG for the system of interest along a spectrum that corresponds to the 'maturity' of CSG. Implications

Table 6 Escalating complex system governance readiness Level classification

GRL	Description	Assessment implications
Nascent	No knowledge of CSG or the systemic worldview essential to understand CSG functions. No representations of the CSG functions, communications channels, or state of CSG. Unknown state of systemic thinking, development constraints, environment, or potential for engagement	• 'Clean sheet' CSG development possibilities • Requires training/education to develop rudimentary understanding of CSG, systems thinking, and the possibilities that development offers • Relevance/utility not apparent
Embryonic	Minimal knowledge of CSG or the systemic worldview necessary to engage in CSG development. No representations of the CSG functions, communications channels, or state of CSG. Unknown state of systemic thinking, development constraints, or potential for engagement	• Some minimal CSG exposure • Requires additional depth/sophistication through generalized education/training on the systems worldview, nature of CSG, and potential for development • Requires training/education to development rudimentary understanding of CSG and the possibilities that development offers • Relevance/utility considered
Forming	First level engagement in CSG and developing the essence of the systemic worldview which underpins CSG. No representations of the CSG functions, communications channels, or state of CSG. Unknown state of systemic thinking, development constraints, or potential for engagement. Utility and value not entirely clear with respect to expectations/potential offered by CSG	• Demonstration of CSG development opportunities through training/education • Understanding the potential that CSG development might offer • First contact looking superficially at the state of CSG and the environment • Requires training/education and application exposure to CSG
Formulated	Engagement in CSG with appropriate level of systemic thinking. Articulation of critical aspects of CSG functions performance. State of systemic thinking and inconsistencies identified, contextual considerations examined	• Systems Thinking capacity understood with respect to environmental demand • Training/education in systems escalated to enhance systemic worldview • Impacts of CSG development utility understood, acknowledged, and supported through realization of CSG state deficiencies

(continued)

Table 6 (continued)

GRL	Description	Assessment implications
Implemented	CSG development methodology deployment initiative planned, resourced, and implemented. Clear designation of preliminary state of CSG, environment complexity, and systems thinking capacity. Expectations for CSG development specified	• CSG development engagement foundations established • Appropriate level of systemic thinking capacity determined and adequate • Clear roles, responsibilities, and accountability are established • Development strategy, timing, resources, and expectations articulated • System performance measures and initiative success defined
Developing	First pass through CSG development methodology accomplished. Framing of context and system of interest established. Composite mapping of the state of CSG constructed, pathologies identified, feasible development initiatives defined and implemented. First generation mapping of the system of interest, CSG functions, communication channels, and environment implemented	• Assessment of development initiative(s) success conducted • Advancement in systemic thinking capacity, state of CSG, and communication channels identified • Clarity in definition of representations for the system of interest, environment, and context established
Developed	Continual passing through CSG development methodology stages. Deepening systems thinking capacity and development endeavors that can be engaged. CSG functions improvement, pathologies reduced, and performance indicators improvement achieved	• Embedding of CSG development into support infrastructure, processes, and context development • CSG ingrained into strategic operations as a 'way of doing business' • Systems language and thinking are routine • CSG development continues to progress in maturity and depth of sophistication
Evolving	CSG is not 'in addition to' the strategic work of the system but becomes the primary work of the system. Additional resources are not allocated specifically for CSG. Instead, CSG is just the approach to accomplishing the work of the system. Continuous evolution of system design, execution, and development are the routine	• CSG is no longer separate from the system of interest • CSG functions and communications drive thinking, decisions, actions, and interpretations for the system • CSG design, execution, and development is a preoccupation of system leadership

(continued)

Table 6 (continued)

GRL	Description	Assessment implications
Evolved	CSG begins to be projected beyond the system of interest. Influence, through demonstration, demands, and expectations become routine to the system. System value is not only in what products/services are produced, but how they are produced. New employees, stakeholders, customers, and entities are exposed to the CSG way of accomplishing the work of the system. The system culture does not distinguish CSG as separate from the operation, but rather it defines the operation of the system	• System identity provides an unambiguous reference point for clarity in decisions, actions, and interpretations for the system • External engagement is transparent, trusted, and effective • CSG is embedded in processes, procedures, support infrastructure, and context without provocation • Development of external entities is achieved by subjecting them to the standards and expectations upon which the system has been designed, executed and evolved

for CSG development, based on the GRL classification, are identified. The implications include identification of the nature and types of feasible activities that can be pursued for CSG development with a reasonable chance for success.

Significant progress has been made in the second stage for CSG. However, it still remains the least mature of the three stages of the development methodology. As this stage continues to develop, there are three important aspects driving future development. First, the processing of the initialization stage results, ultimately framing the state of CSG, captures (measures) a present state of CSG development. There is not a judgment of good or bad, but the state simply remains a depiction of the 'maturity' of CSG and the context for the system of interest at the point of assessment. It offers a reference point that provides a source for dialog as well as a baseline against which the continuing development of CSG can be examined. Second, framing results are a limiting factor as to the nature and types of activities that are appropriate undertakings. Attempts to engage CSG development initiatives that are beyond the capacity of the system to execute are ill-advised. Thus, the activity types that are feasible, given the current state of CSG development, can be identified. This offers a sophisticated 'metering' for the types of activities that are appropriately compatible with the state of CSG development. Over time and with purposeful development initiatives, we would naturally expect the state of CSG to evolve to higher levels of maturity, increasing performance levels, and escalation of the nature and types of development activities that might be pursued.

Third, the results from the development mapping stage provide a direct input into the following governance *development* stage. The nature and scope of activities that might be successfully undertaken to advance the state of CSG are limited by the state of CSG and the context. Therefore, the determination as to whether activities are compatible is important to ensure that expectations for CSG development are consistent with the capacity held by the system. By engaging development activities that are within the capacity of the system to successfully execute, the system reduces the probability of unsuccessful endeavors to enhance CSG.

3.3 Stage 3—Development

The third stage of CSG development, *Development*, is focused on making the first two stages actionable. At a fundamental level *development* is the process through which the meta system is purposefully altered to support future viability. Development is driven by two primary considerations. First, what are the highest priority governance development activities that offer the greatest enhancements. These activities are deemed to be the high priority targets for development. Second, irrespective of priority activities, the selection of engagement initiatives must consider what is feasible. Feasibility is a function of what capacity the system has to engage activity (ies) with a reasonable expectation for successful outcomes (i.e., GRL). This involves purposeful improvement of the system of interest (design, execution, development) as well as context (support infrastructure, leadership, etc.). Ultimately, the purpose of governance development is to enhance system performance through progression of the GRL to more desirable, feasible, achievable, and sustainable levels. As the state of CSG escalates, so too does the level of improvement activities that are advisable to be undertaken. In addition, this stage also attempts to influence the context in ways that will enhance the ability of the system to perform at a higher level.

Table 7 identifies the details of the five interrelated elements that comprise the *development* stage. These five elements include: (1) *Exploration*—concentration on the performance of the metasystem functions with input from the prior two stages, (2) *Innovation*—identification and prioritization of feasible decisions and actions to improve the metasystem functions and context, (3) *Transformation*—implementation of innovation strategies to improve the metasystem functions or context, (4) *Evaluation*—continuous monitoring of the metasystem performance improvements underway, and (5) *Evolution*—monitoring long range system trajectory consistency with a desirable future state for the system of interest, state of governance, and context. It is important that these 5 activities are not intended to be mutually exclusive or independent of one another. Instead, they are continually interactive. Additionally, the development stage becomes a 'cycle within a larger cycle.' This permits the continual evolution of CSG to increasingly higher levels of maturity and ultimately enhanced system performance (Fig. 3).

There are five critical points of consideration for the continuous achievement of the development stage of the CSG development methodology. First, while the different governance development activities are presented as separate, they are not independent or necessarily executed in serial fashion. In fact, they are interrelated and overlapping. Therefore, the consideration and performance of the different activities in the development stage cycle are not mutually exclusive of one another. In essence, they set a frame of reference for a holistic and continuous engagement for CSG development. Second, development of systems is not something that would be entirely

Table 7 Interrelated activities for CSG development

Governance development activity	Purpose	Objectives
Exploration	Holistic analysis and synthesis of metasystem design, execution, and pathologies	• Systemic investigation and self-study of the metasystem initialization profile and development mapping • Identification of set completeness of CSG function mechanisms, effectiveness implementing CSG functions, and effectiveness of individual mechanisms • Conduct systemic inquiry to explore multiple perspectives of the metasystem functions and context • Identify, represent, and prioritize systemic meaning and implications of deficiencies, pathologies, and patterns • Define the metasystem current state and trajectory • Examine models of current system, future system, environment, and context
Innovation	Definition of compatible and feasible metasystem development priorities, strategies, initiatives, and actions	• Develop the high level strategy for systemic modifications to the metasystem • Identification, evaluation, and prioritization of compatible and feasible (contextually) first order (correction within existing system) and second order (correction by system redesign) initiatives to advance governance of the metasystem • Definition of capabilities and competencies (individual and organizational) necessary to engage systemic innovations to advance the GRL and evolve context • Definition of system capacity (resources, infrastructure) and compatibility for engagement of systemic innovation initiatives

(continued)

Table 7 (continued)

Governance development activity	Purpose	Objectives
Transformation	Implementation of systemic metasystem governance strategy, actions, and initiatives to influence system trajectory, GRL advancement, and contextual development	• Holistic deployment planning and resource allocation for initiatives in support of metasystem governance development • Assignment of responsibilities and accountabilities for achievement of transformation initiatives • Exploration of the potential failure modes and mitigation actions necessary to increase probability of success of launched initiatives • Launching of selected initiatives to enhance the metasystem • Assessment of 'rogue' initiatives against the deficiencies, pathologies, and priorities (blueprint) for strategic system governance development
Evaluation	Assessment of the effectiveness of metasystem initiatives and enhanced performance of the metasystem	• Identification of the minimal set of indicators that serve to show progress of the metasystem development efforts • Assess effectiveness of initiatives undertaken for systemic metasystem transformation • Provide feedback for continuing relevance of transformation strategy in light of new system knowledge, understanding, and contextual change

(continued)

Table 7 (continued)

Governance development activity	Purpose	Objectives
Evolution	Setting and monitoring the maturation and trajectory of metasystem governance and system identity	• Assure the long range purposeful trajectory of the system in response to internal and external shifts • Enhance the continuing maturity (GRL advancement) of the system of interest, taking the long view and not corrupted by short-term aberrations • Ensure continuity, sustainability, and viability of the system in relationship to changes in context and the environment • Prevent system erosion through methodical development consistent with shifting demands on the system of interest • Updating of the initialization and development mapping stage products based on evolving CSG

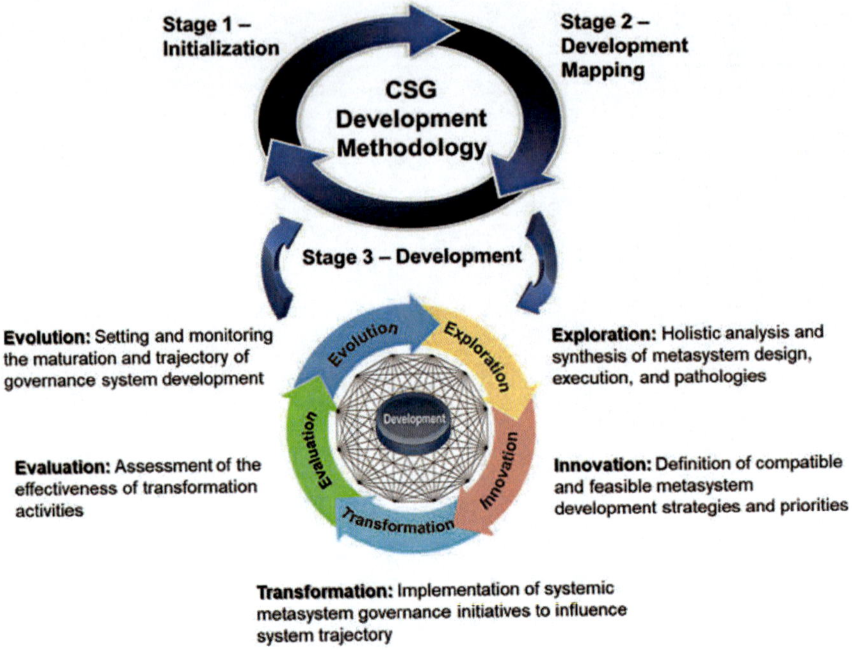

Fig. 3 Development cycle within the CSG development methodology

'new' to a system. On the contrary, systems are always undergoing different 'development activities' that enhance viability (continued existence) prospects. Unfortunately, development is frequently not achieved in an integrated, holistic, or purposeful fashion. The result is a 'hodgepodge' of activities that while well-intentioned individually, in total they are a fragmented aggregation of apparently unrelated activities. In contrast, while some system development might accrue from fragment activities, CSG development is targeted, integrated, and conducted in a purposeful manner. In this sense, the highest priority activities to address pressing deficiencies and the overall fit of activities to the 'whole of development' are key to CSG development. Third, the actions invoked in this stage of CSG development are directed to enhance the maturity of CSG. This becomes 'objectively' measured through activities targeted to make improvements in the GRL and context for the system of interest. By focusing on the input from the *initialization* and *development mapping* stages, the *development* stage proceeds from a more informed set of insights. This advantage stems from the clarity in focus from the identification of CSG gaps, setting of priorities based on the most pressing needs for CSG development. Targeted development can be pursued based on identification of what is feasible given the current state of CSG.

A fourth critical consideration involves the continuing cycling of the *development* stage. Thus, increasingly difficult activities can be taken on as the state of governance matures, the system of interest improves performance, and the context is modified to reduce constraints and amplify enabling factors. Fifth, development also serves as a 'litmus test' to understand the relationship of existing initiatives to the governance development priorities. Thus, decision makers are provided actionable intelligence concerning the contribution of different 'well meaning' activities currently underway or being contemplated to improve the state of CSG. If development initiatives, either ongoing or being considered, cannot be 'justified' as to their relevance to the most pressing needs for improving the GRL and context, they should be called into question. Additionally, the feasibility of different ongoing or planned activities should be considered with respect to their probability of success. What decision makers are ultimately provided through CSG development is: (1) a landscape of system development needs, (2) a prioritization of those needs that also considers feasibility of addressing those needs given the state of CSG and context, and (3) a sound foundation to suggest 'reallocation' of existing resources or 'redirection' of future system development resources to more productive development activities.

CSG development has not been conceived as an easy approach to system improvement. On the contrary, the development path is difficult. The approach presented is comprehensive, theoretically/conceptually grounded, and resource intensive. It requires a sophistication and capacity for systems thinking if it is to be properly engaged. It also requires a supportive context. While a comprehensive application of CSG development is preferred, this does not preclude more limited and modest CSG development activities. For example, upon discovering limitations in systems thinking capacity, there may be an initiative launched to enhance the systems thinking capacity across the workforce. The path to CSG is fraught with potential obstacles that should be considered by individuals or entities contemplating a CSG effort. In the following section, we provide some challenges that should be considered by practitioners considering pursuit of CSG development.

Vignette
The Power and Pull of the Status Quo

The full engagement of CSG development is difficult at best and ill-advised at worst. In recounting a particular attempt to engage CSG, there were several instructive points discovered. As a background, in this situation an organization was 'interested' in what CSG might have to offer and was willing to engage in a brief overview and introductory entry exercise. The entry exercise consisted of establishing a 'snapshot' of the level of systemic thinking capacity in the group, the demands being placed on the organization by the environment, and the current state of CSG in place to govern the organization (system). The results of the 'snapshot' were less than stellar for an organization that considered itself to be 'on the top of their game and industry leaders'. Three instructive points are offered. First, a clear development path forward from the initial results was desirable, but not explicitly provided. There is a clear expectation that regardless as to how CSG is presented, if the mindset is one of 'being told what to do next,' there is great difficulty in thinking that CSG will be embraced, much less successful. Second, organizations (systems) are often focused on immediate problems. CSG requires a focus on the 'long view' and development versus solving direct problems. The linkage of system development as a source of long range problem dissolution was difficult in this case, if not impossible. A mindset dominated by a focus on near term, local problems, and limited engagement expectations left little room for CSG consideration that is focused on long-term, global understanding, and more comprehensive examination. The look for the 'quick fix' placed the nature, thinking, and contributions of CSG in question. Third, when participants returned from the momentary level of thinking/exploration invited by CSG to their 'routine jobs/patterns', any progress quickly returned to a focus on the status quo. The preoccupation with immediate, here and now, issues (urgent/important) could not be suspended for engagement in the out there and future (not urgent/important) that is called for by CSG. Fourth, the very nature of CSG development is threatening to existing power structures. These structures were certain to be called into question through further exploration and immersion in CSG development. In this particular case, the introspection, transparency, and identification of governance deficiencies were too great for the status quo to risk engaging. After all, deficiencies in CSG design, execution, and development can be perceived as a threat to a leadership that has responsibilities for the CSG functions that appear to be questionable. Leaving the level of uncertainty 'as it' was considered much more palatable and less risky than uncovering governance/leadership deficiencies overseen by responsible executives. In effect, better not to ask questions that would likely produce uncomfortable, threatening, or divisive answers. Irrespective of how much there was recognition of the value that might be possible in developing CSG, the powerful 'pull of the status quo' was successful in diverting the momentary engagement in CSG back to the day-to-day issues and crises.

4 Challenges for CSG Development Methodology Deployment

CSG is a *systems-based, holistic, and purposeful approach to complex system development*. CSG offers significant value to help address some of the most vexing problems faced by practitioners (owners, operators, designers, performers) responsible for governance of modern complex systems. However, implementation of CSG development is certainly not free of difficulties and challenges. Despite the offerings of the approach, there are trepidations for engaging CSG which suggest that development should not be taken lightly. CSG development is an approach that requires continuous and purposeful design, execution, and evolution of metasystem functions. The CSG development methodology is an approach to address CSG development.

Success in CSG is not contingent on good will, noble intentions, or strength of desire. On the contrary, success in CSG will be mediated by several factors, several of which have nothing to do with the current state of CSG for the system of interest. Among these exogenous success factors are: (1) the *evolutionary design path* that has brought the system to its current state, (2) *the level of individual capacity* to engage in holistic systemic thinking and action necessary to implement CSG, (3) *organizational competency* for governance that focuses on having a requisite level of knowledge, skills, and abilities to effectively engage CSG, and (4) *support infrastructure enabling/constraining* impacts on CSG development. Each of these factors are examined below.

The *evolutionary design path* is a potential limiting factor in CSG development. The complex system design path defines how a system has come into being. System design (configuration) has three primary paths that might have been pursued. Each path has influences on how CSG development might progress. The first design path is a system coming about by *self-organization.* The *self-organization* approach to complex system design is centered on permitting the relationships and activities undertaken for system benefit to 'take their own unfettered' course of development. This is basically unconstrained design where the structure, behavior, and patterns of relationship are permitted to emerge without constraint. Self-organization is the 'least energy' approach to design. However, this system design approach is particularly troublesome when the resulting 'low energy' design falls short of producing desirable levels of system performance. The result of self-organization design is that *we get what we get, nothing more and nothing less.* As long as the self-organized system design provides performance that remains at an acceptable level, this approach to design must be deemed to be adequate. It has required the least investment of scarce resources for system design. However, as systems become more complex, it is doubtful that unfettered self-organization will produce the levels of sustainability sought in response to internal flux and external turbulence. In this case, self-organization of design for CSG fails to provide sufficient constraint necessary to maintain desired performance levels.

A second approach to system design and development is through *accretion*—where new elements, activities, or modifications occur in a piecemeal or ad hoc fashion. The result of an accretion approach to system design is that systems are fragmented. They are absent an organizing logic that can explain how the system development 'makes sense'. Instead, a disorganized and uncoordinated set of development initiatives are undertaken. Thus, in accretion, additions are made to the system without consideration as to their holistic fit to the larger system. While individually these additions might seem beneficial, incorporation into the larger system might produce unintended consequences that will negate the anticipated benefits upon which their inclusion was perhaps based. At some point a system designed and developed through accretion will cease to make sense. The logic and structure of the design and development are neither apparent nor capable of effectively sustaining the system.

Too often the development of modern complex systems follows development by accretion or self-organization. CSG calls for a third alternative for system development. This alternative is for *purposeful and holistic* development. Purposeful development requires that system development be holistic (considered as an integrated unity) and achieved in a deliberate fashion (purposeful). This is not to suggest that purposeful development does not deviate from the initial formulation. On the contrary, purposeful development is constantly adjusted to shifts in knowledge, understanding, and interpretation of ongoing development results. *Purposeful and holistic* development stands in stark contrast to the pattern observed for development of many modern complex systems. Understanding how a system design has developed (self-organization, accretion, or purposeful design) is influential in how a CSG development endeavor might proceed and what might be reasonable with respect to expectations. If a system is not meeting desirable performance levels, with design having occurred by self-organization or accretion, this might suggest difficulties in instituting CSG development. This does not suggest that CSG cannot be undertaken but rather may forecast a difficult development path ahead.

The *level of individual (participant) systems thinking capacity* has a direct effect on the planning, execution, and expectations for a CSG development endeavor. A development effort is contingent upon the level of systems thinking held by individual participants. The level of systems thinking capacity has nothing to do with strong will, desire, or good intentions. On the contrary, while those elements might contribute to success, they are not indicators of the level of sophistication in system thinking held by the individuals or the aggregate group. Lower levels of systems thinking capacity will limit the types of activities that might feasibly be undertaken to develop CSG. In fact, a low level of systems thinking capacity (individual, group, or both) might indicate that until that is elevated to an acceptable level, it might be a focus for CSG development. This might be through education, structured application, or training programs to enhance systems thinking.

Beyond systems thinking capacity, *organizations have a level of competency (knowledge, skills, and abilities) for engaging CSG.* This competency level will be influential in how CSG development might be performed and what results might be expected for engagement. Competencies might include not only systems thinking skills, but can extend to such reinforcing competencies as leadership, modeling, communication, etc. The array, distribution, and development relevance of competencies are important considerations with respect to the design, execution, and expectations of a CSG development endeavor. It is important to acknowledge that competency development can be built into a CSG development endeavor. However, this should be deliberate rather than an after the fact acknowledgment that competencies are limiting the nature and type of CSG development activities that can be engaged.

Support infrastructure is an important consideration for CSG development endeavors. Support infrastructure includes such aspects as facilities, instrumental policies/processes, procedures for implementing system changes, and management directives. Support infrastructure can be enabling or disabling for a CSG effort. If the support infrastructure can assist in achieving implementation of system development initiatives, it should be utilized to maximum effectiveness. Likewise, if the support infrastructure is a limiting factor to development initiatives, it must be taken into account and part of the design for CSG must include how it will be changed or otherwise addressed. Support infrastructure should be considered and incorporated accordingly as it impacts CSG development. It is shortsighted to engage support infrastructure at later stages of instituting development initiatives when they should have been considered much earlier.

Although the pursuit of CSG development seems enticing, it should not be entered into lightly. CSG development is difficult and has limitations. However, all systems-based approaches attempting to deal with complex systems and their associated problems have limitations. For realistic caution in pursuing CSG development, we offer an additional set of important points for consideration:

1. ***CSG development must involve the system practitioners (owners, operators, designers, performers) who are accountable and responsible for sustainable***

system performance. CSG development pursuit without engagement of these individuals is unlikely to achieve anticipated results. There is no shortcut for involving system practitioners. The responsibility of CSG development cannot be delegated to others or relegated to the status of 'just another initiative.'

2. ***The design for comprehensive CSG development is fallible and must be continually adjusted.*** It is naïve to engage in CSG development assuming that action outcomes can be known in advance. Instead, care must be taken to understand that the design for CSG development is not prescriptive and cannot be static. CSG development must adjust in response to changes in the system itself, the external environment, and the context within which CSG is embedded. The *rate of change for CSG development* design must minimally keep pace with the rate of change in the system, external environment, and context.
3. ***Systemic worldview is critical to performing CSG development.*** The worldview impacts interpretation and framing of all that is encountered for both individuals and organizations. For systems it can range from reductionist (seeing the world as parts and capable of being discreetly broken down and the system understanding existing in the parts) to holistic (seeing the world as defined by relationships and understanding at the whole rather than part level). CSG implementation relies on a holistic systems worldview. Worldviews short of this expectation portend difficulties at later stages.
4. ***CSG development value can accrue across multiple levels.*** CSG can enhance and add value to individuals, entities, and organizations. Care must be taken not to exceed reasonable expectations and feasible achievements in CSG deployment across any level. Although judgment of value is subjective, CSG efforts/expectations should be specified in ways that can be supportive of conclusions regarding the provision of that value.
5. ***The nature of CSG development is evolutionary rather than revolutionary.*** Therefore, the implementation of CSG development requires 'the long view' and patience. Expectations for CSG development must be appreciative of the current state of governance for a system of interest and the context for that system. These will dictate what level of system improvement activities might be feasibly engaged over the near and long-term. Initial excitement and enthusiasm should be tempered, particularly early on in a CSG development endeavor.
6. ***There is inherent 'risk' in engaging comprehensive CSG development.*** It is important to recognize that there is the potential to 'fail' in CSG development. This brings personal and professional risk to participants in the design, execution, and development of CSG. The structuring of CSG efforts should shift levels of risk to facilitators, the system, and the process. Emphasis must remain on engaging feasible activities that increase the state of CSG and evolve the context. As CSG value is seen, the perceived level of risk should diminish.
7. ***Deeper explorations into CSG expose deeper levels of deficiencies.*** CSG exploration can discover inconsistencies that cannot be easily remedied under the current system and context limitations. There is certainly the possibility of discovery of deep systemic issues for which the level of CSG maturity is not

capable of handling. This can represent threats to systems stability and must be appropriately managed.

8. ***CSG development is a protracted 'self-study' of the system of interest, enacted through a new set of lenses, corresponding language, methods, and tools.*** New thinking requires new language, which can produce alternative decision, action, and interpretation in route to pursuit of different outcomes (system performance levels). The willingness to engage in protracted self-study is essential for realization of the benefits offered by CSG development. There is no shortcut to the reflective self-study required for CSG development.
9. ***Engaging CSG development is not a trivial endeavor.*** It is hard work, requiring significant investment of resources, patience to take the 'long view', and sacrifice of instant gratification for sustainable longer term performance improvement. Superficial CSG efforts are not likely to produce desirable or sustainable results, and in fact may make matters worse. Outcome-expectation desires that are incongruent with investments of time, energy, and resources are likely to produce less than desirable results.

The challenges facing CSG development are certainly not insurmountable. They are provided to ensure that practitioners considering CSG development are aware of what CSG development entails. This does not suggest that elements of CSG development (e.g., improvement in individual systems thinking capacity) will not be beneficial or that the deployment of CSG development is a binary 'all or nothing' proposition. On the contrary, there are certainly benefits to be derived from more limited applications of CSG development. However, what can be achieved by CSG development must be consistent with the commitment invested in development efforts. There must be a tempering of expectations based on the multitude of factors that must be taken into account. Ultimately, CSG development is about shifting the governance landscape for a system of interest.

5 Application-Based Insights for Advancing CSG Development Methodology

Thus far, this chapter has provided a grounding background for CSG, a development methodology for deployment of CSG, and a set of concerns for deployment of CSG. The application of CSG has produced many insights from initial efforts. To push the CSG development methodology forward, several of the key insights from application efforts and their implications include:

1. INSIGHT: *Systems worldview is a limiting factor for CSG deployment.* CSG is a systems-theory-based approach to development of the governance functions for a complex system. Engagement for CSG development requires a sufficient grounding in the systems worldview to secure potential gains from deployment. The systems worldview embraces a nonlinear and holistic perspective of all that is encountered. The absence of this requisite systems worldview in those

participants for a CSG deployment is problematic. It is naïve to think that CSG development methodology can be deployed as intended, or achieve the expected results, absent a requisite systems worldview. In response, the CSG development methodology includes, as an upfront effort, the establishment of the state of systems thinking capacity for individual participants as well as the aggregate of participants.

2. INSIGHT: *CSG itself is not a viable entry point for engaging in a CSG development effort.* Although CSG has much to offer for improved system performance, realistically it is not the highest priority for those who might be considering engagement. Those practitioners and entities that stand to gain the most from CSG initiatives are instead focused on 'their problems' and maintaining viability (existence) of their system. Thus, the more appropriate entry point for CSG is to first understand their problems and then draw the linkage to potential CSG value contributions. By engaging in *initialization* activities (e.g., context definition) the direct linkage to the system and utility of further examination through the CSG lenses can be demonstrated. Making this connection between ongoing problems and CSG is critical to draw attention to the possibilities that CSG might bring related to their most vexing issues.
3. INSIGHT: *Starting 'shallow and slow' is preferable to 'deep and fast' to build momentum for CSG.* Engaging CSG is difficult at best and potentially overwhelming at worst. Comprehensive CSG is fraught with difficulties. Completing a marathon is not a short or trivial matter. Capacity must be slowly built as endurance increases as do the prospects for successful completion. CSG engagement is similar. CSG is not a binary (all or nothing) proposition. Instead, there are a spectrum of activities (training, development, modeling, etc.) and focal levels (practitioner, system, enterprise, problem) that might be pursued in the development path to enhance CSG. Through the successive building of confidence and depth of activities, sufficient momentum can be created to engage CSG at increasingly sophisticated levels.
4. INSIGHT: *CSG functions, in an existing system, are already being performed, and thus it is not an 'in addition to' endeavor.* Unlike more traditional system interventions that seek to address a new concern by introduction of a totally new initiative (e.g., Lean, Six Sigma, TQM, BPR, Agile, etc.), CSG functions are already being performed by a system that is viable (exists). Thus, CSG is focused on understanding and potentially improving that which is already being performed. Therefore, the language, thinking, and explorations of CSG are applied to an existing system to improve execution of CSG functions which are already being 'tacitly' performed.
5. INSIGHT: *For CSG engagement, the initial risk should be borne external to the system and participants.* CSG endeavors, irrespective of scope, take resources and present risks to participants and their system of interest. It is unrealistic to expect participants to totally shoulder the 'risk of failure.' Instead, the CSG facilitator should bear the burden of time and risk until the value of investment (time) and utility of CSG engagement (valued results) meet an acceptable *risk-value-cost* trade-off. In effect, CSG should be conducted in a 'safe to

fail' mode. In this sense, initial CSG engagement should offer prospects for an approach that provides high value, low investment, and low risk of failure. This was the concept behind the 'CSG entry' crafted to introduce CSG to potential participants (Keating and Katina [5]).

This set of insights for CSG deployment has been drawn from initial experiences with various applications referenced in the work of (Keating and Katina [5]). While this listing is not all inclusive, it does provide a starting point of considerations for the deployment of CSG development methodology.

6 Summary

In this chapter, we have provided an examination of the CSG development methodology. The examination provided a background into the problem domain that CSG is designed to address. Central to this problem domain are the characteristics of *ambiguity* (lack of clarity in the system and its context), *uncertainty* (the breakdown of explanations rooted in cause-effect relationships), *holism* (loss of meaning from reduction to the component level), *complexity* (excessive number of elements, rich interrelationships, dynamic interaction, and emergence), and *contextually embedded circumstances* (factors, and conditions impacting and impacted by the system). Given this problem domain, CSG was offered as a theoretically grounded, systems-based approach to enhance system performance through the purposeful development activities targeted to improve performance of the system of interest and address contextual issues. Since the earlier work in CSG development methodology [8, 14], our knowledge, understanding, and perspectives have matured. This chapter represents the current state of knowledge for CSG development. While most of the previous work is still consistent with our viewpoint on CSG development, there have been some significant advances in the methodology. This current work represents our most recent state of knowledge. While that state is sure to change, as we continue to explore and learn more about CSG and its related phenomena, we are confident that this work represents a significant movement forward.

CSG development was presented as occurring in three primary stages (Fig. 4), including *initialization, development mapping,* and *development*. *Initialization* includes fixing the state of CSG for the system of interest and elaborating the context for that system. *Development mapping* targeted the setting of priorities for the greatest impact and feasibility of successful achievement for development areas. In this stage, the state of CSG is captured across a nine phased spectrum. This spectrum provides a notional limitation as to the types of activities that might be successfully (feasibly) engaged in development of CSG based on the classification. The final stage, *development*, is focused on the identification, planning, execution, and evaluation of activities selected to undertake in development of the state of CSG.

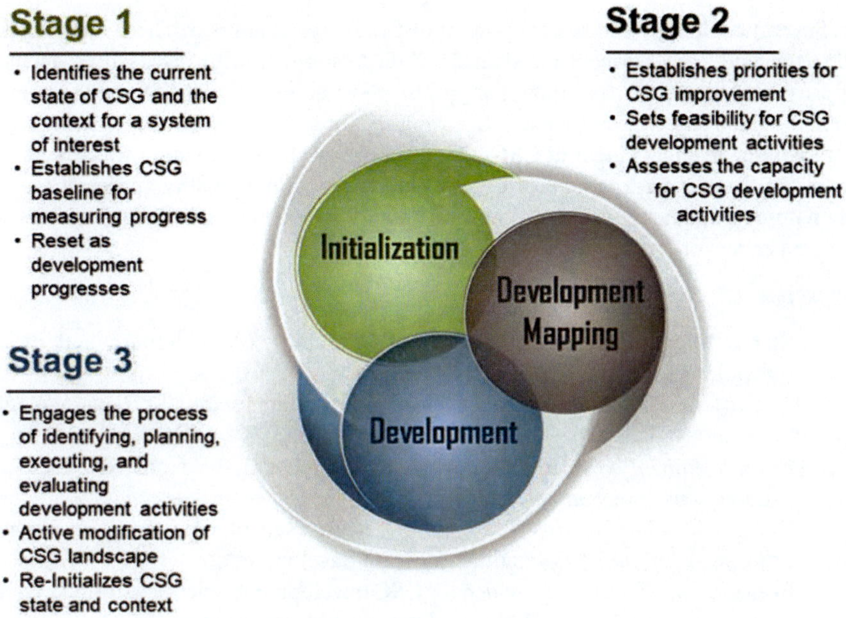

Fig. 4 CSG development methodology stages

CSG development was identified as a continuous cycling that provides the purposeful development of CSG. This purposeful development is steeped in selection of priority feasible activities that can be undertaken to improve the state of CSG as well as the context. Thus, the successive cycling is a continuous re-initialization, shifting of the development mapping, and increasingly deep selection of development activities.

Several success limiting factors were explored. These factors were identified as having a high level of impact as to what can reasonably be expected with respect to CSG development. The first factor dealt with the *evolutionary design path* that has brought the system to its current state. This path was presented as ranging from self-organization to accretion. Ultimately, CSG suggested a difficult path of purposeful design be preferable for increasingly complex systems. The second limiting factor was focused on the *level of individual capacity* existing within the system. Lacking a robust systems thinking capacity, the development of CSG would be limited. A third factor was the *organizational competency* for governance. This competency is focused on the level of requisite knowledge, skills, and abilities to effectively engage CSG. Lacking these skills, although CSG could be engaged, it is doubtful it could have the desired developmental impacts sought. A final factor identified was the *support infrastructure enabling/constraining* impacts on development. Without adequate support infrastructure, a CSG development endeavor would experience limitations in execution. The sum total of these factors was provided to interject a realistic sense of CSG development considerations and limitations.

Several challenges to the deployment of CSG were examined. These challenges serve to guide practitioners considering CSG to more fruitful discussions on the approach, expectations, and potential pitfalls. Several insights for CSG engagement were provided based on experiences from applications. The purpose of this listing was not to dissuade engagement of CSG. On the contrary, the examination was intended to ensure that the engagement of CSG would begin with a healthy appreciation of the nature, scope, and considerations that should be contemplated before taking on such an endeavor.

Exercises

1. Discuss three aspects of the complex system problem domain and their implications for conducting CSG development.
2. Which of the methodology attributes is most important for the CSG development methodology? Why?
3. For the *initialization stage,* discuss how that stage might be expected to change in subsequent development cycles for CSG.
4. Discuss the three forms of system design (self-organization, accretion, purposeful) and their implications for CSG development.
5. Given the considerations provided for CSG development, select and discuss your top three considerations. What guidance would you suggest for practitioners with respect to the considerations you identified?

References

1. Jaradat R, Keating C (2014) Fragility of oil as a complex critical infrastructure problem. Intern J Crit Infrast Prot 7(2):86–99
2. Keating CB, Katina PF (2011) System of systems engineering: Prospects and challenges for the emerging field. Intern J Syst Syst Eng 2(2/3):234–256
3. Keating CB, Katina PF (2012) Prevalence of pathologies in systems of systems. Intern J Syst Syst Eng 3(3/4):243–267. https://doi.org/10.1504/IJSSE.2012.052688
4. Keating CB (2014) Governance implications for meeting challenges in the system of systems engineering field. IEEE System of Systems Engineering Conference, Adelaide, Australia June 9–14
5. Keating CB, Katina PF (2019) Complex system governance: Concept, utility, and challenges. Syst Res Behav Sci 36(5):687–705
6. Kurtz CF, Snowden DJ (2003) The new dynamics of strategy: Sense-making in a complex and complicated world. IBM Syst J 42(3):462–483
7. Mitroff I (1997) Smart Thinking for Crazy Times: The Art of Solving the Right Problems. Berrett-Koehler
8. Keating CB, Calida BY, Jaradat R, Katina PF (2016) Systems thinking. Chap. In Farr JV, Gandhi SJ, Merino DN (eds) The Engineering Management Handbook, 2nd Edition (pp. 281–316). Huntsville, AL, ASEM Press
9. Jackson MC (2019) Critical Systems Thinking and the Management of Complexity. Wiley
10. Checkland P (1999) Soft systems methodology: a thirty year retrospective. Syst Res Behav Sci 17(S1):S11–S58

11. Keating C, Rogers R, Unal R, Dryer D, Sousa-Poza A, Safford R, Peterson W, Rabadi G (2003) System of systems engineering. Eng Manag J 15(3):36–45
12. Ackoff RL (1997) Systems, messes and interactive planning. Soc Engag Soc Sci 3(1997):417–438
13. Rittel HW, Webber MM (1973) Dilemmas in a general theory of planning. Policy Sci 4(2):155–169
14. Keating CB, Katina PF (2016) Complex system governance development: a first generation methodology. Int J Syst Syst Eng 7(1/2/3):43–74
15. Keating CB, Katina PF, Bradley JM (2015) Challenges for developing complex system governance. In: IIE Annual Conference Proceedings (p. 2943). Institute of Industrial and Systems Engineers (IISE)
16. Klir G (1972) Trends in General Systems Theory. Wiley
17. Laszlo A, Krippner S (1998) Systems theories: Their origins, foundations, and development. Advan Psychol 126:47–76
18. Keating CB, Bradley JM (2015) Complex system governance reference model. Int J Syst Syst Eng 6(1/2):33–52
19. Adams KM, Keating CB (2011) Overview of the systems of systems engineering methodology. Int J Syst Syst Eng 2(2/3):112–119
20. Keating CB, Katina PF, Bradley JM (2014) Complex system governance: concept, challenges, and emerging research. Intern J Syst Syst Eng 5(3):263–288
21. Keating CB, Katina PF (2015) Foundational perspectives for the emerging complex system governance field. Int J Syst Syst Eng 6(1/2):1
22. Keating C, Hester P, Meyers T, Calida B (2010) R&D System Governance: An Emerging Perspective. In: Proceedings of R&D Management Conference: Information Imagination & Intelligence in R&D Management. Manchester, London: University of Manchester

Systems Thinking for Complex System Governance

Raed M. Jaradat, Charles B. Keating, Fatine Elakramine, Safae El Amrani, and Larif Oumaima

Abstract For effective implementation of Complex System Governance (CSG), a necessary condition is a high level of systems thinking capacity. Being able to 'think in systems' is essential if effective design, execution, and development are to be undertaken for CSG. An individual must be capable of understanding the complex system's components and how they comprise the whole system. This chapter focuses on better understanding systems thinking in relation to CSG and establishing the level of systems thinking held by an individual/group. Three primary development objectives are pursued for this exploration. First, following an introduction to the chapter, systems thinking is examined. The focus is on providing a rigorous treatment of systems thinking. Second, the role of systems thinking in CSG is examined. This examination is targeted to examine how systems thinking is a fundamental and vital aspect of CSG. Third, a systems thinking instrument is introduced as a vehicle to establish the state of systems thinking of an individual/group. This instrument provides a 'snapshot' of both individual and aggregate systems thinking capacity. The chapter closes with implications that systems thinking holds for the emerging CSG field.

Keywords Complex systems governance · Systems thinking · Systems thinking capacity

---

R. M. Jaradat (✉) · F. Elakramine · S. El Amrani · L. Oumaima
Mississippi State University, Starkville, USA
e-mail: jaradat@ise.msstate.edu

F. Elakramine
e-mail: fe92@ise.msstate.edu

S. El Amrani
e-mail: sa92@ise.msstate.edu

L. Oumaima
e-mail: ol127@ise.msstate.edu

C. B. Keating
Old Dominion University, Norfolk, USA
e-mail: ckeating@odu.edu

C. B. Keating et al. (eds.), *Complex System Governance*, Topics in Safety, Risk, Reliability and Quality 40, https://doi.org/10.1007/978-3-030-93852-9_14

1 Introduction

Systems thinking is touted as an important future capability for society and is frequently suggested as essential to prepare individuals to deal with increasingly complex systems [5]. At a fundamental level, complex systems have several defining characteristics, including:

1. *Large number of entities*—complex systems invariably have a large number of interconnected entities that exponentially increases the number of states the system can occupy and the number of interrelations between elements that are possible.
2. *Richly interrelated*—beyond just the large number of entities and interrelationships among them, complex systems interconnections are 'rich.' This suggests that there are large degrees of 'different' interconnections. These connections are dynamic in nature, with a degree of instability, difficult to completely grasp or understand, and lacking any precision that can provide a definitive description of the interconnection attributes.
3. *Dynamic shifts*—over time, a complex system is subject to change. These changes may follow from elaboration of interconnections or from adjustments in response to environmental change/shifts. Complex systems are not static and defy approaches that assume a 'static/stable' state for purposes of analysis.
4. *Emergence*—the structure, behavior, and performance of complex systems cannot be fully known, or calculated, in advance of the system being operated. The behavior/performance is not known until the system operates. This is where emergence, generated from the interactions among the elements, produces behavior or performance that cannot be known or predicted in advance.

The landscape and need for engaging in CSG has changed substantially as we move firmly into the twenty-first century. And with near certainty, the complex system attributes described above are not going to abate in the near future. In fact, we can posit that the conditions will be exacerbated by multiple confounding factors. Among these factors, we would include (Fig. 1): (1) information- and technology-driven aspects of systems continuing to proliferate as systems become more complex and technology driven, (2) increasing stakeholders who have a perceived interest in a complex system and likely harbor potentially divergent perspectives and varying degrees of politically driven agenda projected onto complex systems, (3) scarce resources that are continually uncertain and subject to shift with little to no notice, resulting in disruption to stable resource planning, (4) constant shifts in performance expectations impacting the ability of a system to effectively respond, (5) advancing technology that can be disruptive and have difficulty integrating/interoperating with existing system technologies, potentially rendering once compatible support infrastructures obsolete, (6) stresses for urgency and near immediate responses to shifting system, context, or environmental demands, (7) emphasis on near term actions and results while surrendering the pursuit of long term stability in deference to immediate emerging crises, rendering long-term planning irrelevant, (8) the constant escalation of complexities and associated uncertainties that become status quo instead of more

Fig. 1 Factors influencing complex systems

limited influences on system operation, and (9) emergence generating unpredictable patterns, structures, and behaviors that result in instabilities in planning, operation, and development of systems. These factors may seem somewhat unsettling and invite questions as to whether or not we will ever have the capabilities to effectively master this domain.

These factors are not likely to subside in the near future. Instead, they are more likely to escalate in frequency of occurrence and severity of impact. Future success in dealing with complex systems will be depending on the degree to which we can effectively mount a response in the face of these factors.

In short, the emerging 'systems world' is an increasingly ambiguous, complex, emergent world of interdependent systems fraught with instabilities and uncertainties [48–50]. There is a pressing need for a new mindset, capabilities, and skills that will increase the probability of success in dealing with complex systems. Unfortunately, 'doing nothing' is as unpalatable as continuing to suffer through the current state of complex system affairs.

Systems thinking can provide a valuable capability for addressing the factors laid out above. Although systems thinking is not the 'silver bullet', 'magic elixir', or 'utopian solution', it can add to the arsenal of weapons available to drive a different level of thinking, decision, action, and interpretation of responses.

This chapter is focused on the use of systems thinking to more effectively deal with increasingly complex systems and their problems. In general, systems thinking has been captured by Haines [33, p. vi] as 'A new way to view and mentally frame what we see in the world,a worldview and way of thinking whereby we see the entity or unit first as a whole, with its fit and relationship to its environment as primary concerns; the parts secondary.'

At a fundamental level, systems thinking involves engaging complex systems from a worldview marked by a focus on the whole and relationships rather than the individual entities of a system. Thus, the behavior, structure, or performance outputs of a system are attributed to the interactions and interrelationships between entities rather than properties held by the component entities. This is fundamental to the perspective of systems thinking for CSG. Having a high level of systems thinking capacity is essential to deal with the conditions that define the nature of complex systems. If CSG is to achieve success, then systems thinking is fundamental to achievement of that success. Arguably systems thinking is the most critical enabling factor for CSG. If an appropriate level of systems thinking does not exist, it is doubtful that the expected contributions of CSG will be realized. Instead, the application of CSG with an inconsistent (nonsystemic/reductionist) worldview is not likely to secure the gains intended by a CSG endeavor. In fact, engaging CSG from a nonsystemic perspective may very likely do more harm than good.

In this chapter, the primary purpose is to explore the nature, role, and implications that systems thinking holds for CSG. The chapter is organized to achieve this purpose by exploring four fundamental points (Fig. 2).

First, systems thinking is examined. This examination is focused on providing an overview of systems thinking and setting the stage for application to CSG. Second, the specific role and relationship of systems thinking to CSG is discussed. The discussion

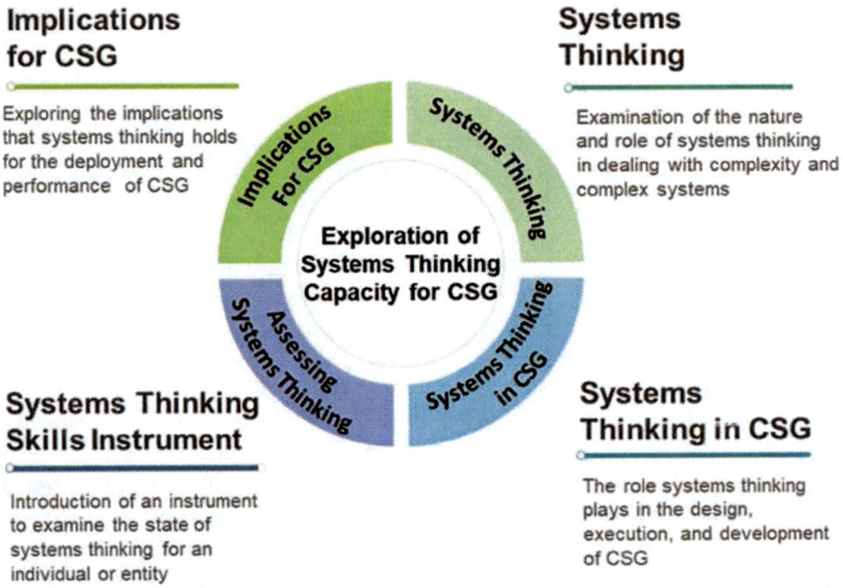

Fig. 2 Organization of the chapter

emphasizes the critical role that systems thinking plays in the design, execution, and development of CSG. Third, an instrument and approach for assessing the capacity of individuals and entities to engage in systems thinking is explored. The influence of this state of systems thinking is examined in relation to performance of CSG functions. Fourth, the implications of systems thinking for CSG are established. The chapter includes a section that provides the CSG instrument for application to assess the level of systems thinking capacity for an individual or entity.

2 Overview of Systems Thinking

Systems thinking is a high-level thinking skill that enables individuals to more effectively engage complex systems [65]. While much has been written about various perspectives of system governance [57], there is limited acknowledgement in the literature that addresses some of the most basic questions related to systems thinking in relation to CSG. The system thinking field has made significant contributions to the advancement of society and our ability to address complex issues. Systems thinking has been applied in such wide-ranging fields as organizational, biological, managerial, economic, and social [1, 12, 18, 21, 32, 36, 46, 58–60, 68, 73, 75, 76, 87, 90]. While the fields of application are different, the same underlying fundamental aspects of systems thinking are applicable across the fields. Again, focus on the whole as exhibiting properties not deducible or existing in the parts of making up the system.

Systems thinking is not entirely a new area. The concept has been in existence for some time, drawing on the earliest explorations found in the Chinese work *The I Ching or Book of Changes* [93]. This work dates to approximately 400 B.C. The *I Ching* noted the dynamic nature of changing relationships among elements, which is consistent with the most basic notions of systems thinking. The most fundamental tenet of systems thinking is found in *holism,* which recognizes the relationship between parts and wholes of systems. Holistic thinking can be found in some of the earliest writings of Aristotle (384–322 BC), who postured that there is more to the whole than that found in the parts [6]. Therefore, the essence of 'systems' and 'systems thinking' is found in interconnectedness and the whole-part distinction. Although there have been advances in society and systems thinking since the earliest works, the quest to understand interrelationships and behavior of wholes has certainly not waned. However, the debates surrounding complex 'systems' and our apparent limited ability to effectively and consistently address them remain unresolved, despite advances in our understanding of the related phenomena. Still yet, the challenges of effectively dealing with systems continue to persist. In Table 1, we list a representative set of perspectives that demonstrates a variety of viewpoints for systems thinking. This is not to suggest the superiority or preference of one perspective over another. Instead, what we can draw from this glimpse is the breadth of perspectives that exist for systems thinking.

Table 1 Multiple perspectives of systems thinking

Author	Perspective
Flood and Carson [28, p. 4]	'a framework of thought that helps us to deal with complex things in a holistic way.'
Checkland [20, p. 318]	'makes conscious use of the particular concept of wholeness captured in the word 'system', to order our thoughts.' 'An *epistemology* which, when applied to human activity, is based upon the four basic ideas: *emergence, hierarchy, communication, and control* as characteristics of *systems*. When applied to *natural* or *designed systems* the crucial characteristic is the *emergent properties* of the whole.'
Gharajedaghi [31, p. 15]	'puts the system in the context of the larger environment of which it is a part and studies the role it plays in the larger whole.'
O'Connor [67, p. 1]	'seeing beyond what appears to be isolated in independent incidents to deeper patterns.'
Haines [33, p. vi]	'A new way to view and mentally frame what we see in the world; a worldview and way of thinking whereby we see the entity or unit first as a whole, with its fit and relationship to its environment as primary concerns; the parts secondary.'
Senge [75, p. 89]	'a discipline for seeing wholes. It is a framework for seeing interrelationships rather than things, for seeing patterns of change rather than static snapshots.' 'encompasses a large and fairly amorphous body of methods, tools, and principles, all oriented to looking at the interrelatedness of forces, and seeing them as part of a common process.'
Capra [17, p. 29]	'a new way of thinking … in terms of connectedness, relationships, context.'
http://www.opbf.org/open-plant-breeding/glossary/so-sz	'A system cannot be understood by an analysis of its parts. Systems thinking concerns the organisation of those parts, as a single system, and the emergent properties that emanate from that organisation.'
Richmond [73, p. 139]	'the art and science of making reliable inferences about behavior by developing an increasingly deep understanding of underlying structure.'

(continued)

Table 1 (continued)

Author	Perspective
ESD Symposium Committee [27], p. 8)	'includes holism, an ability to think about the system as a whole; focus, an ability to address the important system level issues; emergence (see below), recognition that there are latent properties in systems; and trade-offs, judgment and balance, which enable one to juggle all the various considerations and make a proper choice'
Davidz [26, p. 44]	'analysis, synthesis and understanding of interconnections, interactions, and interdependencies that are technical, social, temporal and multi-level'
Ackoff et al. [2, p. 6]	'looks at relationships (rather than unrelated objects), connectedness, process (rather than structure), the whole (rather than just its parts), the patterns (rather than the contents) of a system, and context.'
Arnold and Wade [7, p. 675]	'a set of synergistic analytic skills used to improve the capability of identifying and understanding systems, predicting their behaviors, and devising modifications to them in order to produce desired effects.'

Essential to understanding systems thinking is to understand the essence of systems theory from which systems thinking is drawn. This relationship shows that systems thinking draws from systems theory and is projected to application (Fig. 3).

Following the earlier work of Keating et al. [55], the following development of systems theory is suggested. Systems theory provides a strong conceptual foundation that can influence design, execution, and development of complex systems. Following works on systems theory [3, 52, 91], at a basic level systems theory can be described as a set of *axioms* (taken for granted truths about systems) and *propositions* (principles, concepts, and laws serving to explain system phenomena). Systems theory suggests several central tenets concerning the capacity to deal with environments marked by increasing complexity, instabilities, and ambiguity. These tenets include:

1. All systems are subject to the propositions of systems theory. These propositions define and serve to explain the structure, behavior, and performance of systems,
2. All systems perform a set of systems theory based system CSG functions that, subject to propositions, determine system performance,
3. Violations of system propositions in design, execution, or development of systems have consequences that degrade performance and produce failures in systems, and
4. Systems theory-based proposition violations can provide novel insights for better understanding the relationship of systems theory to inform systems thinking and CSG.

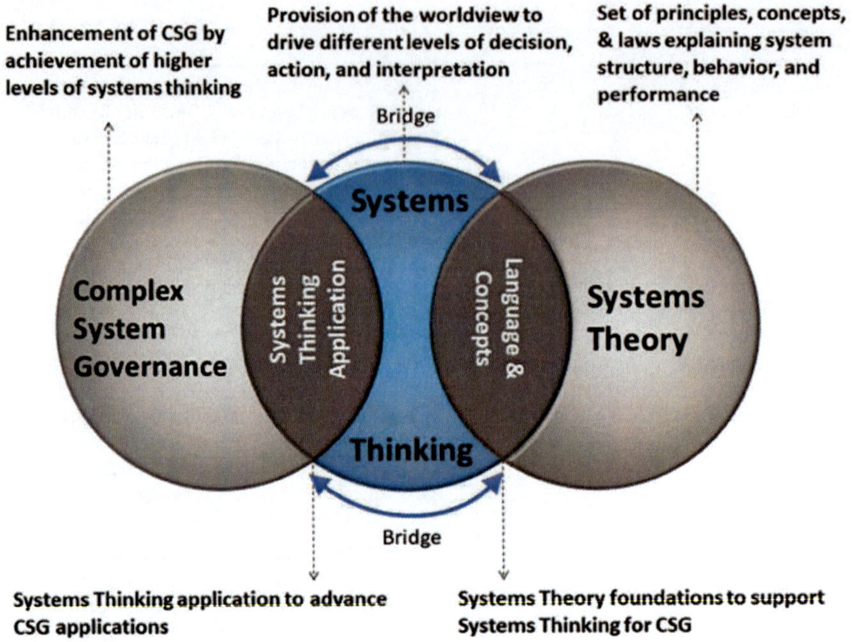

Fig. 3 Relationship of systems theory to systems thinking and CSG application

Further examination of systems theory inevitably leads to general systems theory (GST) to find its genesis. GST does not have a single common accepted definition. However, GST emerged in the 1940s as an attempt to provide an alternative to *reductionism. Reductionism* (focus on the successive 'breaking apart' to produce understanding) is closely aligned with the scientific method, which holds that a complex organism is understood as the sum of its parts, and therefore, they can be reduced to constituent elements [34, 55, 87]. In contrast to reductionism, *'holism'* suggests that understanding of a systems comes from the interrelationships between entities and cannot be ascertained from the properties of the parts. In essence, production properties that exist beyond those held by parts of the system.

An important delineation of systems thinking is the distinction between 'hard' and 'soft' systems thinking modes. Unfortunately, the distinctions between 'hard' and 'soft' set up the potential to view the classification as a binary either/or choice. However, in actuality, a complex system can have both hard and soft elements and arguably will invariably include different degrees of both. While the complete separation of hard and soft distinctions might be false, understanding their implications for systems thinking is important. This is particularly the case when there is the desire to 'rush to judgment' as to the source of deficient system behavior or performance in attributing the deficiency to either hard failure or soft failure. Table 2 provides a set of distinguishing system perspectives [40] that delineates the distinctions between the extremes of hard and soft thinking perspectives.

Table 2 Hard and soft systems thinking

Attribute	Hard system thinking	Soft system thinking
Understanding paradigm	*Reductionism*—focused on understanding through breaking apart (analysis). Performance of a whole can be understood as an aggregation of the properties of the parts	*Holism*—a system is only understood at the (irreducible) whole system level. The behavior of the whole cannot be ascertained simply from understanding the parts. Instead, understanding the system must include the interactions among parts that produce properties beyond those of the parts
Objective	*Optimization*—there is one best solution (optimal) for system performance. This is the solution or configuration which is sought	*Learning*—the primary function of system exploration is to learn about the system and be capable of mounting appropriate response(s) based on that learning to improve a situation
Methodology	*Systematic*—approach is defined by prescribed processes that can be replicated independently of context—prescriptive	*Systemic*—approach is a high-level guide that provides a general set of malleable directions—non-prescriptive and tailorable to circumstances and conditions
Goal/Objectives	*Clearly defined and agreed upon*—goals and objectives are assumed to be clear, unambiguous, defined, and stable	*Ambiguous and shifting*—clarity is not assured, and goals/objectives are subject to multiple interpretations and can be unstable
Perspectives	*Unitary*—assumes that there is an alignment of perspectives for the problem domain	*Pluralist*—there exist multiple, potentially divergent, perspectives on the problem domain
Contextual influences	*Low*—contextual influences are assumed to be 'minimized' by tight bounding of the problem	*High*—contextual influences are seen as integral, systems/problems are not easily separable from context, and boundaries are flexible
Environment	*Stable*—disturbances from the environment are minimal and the rate/depth of resulting changes are not considered overbearing on system solution	*Turbulent*—disturbances are potentially extensive and influence the ability to develop system solutions
Systems-of-interest	*Simple*—low number of variables, interactions well understood, behavior somewhat static or deterministic, environment stable	*Complex*—high number of variables, rich interactions not well understood, dynamic and uncertain, (emergent) pattern/behaviors, environment unstable
Modeling preference	*Mathematical/quantitative*—exact relationships and behavior mathematically predictable	*Non-mathematical/qualitative*—forms of representation non-quantitative in nature. Behavior is not precisely predictable

(continued)

Table 2 (continued)

Attribute	Hard system thinking	Soft system thinking
Boundaries	*Clearly delineated*—boundaries are definitive, stable, and understood	*Unclear and shifting*—boundaries are evolving, unstable, and ambiguous
Worldview	*Aligned*—divergence in worldviews not made explicit or considered central to understanding	*Potentially divergent*—divergence is considered highly probable, with an understanding of divergence sources critical to understanding
Defining metaphor	*Mechanistic*—clear understanding of predictable interrelationships	*Contextual*—lack of clarity in nature of interrelationships and external influences
Behavior	*Predictable*—system behavior is deducible from understanding historical patterns/trends and system interactions	*Emergent*—system behavior cannot be known in advance. Patterns of behavior/performance emerge through the operation of the system

For systems thinking, the inclusion of both hard and soft system thinking modalities is important. Both hard and soft systems thinking are required to engage in treatment of complex systems in a holistic fashion. Thus, the hard system aspects (generally technical, objective, and certain) as well as soft system aspects (generally non-technical, subjective, and uncertain) are necessary for a more holistic perspective of a complex system. Care must be taken not to treat a complex system existing as either

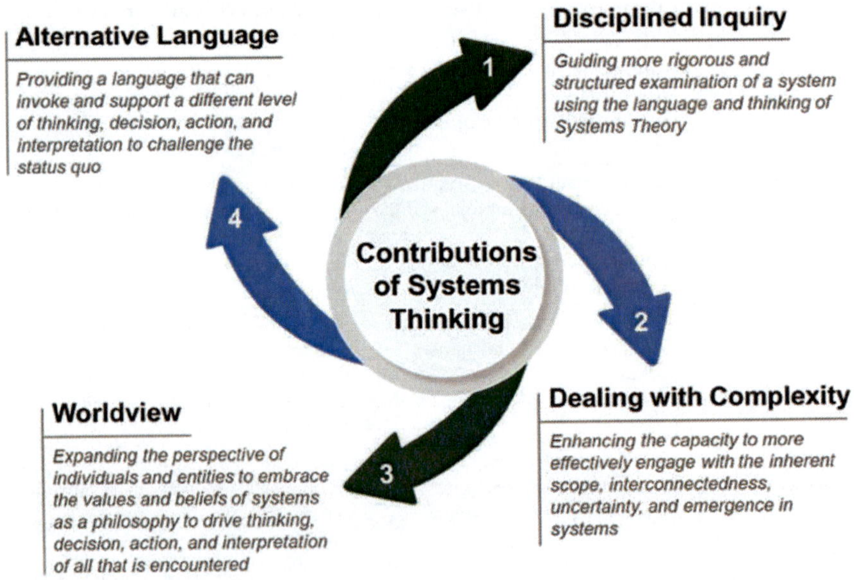

Fig. 4 Contributions of systems thinking

a binary hard or soft entity. In reality, complex systems include both perspectives and formulations. Additionally, the hard system elements influence the soft system elements and vice versa. Most assuredly, CSG must involve both forms of systems thinking and their interactions. To treat CSG otherwise would be shortsighted.

Since present and future environments of CSG are ever-changing, increasingly unstable, and complex, improved systems thinking abilities will be necessary to achieve increased capacity for dealing with complexity. In response, we suggest that systems thinking offers four important foundational contributions for CSG (Fig. 4).

These contributions are the essential offering of systems thinking, drawn from the underlying conceptual foundations of systems theory, including:

- **Disciplined Inquiry**—the application of systems thinking can be helpful in providing approaches, grounded in underlying systems axioms and propositions, to conduct more disciplined inquiry. The very nature of systems thinking suggests a more structured and orderly progression for understanding and holistic framing of complex interrelationships. This is an important point for individuals and entities that must develop effective understanding and governance for complex systems and their problems.
- **Dealing with Complexity**—dealing with increasing complexity is perhaps the greatest future challenge facing society. Systems thinking offers multiple approaches, based in underlying systems theory foundations, to more effectively address complex problems. Complexity is a fact of life for modern systems and their practitioners. The approaches to deal with complexity, rooted in a systems thinking frame of reference, can provide increased effectiveness in addressing increasingly complex systems and their constituent problems. CSG is one such approach.
- **Worldview**—worldview encompasses the values and beliefs that inform a philosophy for how the world functions and establishes the basis for making sense of what we perceive in the world. Aerts et al. (4, p. 9) suggest that 'a worldview is a system of coordinates or a frame of reference in which everything presented to us by our diverse experiences can be placed.' Thus, worldview is what Checkland [19] refers to as *weltanschauung*, the image or model of the world that provides meaning. In effect, the systems worldview offers a particular way of thinking that allows us to give meaning to actions, decisions, and events as they unfold. In effect, systems thinking can provide practitioners with increased capacity to 'make sense' of their circumstances and enhance the capability to mount more effective responses.
- **Alternative Language**—we think through language. Therefore, new language provides avenues for new thinking to more effectively deal with complex systems. The language of systems thinking can be instrumental in coming to a new understanding of familiar issues. Thus, new and different courses of action to address complex problems, both new and old, can be established.

The foundations for systems thinking are fundamental to CSG. Although there are many different formulations of 'systems thinking' in the literature, the importance this field holds for effectiveness in dealing with complex systems is clear. Having established the perspective and importance of systems thinking for CSG, we can now shift focus to developing greater detail into the specific nature and role that systems thinking plays in CSG.

3 Systems Thinking in CSG

Complex system governance (CSG) is a recently created field in a very early stage of development. CSG has been welcomed into the environment of theoretical complex system management due to its potential explanation of systems performance [53]. CSG is fundamentally identified as 'the design, execution, and evolution of the meta-system functions necessary to provide control, communication, coordination, and integration of a complex system' [53, p. 274]. CSG is a system-based approach that aims to enhance governance capabilities to achieve an effective management of complex systems. In other words, CSG is a capability that involves a set of actions required to improve the control of complex systems. Complex systems denote any system that contains an interrelated set of problems that cannot be easily resolved, understood, or formulated. Complex systems are an essential part of many diverse disciplines including economics, meteorology, biology, sociology, etc. [29, 30, 43, 71]. Therefore, enhancing governance of these systems is necessary to advance knowledge across disciplines beholden to complex systems.

The domain challenging present and future complex systems is becoming increasingly difficult to effectively govern (control). While the debate may ensue as to the source of the increased difficulty, the state of the current condition is much less questioned. A capture of the difficulties [51] that must be addressed in this domain is summarized by the following conditions and their impact:

(1) ***Hyperturbulent conditions***—the environment for complex system practitioners is highly dynamic, uncertain, and rapidly changing. This flux places traditional forms of dealing with complex systems in question. Stable platforms for providing control of complex systems from rational/logical designs are relegated to the past. A new era of complexity is here and not likely to subside in the future. This suggests that keeping up with these new conditions will require surrender of the 'illusion' of precision and metered approaches to greater reliance on approaches based in systems thinking-driven agility and resiliency.

(2) ***Ill-defined problems***—the luxury to precisely define problems enroute to developing cogent solutions is diminishing. The circumstances and conditions surrounding problems are potentially in dispute, not readily accessible, or lack sufficient consensus for initial problem formulation and bounding. This throws doubt on the capabilities of traditional approaches to adequately address these poorly understood, dynamic, and emergent problems. The ability to engage a 'systemic perspective' to frame problems is an important contribution that can be made by systems thinking.

(3) ***Contextual Dominance***—expectations must include that the technical 'hard' aspects of a problem may be overshadowed by the contextual 'soft' aspects. Contextual (soft) aspects are those circumstances, factors, conditions, trends, or patterns that influence the framing of the problem, solution development form, solution deployment, and interpretation of deployment results. Complex systems require more holistic treatment called for by systems thinking. Since in many cases, for highly volatile problem domains, context can play a more significant role than the technical aspects, a more holistic perspective provided by systems thinking is necessary.

(4) ***Uncertain approach***—increasingly, complex systems lack clarity on what the appropriate path forward might be to address problems. This calls into question standard approaches which have been successfully applied in the past. No longer are these approaches assured of success. Instead, agreement on the nature of the problem (framing), appropriate approach, and expectations for 'successful' resolution are potential sources of divergence. Systems thinking can be instrumental in better understanding the potential for more informed approaches.

(5) ***Ambiguous expectations and objectives***—given increasingly complex systems, the ability to establish measures of success, system objectives, or requirements is questionable. The conditions of inadequate understanding, incompatible perspectives, or lack of technical competence to proceed with the effort all serve to highlight ambiguity. Traditional thinking, mired in such monikers as 'do a better job', 'do more of the same', or 'work harder', is unlikely to realize any substantiation gains. Systems thinking can be instrumental in finding alternative frames of reference and corresponding paths forward in response to ambiguity.

(6) ***Excessive complexity***—complexity denotes a situation that is highly dynamic, uncertain, emergent, and containing a high number of richly interconnected elements/factors/variables. Dealing with this complexity lies beyond traditional approaches mired in a linear, reductionist perspective for their approach. Proceeding with traditional approaches would require significant reduction, assumptions, and potential oversimplification of the system and its problem domain. This creates the conditions for failure to meet expectations for problem resolution. In no sense should this be taken as a criticism of traditional approaches. On the contrary, traditional approaches, when applied to the right problem/context, have proven successful. However, given the evolving context of complex systems, the utility of traditional approaches must be questioned. Systems thinking brings a different, more holistic, perspective for framing complex system problems and offering alternative paths forward.

(7) ***Pluralist Perspectives***—a characteristic aspect of complex systems is the plurality of different perspectives. Many traditional approaches assume a 'unitary' perspective, where there is agreement on the nature of the problem, appropriate path forward, and expectations. However, for complex systems and stakeholders, the 'unitary' perspective assumption may be questionable. Thus, from a systems thinking perspective, complex systems must be

engaged with the understanding that multiple, potentially divergent, perspectives may well exist. Through the reframing of perspectives brought by systems thinking, appreciation of differences, true differences, and their joint exploration can be obtained.

(8) ***Extended Stakeholders***—stakeholders are individuals or entities that have a 'perceived' interest in the system or problem of interest. This suggests that complex systems can be fraught with a spectrum of stakeholders with different viewpoints. These perspectives may be convergent or divergent. However, systems thinking suggests that all perspectives should be included in the dialog. Thus, insights into the system/problem will not be unnecessarily bounded out.

(9) ***Emergence***—this suggests that until a system operates, the behavior, structure, and performance cannot be precisely known or predicted in advance. This implies that a different level of appreciation and understanding for complex systems must be invoked. In essence, the emphasis must shift from prescriptive detailed planning to planning that appreciates the continuous evolution of a system, its environment, and context. This dynamic 'emergence' of events and patterns cannot be predicted in advance. Thus, systems thinking suggests that designs be resilient (capable of returning to a level of performance following disruption) and robust (designed to accommodate a wide range of 'potential' disturbances without sacrificing performance).

(10) ***Ambiguous Boundaries***—boundaries for complex systems are arbitrary and subject to shifting with new knowledge or understanding. Likewise, the criteria for inclusion/exclusion can also dynamically shift over time. This does not suggest that establishment of boundaries is a fruitless endeavor. On the contrary, the establishment of boundary conditions is important to establish a reference baseline from which changes can be purposefully engaged. From a systems thinking perspective, the boundary conditions must be considered fallible, dynamic, and not absolute.

(11) ***Unstable Planning Foundations***—a high degree of uncertainty in complex systems renders traditional forms of planning of limited utility. For instance, there may be continual shifts in scarce resources, policies, directives, initiatives, and scenarios that create instabilities and make traditional planning difficult. Systems thinking suggests that planning is undertaken with the understanding that complex systems can, and will, shift over time and with new knowledge gained through operation.

(12) ***Information Saturation***—the proliferation of data and information continues to rise exponentially for complex systems. Knowledge of the design for data and information flows is critical to proper functioning of a system. As information is the lifeblood of a complex system, systems thinking is focused on ensuring that communications (flow and interpretation of information) are effectively designed, executed, and evolved to serve the needs of the system.

(13) ***Identity Coherence***—identity provides a reference point for a complex system. This reference point is the set of fundamental values, patterns, and attributes that define the essence of a complex system. Identity is the source

for consistency in decisions, actions and interpretations. Systems thinking is focused on supporting a strong identity such that there is a durable basis for understanding the attributes that are 'non-negotiable' in defining the essence of the system.

(14) ***Generational Shifts***—the generational shift in modern complex systems is well underway. This shift is predicated by the inevitable coming dominance of the millennial generation. For complex systems, this introduces additional strains and burdens in already overly complex systems. From a systems thinking perspective, this generational shift suggests that the incorporation of design modifications should be introduced to lessen the burden of the generational differences while capitalizing on designs that can take advantage of those differentials.

(15) ***Non-ergodicity***—complex systems are characterized by conditions of having no clearly defined states or discernible transitions between system states (Souza-Poza, et al. 2008). This suggests that systems are subject to inherent difficulties in understanding and provoking continuous transition to identified goals. Thus, traditional approaches based on a level of certainty and knowledge of cause–effect relationships is tenuous. Systems thinking suggests that operation in these conditions requires a degree of flexibility, being able to reconfigure as necessary to reset directions based on shifting conditions and understanding.

(16) ***Non-monotonicity***—the reality for complex systems is that increases in knowledge are not reciprocated by increases in understanding. Actions and decisions are always tentative [85]. This suggests that there must be a recognition that knowledge is necessarily provisional, incomplete, and fallible. The systems thinking perspective suggests the need to accept that the system is in continual flux, and allowances for perturbations from external and internal sources must be designed into the execution and development of a system.

The attributes above are certainly not suggested as an absolute or complete set. However, there are three important themes that are suggested for systems thinking for CSG. First, dynamic shifting in knowledge and understanding emerges over time as the system operates. Precise system knowledge designation the precise form, location, duration, or impact of emergent conditions cannot be known in advance. Second, our knowledge of a complex system must always be considered to be incomplete, fallible, and subject to change. This implies that design, execution, and development of a complex system cannot be precisely prescribed in advance. Care must be taken to limit the naïve assumption that the present state of knowledge/understanding will remain static and sufficient for future direction. Third, systems thinking provides a language, way of thinking, and alternative paradigm for dealing with increasingly complex systems and their problems. Following work in systems theory [3, 45, 54, 47, 91], we examine axioms and propositions that provide the language to inform systems thinking with implications for CSG. An axiom is taken as an accepted statement/concept that has been established over time and accepted without further explanation. Propositions are widely held concepts, laws, and principles that have

Table 3 Systems theory axioms, propositions, and implications for systems thinking in CSG

Proposition	Explanation	Implications for systems thinking in CSG
Centrality Axiom: Central to all systems are two pairs of propositions: emergence and hierarchy, and communication and control		
Communication [77, 78, 83]	Communication is a transaction between the information source terminal and the destination terminal, with the sole aim of generation and reproduction of symbols. Information is transmitted as a selection along possible alternative states	All CSG endeavors rely on communications—within designed systems, between systems and their environment, within the CSG function team, and with stakeholders. More rigorous understanding of communications can enhance CSG practice
Control [20]	The process by means of which a whole entity retains its identity and/or performance under changing circumstances	Control entails establishment of constraints and regulatory capacity in systems as well as their development processes. Design for system control impacts system robustness, resilience, viability, and fragility
Emergence [20, 6]	Whole entities exhibit properties and patterns that are meaningful only when they are attributed to the whole, not its parts	Emergence results in system patterns, behavior, or performance that cannot be predicted in advance. However, CSG practitioners can use emergence concepts to more effectively design, execute, and develop systems more adept at dealing with emergent conditions
Hierarchy [70, 20]	Entities meaningfully treated as wholes are built up of smaller entities which are themselves, wholes. In a hierarchy, emergent properties denote the levels	Appreciation of hierarchy implies CSG practice should organize large sets (e.g., data, requirements) into meaningful 'categorizations' which permits greater organization and actionable understanding
Contextual Axiom: System meaning is informed by the circumstances and factors that surround the system		
Complementarity [13]	Two different perspectives or models about a system will reveal truths regarding the system that are neither entirely independent nor entirely compatible	Considering multiple viewpoints and perspectives strengthens CSG practice. Accepting that there is a logic and assumptions that make alternative viewpoints correct, challenging logic/assumptions can enhance CSG dialogs and reduce unnecessary conflict

(continued)

Table 3 (continued)

Proposition	Explanation	Implications for systems thinking in CSG
Incompressibility [24, 72]	Each element in the system is ignorant of the behavior of the system as a whole and only responds to information that is available to it locally. As such, the best representation of a complex system is the system itself and that any representation other than the system itself will necessarily misrepresent certain aspects of the original system	All representations (models) are abstractions of a complex system, inevitably subject to abstraction error. Thus, CSG practice is enhanced by accepting incomplete and fallible knowledge and continually questioning appropriateness and evolving formulations based on new knowledge
Holism [84]	A system must be considered as a whole, rather than a sum of its parts	CSG practice must be deployed across the spectrum of technology, social, human, organizational, managerial, policy, and political dimensions. System failure can occur across this holistic spectrum, not just technology
Boundary [87, 83]	The abstract, semi-permeable perimeter of the system defines the components that make up the system, segregating them from environmental factors and may prevent or permit entry of matter, energy and information	Boundaries determine what is included/excluded for CSG efforts and should be made explicit through criteria for inclusion/exclusion. Compensation should be made for their change over time and shifts in conditions/understanding

(continued)

Table 3 (continued)

Proposition	Explanation	Implications for systems thinking in CSG
Design Axiom: system design is a purposeful imbalance of resources and relationships		
Minimal critical specification [22, 23]	This principle has two aspects, negative and positive. The negative implies that no more should be specified than is absolutely essential; the positive requires that we identify what is essential	Excessive constraint to regulate systems reduces autonomy and wastes scarce system resources. Care must be made to only constrain a system to the degree necessary to preserve outputs and outcomes desired
Power law [66]	The probability of measuring a particular value of some quantity varies inversely as a power of that number	CSG practice must anticipate nonlinear (exponential) relationships in complex systems. Tradeoffs in system performance are not in direct correspondence to either resources invested or effort expended
Requisite Parsimony [63, 82]	The capacity of human short-term recall is no greater than seven plus or minus two items	CSG practitioners should note human limitations to simultaneously focus on multiple factors. This suggests striving for the greatest simplicity possible in systems that are designed, operated, and evolved by people
Requisite saliency [14]	The factors that will be considered in a system design are seldom of equal importance. Instead, there is an underlying logic awaiting discovery in each system design that will reveal the significance of these factors	All characteristics (e.g., design parameters) are not of equivalent importance. CSG practice should provide clarity in criticality and priorities for making effective tradeoff decisions throughout the system lifecycle

(continued)

Table 3 (continued)

Proposition	Explanation	Implications for systems thinking in CSG
Goal Axiom: **systems achieve specific goals through purposeful behavior using pathways and means**		
Equifinality [86]	If a steady state is reached in an open system, it is independent of the initial conditions and determined by the system parameters, e.g., rates of reaction and transport	CSG practice must accept that there are alternative approaches that can produce equivalent results in CSG efforts. Execution of CSG is not achieved by one 'optimal' approach
Multifinality [15]	Radically different end states are possible from the same initial conditions	There are invariably different results that can occur in execution of CSG. Design for CSG must anticipate that circumstances and conditions can influence initial designs
Purposive Behavior [74]	Purposeful behavior is meant to denote that the act or behavior may be interpreted as directed to the attainment of a goal—i.e., to a final condition in which the behaving object reaches a definite correlation in time or in space with respect to another object or event	Designs in CSG should be clear in desired purpose related to fulfilling a need or addressing a problem. Irrespective of intent, system purpose is a function of utility (what a system provides)
Satisficing [80, 81]	The decision-making process whereby one chooses an option that is, while perhaps not the best, good enough	CSG practices should accept that, for complex systems, 'optimal' is not achievable (multiple possibilities can achieve desirable outputs/outcomes} nor desirable (unreasonable resources can be expended in pursuit of 'perfection')

(continued)

Table 3 (continued)

Proposition	Explanation	Implications for systems thinking in CSG
Information Axiom: Systems create, possess, transfer and modify information		
Information Redundancy [79]	The number of bits used to transmit a message, minus the number of bits of actual information in the message	Information in CSG processes should be designed with multiple pathways/mechanisms (redundancy) necessary to ensure both 'receipt' and intended 'interpretation' are met
Redundancy of Potential Command [62]	Effective action is achieved by an adequate concatenation of information	Decision authority in CSG should reside, to the greatest degree possible, at the point in closest proximity to where decision execution actions will be implemented
Operational Axiom: **Systems must be addressed in situ, where the system is exhibiting purposeful behavior**		
Dynamic Equilibrium [87, 64]	An entity exists as expressions of a pattern of processes of an ordered system of forces, undergoing fluxes and continuing flows of matter, energy and information in an equilibrium that is not static	CSG designs, processes, and activities are in constant flux, experience changing circumstances, and should be adjusted 'on the fly' to maintain stability in the face of change
Homeorhesis [88, 89]	The concept encompasses dynamical systems that return to an acceptable trajectory through adjustments in dynamic equilibrium controlled by interrelated regulation mechanisms	Key to CSG achievement is constant adjustment of the path toward a goal. Changes in external circumstances will require adjustment of trajectory – best by purposeful adjustment of regulatory mechanisms than by chance/crisis encounters
Homeostasis [16]	The property of an open system to regulate its internal environment so as to maintain a stable condition, by means of multiple dynamic equilibrium adjustments controlled by interrelated regulation mechanisms	CSG practice must provide adjustments to assure that key (internal) parameters maintain balance in response to inevitable internal flux that might impact ability to achieve CSG objectives

(continued)

Table 3 (continued)

Proposition	Explanation	Implications for systems thinking in CSG
Redundancy [69]	Means of increasing both the safety and reliability of systems by providing superfluous or excess resources	Sufficient mechanisms and resources should be allocated in a CSG effort to assure that achievement of objectives can be attained—given inevitable variabilities whose specific form cannot be know in advance
Relaxation Time [25, 37]	Systems need adequate time to recover from disorder that disturbs its equilibrium, at which point characteristic behavior resumes	Significant changes in CSG processes, design, or execution should be taken with clear understanding of the intent and determinants of success for the effort. Engaging multiple 'changes' can result in indeterminate sources of system oscillation (positive or negative)
Self-organization [8]	The spontaneous emergence of order out of the local interactions between initially independent components	Self-organization in CSG is an effective (least energy consuming) approach to design and execution. Patterns are permitted to emerge without interference (constraint) being invoked
Sub-optimization [35]	If each subsystem, regarded separately, is made to operate with maximum efficiency, the system as a whole will not operate with utmost efficiency	Integration of system elements requires surrendering of autonomy in favor of system integration. CSG practice should, by design, achieve a balance between subsystem autonomy and system integration

(continued)

Table 3 (continued)

Proposition	Explanation	Implications for systems thinking in CSG
Viability Axiom: Key parameters in a system must be controlled to ensure continued existence		
Circular causality [61]	An effect becomes a causative factor for future 'effects', influencing them in a manner particularly subtle, variable, flexible and of an endless number of possibilities	CSG practice must appreciate that some tightly coupled relationships cannot be broken into simple cause-effect understanding. Instead, these relationships can only be understood as a whole, not by breaking down
Feedback [92]	All purposeful behavior may be considered to require negative feedback. If a goal is to be attained, some signals from the goal are necessary at some time to direct the behavior	There is no perfect system design. Variabilities in deployment context require initiation of feedback to make corrections to account for variability impacting system outputs
Recursion [11]	The fundamental laws governing the processes at one level are also present at the next higher level	Irrespective of a particular system, the same system functions must be performed, subject to the same system propositions. If these functions are understood for one system, they are understood for all systems a CSG practitioner will encounter
Requisite Hierarchy [10]	The weaker in average are the regulatory abilities and the larger the uncertainties of available regulators, the more hierarchy is needed in the organization of regulation and control to attain the same result, if possible at all	CSG practitioners should 'flatten' the hierarchy through design of regulatory capacity to assure consistent performance (outputs/outcomes) and diminish uncertainty through design
Requisite Variety [9]	Control can be obtained only if the variety of the controller is at least as great as the variety of the situation to be controlled	The environment within which CSG is performed will generate disturbances (e.g., stakeholder requirement changes). CSG practitioners must provide a design that matches the types and quantities of disturbances that can be experienced

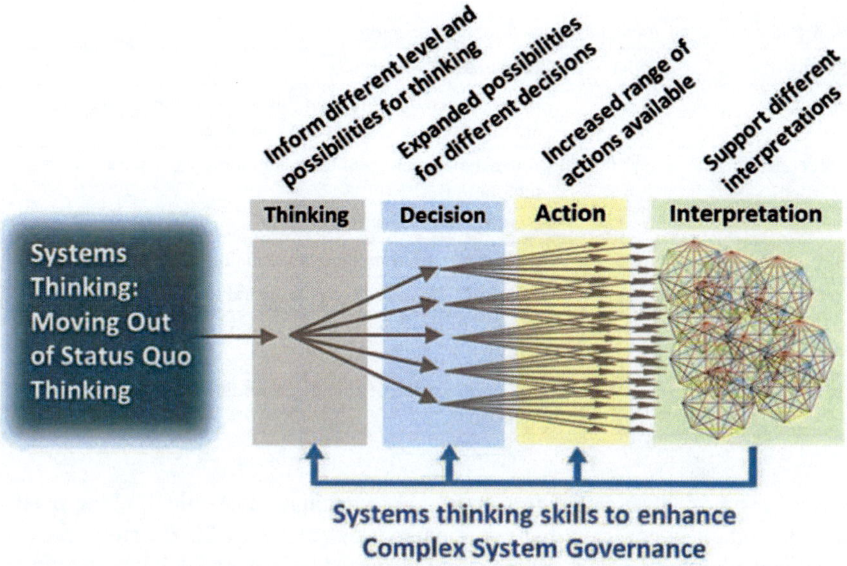

Fig. 5 Systems thinking moving beyond the status quo

been proposed in the systems theory literature, having withstood the test of time. Propositions are organized by associated axioms. Table 3 provides the axiom, associated propositions and implications for systems thinking in CSG. The axioms are drawn from the earlier work of Adams et al. [3] and subsequent amplifying works [54, 55, 91]. The propositions and explanations are drawn directly from the work of Whitney et al. Whitney et al. [91]. Our objective is to extend the propositions and thus bring systems theory closer to systems thinking implications for CSG.

The systems theory propositions provide the language for application of systems thinking for CSG. This language is essential to move to a different (systems) level of thinking. A different level of thinking is a precursor to engaging a corresponding different level of decision, action, and interpretation (Fig. 5). This is consistent with the formulation by Keating et al. [56], calling for movement out of the status quo trajectory pattern by using systems thinking to open the aperture for different thinking, decision, action, and interpretation.

4 Systems Thinking Capacity Assessment

Without doubt, systems thinking is critical to achieving and maintaining effectiveness in CSG.

There is some level of agreement on what systems thinking is for CSG, its importance, and implications. However, there is still the issue of how we go about determining the level of systems thinking for individuals and composite entities who will

Table 4 Seven dimensions of systems thinking capacity

Dimensions	Explanation
Interaction	Interconnectedness in coordination and communication among multiple systems and components
Independence	Balance between local level autonomy versus system integration
Change	Comfort with rapidly shifting systems and situations
Uncertainty	Acceptance of unpredictable situations with limited control
Complexity	Comfort with multidisciplinary and limited understanding
Systems worldview	Understanding system behavior at the whole versus part level
Flexibility	Accommodation of change or modifications in systems or approach

be engaging CSG. The ability to quantify an individual's systemic thinking potential is important in ensuring that participants involved with CSG development can effectively participate. Also, systems thinking capacity is an enabler, or constrainer, as to how prepared individuals may be to fulfill important roles in performing CSG functions. In this section, an approach to determine the level of systems thinking for an individual or entity is developed. This approach is referred to as systems thinking capacity (ST-Cap) instrument.

Based on earlier work by Jaradat [38–41], seven dimensions of systems thinking were developed. The ST-Cap instrument determines an individual's systems thinking capacity by analyzing seven separate dimensions of the individual's natural tendencies (following [40]) in relationship to systems thinking. The seven dimensions of systems thinking capacity are identified in Table 4.

Each of the dimensions has a range, spanning a spectrum from more or less systemic in their disposition. Table 5 identifies the range for each dimension. For each dimension, the left column indicates a propensity for a less systemic perspective. In contrast, the right column indicates a propensity for a more system perspective.

The seven systems thinking dimensions described above are used to establish a person's pre-dispositional thinking and determine their capacity to engage in the degree of systems thinking that is required by CSG. Individuals who exist more in the realm of *holism* are more likely to be more comfortable in dealing with unstable complex environments—this is a hallmark of complex systems and CSG. Although a lack of systems thinking capacity this does disqualify reductionist thinkers from working with complex systems, the high-end systems thinking required of CSG may place a development effort in a difficult situation. Lacking the sufficient systems thinking capacity, the probability for a successful CSG endeavor becomes increasingly doubtful. With respect to CSG, there are opportunities for all 'types' of systems thinking. Those who can focus on the broad 'systemic' aspects of the situation will be better suited to activities, task, and functions that call for heightened systems thinking capacity. Those without a lower level of systems thinking capacity may be better suited, and more comfortable, performing more narrowly defined and task focused activities.

Table 5 Systems thinking characteristics (based on [38, 40])

Systems thinking Dimension	Spectrum of individual systems thinking capacity characteristics	
	Less systemic	More systemic
Interaction	*ISOLATION (N)*—Inclined to local interaction, follow detailed plan, prefer to work individually, enjoy working in small systems, and interested more in cause-effect solutions	*INTERCONNECTIVITY (I)*—Inclined to global interactions, follow general plan, work within a team, and interested less in identifiable cause-effect relationships
Independence	*AUTONOMY (A)*—Preserve local autonomy, tend more to independent decision and local performance level	*INTEGRATION (G)*—Preserve global integration, tend more to dependent decision and global performance level
Change	*RESISTANT TO CHANGE (V)*—Prefer taking few perspectives into consideration, over specify requirements, focus more on the internal forces, like short-range plans and thinking, tend to lock in decisions, and work best in stable environment	*TOLERANT OF CHANGE (Y)*—Prefer taking multiple perspectives into consideration, under specify requirements, focus more on external forces, like long-range plans and thinking, keep decision options open, and work best in changing environment
Uncertainty	*STABILITY (T)*—Prepare detailed plans beforehand, focus on the details, uncomfortable with uncertainty, believe work environment is under control, enjoy objective, and technical problems	*EMERGENCE (E)*—React to situations as they occur, focus on the whole, comfortable with uncertainty, believe work environment is difficult to control, enjoy subjectivity, and non-technical problems
Complexity	*SIMPLICITY (S)*—Avoid uncertainty, work on linear problems, prefer best solution, prefer small scale problems	*COMPLEXITY (C)*—Expect uncertainty, work on multidimensional problems, prefer a working solution, and explore the surrounding environment
Systems worldview	*REDUCTIONISM (R)*—there exist multiple, potentially divergent, perspectives on the problem domain	*HOLISM (H)*—assumes that there is alignment of perspectives for the problem domain
Flexibility	*RIGIDITY (D)*—prefer not to change, like determined plan, motivated by routine	*FLEXIBILITY (F)*—accommodate change, like flexible plans, open to new ideas, unmotivated by routine

The ST-Cap instrument, based on Jaradat's research 38, is a 39 question instrument to establish the capacity for an individual to engage in systems thinking. This instrument represents an important advance in the determination of systems thinking capacity for individuals and entities contemplating engaging CSG. The section below titled 'Systems Thinking Capacity Instrument' provides the full instrument and classification schema.

The ST-Cap instrument provides a 'baseline' snapshot for individuals. In addition, the results of the group of individuals participating in a CSG initiative can be aggregated across their systems thinking profiles. This aggregation provides a locus of systems thinking of the group as well as the distribution of the range of thinking (diversity of thinking) held by the group. In this sense, the aggregate mean and variance for the group distribution across the seven dimensions provide an indicator of the level of systems thinking available in the group. Low levels of systems thinking capacity may indicate that a group needs to first increase the capacity for systems thinking prior to more comprehensive engagement in CSG.

Vignette – We are Agile, but not Flexible?

A production division of a company took the ST-Cap instrument. The division was touted as being based in an agile paradigm. This entailed being able to quickly adapt and improvise to meet shifting demands and changes to products and customer preferences. The group was perplexed when the results of the ST-Cap were tabulated. The group was scored rather low on the flexibility dimension. The result represented the antithesis of what had been continually projected as the central paradigm upon which the system was designed and operated. Through successive exploration, the confounding discovery was examined and the realization that the system was in fact lacking substantial flexibility in being adaptive to shifts in the environment and dealing with internal flux within the system. The effort pointed out the danger of announcing to the world the possession of a particular trait without the substantial evidence to conclude that it actually existed in the system.

There is significant reliance on the systems thinking capabilities of the individuals implementing CSG deployment. These individuals must be capable of mustering the systems thinking capacity to effectively engage CSG at the level of thinking, decision making, and interpretation that is congruent with advancing the state of CSG development. It is important to note that systems thinking and the ST-Cap assessment have been deployed outside of the CSG domain. The implementation of systems thinking is predicated on the (1) systems thinking capacity of the individual acquired naturally or through training, education, or experience, and (2) determination of the predisposition of an individual toward a systemic perspective. This draws a close coupling of an individual's capacities regarding systems thinking with their potential for effectiveness engaging CSG.

Beyond CSG, systems thinking has been implemented in several fields desiring to assess the capabilities of individuals to engage in systems thinking. It is not a foreign concept to assess individuals across dimensions, preferences, and tendencies to identify congruence of individuals to the work they are tasked to perform. As workplace demands for greater systems thinking capabilities increases, the emphasis on conducting assessment of those capabilities also rises. Matching the employees with the demands of assigned tasks and the environment within which those tasks must be performed is critical. It is costly to require individuals to engage in systems thinking based tasks for which they are not adequately suited. Following this perspective, several organizations have used the ST-Cap instrument to better understand the congruence between individual, work, and the environment. For instance, in a recent case study by Karam et al. [44], the ST-Cap instrument was applied in a large U.S. organization to demonstrate how it might aid in selection of future employees to be more compatible with the complexity in their job demands. In the study, two instruments were used to assess individual congruence to work. First, the ST-Cap provided an indicator as to the capability for systems thinking. Second, the complexity demand of the environment was assessed along the same dimensions as the ST-Cap. The result was used to show the differentials along each system's thinking dimension compared to that demanded by the work environment. Thus, the 'fit' of an individual to the tasks they would be performing could be established. The matching of individual ST-Cap and Environmental Complexity Demand was used to assess the match of the employees to the organization's needs [44].

The ST-Cap instrument has also been used in another study by Jaradat et al.[42]. In this study, the ST-Cap instrument was used to compare systems thinking skills across three different sectors, including 'Academic/government' 'Industry/business,' and 'Military.' This study had a sample size of 273 individuals that was comprised of systems engineers and engineering managers, with various levels of expertise, with each of the sectors represented in the sample (100 academic/government, 117 industry/business, and 56 military spanning 25 different organizations). The study findings demonstrated the capability of the ST-Cap instrument to identify and contrast the systems thinking capacity across the three different sectors. Ultimately, the study concluded that the military domain is highly governed by a systemic (holistic)-based profile, while the other two sectors are in between holistic and reductionist profiles [42].

Although the ST-Cap instrument continues to be refined, it has been shown to be a useful indicator of an individual capacity for systems thinking. While this is critical for CSG, systems thinking also has applicability across multiple different contexts where complexity requires heightened capabilities for systems thinking.

5 Implications of Systems Thinking for CSG

Systems thinking is critical to effective deployment of CSG. The language provided by systems thinking (drawn from systems theory) provides the leverage to move to a different level of thinking. This different level of thinking is instrumental in providing the foundations to engage in, and develop, a different corresponding set of decisions, actions, and interpretations in response to increasing complexity.

Determining individual capacity for systems thinking has several important implications for CSG development. First, having an appropriate level of individual/group systems thinking capacity is critical to CSG development. Ultimately, CSG development will be designed, executed, and evolved by individuals. Their capacity to engage in systems thinking will have a profound influence on the speed, depth, and sustainability of CSG development. The CSG development methodology is targeted to *initialization, mapping, and development stages.* At each stage of this methodology, systems thinking is a critical influence. As Fig. 6 indicates, there is a circular causal loop where systems thinking informs CSG Development and CSG Development enhances the level of systems thinking (permitting increasing sophistication in decision, action, and alternative options).

Following Jaradat and Keating [40], systems thinking provides a current baseline of systems thinking held in the system of interest, helps determine what might feasibly be undertaken for CSG development given the current state of systems thinking,

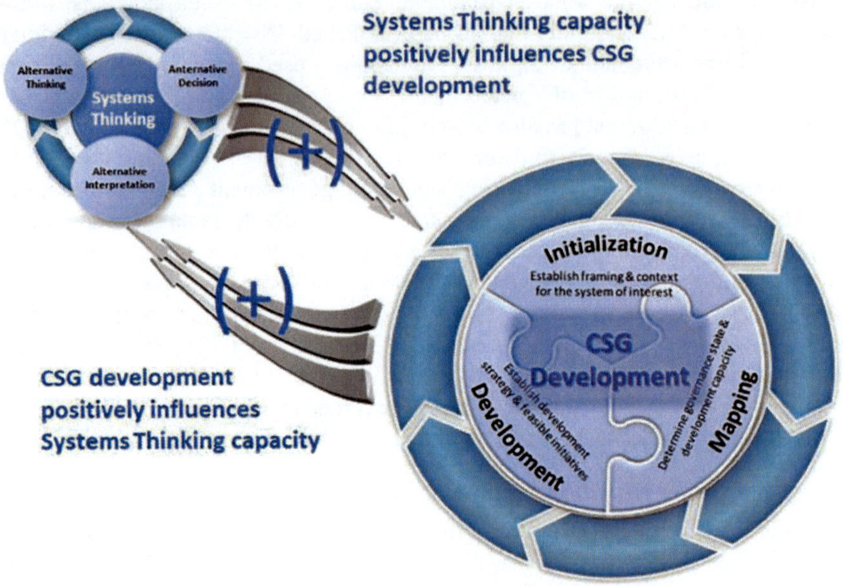

Fig. 6 Circular causal relationship between systems thinking and CSG development

and supports alternative decisions, actions, and interpretations for CSG development. In response, CSG development provides a maturing systems thinking capacity through the development approach, suggests enhancements to systems thinking identified during development, and benefits from enhanced systems thinking supporting increasingly sophisticated CSG development initiatives.

A second implication of systems thinking for CSG involves helping to determine what can feasibly be undertaken with respect to system development. System development activities are constrained by the degree of sophistication (maturity) that exists in the systems thinking capacity held by individuals and the system of interest. If sufficient systems thinking capacity does not exist, then the probability of success for development initiatives requiring systems thinking capacity beyond that available, will likely fall short of expectations. Take for instance a case where a CSG development initiative requires a 'high' level of individual capacity for systems thinking. Further, that the 'high' level is not held by individuals, or within the group, who will assume responsibility for execution of the initiative. The conclusion is that the probability for success in the execution of the initiative is suspect. This does not suggest that at later points in CSG development, the systems thinking capacity will remain at an insufficient level. As CSG development escalates, so too does the increasing capacity for systems thinking. Thus, additional development possibilities to engage 'higher' level CSG development initiatives advances as systems thinking capacity advances.

There are several system thinking challenges that practitioners should consider for deployment of CSG. Each of these challenges represents areas of consideration that might enhance effectiveness of CSG development—but are certainly not prescriptive as a recipe for success in CSG. Systems thinking, while presented as essential to be assessed in the front-end of a CSG development effort, in actuality operates throughout the design, execution, and development stages of CSG. Thus, the degree of systems thinking limits or enhances every aspect of the design, execution, and evolution of CSG. To capsule the system thinking challenges inherent in CSG, the following six challenges, stemming from earlier work of Jaradat and Keating [40] are provided to better appreciate systems thinking considerations for CSG. Among these challenges and their implications for systems thinking in CSG development are:

(1) *Systems Thinking Capacity*—for a system of interest, there is a level of systems thinking capacity that exists for both individuals and the collective group responsible for CSG functions. For individuals, Jaradat's [38] Systems Thinking Capacity instrument is useful as a 'snapshot' of the individual capacity for systems thinking. Although a good start, it cannot be taken as more than an invitation to deeper examination of the prevalence of systems thinking. Beyond the individuals, the larger system should also be considered as to the level of systems thinking that exists at the aggregate level. This aggregate should also include the variability (diversity) in thinking across the aggregate. Ultimately, the success of endeavors to enhance CSG are subject to the match between the degree of systems thinking existing in the system of interest and the level of systems thinking demanded by the development activities undertaken. A mismatch between CSG development activities and the required level of systems thinking capacity is not likely to produce the results desired.
(2) *Support Infrastructure Compatibility with Systems Thinking*—CSG development is unlikely to ever be a 'clean sheet' effort. Instead, it is more likely that CSG development will be deployed for an existing system of interest. Thus, there is already an existing infrastructure in place (e.g., processes, support systems, functions, mores, paradigm) that constrains/enables the system of interest. Engaging systems thinking cannot be accomplished independent of the constraining, as well as enabling, capacity that the system infrastructure provides for CSG development. Compatibility is always necessary between the support infrastructure and the systems thinking capacity required for initiatives undertaken to achieve success in CSG development initiatives. Without this compatibility, limiting infrastructure must be evolved such that sufficient compatibility exists to ensure that CSG development initiatives are feasible given the support infrastructure. Irrespective of the 'goodness' of ideas stemming from systems thinking engagement, failure to account for the impacts of supporting infrastructure is shortsighted. The result of incompatible infrastructure can be failure to achieve the full potential of CSG initiatives.

(3) *Continual Development of Systems Thinking Capacity*—in the initial stages of CSG, development a 'snapshot' of system thinking capacity is constructed. However, systems thinking does not exist in a static state. Care must be taken to actively advance the state of systems thinking at both the individual as well as the system levels. Active development of systems thinking capacity should be by purposeful design as well as systems thinking that emerges through the routine application of CSG development. Simply relying on a 'snapshot' of systems thinking capacity is not sufficient. Complacency with the 'snapshot' of systems thinking capacity is an invitation to make faulty conclusions concerning the nature and types of CSG development activities that can be undertaken. Also, it is not sufficient to simply take a 'snapshot' of systems thinking and conclude that no further development is necessary. Continual development and advancement of systems thinking should be a central element for holistic CSG development.

(4) *Building Sustainable Foundations for Systems Thinking*—building systems thinking capacity lies beyond the shortsighted perspective of learning new tools or attending training events. Getting the most from systems thinking requires practitioners to continually be immersed in the language and thinking stemming from foundational propositions of systems theory. There will always be new tools and techniques that emerge to support systems thinking. However, when systems thinking is grounded in underlying foundations (e.g., systems theory propositions), the lasting impacts will be more pronounced, widely applicable, and sustainable. In this sense, systems thinking will withstand the tests of time and better inform CSG development.

(5) *Appreciation of Uniqueness, not Prescriptive Application*—systems thinking is largely about generating shifts in thinking such that alternative decisions, actions, and interpretations can be developed. Thus, for achievement of this shift, there is not a prescriptive formula, approach, or universal method that can assure repeatable success. Instead, each complex system, context, and set of practitioners are unique for each system of interest. Therefore, this uniqueness invokes the necessity to engage in an equally unique approach to development of systems thinking capacity. In short, there is not a 'one size fits all' approach that can be taken to develop systems thinking capacity. Practitioners must appreciate that every complex problem is unique and exists in a unique context. Thus, both systems thinking, and CSG informed from systems thinking, will miss the mark if they are deployed prescriptively. Instead, they must be tailored to the uniqueness of the system of interest, context, and set of practitioners deploying CSG.

(6) *Shifting from Thinking to Action*—systems thinking invites CSG practitioners to explore complex situations in a way that allows for framing and reframing from different perspectives. The most fundamental system questions are explored from an underlying system thinking perspective. Questions about boundaries, relationships, and behavior are cast from the perspective of the language and thinking of systems. These questions, engaged from a systems thinking perspective, gain the advantages that non systems-based approaches

and perspectives cannot bring to complex systems. However, there is a distinction between thinking and action. Thinking only creates the potential for initiation of action, not the action itself. Care must be taken to appreciate the gap between what can be generated at a 'thinking'-level versus what can be 'feasibly engaged' at the action level for CSG development. For CSG, active management of this gap is essential. Thinking without action is insufficient to develop CSG. 'Thinking alone' will only create the potential for corresponding action. Additionally, caution must be taken as to not limit thinking. However, there must also be an emphasis on not creating unrealistic expectations concerning what can be feasibly achieved in CSG development.

Systems thinking is not a cure-all that can address all problems and guarantee successful CSG endeavors. However, systems thinking invites practitioners to engage CSG development from a different level of thinking. The different level of thinking is appropriate and applicable throughout all stages of CSG development [50]. The criticality of systems thinking, as well as the systems thinking capacity for both individuals and system of interest participants, cannot be overstated for CSG development. The crux of systems thinking lies in the entirely different range of decisions, actions, and interpretations that become accessible in pursuit of CSG development.

6 Systems Thinking Capacity Instrument

This instrument includes 39 items. Each item should be completed by circling the response that best describes your preference. Following the last item, complete the scoring sheet.

1. To address system performance focus should be on
   (a) individual members of the system
   (b) interactions between members of the system

2. Do you prefer to work with
   (a) few systems or people
   (b) many systems or people

3. Are you most comfortable developing a
   (a) detailed plan
   (b) a general plan

4. Do you prefer to
   (a) work individually on a specific aspect of the problem
   (b) organize a team to explore the problem

5. With respect to system interactions, at which level would you prefer to focus
   (a) locally
   (b) globally
6. Do you feel more comfortable working
   (a) individually
   (b) in a group
7. Which is more important to preserve
   (a) local autonomy
   (b) global integration
8. Decisions should be made
   (a) independent of the system
   (b) dependent on the system
9. Parts in a system should be more
   (a) self-reliant
   (b) dependent
10. Giving up local decision authority should be
   (a) resisted
   (b) embraced
11. Performance is determined more by actions at the
   (a) local level
   (b) global level
12. Do you prefer to think about the time to implement change in a system as
   (a) short
   (b) long
13. Change in a system is most likely to occur as
   (a) revolutionary
   (b) evolutionary
14. In turbulent environments, planning for system change is
   (a) useful
   (b) wasteful
15. Forces for system change are driven more
   (a) internally
   (b) externally

16. To evolve a system, would you prefer to find
    (a) One best approach
    (b) Multiple possible approaches
17. To ensure system performance, it is better to
    (a) overspecify requirements
    (b) underspecify requirements
18. Would you most prefer to work in group that
    (a) prepares detailed plans beforehand
    (b) reacts to situations as they occur
19. You prefer to focus more on the
    (a) specific details
    (b) whole
20. In dealing with unexpected changes, you are generally
    (a) uncomfortable
    (b) comfortable
21. Control of the work environment is
    (a) possible
    (b) not possible
22. I prefer to work on problems for which the solution is
    (a) objective
    (b) subjective
23. I most enjoy working on problems that primarily involve
    (a) technical issues
    (b) non-technical issues
24. Are you more inclined to work on something that follows
    (a) regular patterns
    (b) irregular patterns
25. Once desired performance is achieved, a system should be
    (a) left alone
    (b) adjusted
26. In dealing with a system, would you prefer it io be
    (a) small
    (b) large

27. I prefer to work on problems for which the approach is
    (a) standardized
    (b) unique
28. In solving a problem, I generally try to get opinions from
    (a) a few people
    (b) many people
29. A solution to pProblem should always be
    (a) the best solution
    (b) a working solution
30. A system can be understood by analyzing the parts
    (a) agree
    (b) disagree
31. In thinking about this system, I would prefer to focus on
    (a) particulars
    (b) the whole
32. System performance is primarily determined by individual components
    (a) agree
    (b) disagree
33. A problem should first be addressed at what level
    (a) specific
    (b) general
34. Once successful, a technical solution will result in similar success in other applications
    (a) agree
    (b) disagree
35. I am most comfortable working where circumstances require
    (a) minimal adjustment
    (b) constant adjustment
36. Once a system is deployed, modifications and adjustments indicate that the design was
    (a) inadequate
    (b) flexible

37. In planning for a system solution, plans should be

    (a) fixed
    (b) expected to change

38. With respect to execution of a plan

    (a) I prefer to follow the plan as closely as possible
    (b) I am comfortable with deviating from the plan

39. I would describe my preferred work environment as one for which outcomes

    (a) are predetermined
    (b) emerge

Directions for scoring:

1. For each question, check the ‘a’ or ‘b’ block.
2. Add the total number of ‘a’ answers in the box at the bottom of each column. Do the same for the ‘b’ answers.
3. There are now seven pairs of numbers.
4. Circle the letter below the larger of the numbers of each pair.
5. Place the letter in the six boxes below the pairs (in the order from left to right).
6. These combinations identify the individual’s systemic thinking preference profile.
7. The complete profile is the seven letter combination from the fourteen letters.
8. For the ‘b’ column in each dimension, divide the number of ‘b’ responses by the number of questions (e.g., for Dimension 1, if there were 4 ‘b’ responses, it would be (4 ‘b’ responses)/6 Total Questions = 0.66 or 66%. Record this number on the corresponding ‘radar chart’ for each dimension. For any that are evenly split mark both letters. For example, if Complexity has 3 “a” responses and 3 ‘b’ responses, then the profile letter would be S/C.

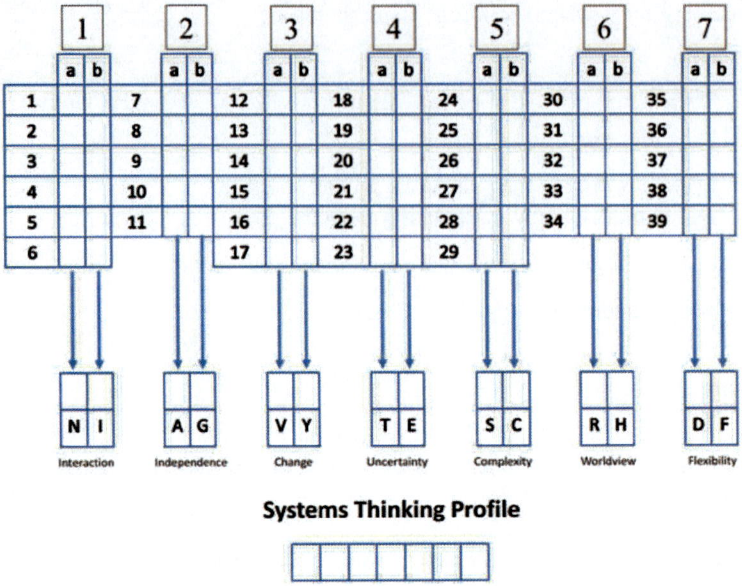

Calculation aid	
5ths	Percentage
1/5	20
2/5	40
3/5	60
4/5	80
5/5	100
6ths	
1/6	17
2/6	33
3/6	50
4/6	67
5/6	83
6/6	100

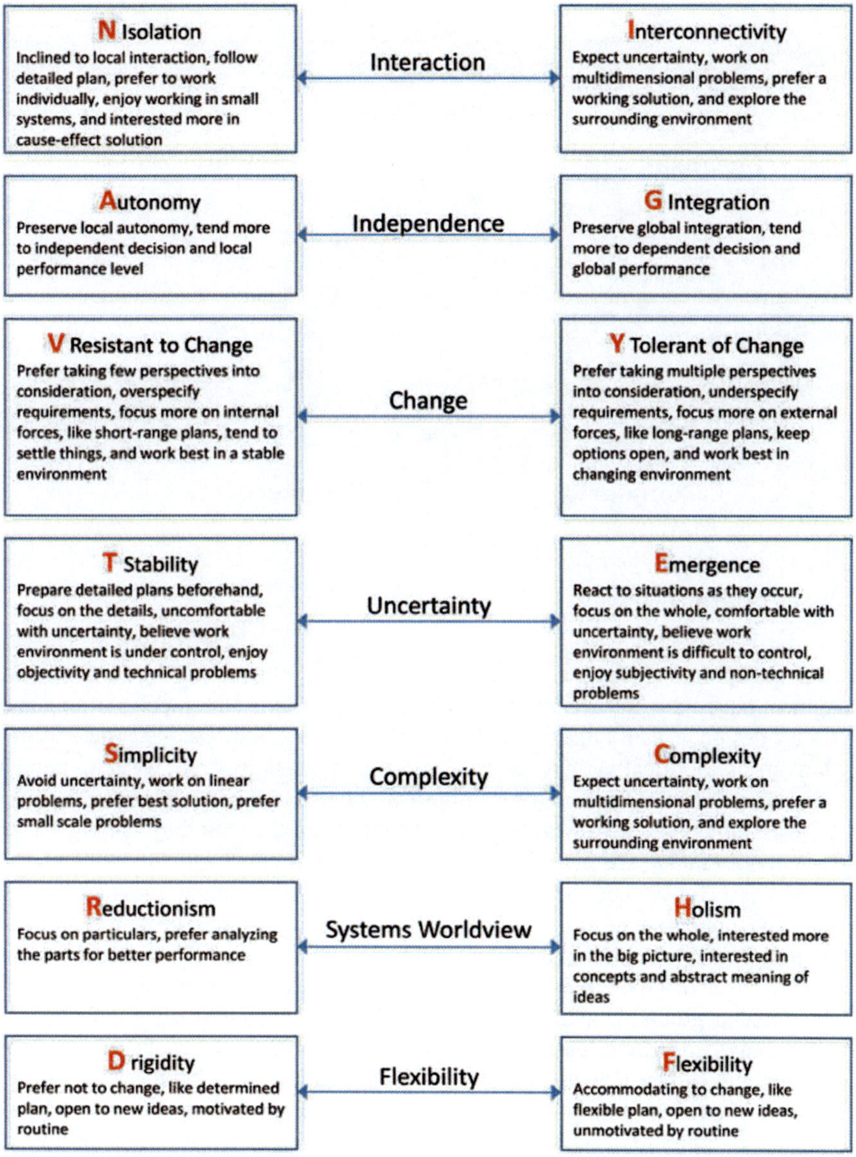
N Isolation
Inclined to local interaction, follow detailed plan, prefer to work individually, enjoy working in small systems, and interested more in cause-effect solution
Interaction
Interconnectivity
Expect uncertainty, work on multidimensional problems, prefer a working solution, and explore the surrounding environment
Autonomy
Preserve local autonomy, tend more to independent decision and local performance level
Independence
G Integration
Preserve global integration, tend more to dependent decision and global performance
V Resistant to Change
Prefer taking few perspectives into consideration, overspecify requirements, focus more on internal forces, like short-range plans, tend to settle things, and work best in a stable environment
Change
Y Tolerant of Change
Prefer taking multiple perspectives into consideration, underspecify requirements, focus more on external forces, like long-range plans, keep options open, and work best in changing environment
T Stability
Prepare detailed plans beforehand, focus on the details, uncomfortable with uncertainty, believe work environment is under control, enjoy objectivity and technical problems
Uncertainty
Emergence
React to situations as they occur, focus on the whole, comfortable with uncertainty, believe work environment is difficult to control, enjoy subjectivity and non-technical problems
Simplicity
Avoid uncertainty, work on linear problems, prefer best solution, prefer small scale problems
Complexity
Complexity
Expect uncertainty, work on multidimensional problems, prefer a working solution, and explore the surrounding environment
Reductionism
Focus on particulars, prefer analyzing the parts for better performance
Systems Worldview
Holism
Focus on the whole, interested more in the big picture, interested in concepts and abstract meaning of ideas
D rigidity
Prefer not to change, like determined plan, open to new ideas, motivated by routine
Flexibility
Flexibility
Accommodating to change, like flexible plan, open to new ideas, unmotivated by routine

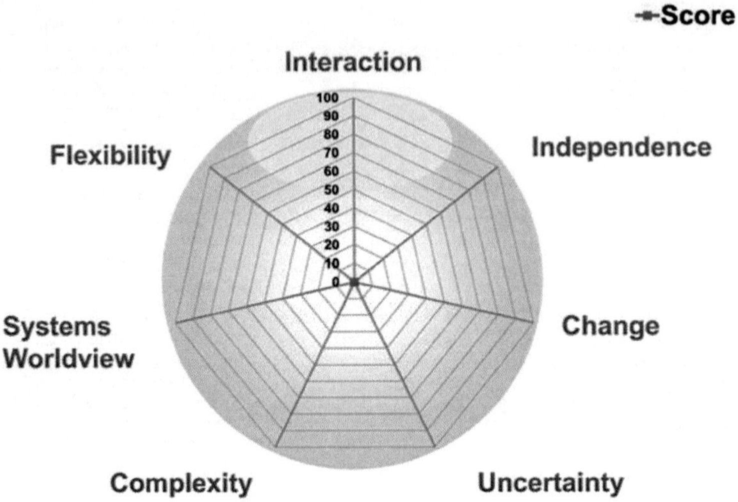

7 Summary

In this chapter, the nature, role, and implications of systems thinking for CSG development have been introduced. Systems thinking provides an important foundation for the design, execution, and development of complex systems. In essence, systems thinking is a critical skill set to have for effectiveness in CSG. Systems thinking was presented from the perspective of providing an individual, and entities, capacity to more effectively deal with the inherent complexity, uncertainty, ambiguity, and emergence that are characteristic of complex systems.

The basis for systems thinking is grounded in the underlying worldview of *holism*. For CSG, holism suggests that pursuit of CSG development must be targeted to understanding system behaviors as a function of not only the entities that comprise the system but also of their interactions. Additionally, the tenets of systems thinking were presented as grounded in the underlying language of systems theory. This language of systems theory provides the basis for moving to a different level of thinking. Subsequently, this different level of thinking enables a different decision space to emerge, different corresponding potentials for actions to enact decisions, and a different range of interpretation accessible to understand 'differently.' Also, systems thinking creates the conditions for engaging in: (1) more disciplined inquiry, grounded in the underlying axioms and propositions of systems theory that enable a different level of understanding of complex system behavior, structure, and performance, (2) support for more effective engagement to deal with increasingly complex systems and their problems, (3) exposure to an alternative language to assist in framing complexity, complex systems, and problems in different ways, and thus opening unforeseen possibilities for resolution, and (4) exposure to a different worldview, inviting movement to a different interaction, interpretation, and understanding to all that is encountered with respect to complex systems. Thus, the reach of systems thinking, while essential to effectiveness in CSG, has a range that far exceeds implications for CSG alone.

Also noted in this chapter is the essence of systems thinking capacity and its potentially enabling, as well as disabling, influence on CSG. Systems thinking plays a critical role in the 'front-end' framing for CSG development. It is noteworthy that systems thinking is not a panacea of effectiveness that can guarantee universal success. On the contrary, it is one aspect, albeit a critical aspect, in foundations essential for effective CSG development. Additionally, the relationship between systems thinking and CSG development was examined as reciprocal in nature. Systems thinking enhances prospects for CSG development. Likewise, CSG development is instrumental in enhancing the state of systems thinking for a system of interest. Each influences, and is influenced by, the other.

Systems thinking capacity was introduced as an approach to establish the level of systems thinking for an individual or set of individuals. The ST-Cap instrument consists of 39 questions that provide an indicator of the level of systems thinking. While the results of the ST-Cap instrument must be considered a 'snapshot' of systems thinking, it does provide an 'indicator' as to the state of systems thinking along the seven dimensions of systems thinking assessed. Also, systems thinking is not static and efforts for CSG development can also include purposeful development of systems thinking.

Systems thinking was suggested as an essential aspect of CSG development. Absent an appropriate level of systems thinking, the prospects for successful CSG endeavors are questionable. Additionally, systems thinking is not a static indicator, but rather something that can and will shift over time as CSG is developed. Implications for systems thinking were proposed, primary among which was the need to move from 'thinking' to 'action' based on the insights and understanding generated from systems thinking. Systems thinking has wide-ranging implications for CSG and exists as a cornerstone for CSG development.

Exercises

1. For complex systems, describe the distinction between 'hard' and 'soft' systems aspects and discuss the implications for 'holistic' treatment for CSG development.
2. Complete the ST-Cap assessment instrument in the section titled 'System Thinking Capacity Instrument.' Identify your 7 letter profile for systems thinking.
3. Based on your ST-Cap assessment profile, discuss the implications of your profile for engaging systems thinking. Where might you have developmental areas or potential 'blind spots' suggested from your profile?
4. Discuss the role that systems thinking plays in CSG and CSG development.

References

1. Ackoff RL (1999) Re-creating the corporation: a design of organizations for the 21st century. Oxford University Press
2. Ackoff R, Gharajedaghi J, Carey A (2010) Systems thinking for curious managers: with 40 new management f-laws. Triarchy Press
3. Adams KM, Hester PT, Bradley JM, Meyers TJ, Keating CB (2014) Systems theory as the foundation for understanding systems. Syst Eng 17(1):112–123. https://doi.org/10.1002/sys.21255
4. Aerts D, Apostel L, De Moor B, Hellemans S, Maex E, Van Belle H, Van der Veken J (1994) World views: from fragmentation to integration. VUB Press. http://theperihelioneffect.com/world-views-from-fragmentation-to-integration/
5. Aoun JE (2017) Robot-proof: higher education in the age of artificial intelligence. The MIT Press
6. Aristotle (2002) Metaphysics: book H-form and being at work (Sachs J, trans.; 2nd edn). Green Lion Press
7. Arnold RD, Wade JP (2015) A definition of systems thinking: a systems approach. Procedia Comput Sci 44:669–678. https://doi.org/10.1016/j.procs.2015.03.050
8. Ashby WR (1947) Principles of the self-organizing dynamic system. J Gen Psychol 37(2):125–128. https://doi.org/10.1080/00221309.1947.9918144
9. Ashby WR (1956) An introduction to cybernetics. Chapman & Hall, Ltd.
10. Aulin-Ahmavaara AY (1979) The law of requisite hierarchy. Kybernetes 8(4):259–266. https://doi.org/10.1108/eb005528
11. Beer S (1979) The heart of the enterprise. Wiley
12. Boardman J, Sauser B (2008) Systems thinking: coping with 21st century problems. CRC Press. https://doi.org/10.1201/9781420054927
13. Bohr N (1928) The quantum postulate and the recent development of atomic theory. Nature 121(3050):580–590
14. Boulding KE (1966) The impact of social sciences. Rutgers University Press.
15. Buckley W (1967) Sociology and modern systems theory. Prentice-Hall
16. Cannon WB (1929) Organization for physiological homeostasis. Physiol Rev 9(3):399–431
17. Capra F (1996) The web of life: a new scientific understanding of living systems. Anchor Books.
18. Checkland PB (1989) Soft systems methodology. Hum Syst Manag 8(4):273–289
19. Checkland PB (1990) Soft systems methodology: a thirty year retrospective. In: Checkland PB, Scholes J (eds) Soft systems methodology in action. Wiley, pp A1–A66
20. Checkland PB (1993) Systems thinking, systems practice. Wiley
21. Chen D, Stroup W (1993) General system theory: toward a conceptual framework for science and technology education for all. J Sci Educ Technol 2(3):447–459
22. Cherns A (1976) The principles of sociotechnical design. Human Relat 29(8):783–792. https://doi.org/10.1177/001872677602900806
23. Cherns A (1987) Principles of sociotechnical design revisted. Human Relat 40(3):153–161
24. Cilliers P (1998) Complexity and postmodernism: Understand complex systems. Routledge.
25. Clemson B (1984) Cybernetics: a new management tool. Abacus Press.
26. Davidz HL (2006) Enabling systems thinking to accelerate the development of senior systems engineers. Thesis, Massachusetts Institute of Technology. https://dspace.mit.edu/handle/1721.1/34200
27. ESD (2002) ESD symposium committee overview: engineering systems research and practice. Working paper ESD-WP-2003–01.20. Massachusetts Institute of Technology. https://dspace.mit.edu/handle/1721.1/102748
28. Flood RL, Carson ER (1993) Dealing with complexity: an introduction to the theory and application of systems science. Plenum Press
29. Foster J (2005) From simplistic to complex systems in economics. Camb J Econ 29(6):873–892. https://doi.org/10.1093/cje/bei083

30. Friedkin NE (2015) The problem of social control and coordination of complex systems in sociology: a look at the community cleavage problem. IEEE Control Syst. https://doi.org/10.1109/MCS.2015.2406655
31. Gharajedaghi J (1999) Systems thinking: managing chaos and complexity: a platform for designing business architecture. Butterworth-Heinemann.
32. Graczyk SL (1993) Get with the system: general systems theory for business officials. Sch Bus Aff 59(2):16–20
33. Haines SG (2000) The managers pocket guide to systems thinking and learning. Human Resource Development Press
34. Hammond D (2002) Exploring the genealogy of systems thinking. Syst Res Behav Sci 19(5):429–439. https://doi.org/10.1002/sres.499
35. Hitch C (1953) Sub-optimization in operations problems. Oper Res 1(3):87–99. https://doi.org/10.1287/opre.1.3.87
36. Hoefler BG, Mar BW (1992) Systems engineering methodology for engineering planning applications. J Prof Issues Eng Educ Pract 118(2):113–128. https://doi.org/10.1061/(ASCE)1052-3928(1992)118:2(113)
37. Holling CS (1996) Engineering resilience versus ecological resilience. In: Schulze P (ed) Engineering within ecological constraints. National Academies Press, pp 31–43
38. Jaradat RM (2014) An instrument to assess individual capacity for system thinking. Ph.D., Old Dominion University, Norfolk, Virginia, USA
39. Jaradat RM (2015) Complex system governance requires systems thinking-how to find systems thinkers. Int J Syst Syst Eng 6(1–2):53–70
40. Jaradat RM, Keating CB (2016) Systems thinking capacity: implications and challenges for complex system governance development. Int J Syst Syst Eng 7(1/2/3):75–94. https://doi.org/10.1504/IJSSE.2016.076130
41. Jaradat RM, Keating CB, Bradley JM (2017) Individual capacity and organizational competency for systems thinking. IEEE Syst J 12(2):1203–1210
42. Jaradat R, Hamilton MA, Dayarathna VL, Karam S, Jones P, Wall ES, Amrani SE, Hsu GSE (2019) Measuring individuals' systems thinking skills through the development of an immersive virtual reality complex system scenarios. In: 2019 ASEE annual conference & exposition, Tampa, Florida. https://doi.org/10.18260/1-2--33090
43. Kaneko K (2006) Life: an introduction to complex systems biology. Springer. http://site.ebrary.com/id/10145135
44. Karam S, Nagahi M, Dayarathna (Nick) VL, Ma J, Jaradat R, Hamilton M (2020) Integrating systems thinking skills with multi-criteria decision-making technology to recruit employee candidates. Expert Syst Appl 160:113585. https://doi.org/10.1016/j.eswa.2020.113585
45. Katina PF (2016) Systems theory as a foundation for discovery of pathologies for complex system problem formulation. In: Masys AJ (ed) Applications of systems thinking and soft operations research in managing complexity. Springer International Publishing, pp 227–267. http://link.springer.com/chapter/https://doi.org/10.1007/978-3-319-21106-0_11
46. Katz D, Kahn RL (1966) The social psychology of organizations. Wiley
47. Keating CB (2015) Complex system governance: theory to practice challenges for system of systems engineering. In: 2015 10th system of systems engineering conference (SoSE), pp 226–231. https://doi.org/10.1109/SYSOSE.2015.7151955
48. Keating CB, Katina PF (2011) Systems of systems engineering: prospects and challenges for the emerging field. Int J Syst Syst Eng 2(2/3):234–256. https://doi.org/10.1504/IJSSE.2011.040556
49. Keating CB, Katina PF (2012) Prevalence of pathologies in systems of systems. Int J Syst Syst Eng 3(3/4):243–267. https://doi.org/10.1504/IJSSE.2012.052688
50. Keating CB, Katina PF (2016) Complex system governance development: a first generation methodology. Int J Syst Syst Eng 7(1/2/3):43–74. https://doi.org/10.1504/IJSSE.2016.076127
51. Keating CB, Bradley JM, Katina PF (2016a) Systemic analysis of complex system governance for acquisition. In: Long S, Ng E-H, Downing C, Nepal B (eds) Proceedings of the thirteenth annual acquisition research symposium. NPS, pp 196–214

52. Keating CB, Calida BY, Jaradat RM, Katina PF (2017) Systems thinking. In: Farr JV, Gandhi SJ, Merino DN (eds) The engineering management handbook, 2nd edn. The American Society of Engineering Management, pp 281–316
53. Keating CB, Katina PF, Bradley JM (2014) Complex system governance: concept, challenges, and emerging research. Int J Syst Syst Eng 5(3):263–288. https://doi.org/10.1504/IJSSE.2014.065756
54. Keating CB, Katina PF, Bradley JM, Pyne JC (2016b) Systems theory as a conceptual foundation for system of systems engineering. Insight 19(3):47–50. https://doi.org/10.1002/inst.12108
55. Keating CB, Katina PF, Hodge R, Bradley JM (2020) Systems theory: bridging the gap between science and practice for systems engineering. INCOSE Int Symp 30(1):1017–1031. https://doi.org/10.1002/j.2334-5837.2020.00769.x
56. Keating CB, Katina PF, Jaradat R, Bradley JM, Hodge R (2021) Systems thinking: a critical skill for systems engineers. INCOSE Int Symp 31(1):522–536. https://doi.org/10.1002/j.2334-5837.2021.00852.x
57. Keohane RO, Nye JS (1977) Power and interdependence: world politics in transition (1977 edition). TBS The Book Service Ltd.
58. Kim DH (1995) Systems thinking tools: a user's reference guide. Pegasus Communications, Inc.
59. King WR, Cleland DI (1972) Management: a systems approach. McGraw-Hill Companies. https://www.biblio.com/book/management-systems-approach-william-r-king/d/1242948028
60. Klir GJ (1985) Architecture of systems problem solving. Plenum Press
61. Korzybski A (1994) Science and sanity: an introduction to non-Aristotelian systems and general semantics. Wiley.
62. McCulloch WS (1965) Embodiments of mind. MIT Press
63. Miller GA (1956) The magical number seven, plus or minus two: some limits on our capacity for processing information. Psychol Rev 63(2):81–97. https://doi.org/10.1037/h0043158
64. Miller JG (1978) Living systems. McGraw-Hill
65. Nagahi M, Jaradat R, El Amrani S, Hamilton M, Goerger SR (2020) Holistic and reductionist thinker: a comparison study based on individuals' skillset and personality types. Int J Syst Syst Eng 10(4):337. https://doi.org/10.1504/IJSSE.2020.112312
66. Newman MEJ (2005) Power laws, Pareto distributions and Zipf's law. Contemp Phys 46(5):323–351. https://doi.org/10.1080/00107510500052444
67. O'Connor J (1997) The art of systems thinking: essential skills for creativity and problem solving. Thorsons
68. O'Connor J, McDermott I (1997) The art of systems thinking. Thorsons, San Francisco
69. Pahl G, Beitz W, Feldhusen J, Grote K-H (2011) Engineering design: a systematic approach (Wallace K, Blessing LTM (eds & trans; 3rd edn). Springer
70. Pattee HH (1973) Hierarchy theory: the challenge of complex systems. Braziller
71. Peters O, Neelin JD (2006) Critical phenomena in atmospheric precipitation. Nat Phys 2(6):393–396. https://doi.org/10.1038/nphys314
72. Richardson KA (2004) Systems theory and complexity: part 2. E:CO 6(4):77–82
73. Richmond B (1993) Systems thinking: critical thinking skills for the 1990s and beyond. Syst Dyn Rev 9(2):113–133. https://doi.org/10.1002/sdr.4260090203
74. Rosenblueth A, Wiener N, Bigelow J (1943) Behavior, purpose and teleology. Philos Sci 10(1):18–24
75. Senge PM (1990) The fifth discipline: the art and practice of the learning organization. Doubleday/Currency
76. Senge PM (1994) The Fifth discipline fieldbook: strategies and tools for building a learning organization. Currency, Doubleday. http://www.loc.gov/catdir/description/random046/93050130.html
77. Shannon CE (1948a) A mathematical theory of communication: part 1. Bell Syst Tech J 27(3):379–423

78. Shannon CE (1948b) A mathematical theory of communication: part 2. Bell Syst Tech J 27(4):623–656
79. Shannon CE, Weaver W (1949) The mathematical theory of communication. University of Illinois Press
80. Simon HA (1955) A behavioral model of rational choice. Q J Econ 69(1):99–118. https://doi.org/10.2307/1884852
81. Simon HA (1956) Rational choice and the structure of the environment. Psychol Rev 63(2):129–138
82. Simon HA (1974) How big is a chunk? Science 183(4124):482–488
83. Skyttner L (2005) General systems theory: problems, perspectives, practice, 2nd edn. World Scientific Publishing Co. Pte. Ltd.
84. Smuts J (1926) Holism and evolution. Greenwood Press
85. Sousa-Poza AA, Kovacic S, Keating CB (2008) System of systems engineering: an emerging multidiscipline. Int J Syst Syst Eng 1(1/2):1–17. https://doi.org/10.1504/IJSSE.2008.018129
86. von Bertalanffy L (1950) The theory of open systems in physics and biology. Science 111(2872):23–29
87. von Bertalanffy L (1968) General system theory: foundations, developments, applications. George Braziller
88. Waddington CH (1957) The strategy of genes: a discussion of some aspects of theoretical biology. Allen and Unwin
89. Waddington CH (1968) Towards a theoretical biology. Nature 218(5141):525–527
90. Waring A (1996) Practical systems thinking. International Thomson Business Press
91. Whitney K, Bradley JM, Baugh DE, Chesterman CW (2015) Systems theory as a foundation for governance of complex systems. Int J Syst Syst Eng 6(1–2):15–32. https://doi.org/10.1504/IJSSE.2015.068805
92. Wiener N (1948) Cybernetics: or control and communication in the animal and the machine. MIT Press
93. Wilhelm H (ed) (1967) The I Ching or book of changes (Baynes CF trans). Princeton University Press

Leadership for Complex System Governance

David Walters

Abstract Recognizing complexity and uncertainty as norms for the environments in which organizations exist, complexity theory, complex systems, and complex adaptive systems have been suggested as appropriate to address these challenges. Further, merging aspects of complex systems and governance to develop a systems-based framework for design, evolution, and implementation of the metasystem which governs a complex system has been proposed. In this chapter, we explore leadership issues associated with governance of complex systems. Topics include the relationship between systems leadership and the Complex System Governance (CSG) reference model; a review of leadership including leadership styles, functions, ethics, complex system leadership, and Complex System Governance leadership; system leadership practice including responsibilities, effects and impacts of leadership, the role of leadership in metasystem design, implementation and operation; and the applicability and integration of systems leadership into the CSG reference model. Exercises will then be addressed.

Keywords Leadership · Complexity theory · Complex systems · Complex system governance

1 Introduction

The extensive body of scholarly literature regarding leadership addresses the subject across a number of domains including nursing, education, academia, economics, politics, military, and business, to name but a few. Much of this literature, developed over a significant period of time, addresses leadership from the perspectives of required personal skills, organizational impact, developmental issues, behavioral characteristics, or ethical issues. Over these many years of study, researchers have developed more than 60 classification systems to help characterize leadership. Despite these efforts, there is no single accepted definition for leadership. However, for the

D. Walters (✉)
Old Dominion University, Norfolk, USA
e-mail: dwalt23321@cox.net

C. B. Keating et al. (eds.), *Complex System Governance*, Topics in Safety, Risk, Reliability and Quality 40, https://doi.org/10.1007/978-3-030-93852-9_15

purposes of this chapter, a definition for leadership, based on concepts from the literature, will be identified and will be used as a basis for further discussions. Leadership functions, responsibilities, effects, and impacts as well as leadership ethics will be also explored.

It is clear from reviewing the literature that the study of leadership from a systems theoretic perspective is a relatively recent endeavor with an emerging emphasis on complexity theory. The primary goal in this chapter is to explore the role of leadership in the governance of complex systems from a systems theory and management cybernetics perspective. The systems theory axioms and propositions which underpin this study were covered in detail in the "Complexity" chapter.

2 The Context for Complex System Governance Leadership

Complex systems exist in environments characterized by uncertainty and complexity. Governance of these systems requires execution of metasystem functions necessary to ensure system viability, or more specifically, control, communication, coordination, and integration of the complex system [1]. CSG is an emerging field based in systems theory, management cybernetics, as well as governance, and has been defined as the '*design, execution, and evolution of the metasystem functions necessary to provide control, communication, coordination and integration of a complex system*' [2] (p. 5). The CSG reference model, previously addressed in "Complex System Governance Reference Model", is a systems-based construct that identifies essential Complex System Governance functions. As a reminder, the functions identified in the CSG reference model include policy and identity (Metasystem 5), system context (Metasystem 5*), strategic system monitoring (Metasystem 5′), system development (Metasystem 4), learning and transformation (Metasystem 4*), environmental scanning (Metasystem 4′), system operations (Metasystem 3), operational performance (Metasystem 3*), and information and communications (Metasystem 2). Figure 1 is a graphical representation of the CSG reference model. Design, implementation, and operation of the system, including the metasystem which provides for these functions, are the responsibility of leadership; therefore, system leadership has a role in each of the aspects of the Complex System Governance reference model. A more detailed discussion of the roles and responsibilities of leadership in ensuring the functions identified in the reference model are executed is provided later in this chapter.

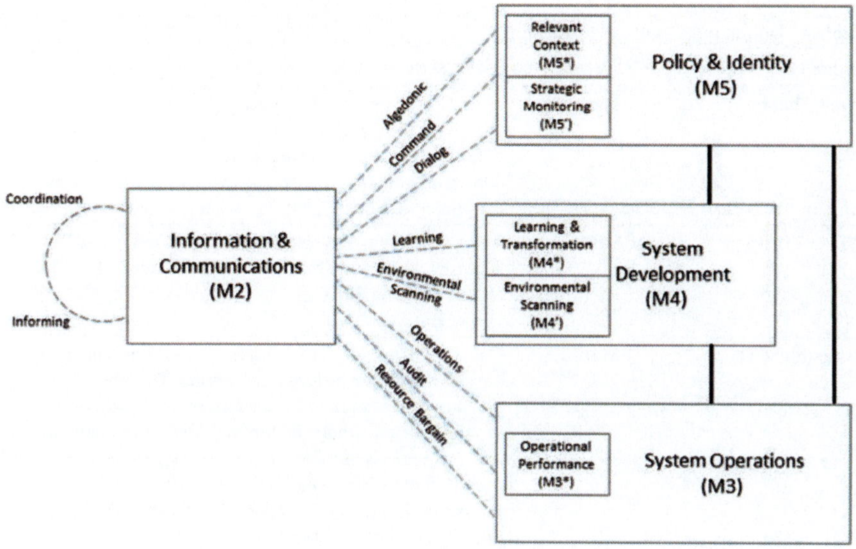

Fig. 1 Complex System Governance reference model

3 Leadership—A Review

3.1 Overview

History is replete with examples of leaders throughout the ages so it is no revelation that leadership has been practiced for many years. Serious study of leadership has also been conducted for quite some time from many points of view. There exists extensive literature regarding leadership addressing the subject across a number of domains including economics, politics, military, education, nursing, academia, and business, to name but a few. Perspectives explored in this literature include required personal skills, organizational impact, developmental issues, traits (behavioral characteristics), or ethical issues associated with leadership. More than 60 classification systems to help characterize leadership [3] have been developed over the years. Eight dominant 'different schools' of contemporary leadership are apparent from the literature. There is also an emerging school. Among the contemporary schools of leadership are those subscribing to *Trait Theory, Skills Theory, Situational Theory, Contingency Theory, Path-Goal Theory, Transformational Theory, Transactional Theory,* and *Servant Theory.* The emerging school of leadership subscribes to *Complexity Leadership Theory.* Table 1 provides a discussion on each of these leadership theories. In spite of this history, no single, universally agreed upon definition of leadership exists [4, 5]. Some generally (but not universally) accepted characteristics of leadership include leadership is a *process*, involving *influence* resulting from *dynamic interaction*, occurring in a *group*, and working toward a *common goal* [4, 6, 7].

Table 1 Dominant schools of leadership

School	Perspective	Discussion
Trait Theory	Traits	This theory suggests that people either are or are not born with the *qualities* that predispose them to success in leadership roles. Intelligence, sociability, and determination are consistently cited as central qualities Proponent: Crowley [11]
Skills Theory	Skills	This theory suggests that learned knowledge and acquired skills and abilities are significant factors in effective leadership Proponent: Katz [12]
Situational Theory	Multiple	This theory suggests that different styles of leadership are warranted for different situations. The ability to adapt or adjust to the circumstances of the situation is required. The primary factors that determine how to adapt are an assessment of the competence and commitment of followers. The assessment of these factors determines if a leader should use a more directive or supportive style Proponent: Hersey and Blanchard [13]
Contingency Theory	Multiple	This theory suggests that effectiveness of a leader is contingent on how well the leader's style matches a specific setting or situation Proponent: Fiedler [14]
Path-Goal Theory	Multiple	This theory suggests that effective leaders can improve the motivation of followers by clarifying the requirements and removing obstacles to high performance and desired objectives. There is an expectation that people will be more focused and motivated if they believe they are capable of high performance, believe their work is worthwhile, and their effort will result in desired outcomes Proponent: Evans [15]
Transformational Theory	Multiple	This theory suggests that leadership is a process of engagement in which a leader is able to create a connection with followers that results in increased motivation and morality in both followers and leaders. The leader must be attentive to the needs and motives of followers in order to help them reach their maximum potential. Transformational leadership may also describe how leaders can initiate, develop, and implement important changes in an organization Proponent: Downton [16]

(continued)

Table 1 (continued)

School	Perspective	Discussion
Transactional Theory	Multiple	This theory focuses on the exchanges that take place between leaders and followers. It suggests that a leader's job is to create structures that make clear what is expected and also the consequences for meeting or not meeting these expectations. Often likened to management Proponent: Weber [17]
Servant Theory	Multiple	This leadership theory suggests that leaders should be servants first, that is, in order to be effective, leaders must place the needs of their followers, customers, and the community ahead of their own interests Proponent: Greenleaf [18]
Complexity Theory	Process	This leadership theory focuses on enabling learning, creativity, and adaptability of complex adaptive systems (CAS) within a context of knowledge-producing organizations Proponentes: Uhl-Bien, Marion, McKelvey [19]

Implied by these characteristics is that leadership involves a relationship between a leader and those led [8, 9]. Marion [10] reminds us the goal to which a group aspires may be temporal or not well understood due to complexity, thus requiring conditions to be created by leadership that enable productive pursuit of the goal. For the purpose of this chapter, leadership will be defined as follows: *a process whereby the conditions that enable a group to productively pursue a common goal is created or fostered.* Creating the conditions to enable pursuit of the goal includes influencing the group through dynamic interactions. A leadership role is not necessarily the sole province of official position, nor is it static. Any individual in a group may take on a leadership role. It is also the case that positional authority does not necessarily result in leadership. For example, a person in a position of authority who might be expected to provide leadership might delegate that role to another person or persons within the team in a particular instance.

3.2 Leadership Styles

Leadership has also been analyzed from the perspective of style. Among the styles often identified are *Autocratic, Democratic, Laissez-Faire, Charismatic, Transactional, Transformational* and *Situational.* Table 2 provides information regarding each of these leadership styles.

Table 2 Leadership styles

STYLE	DESCRIPTION	APPLICATION	EFFECT	SOURCE
Autocratic	Leader makes decisions unilaterally even if input from other team members would be useful	Appropriate for situations in which a decision is needed quickly	This leadership style can have a detrimental effect on team morale, dynamics, and productivity	[20, 21]
Democratic	Leader makes the final decisions but includes members of the team in the decision-making process	Appropriate for leaders to foster an environment in which the team is significantly engaged in projects, participate in decision making, and are encouraged to be creative	This style of leadership may not always be effective, especially when a quick decision is required	[20, 21]
Laissez-Faire	Leader gives their team a good deal of autonomy in determining how work will be accomplished and how deadlines are set	Advice and resources are provided, if required, but otherwise the leader does not get involved. Inappropriate for teams not able to self-manage	May result in high job satisfaction	[21]
Charismatic	Leader inspires and motivates team members	Leaders who employ this style of leadership often focus on themselves and their own ambitions	Depending on the leader's motivations, this may affect the team positively or negatively	[10, 21]
Transactional	Leaders and those led operate under the premise that assigned team members agree to obey the leader when they accept a job as they are being paid for their effort and compliance on a short-term task	This type of leader–follower relationship makes clear everyone's roles and responsibilities	Does not encourage creative tension within the team. Rewards (positive or negative) are contingent upon performance	[21, 22]

(continued)

Table 2 (continued)

STYLE	DESCRIPTION	APPLICATION	EFFECT	SOURCE
Transformational	Leader motivates teammates with a shared vision of the future which is communicated well, and clear goals for the achievement of that future are set	For environments where change in individuals or the system is desirable	This style of leadership typically results in high productivity and engagement on the part of the team members	[21, 22]
Situational	A leadership style in which the leader uses different leadership styles depending on the maturity of assigned team members	Differing approaches used depending on maturity of teammates. Directing for immature, participative, or delegating for mature teammates	Works if the leader is attuned to the needs of the people led, otherwise can demotivate or not provide sufficient support	[21, 22]

3.3 Leadership Functions

A definition for leadership has been provided, a number of common leadership styles have been addressed, and we now turn our attention to the functions of leadership which should be performed regardless of the style employed. Although by no means a categorical list, functions associated with leadership may include *create a vision* [5, 10, 23]; *build trust* [9, 24, 25]; *enable* [10, 23]; *adapt* [9, 10, 26]; *communicate* [23, 25, 27]; *protect* [10, 25]; and *influence* [5, 6, 9]. Figure 2 illustrates these leadership functions and some of the actions which support those functions.

Creating a vision entails envisioning a future desired or possible state or condition. When developing a vision, beginning with a broad perspective helps avoid exclusion of potentially desirable opportunities or alternatives. Developing a vision is a necessary part of building a plan or roadmap to channel a team's effort toward a shared goal [23] making it easier for teammates to appreciate their fit in the organization and guiding behavior [4]. It is also an opportunity to question accepted assumptions [27] as part of an effort to reframe the future. Once a vision has been created, it must be communicated if pursuit of that vision is expected. Communication will be addressed in more detail later in this chapter.

Building trust within a team is an important function of a leader. For the purposes of this chapter, we will use the definition of trust provided by the Merriam-Webster online dictionary—*trust* is defined as *the belief that the person is reliable, good, honest, and effective* [28]. Trust in a leader is, therefore, the belief that the leader is reliable, good, honest, and effective. Reliability suggests the leader's actions are consistent and the team is not having to be wary of the leader's mood or guess their next move. Good suggests the leader's actions, and behaviors are sufficiently

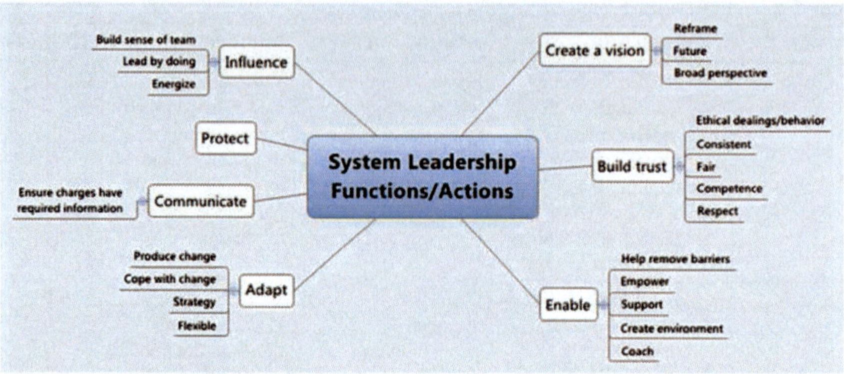

Fig. 2 System leadership functions/actions

transparent and ethical. Fair means the leader treats teammates as well as others justly and respectfully. Honest suggests the leader does not engage in purposeful deception. Effective suggests the leader is competent and can lead the team to accomplish goals. Maccoby suggests '*People follow a leader either out of fear or for a mix of positive reasons such as hope for success, trust in the leader, excitement about a project or mission, or the opportunity to stretch oneself to the limit*' [29] (p. 57). Fear does not lead to trust. It is far better for long-term success for teammates to follow a leader for positive reasons rather than fear. Given a leader can build trust within the team, that leader will likely enjoy the respect of those teammates.

Another function of leadership is *enabling* the team to pursue accomplishment of goals. Included in this function may be creating conditions for team success, barrier removal, empowering team members, supporting them, and coaching them. The path from current conditions to the goals envisioned can be challenging and circuitous. Marion [10] suggests it is up to a leader to create conditions for a creative environment within which the team can succeed. Empowering team members encourages the exercise of initiative and self-sufficiency [4] which can enable quicker response to issues and lessen the leader's burdens. There are times, however, when situations require the engagement of a leader with positional authority. In these cases, the team members do not have the capability to overcome the barriers that have been encountered. It is incumbent upon the leader to recognize these situations and to take appropriate and timely action. Inadequacy of team members' capabilities (knowledge, skills, or abilities) may also be a barrier to success for the team. In these cases, the leader must recognize the issues, develop remediation options, and help the team overcome those shortcomings.

Toor [5] suggests leadership is about *adapting* to as well as producing change. Developing strategies to attain the goals of the vision which in turn allow plans for execution to be developed are the responsibility of leadership. The environments within which systems operate are not static; therefore, it should be expected that execution plans will be constantly reviewed and adapted, as necessary, to extant conditions. Leaders, therefore, must be flexible in order to cope with the inevitable changes in conditions that will most likely occur. Leadership is responsible for ensuring the system is capable of meeting current expectations and is on a trajectory for continuing to meet future goals.

Communicating is an essential function of leadership which involves presence as well as listening [23]. Leadership communication is the essential information to the right audience in a timely manner [30]. Ensuring lines of communications exists across all levels of an organization and is important to ensure requisite information is available when and where needed [10]. Effective communications can catalyze individual and group participation while enabling leaders to 'exert their influence through the stories they tell and the ways they embody those stories' [31] (p. 155).

The welfare of the team is the responsibility of the leader. There may be times when the team or team members need protection from spurious threats or conditions from within or outside the organization (system). This does not mean protecting poor performers as this would be counter to the welfare of the team. Endeavors associated with change have some level of risk. Failure, when prudent risks were assumed, is not unreasonable; however, in some organizations, no failure is acceptable. It is in these cases that it is important that the leader provides protection for team members involved.

Kaiser [6] suggests leaders have an effect (either a positive or negative) on the performance of an organization by *influencing* individuals to contribute to the attainment of the organization's goals. Good leadership takes strength of character and a firm commitment to doing the right thing, at the right time, for the right reason including following through on commitments [21, 30].

Figure 2 is an illustration of leadership functions and actions.

3.4 Leadership Ethics

In spite of ethical behavior being important to effective leadership [32], a significant body of research literature has yet to be published regarding the theoretical foundations of ethical leadership [4].

Ethics is concerned with the character of individuals or groups (institutions, societies) which includes the values and morals which are recognized to be desirable [4, 33]; essentially: what is right or wrong, good or bad. Leadership, from the perspective of ethics, is about what a leader does and who they are as a person.

Northouse [4] suggested a set of principles for ethical leadership which include *respect, service, justice, honesty, and community*. How a leader addresses these principles is indicative of their character.

Respect for the individual is often identified as an issue that is important to ethical leadership [4, 34, 35]. Ethical leaders treat individuals with dignity and as valuable team assets by empowering them to be creative [35], not simply as means to an end [4]. In the context of leadership, respect is having consideration for a person's perspective, conviction, or position regarding an issue under consideration. Respect also includes listening attentively to members of the team and being tolerant of varying points of view [4].

Service is about attending to others, which might include mentoring, empowering, coaching, or team building [4, 35]. Service is about putting the other's welfare foremost in a leader's effort while foregoing personal gain for the sake of team members or the organization.

Northouse [4] suggests *justice* is about treating people fairly. Transparency in actions can be important since the perception of fair treatment is a subjective judgment made by affected or observing individuals [34]. Transparency can help the affected person(s) or observers be more confident in the motives of the leader. Justice is not necessarily about treating everyone the same, it is more about giving everyone an equal opportunity. For example, there may be situations in which the team leader might need to allow individual team members more latitude during specific periods of time to help them succeed.

Honesty is about not purposely deceiving others by not telling the truth or by not representing reality as fully as possible [4]. A leader can find themselves in a difficult position when they cannot share information fully with team members. In this case, a delicate balance is required between guarding privileged information that has been shared with the leader and openness. Trustworthiness, an important issue in the relationship between leaders and their team, is influenced by the perception of their honesty.

Northouse [4] (p. 437) observes that '*an ethical leader takes into account the purpose of everyone involved in the group and is attentive to the interests of the community and the culture.*' Ethical leadership seeks a common goal for the team, one that each of the team members can embrace as their own. The goal will be one that is beneficial to each of the team members.

In summary, good leaders show respect to the individuals, they put the welfare and interests of the group ahead of their own self-interests, treat people fairly and justly, are honest with their team, and consider the greater good for the community [4].

A definition of leadership, leadership styles, a number of important functions of leadership, and leadership ethics have been discussed. It is now time to address complex systems leadership.

3.5 Complexity Theory Perspective of Leadership

It has been recognized that complexity and uncertainty are the norms for the environments in which organizations exist [36]. As a result, complexity theory, complex systems, and complex adaptive systems are being suggested as appropriate to address these conditions [7, 36–38]. *Complexity theory* is the 'study of complex and chaotic systems and how order, pattern and structure can arise from them' [39]. Johnson [40] (p. 17) observed that '*complexity looks at the complicated and surprising things which can emerge from the interaction of a collection of objects which themselves may be rather simple.*' Complex systems were introduced earlier in this work. Among the characteristics of complex systems discussed were the large number of elements, loosely organized, with many interactions between those elements. In addition, these systems evolve over time and are open to their environment [41]. Hazy [37] observes that complex adaptive systems (CAS) have a capacity to adapt to changes in the environment within which they exist. Uhl-Bien, et al. [19] (p. 302) credit Dooley [42] with describing CAS as '*an aggregate of interacting agents that behave according to three principles: order is emergent as opposed to predetermined, the system's history is irreversible, and the system's future is often unpredictable.*' Marion and Uhl-Bien [10] suggest that organizations are complex adaptive systems composed of agents who interact with one another, mutually affecting each other, which generates new behavior for the system as a whole. Systems theory is being used to help in the understanding of complex and complex adaptive systems leadership (implicitly if not explicitly) with a limited set of propositions considered and is tending away from strictly considering hierarchical, positional, or character/characteristic-based perspectives to also include viewing leadership as an emerging event resulting from interactions between people [7] where leaders enable conditions appropriate to success of the endeavor. Implicit in this perspective is leadership can emerge from any part of an organization for a specific instance. In summary, *leadership* in complex systems is an emergent, interactive dynamic that produces outcomes while *leaders*, who may emerge from any part of the organization or team, are individuals who influence these interactions and outcomes [19].

3.6 Complex System Governance Leadership

Turning now to *Complex System Governance leadership*, a review of the roles and responsibilities of leadership with respect to the requisite functions identified in the complex systems governance reference model will be explored. Complex System Governance has been defined as '*the performance of (metasystem) functions necessary to provide direction, communication, control, and change necessary to ensure continuing viability of a system*' [43] (p. 156). The complex system reference model discussed in "Complex System Governance Reference Model" identifies nine functions that must be performed by the metasystem to ensure system viability. Recall those functions included policy and identity (M5); system context (M5′); strategic system monitoring (M5*); system development (M4); environmental scanning (M4′); learning and transformation (M4*); system operations (M3); operational performance monitoring (M3*); and information and communications (M2). The metasystem roles and responsibilities of leadership from a high-level perspective are shown in Table 3. A more detailed discussion of these roles and responsibilities is provided later in the practice section of this chapter.

The next section of this chapter will discuss systems leadership issues with implications for practice.

Table 3 Complex System Governance reference model leadership roles and responsibilities

Metasystem	Function	Leadership Role	Leadership Responsibility
M5	Provide direction, oversight, accountability, and evolution of the System	Enable provision of direction, oversight, accountability, and evolution of the system	Ensure direction, oversight, accountability, and evolution of the system are provided and occur
M5′	Monitor the system context (the circumstances, factors, conditions, or patterns that enable or constrain the system)	Enable the system context to be monitored	Ensure the system context is monitored
M5*	Maintain system context and monitor measures for strategic system performance	Enable maintenance of system context and monitoring of strategic system performance measures	Ensure maintenance of system context and monitoring of strategic system performance measures
M4	Analyze and interpret implications and potential impacts of trends, patterns, and precipitating events in the environment in order to develop future scenarios, design alternatives, and future focused planning to position the system for future viability	Enable analysis and interpretation of trends, patterns, and events in the environment to enable planning for future system viability	Ensure environmental analysis and interpretation in support of planning for future system viability
M4′	Provide the design and execution of scanning of the environment with focus on patterns, trends, threats, events, and opportunities for the system	Enable the design and execution of environmental scanning	Ensure environmental scanning occurs
M4*	Provide for identification and analysis of metasystem design errors (second-order learning) and suggest design modifications and transformation planning for the system	Enable identification of metasystem design errors	Ensure identification of metasystem design errors

(continued)

Table 3 (continued)

Metasystem	Function	Leadership Role	Leadership Responsibility
M3	Maintain operational performance control through the implementation of policy, resource allocation, and design for accountability	Enable operational performance control	Ensure operational performance control
M3*	Monitor measures for operational performance and identity variance in system performance requiring system-level response	Enable operational performance measures to be monitored and identify variance in system performance requiring system-level response	Ensure operational performance measures are monitored and system performance variances requiring system-level responses are identified
M2	Design and implement the architecture for information flow, coordination, transduction and communications within the metasystem and between the metasystem, the environment, and the governed system	Enable design and implementation for information flow, coordination, transduction and communications within the metasystem and between the metasystem, the environment and the governed system	Ensure design and implementation for information flow, coordination, transduction and communications within the metasystem and between the metasystem, the environment and the governed system

4 Applicability of Leadership to Complex System Governance

'Leadership is a highly sought-after and highly valued commodity' [4] (p. 1). Whether positional or emergent, teams coalesce around good leadership which affects team members' satisfaction, attitude, performance, and system (organization) success in a positive manner [44]. Leadership also affects the culture of an organization—by talk or action, the corporate culture is communicated to team members, either good or bad, ethical or unethical. In this next section, the responsibilities, effects, and impacts of leadership will be explored.

4.1 Leadership Responsibilities, Effects, and Impacts

Earlier in this chapter, the growing recognition that complexity and uncertainty prevail in the environments in which organizations exist [36] was discussed as was the appropriateness of complexity theory, complex systems, and complex adaptive systems to address these conditions [7, 36–38]. Also discussed was that *leadership* in complex systems is an emergent, interactive dynamic that produces outcomes while *leaders*, who may emerge from any part of the organization or team, are individuals who influence these interactions and outcomes [19]. Regardless of the source of leadership, positional or emergent, the leadership functions discussed earlier in this chapter need to be performed. The reader will recall the functions identified include creating a vision, building trust, enabling, adapting, communicating, protecting, and influencing. It is the responsibility of the leader to contribute to creating a vision, which entails envisioning a desired or possible state or condition in the future. This may be accomplished by the leader alone or by bringing together appropriate team members and leading an effort to define the vision, or by creating the environment in which the vision can be developed by team members. Engaging team members in the development of a vision, although possibly taking more effort and time consuming, has advantages including exploring multiple perspectives (complementarity), beginning the process of disseminating the information related to the vision (communications), and jump-starting buy-in by those participating. Regardless, the vision must be communicated to the whole team in a way in which each member can embrace the vision. Along with creating a vision, an appropriate organizational design to pursue the vision must be identified and implemented which is also a leadership responsibility. Care should be taken to ensure the resulting organization enables team success, supports emergent networks, and does not make it more difficult for the team to succeed. As the environment within which the organization (system) exists is not static, organizational (system) design is not a ‘one and done’ effort, it must be flexible and requires continual review to ensure requisite functions are being performed. This does not imply continuous organizational change which can be disruptive (relaxation time). The effects on the organization of having established goals to which the team can aspire, a construct within which to operate, and a course to follow include diminished uncertainty and a sense of purpose. As a result, the impact on the team members may include diminished stress and a sense of purpose.

As previously discussed, building trust within a team is an important function of leadership. For a leader to build trust, team members need to believe the leader is reliable, good, honest, and effective. ‘Actions speak louder than words’ may be trite but without a doubt true in this area. Trust is earned over a period of time but can be lost in an instant. It is important, therefore, that the leader be consistent, ethical, forthright, and effective. The effects may include establishing an environment within the organization in which the leader can inspire team members to embrace the vision as their own [23]. Impacts include a team working toward a goal with more enthusiasm and being more willing to sacrifice to achieve the goal.

Enabling is about creating conditions for team success, helping to remove barriers, empowering team members, supporting them, or coaching. There are situations that arise that only a leader with positional authority can resolve. In these situations, it is the responsibility of that leader to act in a timely manner to resolve the issue thus allowing the team to devote their efforts to attaining the goal. Impact includes improved team morale and a sense of being able to accomplish tasking.

As previously discussed, leadership is about *adapting* to as well as producing change [5]. Leaders are charged with developing strategies to attain the goals of the vision which in turn allow plans for execution to be developed. The environments within which systems operate are complex, therefore not completely understood (system darkness), and are not static. It should be, therefore, no surprise that execution plans must be constantly reviewed and adapted to extant conditions as currently understood. This requires leaders to be flexible in order to cope with the inevitable changes in conditions that will occur. Ensuring the system is capable of meeting current expectations and is on a trajectory for continuing to meet future goals (homeorhesis) is a responsibility of leadership. The effect is relevant planning and organizational constructs for team operations. Impact is a relevant team effort toward achieving goals.

Visions, goals, directives, or plans are worth little if the information contained therein is not shared with appropriate agents. Communicating the right information, at the right time, to the right people is important to the success of any endeavor. Leadership's role is to ensure means are available to facilitate communications as well as using those means to ensure team members have the opportunity to be properly informed. Communicating, that is transmitting information accurately and reliably from a source to a recipient, is not an easy task. What the source thought they said and what the recipient interpreted from what they heard may be significantly different even if there was minimum noise in the transmission channel. For example, in the 2013 Federal Viewpoint Survey, nearly 53 percent of the survey respondents indicated dissatisfaction with 'the information you receive from management on what's going on in your organization' [45]. Similar concerns can be found in other domains as well. The point is good communication, that truly informs, is a difficult but necessary undertaking that starts with leadership. Communications can affect the climate of an organization with good communications having a positive effect while poor communications can encourage a negative organizational climate. The impact can be seen in efficiency and morale.

A leader has a responsibility to help ensure the welfare of the team. Sometimes this requires the leader to protect the team or its members. Team members need to know that their leader 'has their back' when they perform their duties in good faith. In the end, the leader is responsible for the performance of the team. This concept is no more evident than in the sign that was displayed on President Harry Truman's (1945–1953) desk in the White House that read 'The Buck Stops Here' [46]. This does not, however, mean protecting poor performers or illegal activity. A team that has confidence in the leader backing their efforts and deflecting spurious threats may be more willing to take (prudent) risks.

As previously discussed, leaders can have either a positive or negative effect on the performance of an organization by *influencing* individuals to contribute to the attainment of the organization's goals [6]. Influence can be for good or bad and is not always overt and obvious. On occasion, results of influence can be immediate and obvious as in the case where a leader, by word or actions, inspires current team performance. There are other instances where influence is more subtle and takes longer to manifest the results. For example, a leader can impact the perceptions and character of a team and its members by consistent performance of their duties in an ethical and effective manner. This can lead to an understanding that there is an expectation that team members will act in a similar manner. The converse is also true. If the leader's actions are continually unethical, it should be of no surprise that team members follow the leader's lead. Leadership, at least good leadership, takes strength of character and a firm commitment to doing the right thing, at the right time, for the right reason including following through on commitments [21, 30]. This is the essence of leading by example. The effect is an established environment of expected behavior. The team members' conduct is among the impacts of leadership influence.

4.2 Role of leadership in Metasystem design, Implementation, and Operation

In "Complex System Governance", the concept of *metasystem* was introduced. Beer defined metasystem as *"...a system over and beyond a system of lower logical order, and therefore capable of deciding propositions, discussing criteria, or exercising regulation for systems that are themselves logically incapable of such decisions and discussions or self-regulation...*' [47] (p. 402). The purpose of the metasystem is to provide the functions necessary for organizational (system) viability. This means the functions are necessary for the organization (system) to continue to exit. The metasystem is the means by which Complex System Governance is exercised. Although these functions could emerge from a self-organization effort, an explicitly designed metasystem (purposeful design) is preferable. As previously discussed, the organizational construct within which the organization operates is the responsibility of leadership; therefore, leadership is responsible for design, implementation, and operation of the metasystem which is part of that construct. Exercising purposeful design over the metasystem affords leadership the opportunity to influence how the necessary functions are performed and by whom as well as influencing the efficiency with which they are performed.

The next section will explore the relationships between systems leadership and the Complex Systems Governance reference model discussed in "Complex System Governance Reference Model".

4.3 Applicability and Integration into the Complex System Governance Reference Model

"Complex System Governance Reference Model" introduced the Complex System Governance reference model which is an organizing systems-based construct for facilitating development of essential Complex System Governance functions. In this section, the relationship between leadership and each of the aspects of the Complex System Governance reference model will be explored. Table 2 provided an introduction to the roles and responsibilities of leadership in relation to the governance functions identified in the Complex System Governance reference model. Table 4 discusses each of the roles and responsibilities of leadership in regard to the identified governance functions including potential leadership actions and mechanisms that might be used.

5 Summary

The purpose of this chapter was to explore leadership for Complex Systems Governance. In the process, a definition of leadership was provided, the relationship between systems leadership and the Complex System Governance reference model was explored; a review of leadership including leadership styles, functions, ethics, complex system leadership, and Complex System Governance leadership was completed; system leadership practice including responsibilities, effects and impacts of leadership, the role of leadership in metasystem design, implementation, and operation was addressed; and the applicability and integration of systems leadership into the CSG reference model were explored.

In the end, CSG leadership adds to the mosaic of understanding of leadership that has been developed over years of research that resulted in the various perspectives of leadership previously discussed.

The next section will address exercises.

6 Exercises

1. An organization has recognized a trend in their customer base to which a response is required. A team is established to analyze the conditions and determine organizational changes, in function or configuration, required to meet this challenge. Discuss the role of leadership in this effort specifically identifying systems propositions that may be in play.
2. Creating a vision for an organization (system) has been identified as a responsibility of leadership. Discuss the role of leadership in creating a vision from your perspective. Identify the system propositions that might influence the design of the vision and explain what the influence of each is.
3. Discuss the advantages and disadvantages of a purposefully designed metasystem as contrasted with an emergent design resulting from self-organization. Include in the discussion the systems propositions that may influence these efforts.
4. In the framing vantage of the Complex Systems Governance development reference model, one of the issues addressed is the bounding of the system of interest. Discuss what this entails, the importance of bounding the system, the role of leadership, and identify supporting system propositions.

Table 4 Leadership roles and responsibilities

Leadership Role	Actions	Mechanisms	Leadership Responsibility	Actions	Mechanisms
System: M5—Policy and identity Function provides direction, oversight, accountability, and evolution of the system. Focus includes policy, mission, vision, strategic direction, performance, and accountability of the system such that (1) the system maintains viability, (2) identity is preserved, and (3) the system is effectively projected both internally and externally					
Enable system identity to be established and maintained in the face of changing environment and context	Provide guidance, communications channels, and resources to establish and maintain system identity in the face of changing environment and context	- guidance documents - internal communications - meetings - resource allocations	Ensure system identity is established and maintained in the face of changing environment and context	Monitor how well the system identity reflects the desired characteristics and culture of the system as well as how well it is distributed throughout the system. Increase engagement as necessary to achieve desired results and be prepared to reallocate resources if required	- direct personal interactions - meetings - internal correspondence - resource reallocation
Enable the system vision, strategic direction, purpose, mission, and interpretation to be defined, clarified, and propagated	Provide guidance and resources required to define, clarify, and propagate the system vision, strategic direction, purpose, mission, and interpretation	- guidance documents - internal communications - meetings - resource allocations	Ensure the system vision, strategic direction, purpose, mission, and interpretation are defined, clarified, and propagated	Monitor how well the system vision, strategic direction, purpose, mission, and interpretation are defined, clarified, and propagated throughout the system. Increase engagement as necessary to achieve desired results and be prepared to reallocate resources if required	- direct personal interactions - meetings - internal correspondence - resource reallocation

(continued)

Table 4 (continued)

Leadership Role	Actions	Mechanisms	Leadership Responsibility	Actions	Mechanisms
Enable the system focus to be determined and balanced between present and future on an active basis	Provide guidance and requisite resources to actively determine and balance the system focus between the present and the future	- guidance documents - internal communications - meetings - resource allocations	Ensure the system focus is determined and balanced between present and future on an active basis	Monitor how well the system focus is determined and balanced between present and future on an active basis. Increase engagement as necessary to achieve desired results and be prepared to reallocate resources if required	- direct personal interactions - meetings - internal correspondence - resource reallocation
Enable strategic plan dissemination	Provide guidance, venue, and resources required to disseminate the strategic plan	- guidance documents - internal communications - meetings - resource allocations	Ensure the strategic plan is disseminated	Monitor how well the strategic plan is being disseminated. Increase engagement as necessary to achieve desired results and be prepared to reallocate resources if required	- direct personal interactions - meetings - internal correspondence - resource reallocation
Enable oversight of strategic plan execution	Provide guidance and resources to oversee execution of the strategic plan	- guidance documents - internal communications - meetings - resource allocations	Ensure strategic plan execution is afforded oversight	Oversee execution of the strategic plan	- meetings - internal correspondence - resource reallocation
Enable provision of capital resources necessary to support the system	Provide requisite resources to support the system	- resource allocations	Ensure capital resources necessary to support the systems are provided	Monitor availability of requisite capital for system support. Be prepared to reallocate resources as required	- meetings - internal correspondence - resource reallocations

(continued)

Table 4 (continued)

Leadership Role	Actions	Mechanisms	Leadership Responsibility	Actions	Mechanisms
Enable setting of present and future problem space for focus of product, service, and content development and deployment	Provide guidance and resources necessary to define focus for present and future products, services, and content development and deployment	- guidance documents - internal communications - meetings - resource allocations	Ensure setting of present and future problem space for focus of product, service, and content development and deployment	Monitor focus of present and future products, services, and content development and deployment. Be prepared to provide additional guidance and reallocate resources if required	- meetings - internal correspondence - resource reallocations
Enable establishment of strategic dialog forums	Provide guidance, venues, and resources for the establishment of strategic dialog forums	- guidance documents - internal communications - meetings - resource allocations	Ensure establishment of strategic dialog forums	Monitor the establishment of and engage in strategic dialog forums	- direct personal interactions - meetings - internal correspondence
Enable autonomy-integration balance to be preserved in the system	Provide expectations and guidance to preserve the autonomy-integration balance in the system	- guidance documents - internal communications - meetings	Ensure autonomy-integration balance in the system is preserved	Monitor autonomy-integration balance in the system. Be prepared to provide additional guidance as required to attain and maintain desired results	- direct personal interactions - meetings - internal correspondence
Enable system products, services, content, and value to be marketed	Provide expectations, guidance, and resources for marketing of system products, services, content, and value	- guidance documents - internal communications - meetings - resource allocations	Ensure system products, services, content, and value are marketed	Monitor marketing of system products, services, content, and value. Be prepared to provide additional guidance and to reallocate resources as required to attain and maintain desired results	- meetings - internal correspondence - resource reallocations

(continued)

Table 4 (continued)

Leadership Role	Actions	Mechanisms	Leadership Responsibility	Actions	Mechanisms
Enable planning and execution of public relations	Provide expectations, guidance, and resources for planning and execution of public relations	- guidance documents - internal communications - meetings - resource allocations	Ensure planning and execution of public relations are accomplished	Monitor public relations planning and execution. Be prepared to provide additional guidance or engage as necessary as well as to reallocate resources to attain and maintain desired results	- meetings - internal correspondence - resource reallocations
Enable external mentorship to be developed	Provide guidance, expectations, and resources for mentorship development	- guidance documents - internal communications - meetings - resource allocations	Ensure external mentorship is developed	Monitor mentorship development. Engage as necessary. Be prepared to provide additional guidance and to reallocate resources as required to attain and maintain desired results	- meetings - internal correspondence - resource reallocations
Enable system policy direction to be established	Provide guidance and resources necessary for establishment of system policy direction	- policy documents - internal communications - meetings - resource allocations	Ensure system policy direction is established	Monitor system policy direction. Be prepared to engage, provide additional guidance, and/or reallocate resources as necessary to attain and maintain desired results	- meetings - internal correspondence - resource reallocations
Enable system identity to be maintained	Provide guidance and resources necessary to maintain system identity	- guidance documents - internal communications - meetings - resource allocations	Ensure system identity is maintained	Determine system elements' perspectives of system identity. Be prepared to engage, provide additional guidance, and/or reallocate resources as required to maintain system identity	- direct personal interactions - meetings - internal correspondence - resource reallocation

(continued)

Table 4 (continued)

Leadership Role	Actions	Mechanisms	Leadership Responsibility	Actions	Mechanisms
Enable system interests to be represented to external constituents	Provide guidance and resources for system interests to be represented to external constituents	- guidance documents - internal communications - meetings - resource allocations	Ensure system interests are represented to external constituents	Monitor representation of system interests to external constituents. Be prepared to engage, provide additional guidance and/or reallocated resources as required to attain and maintain desired results	- external correspondence - internal correspondence - meetings - resource reallocations
Enable the expanded network for the system (strategic partners) to be defined and integrated	Provide guidance and resources to define and integrate the expanded system network	- guidance documents - internal communications - meetings - resource allocations	Ensure the expanded network for the system (strategic partners) is defined and integrated	Monitor the defining and integrating of the expanded network for the system. Be prepared to engage, provide additional guidance, and/or reallocate resources to attain and maintain or desired results	- external correspondence - internal correspondence - meetings - resource reallocations
Enable scenarios for system transformation to evolve	Provide guidance and resources for system transformation evolution	- guidance documents - internal communications - meetings - resource allocations	Ensure scenarios for system transformation to evolve are developed	Monitor development of scenarios for system transformation. Be prepared to engage, provide additional guidance, and/or reallocate resources as required to attain or maintain desired results	- meetings - internal correspondence - resource reallocations
Enable strategic transformation direction to be implemented	Provide expectation, guidance, and resources for implementation of strategic transformation direction	- guidance documents - internal communications - meetings - resource allocations	Ensure strategic transformation direction is implemented	Monitor implementation of strategic transformation direction. Be prepared to engage, provide additional guidance, and /or reallocate resources required to attain or maintain desired results	- direct personal interactions - meetings - internal correspondence - resource reallocation

(continued)

Table 4 (continued)

Leadership Role	Actions	Mechanisms	Leadership Responsibility	Actions	Mechanisms
System M5′—System context					
Function: Monitor the system context (the circumstances, factors, conditions, or patterns that enable or constrain the system)					
Enable system context to be identified	Provide guidance and resources to identify system context. Participate in effort to identify system context	- guidance documents - internal communications - meetings - resource allocations	Ensure system context is identified	Monitor identification of system context. Be prepared to increase engagement and/or reallocate resources if necessary to attain or maintain desired results	- meetings - internal correspondence - resource reallocations
Enable contextual impacts on system performance (constraining and enabling) to be assessed	Provide guidance, expectations, and resources for assessment of contextual impacts on system performance	- guidance documents - internal communications - meetings - resource allocations	Ensure contextual impacts on system performance (constraining and enabling) are assessed	Monitor assessment of contextual impacts on system performance. Be prepared to provide additional guidance and/or reallocate resources to attain or maintain desired results	- meetings - internal correspondence - resource reallocations
Enable context to be actively managed	Provide guidance and resources to actively manage context	- guidance documents - internal communications - meetings - resource allocations	Ensure context is actively managed	Monitor context management. Be prepared to engage, provide additional guidance, and/or reallocate resources to attain or maintain desired results	- meetings - internal correspondence - resource reallocations
Enable boundary spanning to determine the boundary conditions, values, and judgments for the system to be conducted	Provide guidance and resources to conduct boundary spanning to determine the boundary conditions, values, and judgments for the system	- guidance documents - internal communications - meetings - resource allocations	Ensure boundary spanning to determine the boundary conditions, values, and judgments for the system is conducted	Monitor effort to determine the system's boundary conditions, values, and judgements. Be prepared to engage, provide additional guidance, and/or reallocate resources if necessary	- direct personal interactions - meetings - internal correspondence - resource reallocation

(continued)

Table 4 (continued)

Leadership Role	Actions	Mechanisms	Leadership Responsibility	Actions	Mechanisms
Enable inquiry into contextual barriers to system execution or development to be conducted	Provide guidance and resources to conduct inquiry into the contextual barrier to system execution or development	- guidance documents - internal communications - meetings - resource allocations	Ensure inquiry into contextual barriers to system execution or development is conducted	Monitor progress in determining contextual barriers to system execution or development. Be prepared to provide additional guidance, engage, and/or reallocate resources as necessary	- direct personal interactions - meetings - internal correspondence - resource reallocation
Enable the influence of contextual aspects for the system to be monitored and assessed	Provide guidance and resources to monitor and assess the influence of contextual aspects for the system	- guidance documents - internal communications - meetings - resource allocations	Ensure the influence of contextual aspects for the system is monitored and assessed	Review effort to monitor and assess the influence of contextual aspects for the system. Be prepared to engage, provide additional guidance, and/or reallocated resources if necessary	- direct personal interactions - meetings - internal correspondence - resource reallocation
System M5*—Strategic system monitoring Function: Maintain system context and monitor measures for strategic system performance and identify variance requiring metasystem-level response with emphasis on variability that may impact future system viability					
Enable tracking of ongoing system performance based on dashboard measure of performance for operations	Provide guidance and resources to develop/maintain dashboard for system performance measures as well as tracking of ongoing system performance based on dashboard information	- guidance documents - internal communications - meetings - resource allocations	Ensure tracking of ongoing system performance based on dashboard measure of performance for operations is accomplished	Monitor dashboard information. Be prepared to provide additional guidance and/or reallocate resources if necessary	- meetings - internal correspondence - resource reallocations

(continued)

Table 4 (continued)

Leadership Role	Actions	Mechanisms	Leadership Responsibility	Actions	Mechanisms
Enable system performance to be disseminated throughout the system	Provide guidance, communications channels, and resources to disseminate system performance throughout the system	- guidance documents - internal communications - meetings - resource allocations	Ensure system performance is disseminated throughout the system	Monitor system performance dissemination throughout the system. Be prepared to provide additional guidance, communication channels, and/or reallocate resources if necessary	- direct personal interactions - meetings - internal correspondence - resource reallocation
Enable inquiry into performance aberrations to be conducted	Provide expectations, guidance, and resources to conduct inquiry into system performance aberrations	- guidance documents - internal communications - meetings - resource allocations	Ensure inquiry into performance aberrations is conducted	Monitor inquiries into system performance aberrations. Be prepared to engage, provide additional guidance, and/or reallocate resources to attain/maintain desired results	- direct personal interactions - meetings - internal correspondence - resource reallocation
Enable operational performance measures to be monitored and assessed for continuing adequacy	Provide guidance and resources to monitor and assess continuing adequacy of operational performance measures	- guidance documents - internal communications - meetings - resource allocations	Ensure operational performance measures are monitored and assessed for continuing adequacy	Review effort to monitor and assess operational performance measures for continuing adequacy. Be prepared to engage, provide additional guidance, and/or reallocate resources if necessary	- direct personal interactions - meetings - internal correspondence - resource reallocation

(continued)

Table 4 (continued)

Leadership Role	Actions	Mechanisms	Leadership Responsibility	Actions	Mechanisms
System M4—System development Function: Analyze and interpret implications and potential impacts of trends, patterns, and precipitating events in the environment in order to develop future scenarios, design alternatives, and future focused planning to position the system for future viability					
Enable the analysis and interpretation of environmental scanning results for shifts, their implications, and potential impacts on system evolution	Provide resources necessary to analyze and interpret environmental scanning results as well as means to communicate the results to appropriate parts of the system	- resource allocations - internal correspondence	Ensure environmental scanning results are analyzed and interpreted for shifts, their implications, and potential impacts on system evolution	Monitor the results of environmental scanning. Be prepared to make requisite changes to existing operations and development plans based on information resulting from the analysis and interpretation of the environmental scan	- internal correspondence - meetings - resource reallocations
Guide the development of the system strategic plan and system development map	Provide clear expectations regarding the development of the strategic plan and system development map. Provide resources to develop the strategic plan and system development map	- guidance documentation - meetings - resource allocations	Ensure the system strategic plan and system development map are developed	Monitor the development of the system strategic plan and system development map. Be prepared to provide additional guidance and/or reallocate resources as necessary	- direct personal interactions - internal correspondence - meetings - resource reallocations
Enable the use of information and knowledge resident in M4 to inform strategic planning	Make system development information available to inform strategic planning. Devote resources necessary	- meetings - internal correspondence - resource allocations	Ensure information and knowledge resident in M4 are used to inform the strategic plan	Monitor the use of system development information to inform the development of the strategic plan. Be prepared to provide guidance and/or resources as required	- direct personal interactions - internal correspondence - meetings - resource reallocations

(continued)

Table 4 (continued)

Leadership Role	Actions	Mechanisms	Leadership Responsibility	Actions	Mechanisms
Enable future product, service, and content development to be guided by information and knowledge resident in M4	Share the expectation that future product, service, and content development will be guided by information and knowledge resident in M4. Provide necessary resources to allow M4 to participate in development as appropriate	- guidance documentation - meetings - resource allocations	Ensure information and knowledge resident in M4 are used to guide future product, service, and content development	Monitor the use of information and knowledge resident in M4 in development plans for future products, service, and content development. Be prepared to provide guidance if not the case	- meetings - internal correspondence - resource reallocation
Enable investment priorities to be guided by information and knowledge resident in M4	Provide clear expectations that investment priorities are to be guided by information and knowledge resident in M4. Provide resources and communications channels necessary to inform the system	- guidance documentation - meetings - resource allocations	Ensure information and knowledge resident in M4 are used to guide investment priorities	Monitor the use of information and knowledge resident in M4 in determining investment priorities. Be prepared to provide guidance if not the case	- meetings - internal correspondence
Enable future relationships critical to system development to be identified	Provide criteria for critical future relationships. Provide resources for development of candidates. Provide a venue for review	- guidance documentation - meetings - resource allocations	Ensure future relationships critical to system development are identified	Monitor identification and review of critical future relationships. Engage as necessary and be prepared to reallocate resources as required	- meetings - internal correspondence - resource reallocation
Enable future development opportunities and targets that can be pursued in support of mission and vision of the system to be identified	Provide resources to identify future development opportunities and targets aligned with mission and vision for the system. Provide a venue for review	- resource allocations - meetings	Ensure future development opportunities and targets that can be pursued in support of the mission and vision of the system are identified	Monitor identification and review of future development opportunities and targets. Engage as necessary. Be prepared to reallocate resources as required	- meetings - internal correspondence - resource reallocation

(continued)

Table 4 (continued)

Leadership Role	Actions	Mechanisms	Leadership Responsibility	Actions	Mechanisms
System M4′—Environmental scanning					
Function: Provide the design and execution of scanning of the environment with focus on patterns, trends, threats, events, and opportunities for the system					
Enable environmental scanning for the entire system to be designed	Provide expectations and resources for design of environmental scanning	- guidance documents - internal correspondence - meetings - resource allocations	Ensure environmental scanning for the entire system is designed	Monitor design of environmental scanning. Engage as necessary. Be prepared to reallocate resources as required	- meetings - internal correspondence - resource reallocation
Enable the environmental scanning designs to be executed	Provide guidance and resources necessary for execution of environmental scanning designs	- guidance documents - internal correspondence - meetings - resource allocations	Ensure the environmental scanning designs are executed	Monitor execution of designs of environmental scanning. Engage as necessary. Be prepared to reallocate resources as required	- meetings - internal correspondence - resource reallocation
Enable a model of the metasystem environment to be maintained	Provide expectations and resources for maintenance of a model of the metasystem environment	- guidance documents - internal correspondence - meetings - resource allocations	Ensure a model of the metasystem environment is maintained	Monitor execution of the maintenance of a model of the metasystem environment. Engage as necessary. Be prepared to reallocate resources as required	- meetings - internal correspondence - resource reallocation
Enable emergent environmental conditions and events to be captured	Provide expectations and resources for capturing emergent environmental conditions and events	- guidance documents - internal correspondence - meetings - resource allocations	Ensure emergent environmental conditions and events are captured	Monitor efforts to capture emergent environmental and events. Engage as necessary. Be prepared to reallocate resources as required	- meetings - internal correspondence - resource reallocation

(continued)

Table 4 (continued)

Leadership Role	Actions	Mechanisms	Leadership Responsibility	Actions	Mechanisms
Enable results from environmental scanning to be consolidated and synthesized	Provide resources to consolidate and synthesize the results from environmental scanning	- resource allocations - internal correspondence - meetings	Ensure results from environmental scanning are consolidated and synthesized	Monitor efforts to consolidate and synthesize the results from environmental scanning. Engage as necessary. Be prepared to reallocate resources as required	- meetings - internal correspondence - resource reallocation
Enable essential environmental information and shifts to be disseminated throughout the system	Make essential environmental information and shifts available throughout the system. Devote resources necessary	- guidance documents - meetings - resource allocations	Ensure essential environmental information and shifts are disseminated throughout the system	Monitor efforts to disseminate essential environmental information and shifts throughout the system. Engage as necessary. Be prepared to reallocate resources as required	- meetings - internal correspondence - resource reallocation
System M4*—Learning and transformation Function: Provide for identification and analysis of metasystem design errors (second-order learning) and suggest design modifications and transformation planning for the system					
Enable processing of inputs for system wide implications	Provide resources required to process metasystem design errors for system wide implications	- resource allocations	Ensure inputs for system wide implications are processed	Monitor processing of inputs of metasystem design errors. Engage as necessary and be prepared to reallocate resources as required	- meetings - internal correspondence - resource reallocation
Enable identification of mechanisms for double-loop learning	Foster a culture of learning and provide resources required to identify mechanisms for double-loop learning	- guidance documents - internal communications - meetings - resource allocations	*Ensure mechanisms for double-loop learning are identified*	Monitor progress in identifying mechanisms for double-loop learning. Engage as necessary and be prepared to reallocate resources as required	- meetings - internal correspondence - resource reallocation

(continued)

Table 4 (continued)

Leadership Role	Actions	Mechanisms	Leadership Responsibility	Actions	Mechanisms
Enable designing of objectives, measures, and accountability for second-order learning in the system	Provide clear expectations and resources for designing objectives, measures, and accountability for second-order learning in the system	- guidance documents - internal communications - meetings - resource allocations	Ensure objectives, measures, and accountability for second-order learning in the system are designed	Monitor design of objectives, measures, and accountability for second-order learning in the system. Engage as necessary and be prepared to reallocate resources as required	- meetings - internal correspondence - resource reallocation
Enable future transformation analysis	Provide resources for future transformation analysis. Articulate expectations for outputs and outcomes	- guidance documents - internal communications - meetings - resource allocations	Ensure future transformation analysis is accomplished	- engage as required and reallocate resources if needed	- meetings - internal correspondence - resource reallocation
Enable provision of future focused input to strategy development	Provide a venue for inclusion of future focused input to strategy development as well as required resources	- guidance documents - internal communications - meetings - resource allocations	Ensure future focused input to strategy development is provided	Monitor strategy development, engage if necessary, and be prepared to reallocate resources if required	- meetings - internal correspondence - resource reallocation

(continued)

Table 4 (continued)

Leadership Role	Actions	Mechanisms	Leadership Responsibility	Actions	Mechanisms
System M3—Systems operations Function: Maintain operational performance control through the implementation of policy, resource allocation, and design for accountability					
Enable oversight of products, services, value, and content delivery	Develop and communicate expectations. Establish clear authority and responsibility for operational performance oversight and provide requisite resources for accomplishment. Foster culture that encourages recognizing and addressing emergent issues by the system at the location closest to the issue (self-organization)	- policy documents - operational plans - meetings - resource allocations	Ensure oversight of products, services, value, and content delivery is provided	Maintain visibility of operational performance. Use performance aberrations to vector additional scrutiny to specific areas. Be prepared to clarify guidance, reassign authority and responsibility, or reallocate resources in response to aberrations or emergent conditions	- meetings - direct personal interactions
Enable system planning and control for ongoing day-to-day operational effectiveness	Develop and communicate expectations (minimal critical specifications). Recognize time constraints for decisions associated with day-to-day operations (emergence). Delegate planning and control authority to the lowest level (closest to the issue) as appropriate (autonomy). Provide resources for accomplishment	- policy documents - meetings - resource allocations	Ensure system planning and control for ongoing day-to-day operational effectiveness are provided	Maintain visibility of planning and control for ongoing day-to-day operational effectiveness. Use performance aberrations to determine which areas might need additional scrutiny. Be prepared to clarify guidance, reassign authority and responsibility, or reallocate resources in response to aberrations or emergent conditions	- meetings - direct personal interactions

(continued)

Table 4 (continued)

Leadership Role	Actions	Mechanisms	Leadership Responsibility	Actions	Mechanisms
Enable development of near-term system design response to evolving operational issues	Recognize that parts of the system closest to the evolving operational issues may have best perspectives on required system design responses to those issues. Empower these parts of the system to address near-term system design responses to evolving operational issues. Provide requisite resources for accomplishment	- policy documents - meetings - resource allocations	Ensure near-term system design response to operational issues is developed	Maintain visibility of evolving operational issues and responses there to. Be prepared to provide guidance or reallocate resources in response to emergent conditions	- direct personal interactions - internal correspondence - meetings - resource reallocation
Enable monitoring of operational performance measures	Provide resources to monitor operational performance impacted by evolving operational issues	- resource allocations	Ensure operational performance measures are monitored	Maintain visibility of performance. Be prepared to provide guidance or reallocate resources in response to emergent conditions	- direct personal interactions - internal correspondence - meetings - resource reallocation
Enable interpretation and implementation of system policies and direction from an operational perspective	Provide operational policies and direction that are clearly communicated. Provide clear expectations regarding implementation as well as resources required for monitoring	- meetings - internal correspondence - resource allocations	Ensure system policies and direction are (correctly) interpreted and implemented from an operational perspective	Maintain visibility of performance. Be prepared to provide guidance or reallocate resources in response to emergent conditions	- direct personal interactions - internal correspondence - meetings - resource reallocation

(continued)

Table 4 (continued)

Leadership Role	Actions	Mechanisms	Leadership Responsibility	Actions	Mechanisms
Enable determination of resources, expectations, and performance measurement for operational performance	Provide clear guidance regarding expectations for operational performance. Provide resources necessary to determine how to meet those expectations as well as to measure that performance	- policy documents - meetings - resource allocations	Ensure resources, expectations, and performance measurement for operational performance have been determined	Monitor effort to determine resources, expectations, and performance measures for operational performance. Be prepared to provide additional guidance or reallocate resources	- meetings - internal correspondence - resource reallocation
Enable design for accountability and performance reporting for operations	Provide clear expectations regarding accountability and performance reporting for operations. Provide resources required to design and execute the reporting process (or system?)	- policy documents - meetings - resource allocations	Ensure accountability and performance reporting for operations	Monitor accountability and performance reporting. Be prepared to provide additional guidance or reallocate resources if reporting is inadequate	- internal correspondence - meetings - resource reallocation
System M3*—Operational performance monitoring Function: Monitor measures for operational performance and identity variance in system performance requiring system-level response. Emphasis is on variability and performance trends that may impact system viability					
Enable tracking of ongoing performance of the system based on dashboard measures of performance for operations	Provide guidance for dashboard development and measures of performance for operations to be monitored. Provide resources to develop, operate, and maintain a dashboard	- guidance documents - meetings - resource allocations	Ensure ongoing performance of the system is tracked based on measures of performance for operations	Monitor ongoing system performance and use of dashboard. Be prepared to provide additional guidance or reallocate resources if the dashboard is not providing requisite information for tracking of ongoing operations, or if the dashboard is not being used	- direct personal interactions - meetings - internal correspondence - resource reallocation

(continued)

Table 4 (continued)

Leadership Role	Actions	Mechanisms	Leadership Responsibility	Actions	Mechanisms
Enable dissemination of system performance information throughout the system	Provide clear expectations regarding dissemination of system performance information throughout the system. Provide resources required to enable information dissemination	- policy documents - meetings - resource allocations	Ensure system performance information is disseminated throughout the system	Monitor availability of system performance information throughout the system. Be prepared to provide additional guidance or reallocate resources if necessary	- direct personal interactions - meetings - internal correspondence - resource reallocation
Enable the conduct of inquiry into performance aberrations	Allocate resources necessary to conduct inquiry into performance aberrations	- resource allocations	Ensure inquiries into performance aberrations are conducted	Monitor and/or engage in efforts to address performance aberrations as appropriate. Be prepared to allocate resources as required to support this effort	- direct personal interactions - meetings - internal correspondence - resource reallocation
Enable the monitoring and assessment of the continuing adequacy of operational performance measures	Allocate resources necessary to monitor and assess the continuing adequacy of operational performance measures	- resource allocations	Ensure the continuing adequacy of operational performance measures are monitored and assessed	Maintain visibility of and/or engage in efforts to monitor and assess the continuing adequacy of operational performance measures	- direct personal interactions - meetings - internal correspondence - resource reallocation

(continued)

Table 4 (continued)

Leadership Role	Actions	Mechanisms	Leadership Responsibility	Actions	Mechanisms
System M2—Information and communications					
Functions: Design and implement the architecture for information flow, coordination, transduction and communications within the metasystem and between the metasystem, the environment, and the governed system					
Enable the design and maintenance of the architecture for information flows and communications between the metasystem and environment, and between the metasystem and the governed system	Provide vision/guidance regarding expected outputs/outcomes as well as requisite resources to design and maintain the information and communication architecture. Recognize that the initial vision/guidance will be incomplete and that emergence will influence the results thus requiring this to be a continuous effort	- meetings - internal correspondence - guidance documents - resource allocations	Ensure the architecture for information flows and communications between the metasystem and environment, and between the metasystem and the governed system is designed and maintained	Stay abreast of the status of the architectural products. Provide oversight by meeting with the architecture team as required to review products and processes resulting from this effort. Continue providing requisite resources	- meetings - product/project reviews - resource reallocation
Enable accessibility of coordinating information within the system	Insist that requisite communication channels be available to support access to coordinating information and foster a culture of information sharing. Expect emergent requirements. Provide requisite resources to support communication channels	- meetings - internal correspondence - resource allocations	Ensure coordinating information within the system is accessible	Query personnel to determine success of gaining access to coordinating information. Be prepared to 'encourage' removal of roadblocks and bottlenecks	- meetings

(continued)

Table 4 (continued)

Leadership Role	Actions	Mechanisms	Leadership Responsibility	Actions	Mechanisms
Enable identification of standard processes and procedures necessary to facilitate transduction and provide effective integration and coordination of the system	Provide vision/guidance regarding expected outputs/outcomes of effort to identify standard processes and procedures as well as requisite resources. Be mindful of minimal critical specification and emergence	- guidance documents - internal correspondence - meetings - resource allocations	Ensure standard processes and procedures necessary to facilitate transduction and provide effective integration and coordination of the system are identified	Maintain visibility into system operations from the perspective of availability of guidance provided by standard processes and procedures. Be prepared to add, delete, or modify processes or procedures when required to improve integration and coordination	- direct personal interactions - meetings - guidance document revisions - resource reallocations
Enable forums to identify and resolve emergent conflict and coordination issues within the system	Encourage a culture of cooperation and open-minded, straight forward discussion regarding conflict and coordination issues (complementarity). Set expectations that conflict and coordination issues will be resolved at the lowest level possible (closest to the issue)	- personal example - meetings - guidance documents	Ensure forums to identify and resolve emergent conflict and coordination issues within the system are created as needed	Observe the system's ability to engage in open, straight forward discussion regarding conflict and coordination issues. Query personnel regarding conflict resolution	- direct personal interactions - meetings

5. Discuss how an appropriate leadership style for an organization might be determined.
6. From your perspective, is there a need for leadership style within an organization to change over time? If so, under what circumstances?
7. What are some of the consequences of changing leadership style? What consequences might result if the leadership style within an organization is not congruent with the needs of the organization?

References

1. Keating C, Katina P, Bradley J (2014) Complex system governance: concept, challenges, and emerging research. Int J Syst Syst Eng 5(3):263–288
2. Keating C, Katina P (2015) Editorial: foundational perspectives for the emerging complex system governance field. Int J Syst Syst Eng 6(1/2):1–14
3. Fleishman E, Mumford M, Zaccaro S, Levin K, Hein M (1991) Taxonomic effort in the description of leader behavior: a synthesis and functional interpretation. Leadersh Quart 2(4):245–287
4. Northouse PG (2013) Leadership theory and practice, 6th edn. SAGE Publications Ltd., Thousand Oaks, CA
5. Toor S (2011) Differentiating leadership from management: an empirical investigation of leaders and managers. Leadersh Manag Eng 11:310–320
6. Kaiser R, Hogan R, Craig SB (2008) Leadership and the fate of organizations. Am Psychol 63(2):96–110. https://doi.org/10.1037/0003.0066X63.2.96
7. Lichtenstein B, Uhl-Bien M, Marion R, Seers A, Douglas J (2006) Complexity leadership theory: an interactive perspective on leading in complex adaptive systems. Emerg Complex Orga 8(4):2–12
8. Kupers W, Weibler J (2008) Inter-leadership: why and how should we think of leadership and fellowship integrally? Leadership 4:443–475. https://doi.org/10.1177/1742715008095190
9. Maccoby M, Scudder T (2011) Strategic intelligence: a conceptual system of leadership for change. Perform Improv 50(3):32–40. https://doi.org/10.1002/pfi.20205
10. Marion R, Uhl-Bien M (2001) Leadership in complex organizations. Leadersh Quart 12:389–418
11. Crowley W (1931) The traits of face-to-face leaders. Psychol Sci Public Interest 26(3):304–313
12. Katz R (1955) Skills of an effective administrator. Harv Bus Rev 33(1):33–42
13. Hersey P, Blanchard K (1969) Management of organizational behaviour—utilization of human resources. Prentice Hall, New Jersey
14. Fiedler F (1964) A Contingency model of leadership effectiveness. In: Berkowitz L (ed) Advances in experimental social psychology, vol 1. Academic Press, New York
15. Evans M (1970) The effects of supervisory behavior on the path-goal relationship. Organ Behav Hum Perform 5(3):277–298
16. Downton JV (1973) Rebel leadership, New York
17. Weber M (1947) The theory of social and economic organization. The Free Press, New York
18. Greenleaf R (1977) servant leadership a journey into the nature of legitimate power and greatness. Paulist Press, New Jersey
19. Uhl-Bien M, Marion R, McKelvey B (2007) Complexity leadership theory: shifting leadership from the industrial age to the knowledge era. Leadersh Q 18(4):298–318
20. Gonos J, Gallo P (2013) Model for leadership style evaluation. Management 18(2):157–168
21. MindTools (2014) Core leadership theories learning the foundations of leadership. www.mindtools.com

22. Giltinane C (2013) Leadership styles and theories. Nurs Stand 27(41):35–39
23. Prieto B (2013) Establishing and building leadership skills. Leadersh Manag Eng 13:209–211
24. Katzenbaugh J, Santamaria J. (1999) Firing up the front line. Harv Bus Rev 107–117
25. Kotter J (1990) What leaders really do. Harv Bus Rev 68(3):103–111
26. Osborn R, Hunt J, Jauch L (2002) Toward a contextual theory of leadership. Leadersh Q 13:797–837
27. Vera D, Crossan M (2004) Strategic leadership and organizational learning. Acad Manag Rev 29(2):222–240
28. Merriam-Webster (ed) (2014) Merriam-Webster on-line dictionary. Merriam-Webster, Inc., Springfield, MA
29. Maccoby M (2000) Understanding the difference between management and leadership. Res Technol Manag 43(1):57–59
30. Lowder R (2009) The most important attributes of leadership—a personal skills improvement guide. Outskirts Press, Inc., Denver
31. Gardner H, Laskin E (1995) Leading minds an anatomy of leadership. BasicBooks, New York, NY
32. Thompson K, Thach E, Morelli M (2010) Implementing ethical leadership: current challenges and solutions. Insights Chang World J 2010(4):107–130
33. Gabriele E (2011) Ethics leadership in research, healthcare and organizational systems: commentary and critical reflections. J Res Adm 42(1):88–102
34. Caldwell C, Hayes L, Karri R, Bernal P (2008) Ethical stewardship—implications for leadership and trust. J Bus Ethics 78:153–164
35. Sankar Y (2003) Character not charisma is the critical measure of leadership excellence. J Leadersh Organ Stud 9(45–55)
36. Coffey G (2010) A systems approach to leadership how to create sustained high performance in a complex and uncertain environment. Springer, London
37. Hazy J, Goldstein J, Lichtenstein B (eds) (2007) Complex systems leadership theory: new perspectives from complexity science on social and organizational effectiveness, vol 1. ISCE Publishing, Mansfield, MA
38. Onyx J, Leonard R (2010) Complex systems leadership in emergent community projects. Community Dev J 46(4):493–510
39. Dictionary.com (2014) http://dictionary.reference.com/
40. Johnson N (2009) Simply complexity. Oneworld Publications, London, UK
41. Skyttner L (2001) General systems theory: ideas and applications, 1st edn. World Scientific Publishing Co., Pte. Ltd., River Edge, NJ
42. Dooley K (1996) Complex adaptive systems: a nominal definition. www.researchgate.net
43. Keating C (2014) Governance implications for meeting challenges in the system of systems engineering field
44. Bass B (1990) Bass and Stogdill's handbook of leadership—theory, research, and managerial applications, 3rd edn. The Free Press, New York, NY
45. OPM (2013) 2013 Federal employee viewpoint survey. OPM, Washington, DC. http://www.fedview.opm.gov/2013/Reports
46. Truman_Library (2014) 'The Buck Stops Here' Desk Sign. http://www.trumanlibrary.org/buckstop.htm
47. Beer S (1994) Brain of the Firm. John Wiley & Sons, New York, NY

Introducing Complex Systems Governance to Practitioners

James C. Pyne, Charles B. Keating, and Polinpapilinho F. Katina

Abstract Complex System Governance (CSG) is an emerging field with the potential to enhance capabilities for the design, execution, and evolution of complex systems. CSG offers a theoretically grounded, model informed, and methodologically driven approach to more effectively deal with complex systems and their problems. However, initial CSG applications have identified multiple impediments to systemic intervention to deploy this new and novel field. In this chapter, we discuss strategies to effectively deploy systemic intervention in support of CSG. Four primary objectives are pursued, including: (1) identification of major forms of systemic intervention for complex systems in general and a corresponding classification schema, (2) presentation of a dynamic and tailored approach (CSG Entry) to improve prospects for introductory systemic intervention for CSG, (3) results from an initial application of CSG Entry in a field setting, and (4) suggestion of lessons learned from initial applications of CSG Entry in relationship to systemic intervention. The chapter concludes with examination of future development directions for systemic intervention to advance CSG performance.

1 Introduction

Landscape of the Modern Complex System Practitioner

Practitioners continue to be besieged with complex systems and their problems that at first glance appear increasingly intractable. This is especially true of organizations. For the purposes of this chapter, we use the term 'enterprise system' to denote an organizational complex system. The shifting landscape of the systems engineering

J. C. Pyne · C. B. Keating (✉)
Old Dominion University, Norfolk, VA, USA
e-mail: ckeating@odu.edu

P. F. Katina
Department of Informatics and Engineering Systems, University of South Carolina Upstate, Spartanburg, SC, USA
e-mail: pkatina@uscupstate.edu

C. B. Keating et al. (eds.), *Complex System Governance*, Topics in Safety, Risk, Reliability and Quality 40, https://doi.org/10.1007/978-3-030-93852-9_16

practitioner might be characterized by several dominant characteristics. Following previous recitations of this landscape from recent works [4, 10, 11, 15], the following summary is offered with respect to characteristics and their nature for the domain faced by system practitioners dealing with enterprise systems. These include:

1. *Exponential Rise in Complexity*—the availability, magnitude, and accessibility of information are beyond current capabilities to structure, order, and reasonably couple decisions, actions, and consequences. This, coupled with compression of time and the interconnectedness of 'everything,' is challenging our capacity to mount effective responses.
2. *Dominance of Emergence*—the appearance of structures, behaviors, performance, or consequences that cannot be known in advance renders traditional forms of planning innocuous at best, unsuited to current realities, and potentially detrimental. Current methods are failing to provide practitioners with the necessary capabilities to engage highly emergent situations.
3. *Ambiguity in Understanding*—instabilities in understanding, shifting boundary conditions, and unstable structural patterns create a lack of clarity for decisive action.
4. *Uncertainty as a Norm*—the inability to have any measured degree of confidence in how to proceed to produce desired performance is not the exception but rather the stable state of affairs.
5. *Holistic Satisficing Solution Spaces*—the modern problem space is not limited to simple, absolute, or isolated solution forms. The spectra of technology/technical, organizational/managerial, human/social, and political/policy are in play across special, temporal, and social dimensions.
6. *Contextual Dominance*—unique circumstances, factors, patterns, and conditions permeate all systems. They can be both enabling and constraining to decision, action, and interpretation.

Dealing with these characteristics is not insurmountable. However, effectively dealing with them requires a different level of thinking. While these characteristics are certainly not intended to present an 'absolute' depiction of the landscape, they serve as a reminder of the stark reality faced by practitioners. The domain of the enterprise system practitioner appears to be intractable. Complex System Governance (CSG) is an emerging field designed to address this increasingly hostile landscape, which represents a 'new normal' for system practitioners. A snapshot of the realities facing practitioners in this 'new normal' is shown below in Fig. 1.

Three primary conclusions are offered for this set of realities facing practitioners of enterprise systems. First, the nature of this landscape is not likely to improve in the future. More probable is that these elements will escalate in frequency and severity of their impacts. Second, our current approaches to deal with the systems characterized by these conditions are not having the desired impact. This is evidenced by the increasing number of tools, technologies, and approaches attempting to address complex systems without resolution of associated issues. This is not intended to disparage any of those tools, technologies, or approaches, but rather only recognizes

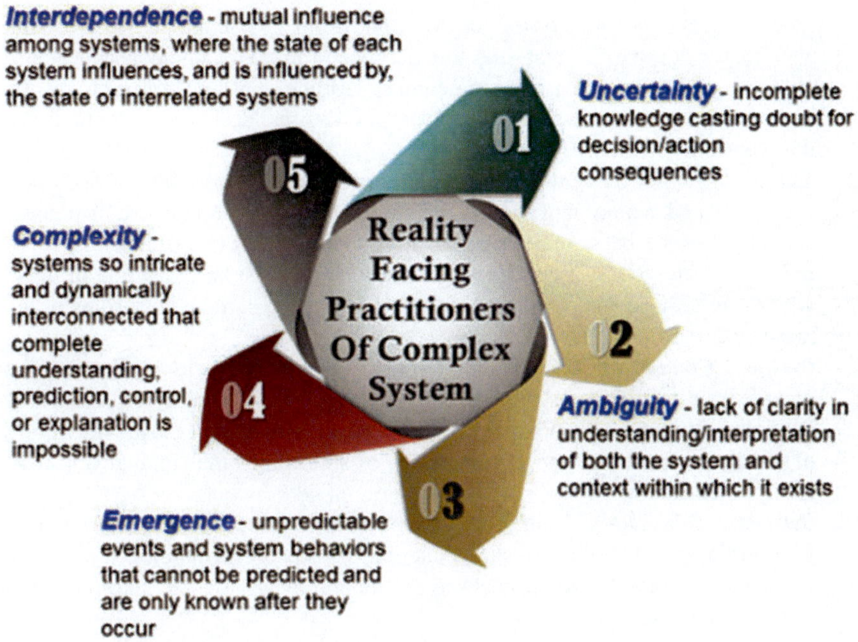

Fig. 1 Five realities for complex system practitioners (adapted from [18])

that the search must continue for more effective approaches. The presented characteristics are representative of a complex system problem domain. Therefore, approaches that are not consistently developed, grounded, or applied in a manner appreciative of 'systems' are not likely to 'match' the complexity demanded by this domain. We now shift to a discussion of intervention to assist enterprise system practitioners in dealing with their new reality from a CSG perspective.

Systemic Intervention

Intervention is certainly not a new concept. Almost every management theory has a related intervention strategy. Facilitators have adapted them to suit their individual practices. At the very essence of intervention is the notion that there is (1) involvement, (2) intention to alter actions/outcomes, and (3) use of some form of leverage (force) to carry out the effort. While this depiction is helpful, systemic intervention has a different connotation. Following [20], we describe systemic intervention as *the purposeful action by an agent, generally human for complex systems, to produce change in a system or situation.* For our perspective of systemic intervention, the following elements are offered [17]:

1. *Purposeful*—engagement in intervention with the intention to achieve some desired aim. The importance of this aspect of systemic intervention is that it requires the outcome (expectations) for the intervention to be specified

(determined) in advance of the intervention. From a systemic perspective, this also must acknowledge that, due to emergence (unpredictable consequences), although there are 'desirable' outcomes, latitude must be given to results and directions not necessarily conforming to desires, design, or intentions for intervention.

2. *Human Agent*—at the center of any systemic intervention are people. The design, execution, and evolution of a systemic intervention are accomplished by people. As such, people become the central driving force behind systemic intervention. So much so that effectiveness in intervention must be a function of those who design, those who conduct, and those who play participatory roles in the intervention effort.
3. *Produce Change*—from a systemic perspective, change in a system may include modifications in structure, behavior, or understanding/interpretation of a system/situation. This point is critically important, since it moves the notion of change beyond the narrow conception of solution as the singular objective for intervention.
4. *Systemic*—this invokes the entirety of the 'systems' perspective for intervention. In contrast to a focus on linear, reduction, or piecemeal inquiry, a systemic orientation to intervention is focused on the nonlinear, holistic, and integrated inquiry into a system.

There are four primary conclusions with respect to the systemic nature of intervention identified for CSG development. First, although the notion of intervention is well known, the nature of 'systemic intervention' introduces a different level of thinking, possibility for different corresponding actions, and can invoke a different level of understanding/interpretation of a situation. Second, systemic intervention does not exist in a binary fashion of 'present' or 'not present.' Rather, it is best to recognize that systemic intervention might be achieved in 'degrees of application.' This opens the possibility of systemic intervention having a spectrum of depth in delivery. Third, the engagement in systemic intervention has real consequences for performance of a given system—introducing an entire spectrum of development possibilities. These developmental 'change' possibilities range across the spectrum of technology, human, social, organizational, managerial, policy, and political dimensions. In addition, although 'everything' cannot change simultaneously for a given system, changes pursued can be assessed for feasibility and their specific fit to the larger landscape of systemic issues identified during intervention inquiry. Each system is unique and must be taken as it is with its own individual culture, peculiar language, available resources, perceived needs, and variety load. Therefore, the associated systemic intervention design, execution, and development expectations must be unique. Fourth, systemic intervention must be engaged by individuals with some level of a 'systems worldview.' In effect, since intervention is undertaken by people, their worldview, and the degree that it is consistent with a systems mindset, will enable or constrain any systemic intervention effort. Thus, while systemic intervention provides an exciting and substantial movement forward for CSG development, it must be engaged with a healthy skepticism.

The focus now shifts to elaboration of the different roles and specific forms of systemic intervention. This elaboration is essential to clearly understand where individuals are placed in a systemic intervention and the particular type (form) being pursued. Both of these aspects require clarity concerning systemic intervention—hopefully at the outset of an initiative.

Observation versus Intervention. As the definition of systemic intervention highlights the characteristic of purposeful action, one may consider what happens when there is inaction and only observing the system. [21] addresses conducting an observation versus conducting an intervention. Midgley defines independent observation as 'observation detached from the values and idiosyncrasies of the observer' [21], p. 9) and suggests that without this independence, an intervention has been conducted. This is not to pass judgment on intervening (or not intervening). In fact, this further highlights a need for a systemic approach to intervention.

In addition to the independence of the observer, the observer's engagement with the system can determine if an intervention is occurring.

Four situations can occur between the observer and the system as follows:

1. Observation performed and *known* by the system results in an intervention as the system has been changed by the knowledge of the observation occurring.
2. Observation performed and *known* by the system does not result in an intervention because the system remains unchanged by the observation occurring.
3. Observation performed and *unknown* by the system results in no intervention because the system is unchanged.
4. No observation—no intervention.

The key takeaway is that there is a distinction between merely observing versus conducting an intervention and that when an intervention occurs, a need for a systemic approach exists.

Roles and Forms of Intervention. In the initiation of intervention, we present four primary forms of intervention and their associated role expectations. It is important to be clear on which of the forms of intervention are being pursued. In addition, each of the different forms requires a specific role to be played by both the interventionist and those enlisting the intervention.

Table 1 summarizes three basic forms of intervention. This is not to say that there might be different configurations or hybrids of the different forms. However, these four basic forms provide an adequate definition of the landscape for intervention.

These three forms of intervention are not intended to define the entire scope of intervention. However, they do provide a survey of the range of intervention possibilities for systems. There are three important conclusions offered with respect to intervention implications. First, there is a range of 'intensity' and corresponding expectations related to the different intervention forms. The simple 'expert advice' intervention is certainly not to the depth or expectations that would be characteristic of the 'participatory' intervention form. Second, there is a range of risk incurred in any intervention. As the intervention moves from 'problem resolution' to 'participatory,' the risk shifts from the interventionist to the client organization. Thus, for

Table 1 Forms and roles for systemic intervention (adapted from [18])

Intervention Form	Nature	Roles	Accountability	Example
Problem Resolution	Engagement for a specific problem to be resolved by the intervention. Expertise is beyond that held by the system in focus	Intervention which brings specific competence not held within the system, or intended to be developed in the future	Risk for proper resolution of a problem is held by the interventionist	Bringing on an expert to solve a specific targeted governance problem in the system (e.g., communications)
Expert Advice	Engagement of an expert for their specific advice concerning a problematic situation	The client system provides data and description of a problematic situation, leaving the interventionist to provide prescriptive advice for resolution	Interventionist has responsibility for the prescription adequacy. Client system holds responsibility for implementation of recommendations	Engaging an expert to make recommendations concerning implementation of a new program (e.g., supply chain logistics)
Participator	Engaging in a shared intervention effort to design, analyze, and improve system to performance	The intervention design, execution, and assessment are shared between interventionist and system actors	The responsibility for conduct and results are shared between all parties in the intervention	Engaging in a comprehensive CSG development effort

holistic intervention, characteristic of the participatory form, there is a sharing of risk for success of the intervention. Third, the ability to make objective determinations with respect to 'success' of the intervention endeavor decreases as the form of intervention moves from 'expert advice' to 'participatory' forms. Fourth, as the depth of intervention increases (from advice to participatory) so too does the risk for failure or falling short of expectations. This is not unexpected, as the nature of problems and their scope, breadth, and depth is increasing with the different intervention forms, with participatory representing the most comprehensive and extreme intervention case. It should be emphasized that the forms of intervention are not binary in nature. Instead, they can exist in different combinations and hybrid forms.

Systemic Intervention for Complex System Governance

CSG is not an easily approachable subject. Instead, it requires commitment to a 'long view,' 'sustainable,' and 'integrated' endeavor. It focuses on the very core of complex system design, execution, development, and maintenance for organizations. However, as with all systemic intervention approaches, it should be met with a healthy skepticism. It would be unrealistic to engage in a comprehensive systemic intervention with little more than a 'promise' of effectiveness. Both the practitioners in the complex (enterprise) system as well as the facilitator will be learning throughout the process, but it is especially important for the facilitator to be observant. For the intervention to be successful, the enterprise has to be engaged as it is, not as the facilitator thinks it should be. CSG intervention cannot be a 'one size fits all' program. It must be adapted to suit the individual enterprise. Of particular interest in the initial phases of the intervention are the enterprise's worldview, context, and language. These are not independent attributes. They develop in a system concurrently and are linked. Changes in any one invoke changes in the others. Worldview can be thought of as a system's outlook or belief about its place in the environment and how it can or should interact with it. Its contexts are factors, conditions, circumstances, and influences, both internal and external to the system, that influence the behavior of the system. Special attention should be given to language. Each complex (enterprise) system has a language that has developed over time in response to its context, environment and the technologies it is involved with. Likewise, CSG has been developed from a systems context in an academic environment and has developed its own specific language. The facilitator must communicate using language readily understandable by the enterprise system practitioners. As the intervention proceeds and the enterprise system practitioners become more familiar with CSG concepts, a common understanding will evolve.

Understanding these provides a foundation for structuring an intervention plan specific to the enterprise. *CSG Entry* (discussed in more detail below) has been developed as a first introduction to CSG to lessen comprehensive engagement hesitation and to provide the facilitator an opportunity to make some initial observations before proceeding with more, in depth, intervention.

Vignette—Worldview, Context, and Language Changes

In an enterprise system, worldview, context, and language are some of the components that make up what some would call its culture. From the early 1970s through the late 1980s, a large governmental utility evolved a specific culture in response to increased regulatory pressure and reinforced by the overall culture of the region that had a large population of active and retired military personnel. This culture influenced how the enterprise system interacted with regulators and other governmental agencies in the region. Although cordial on the surface, the underlying attitudes were adversarial. Indeed, the internal language used to describe the outcome of correspondence and meetings was laced with phrases like 'we really beat them today' or 'we have to fight back against those permit requirements'. The internal departments were effectively siloed and competed for resources with the stronger department heads winning out over the others. Although the enterprise was operating efficiently and was in compliance with all of the regulations, its mission was narrowly defined to meeting regulatory requirements and limiting customer rate increases. Starting in the 1990s and continuing into the new millennium, the culture began to change. The shift coincided with some internal development programs, followed by a change in leadership as older managers began to retire. The enterprise's governing body specifically chose a top executive with a more expansive worldview. As a result, the language used to discuss relations with other agencies and regulators began to soften as did the walls between the siloed departments. Interactions between the enterprise and others in its environment became less adversarial. This led to new initiatives that were targeted not only to improve operations related to the enterprise's traditional mission but expand it to improve the regional environment in ways that were not previously considered.

Intervention Planning and Execution

The intervention requires a rigorous plan. The facilitator must acquire some basic information about the enterprise system in order to develop an intervention plan that is specific to the enterprise considering its context, environment, level of systems thinking, and its current governance condition. The initial phases of the intervention should be designed to introduce the facilitator to the enterprise and for the facilitator to gather the information required for development of a tailored intervention plan that will address specific system weaknesses.

Each enterprise system is unique and will require a strategy tailored to suit its peculiarities. However, there can be a common framework, especially in the early phases of the intervention that can help in developing the structure of the intervention plan and its execution.

Diagnostic Phases of Systemic Intervention

It must be accepted that in the early phases of systemic intervention, the larger portion of the effort will reside with the facilitator. As the intervention progresses, the practitioners will gain insight into systems concepts and methods, in general, and CSG in particular, to be able to accept more of the effort. It must be made clear to the practitioners that the effort they expend is not in addition to their current responsibilities. Instead, governance of the enterprise system is their responsibility. They are already practicing it on a daily basis. The intervention is designed to improve their governance of the enterprise system, especially in its functions.

Early in the diagnostic phase of an intervention it would be beneficial to elicit the system practitioner's concerns about the functioning enterprise system's operation and governance. This can be accomplished utilizing the CSG Entry program methods, discussed below, as well as free form discussions with the system practitioners. It is important to give credence to the concerns that practitioners have. These concerns may be, in reality, symptoms that are the result of deeper systemic problems or pathologies. By first focusing on symptoms brought to light by the practitioners then proceeding to relate those to the deeper systemic issues, the practitioners begin to realize the value of CSG-based methods with respect to enhancing their enterprise system performance.

As the practitioners gain exposure and experience with CSG-based methods, they will be able to take a more active role in the intervention. As their participation increases, the role of the facilitator will begin to decrease as shown in Fig. 2. The result will be that the practitioners will begin to see CSG, its concepts, and methods as not just another 'add on' program but as integral to the functioning of the enterprise system.

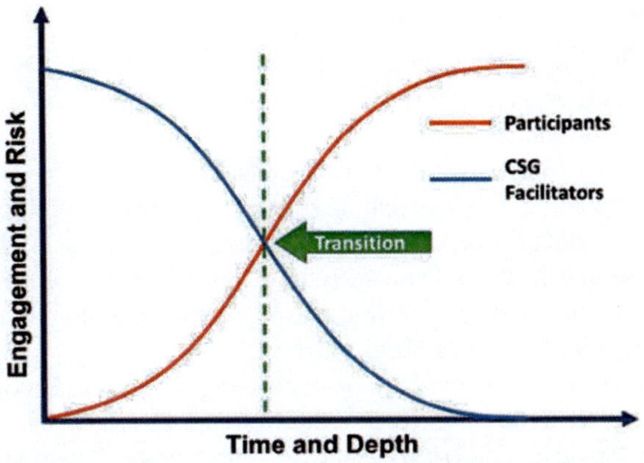

Fig. 2 Relative participatory effort

CSG Entry

The first phase of an enterprise system intervention should consist of an introductory program. CSG Entry has been developed specifically for this purpose. The combination of System Thinking Capacity (ST-Cap), Environmental Complexity Demand (ECD), and Governance System Diagnostic Check (GDC) provides insight and background information for the facilitator and introduces the enterprise system practitioners to systems thinking and CSG concepts. There is a more in-depth examination of CSG Entry later in this chapter.

There is considerable value gained from the CSG Entry alone, so if the enterprise systems practitioners only complete this initial phase, the effort would not have been wasted.

In order to develop a more complete articulation of the enterprise system landscape and its current state of governance, the facilitator and the enterprise system practitioners must investigate deeper. A rigorous system mapping will provide the facilitator and the enterprise system practitioners with insights into the system, how it is designed to function and how its governance functions are integrated.

System Mapping

In many cases, the enterprise structure was developed in a less complex environment and now may need to be updated to suit current complexity, or the structure has been changed on an ad hoc basis to cope using a trial and error-type method not based on a rigorous method but based on some type of systems method. An important part of the systemic intervention is coming to an understanding of the components, relationships, and overall architecture [1] of the enterprise governance system with respect to the metasystem functions as articulated in the CSG reference model [14]. There are many approaches that can be used to perform the system mapping, and the approach should be chosen with respect to what has been discovered about the enterprise system context during the CSG Entry process.

Metasystem Pathology Assessment

The system mapping effort should be followed by an investigation to uncover weaknesses in the governance structure (pathologies) that are inhibiting system performance.

There are currently two possible methods to consider investigating system pathologies. System pathologies are rigorously investigated using the M-path method [9] which is discussed in more detail later in this chapter. The M-Path is a rigorous method that has been fully developed and verified through extensive research. It provides a detailed compilation of the pathologies adversely affecting the nine governance metasystem functions.

Another method for investigating system pathologies has been proposed and utilizes a modified version of failure mode effects and criticality analysis (FMECA) adapted for CSG (FMECA-CSG) and is under development. The five-phase FMECA-CSG approach is being mapped onto corresponding M-Path system pathologies. It will be a less intensive method that can perform an analysis of pathologies similar, although not as in depth or rigorous as M-Path. It may also be used as an initial step to better target the M-Path effort.

CSG Entry, system mapping, and M-Path/FMEC-CSG together comprise a suite of methods that when used at the initial phase of a systemic intervention provide essential information to develop a comprehensive plan to advance the intervention toward the production of favorable results.

In summary, during these initial phases of the intervention, the investigation should concentrate on five aspects of the enterprise system. These include the following:

1. The System Thinking Capacity of the enterprise system.
2. The enterprise system's Environmental Complexity Demand.
3. A diagnostic of the enterprise system's governance (metasystem).
4. A rigorous mapping of the enterprise system and metasystem.
5. An investigation of the enterprise system's pathologies.

2 CSG Entry as an Approach to Begin Systemic Intervention

CSG has not been presented as a 'magic elixir' or 'silver bullet' that can cure all system/organizational ills. CSG development is not a 'sprint,' a 'fad,' 'easy,' or an 'isolated' endeavor. Instead, it requires commitment to a 'long view,' 'sustainable,' and 'integrated' endeavor. It focuses on the very core of complex system design, execution, development, and maintenance for organizations. However, as with all systemic intervention approaches, it should be met with a healthy skepticism. It would be unrealistic to engage in a comprehensive systemic intervention without more than a 'promise' of effectiveness. Thus, *embarking on a comprehensive CSG development effort as a first step is unrealistic.* The associated risks and inherent uncertainties in a comprehensive CSG endeavor are simply too great as a first step.

Therefore, *CSG Entry* has been developed as a first introduction to begin a systemic intervention effort. It represents a 'hands-on' low-risk, efficient, and value-adding introduction to CSG. In a nutshell, CSG has been developed as a systems-based approach that:

1. *Appreciates the 'new normal' for practitioners marked by increasing complexity in their organizations, systems, and environment,*
2. *Offers an alternative perspective and approach to better understand critical system functions directly responsible for performance,*
3. *Is based in the application of fundamental system laws that govern performance of all systems, and*
4. *Enhances capacity to more effectively deal with increasingly complex systems, environments, and problems.*

It is a four-phased *CSG Entry* (Exhibit 2) approach that offers an efficient, convenient, low-risk, and value-added introduction to CSG (Fig. 3).

CSG Entry offers a 'hands-on' first exposure to CSG that is a short-term, efficient, and value-adding endeavor. It can be achieved from start to finish in the four phases with a minimal investment of time and resources spread out over a time period convenient to the enterprise system practitioners. A summary of the four phases of CSG Entry includes the following:

1. *PHASE 1: INVITATION TO CONDUCT CSG ENTRY*—the organization agrees to engage in a CSG Entry effort and is provided a basic overview of the process and expectations. The focal entity (unit, team, organization) is identified, and

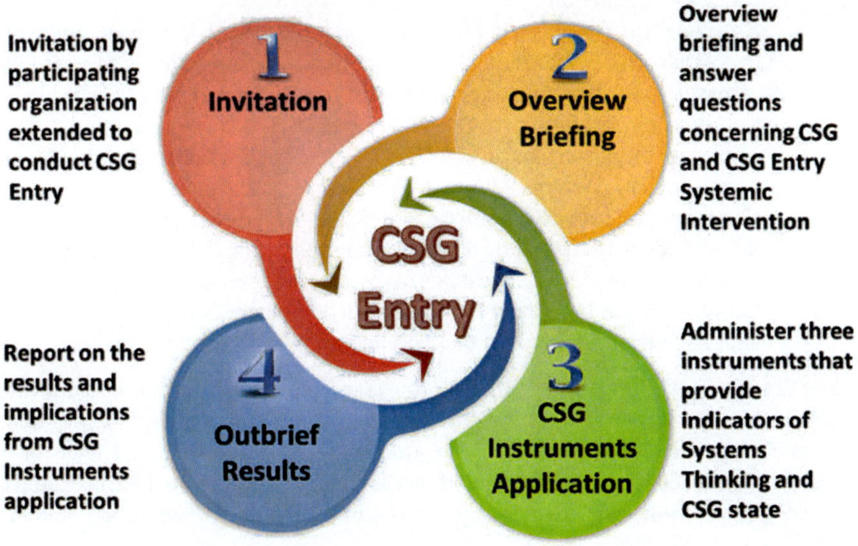

Fig. 3 Four phases for CSG entry (adapted from [18])

prospective participants are selected and a tentative timetable for completion set. This should include a discussion with someone in authority that will act as the sponsor for the intervention to gauge the level of interest and support.

2. *PHASE 2: OVERVIEW BRIEFING*—this briefing is designed to introduce participants to CSG and the *CSG Entry* approach. Questions are answered, expectations are set, and preparations are made to execute *CSG Entry*. In this briefing, the nature of CSG is kept to an overview level, and the emphasis is on instruction and clarification of the three instruments to be completed by the participants. Plan this phase and phase 1 to not only introduce the facilitator to the enterprise but to learn about the enterprise for use in adapting the intervention strategy.
3. *PHASE 3: CSG ENTRY INSTRUMENTS APPLICATION*—this phase is designed around administration of three web-based instruments that provide a set of insights concerning CSG. The total time investment in this phase is approximately 30 min per participant to take the three instruments. The results of these instruments are anonymous, and only aggregate information is compiled. Each instrument provides a snapshot of a different aspect related to CSG for the focal entity (unit, team, organization). In summary, the three instruments are as follows:

   a. ***Systems Thinking Capacity***—examines seven dimensions of systems thinking through a 39-question web-based survey instrument. The instrument determines the relative preference for systems thinking that exist in the participating group. Although each individual has a personal profile for systems thinking preference, only aggregates are collected and reviewed for CSG implications.
   b. ***Environment Complexity Demand***—examines the degree of complexity that exists in the environment of the enterprise system. This is captured by assessment of the seven dimensions of systems thinking in relationship to the environment through a 43-question web-based survey instrument. The aggregate of participant responses is collected and mapped onto the seven dimensions of Systems Thinking Capacity.
   c. ***Governance System Diagnostic Check***—a 45-question web-based survey that guides participants through an examination of CSG function performance to provide a 'snapshot' of performance across the nine essential governance functions. Participant responses are anonymous, and only aggregate data are used for analysis and mapping of the results.

4. *PHASE 4: OUTBRIEF RESULTS*–after completion of the three instruments, results are analyzed and compiled in a technical document provided to help guide interpretation of results. The document is provided in advance of a briefing presentation conducted with participants to explore the interpretations, answer questions, and suggest implications of the results for individuals and the participating entity.

In sum, *CSG Entry* offers an *efficient, low-risk, and value-added* set of activities to introduce CSG. This approach represents a 'hands-on' demonstration of the practical utility of CSG for helping to address some of the most vexing problems facing organizations and practitioners responsible for design, execution, and development of complex systems. CSG development, beyond *CSG Entry*, is not easy, fast, or achievable by following a prescriptive recipe. However, the *CSG Entry* approach outlined above offers an important first step for more comprehensive systemic intervention. To be able to move forward to more, in depth, systemic intervention, the practitioners and participants must perceive that systemic intervention will produce something of value for them. *CSG Entry* is the first step in demonstrating systemic intervention value. However, even if nothing is pursued beyond the *CSG Entry* effort, there is still significant value that can accrue.

Results from an Initial Application of CSG Entry

This section presents results from an initial application of the CSG Entry approach. The objective of this entry was to introduce CSG, through hands-on experience in an operational setting. This discussion is limited to what resulted and what was learned from the application. The presentation is broken down to the corresponding phases of CSG Entry (Invitation, Overview Briefing, CSG Instruments Application, and Outbriefing of the Results).

1. *Phase 1: Invitation*—project sponsors were provided a brief overview of CSG and the potential value that CSG might provide to the organization (system in focus). The expectations with respect to resources necessary to engage were explained. The nature of CSG, coupled with the efficient deployment, limited risk, and potential value were considered as sufficiently reasonable by the sponsor to engage in the effort.
2. *Phase 2: Overview Briefing*—the project sponsors selected participants for the CSG Entry effort. These were mainly leaders from the various organizational departments. The selected participants were briefed on the basics of CSG, the approach to CSG Entry, and the specifics of the instruments that would be deployed to provide data for the CSG Exploration. Timeframes were established, all questions answered, and access to the CSG instrumentation was provided.
3. *Phase 3: Instrument Application*—participants completed the three web-based instruments (System Thinking Capacity, Environment Complexity Demand, and 14-Point Governance Check) consistent with the timetable scheduled. Data were collected and prepared for outbriefing of results.
4. *Phase 4: Outbrief Results*—results of the instruments application were prepared in a technical report, and an outbriefing of the results was conducted with the participants. In this particular project, the participants were not provided with the technical report in advance of the outbriefing. For brevity, we include a snapshot of the representations of the results.

The following are the actual results of the CSG Entry effort as presented to the participants during the outbriefing. The presentation includes the numerical results of the instruments as well as definitions of the parameters measured.

The results of each of the three web-based instruments and their implications are shown below. *Care must be taken not to overreach the results and their implications.* Remember, these results stem from three short instruments and represent only a '*snapshot*' of several aspects of the enterprise. As such, it would be shortsighted to attribute 'absolutes' to the results. Rather, they are indicators that suggest potential implications, can focus further explorations/discussions, and suggest development directions for possible consideration to enhance system governance. We now examine each of the instruments and their results.

Systems Thinking Capacity—this instrument examined seven dimensions of systems thinking through a 39-question web-based survey instrument. The instrument determines the relative preference for systems thinking that exist in the participating group. The results of this instrument are provided in Table 2 and Fig. 4. The % spread provides the degree of variability of thinking in the group (max variability is 100%, indicating the group includes thinking along the entire spectrum of the dimension, where 0% would indicate total uniformity of thinking). The Systems Thinking Capacity % provides the degree to which a preference for systems thinking exists in the group as a whole (100% is maximum systems thinking capacity).

Table 2 Definition of systems thinking capacity dimension and variability of responses

Dimension	Definition	% Spread of group members (max spread 100)	Systems thinking capacity % (max 100, more systemic)
Complexity	Comfort with multidimensional problems and limited system understanding	**53**	**30.3**
Independence	Balance between local-level autonomy versus system integration	**42**	**58.2**
Interaction	Interconnectedness in coordination and communication among multiple systems	**42**	**77.3**
Change	Comfort with rapidly shifting systems and situations	**32**	**59.1**
Uncertainty	Acceptance of unpredictable situations with limited control	**45**	**57.6**
Systems Worldview	Understanding system behavior at the whole versus part level	**34**	**47.3**
Flexibility	Accommodation of change or modifications in systems or approach	**18**	**27.3**

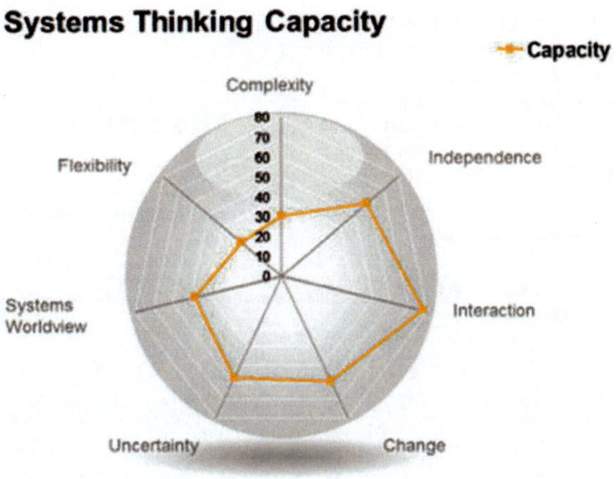

Fig. 4 Systems thinking capacity

The following diagrams break down the definition and results for each of the seven dimensions of Systems Thinking Capacity.

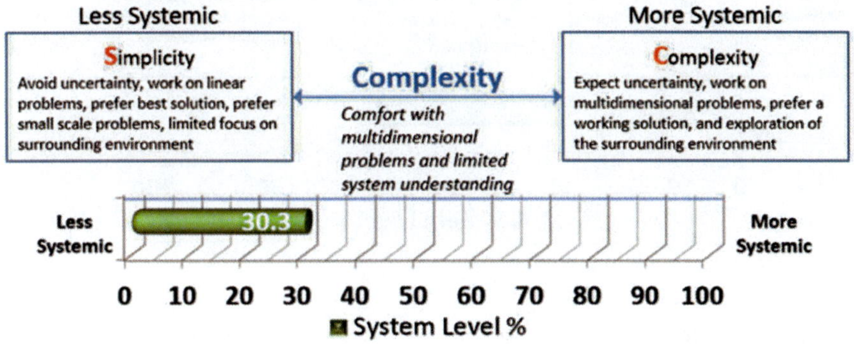

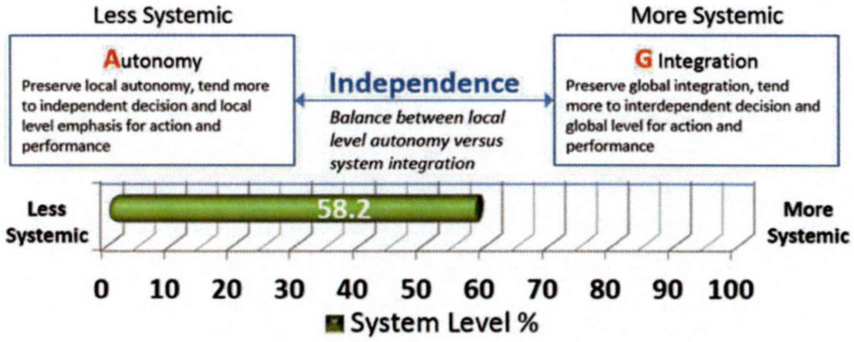

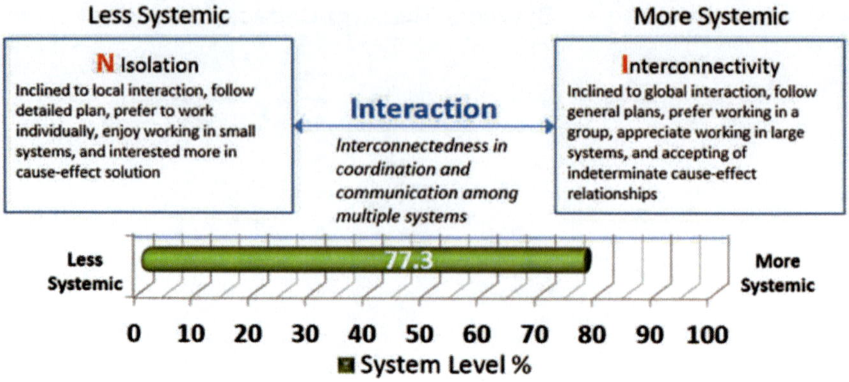
Less Systemic
N Isolation
Inclined to local interaction, follow detailed plan, prefer to work individually, enjoy working in small systems, and interested more in cause-effect solution
Interaction
Interconnectedness in coordination and communication among multiple systems
More Systemic
Interconnectivity
Inclined to global interaction, follow general plans, prefer working in a group, appreciate working in large systems, and accepting of indeterminate cause-effect relationships
Less Systemic
77.3
More Systemic
0 10 20 30 40 50 60 70 80 90 100
System Level %

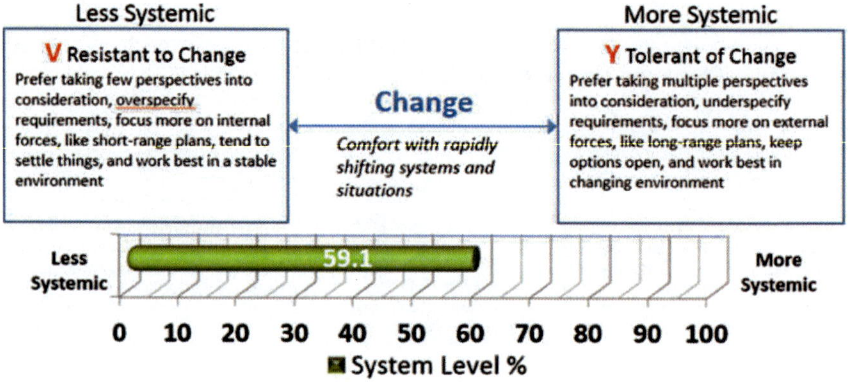
Less Systemic
V Resistant to Change
Prefer taking few perspectives into consideration, overspecify requirements, focus more on internal forces, like short-range plans, tend to settle things, and work best in a stable environment
Change
Comfort with rapidly shifting systems and situations
More Systemic
Y Tolerant of Change
Prefer taking multiple perspectives into consideration, underspecify requirements, focus more on external forces, like long-range plans, keep options open, and work best in changing environment
Less Systemic
59.1
More Systemic
0 10 20 30 40 50 60 70 80 90 100
System Level %

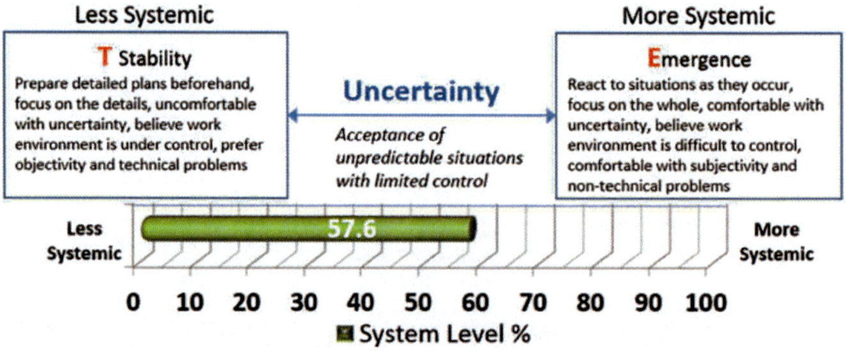
Less Systemic
T Stability
Prepare detailed plans beforehand, focus on the details, uncomfortable with uncertainty, believe work environment is under control, prefer objectivity and technical problems
Uncertainty
Acceptance of unpredictable situations with limited control
More Systemic
Emergence
React to situations as they occur, focus on the whole, comfortable with uncertainty, believe work environment is difficult to control, comfortable with subjectivity and non-technical problems
Less Systemic
57.6
More Systemic
0 10 20 30 40 50 60 70 80 90 100
System Level %

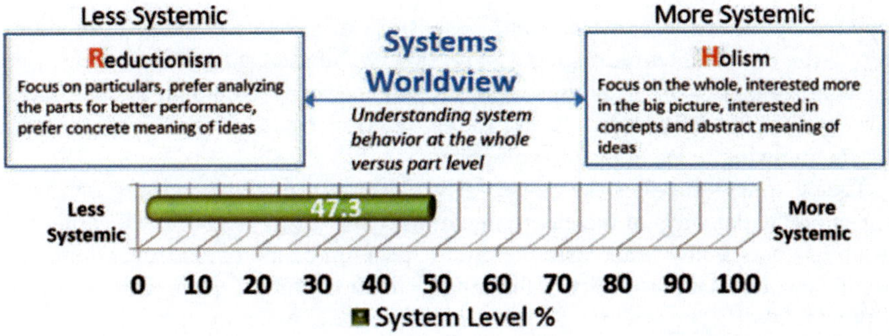

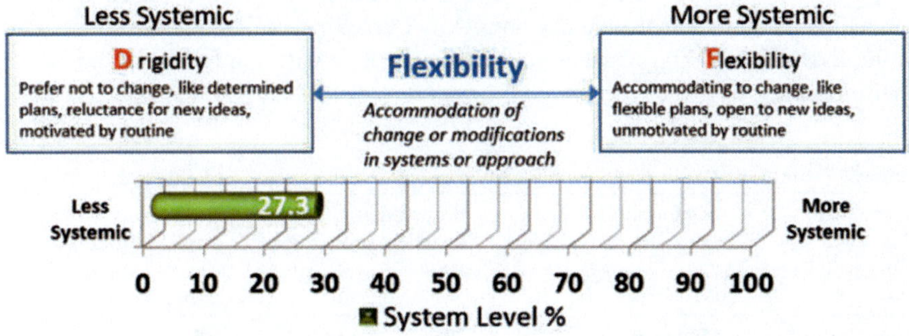

Figure 5 provides a summary overview of the results for the Systems Thinking Capacity identified for the aggregate of the group.

Environment Complexity Demand: This instrument examined the degree of perceived complexity that exists in the environment of the enterprise. This was

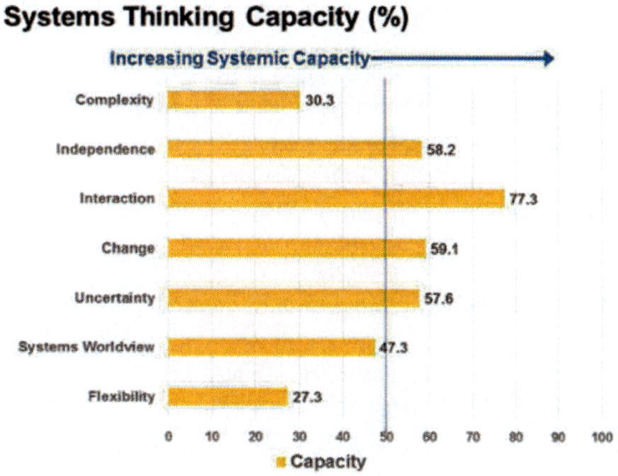

Fig. 5 Systems thinking capacity summary (max 100%, more systemic)

captured by assessment of the seven dimensions of systems thinking in relationship to the environment through a 43-question, web-based survey instrument. The aggregate of participant responses was collected and mapped onto the seven dimensions of Systems Thinking Capacity. Table 3 and Fig. 6 summarize the results with details following in the associated diagrams.

The % spread provides the degree of variability of thinking in the group with respect to the demands of the environment (max variability is 100%, indicating the group has perspectives that span the entire spectrum of the dimension, where 0% would indicate total uniformity of thinking). The Environment Complexity Demand % provides the degree to which the group perceives the complexity in the environment (100% represents a maximum in the complexity which must be responded to by the organization). The results of this instrument are provided in Table 3 and Fig. 7.

The following diagrams break down the definitions and results for each of the seven dimensions of Environment Complexity Demand. Recall that these are parallel to the Systems Thinking Capacity seven dimensions with the focus shifted to the environment.

Table 3 Definition of environment complexity demand dimensions and variability of responses

Dimension	Definition	% spread of group members (max spread 100)	Environment complexity Demand % (max 100, more systemic)
Complexity	Range of multidisciplinary requirements and understanding	**39**	**55.8**
Independence	Balance between local-level autonomy versus system integration	**40**	**45.5**
Interaction	Interconnectedness in coordination and communication among multiple systems	**32**	**54.5**
Change	Rapidly shifting systems and situations	**40**	**54.5**
Uncertainty	*Unpredictable situations with limited control*	**55**	**47.7**
Systems Worldview	*System behavior at the whole versus part level*	**43**	**54.5**
Flexibility	Ease of change or modifications in systems or approach	**42**	**38.2**

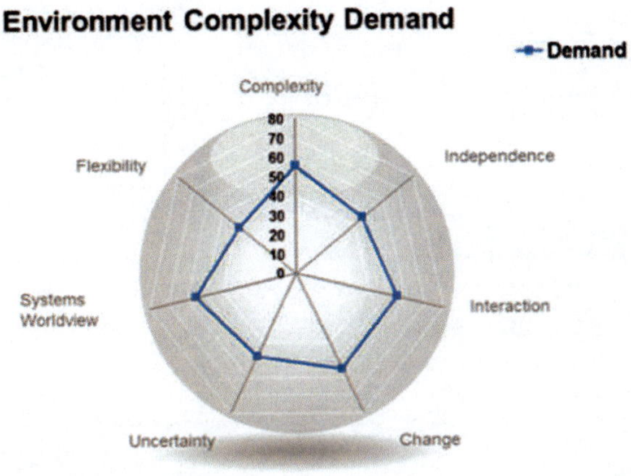

Fig. 6 Environment complexity demand

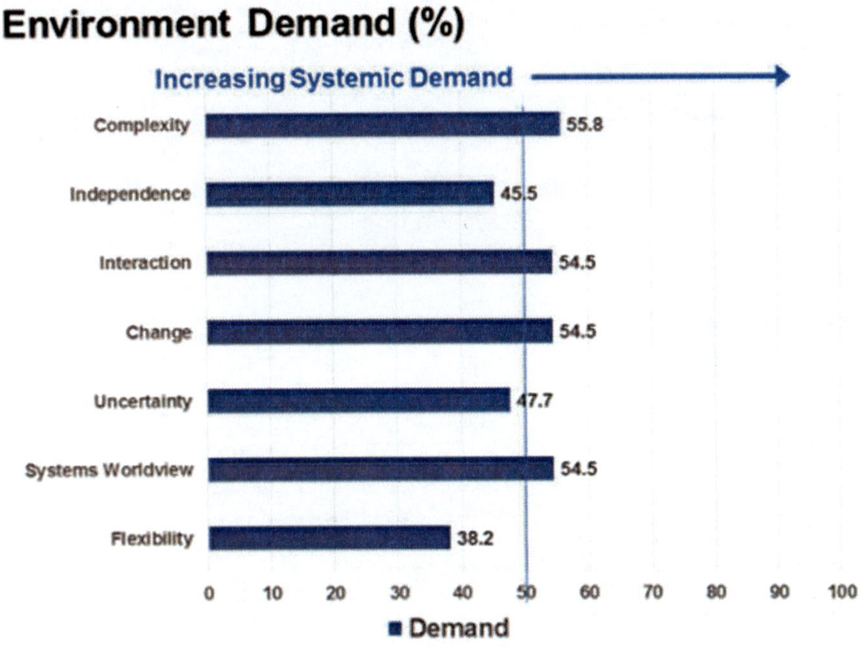

Fig. 7 Environment complexity demand summary (max 100%, more systemic)

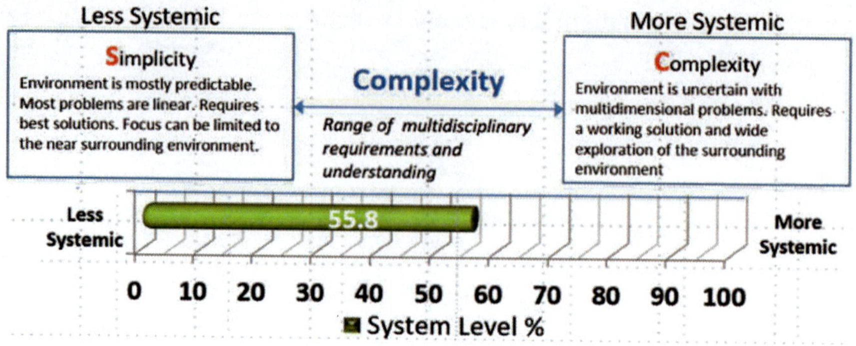
Less Systemic
Simplicity
Environment is mostly predictable. Most problems are linear. Requires best solutions. Focus can be limited to the near surrounding environment.
Complexity
Range of multidisciplinary requirements and understanding
More Systemic
Complexity
Environment is uncertain with multidimensional problems. Requires a working solution and wide exploration of the surrounding environment
Less Systemic
55.8
More Systemic
0 10 20 30 40 50 60 70 80 90 100
System Level %

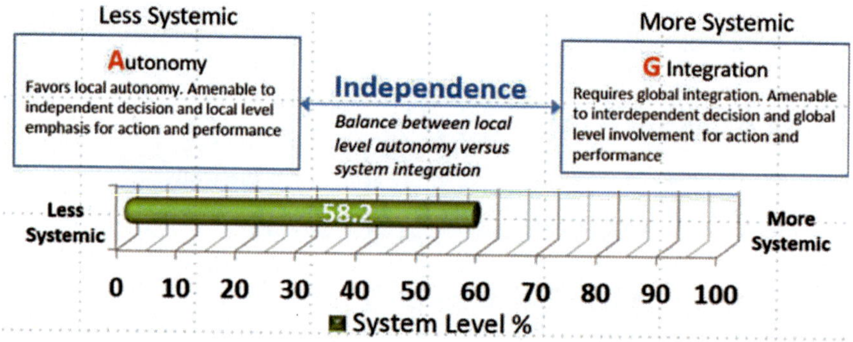
Less Systemic
Autonomy
Favors local autonomy. Amenable to independent decision and local level emphasis for action and performance
Independence
Balance between local level autonomy versus system integration
More Systemic
G Integration
Requires global integration. Amenable to interdependent decision and global level involvement for action and performance
Less Systemic
58.2
More Systemic
0 10 20 30 40 50 60 70 80 90 100
System Level %

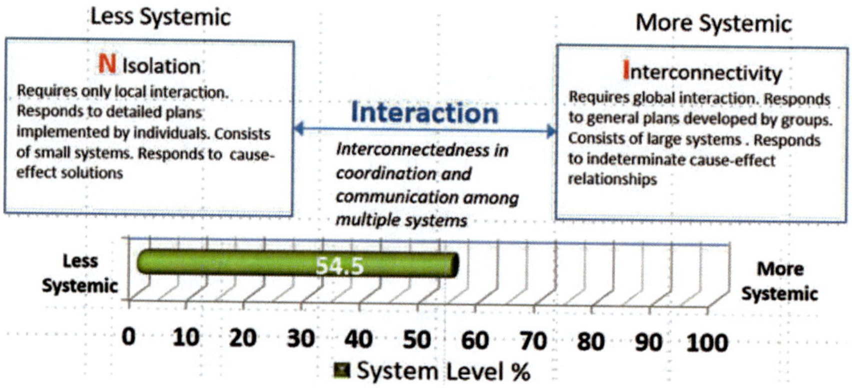
Less Systemic
N Isolation
Requires only local interaction. Responds to detailed plans implemented by individuals. Consists of small systems. Responds to cause-effect solutions
Interaction
Interconnectedness in coordination and communication among multiple systems
More Systemic
Interconnectivity
Requires global interaction. Responds to general plans developed by groups. Consists of large systems . Responds to indeterminate cause-effect relationships
Less Systemic
54.5
More Systemic
0 10 20 30 40 50 60 70 80 90 100
System Level %

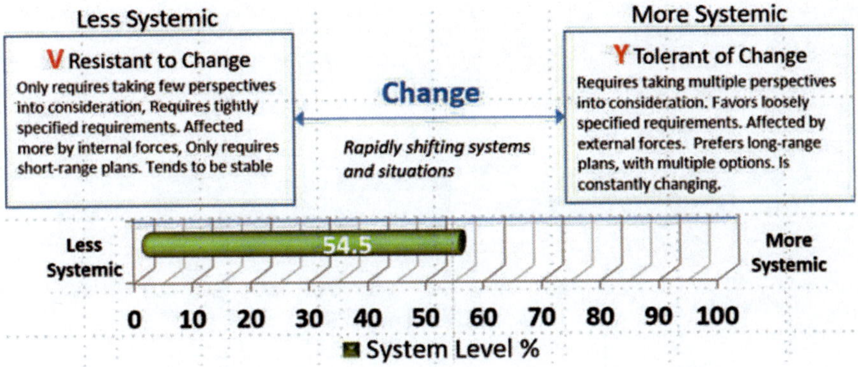
Less Systemic
V Resistant to Change
Only requires taking few perspectives into consideration, Requires tightly specified requirements. Affected more by internal forces, Only requires short-range plans. Tends to be stable
Change
Rapidly shifting systems and situations
More Systemic
Y Tolerant of Change
Requires taking multiple perspectives into consideration. Favors loosely specified requirements. Affected by external forces. Prefers long-range plans, with multiple options. Is constantly changing.
Less Systemic
54.5
More Systemic
0 10 20 30 40 50 60 70 80 90 100
System Level %

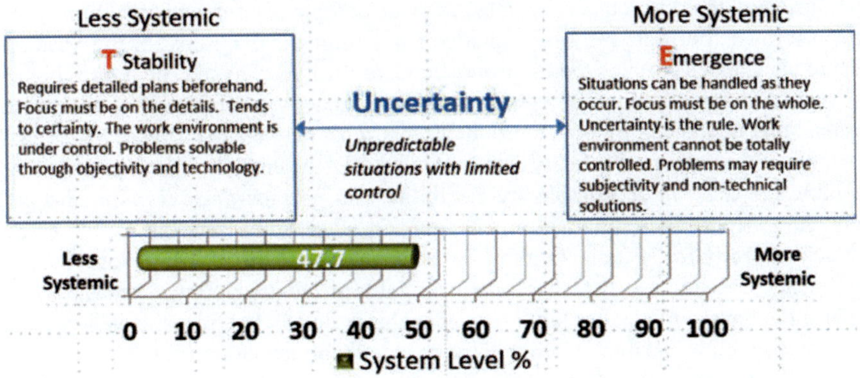
Less Systemic
T Stability
Requires detailed plans beforehand. Focus must be on the details. Tends to certainty. The work environment is under control. Problems solvable through objectivity and technology.
Uncertainty
Unpredictable situations with limited control
More Systemic
Emergence
Situations can be handled as they occur. Focus must be on the whole. Uncertainty is the rule. Work environment cannot be totally controlled. Problems may require subjectivity and non-technical solutions.
Less Systemic
47.7
More Systemic
0 10 20 30 40 50 60 70 80 90 100
System Level %

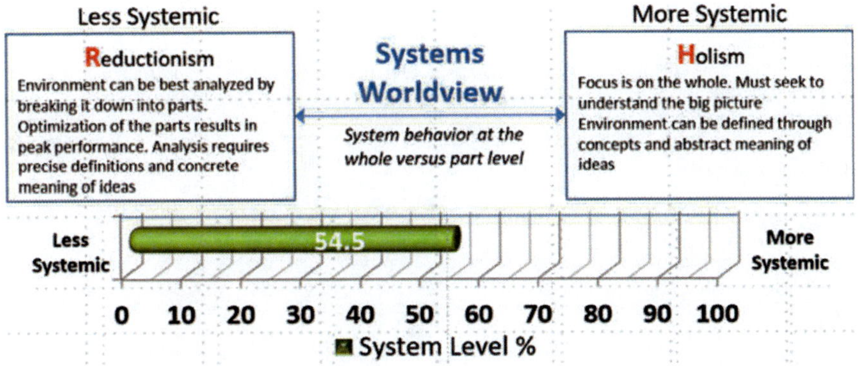
Less Systemic
Reductionism
Environment can be best analyzed by breaking it down into parts. Optimization of the parts results in peak performance. Analysis requires precise definitions and concrete meaning of ideas
Systems Worldview
System behavior at the whole versus part level
More Systemic
Holism
Focus is on the whole. Must seek to understand the big picture Environment can be defined through concepts and abstract meaning of ideas
Less Systemic
54.5
More Systemic
0 10 20 30 40 50 60 70 80 90 100
System Level %

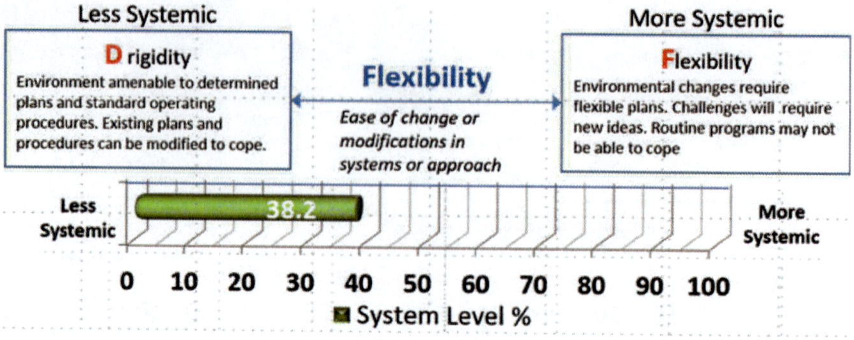

The combination of System Thinking Capacity and Environment Complexity Demand is instructive in understanding the degree to which the organization views its Systems Thinking Capacity in relation to the complexity demands of the environment it must confront. We have provided two composite diagrams for the Systems Thinking Capacity versus the Environment Complexity Demand. The figures below provide a composite view of System Thinking Capacity available in the enterprise versus that demanded by the environment that must be engaged. Figure 8 suggests that the environment complexity demands more than the enterprise's current systems thinking capacity in the complexity, flexibility, and systems worldview dimensions. For the independence, interaction, change, and uncertainty dimensions, the enterprise's capacity for systems thinking exceeds the complexity demands of the environment. This suggests that there are potential targeted development areas that are in need of having the 'gap' closed between what is demanded from the environment and that which is capable of being delivered by the organization (Fig. 8).

System Governance Check—this instrument was a web-based survey that allowed participants to examine 14 elements of system governance. The results provide a 'snapshot' of several aspects of governance. Participant responses were aggregated

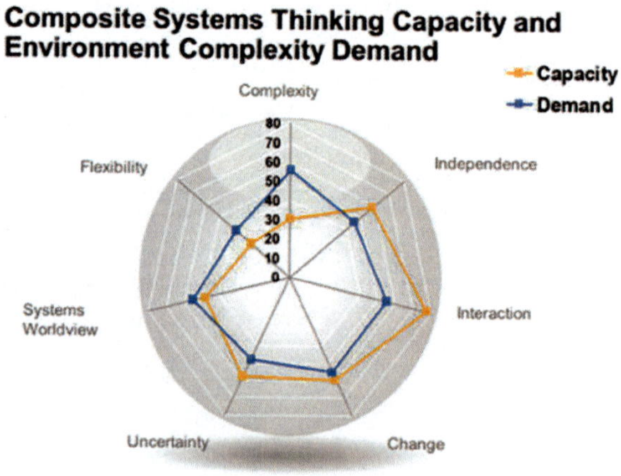

Fig. 8 Composite systems thinking capacity and environment complexity demand summary (max 100%, more systemic)

to provide for the analysis. The results of the 14-Point Governance Check instrument are provided below and detailed in the Table that follows.

Table 4 provides a summary of the raw data numerical results for the governance areas examined in the instrument (Fig. 9).

The results of the 14-Point Governance check suggest that there are developmental areas that might be engaged to enhance system governance. In addition, there was a variability in the range of responses. The governance check is precisely that a check. As such it is a 'snapshot' of indicators and cannot be taken as absolute. Instead, it is an invitation to deeper dialog concerning the state and development of system governance.

Implications of the CSG Entry Effort

The CSG Entry was successful in providing a 'snapshot' of three distinct aspects of how the organization (system) viewed itself along three dimensions related to CSG (Systems Thinking Capacity, Environment Complexity Demand, Preliminary State of Governance). While we hesitated to draw absolutes concerning the results, we offered three high-level indicators for more critical examination from the initial inspection of results:

1. There were three areas of Systems Thinking Capacity that indicated the organization was not at a level demanded by the environment being faced. These included complexity, systems worldview, and flexibility.
2. As an aggregate, the group had several areas that could be considered good focal candidates to enhance Systems Thinking Capacity. In addition, there were several areas of Systems Thinking Capacity that exceed the environment demand—with the greatest difference being observed in the interaction, independence, and uncertainty dimensions. These offered possible areas to exploit with respect to system governance design, execution, and development.
3. The results of the 14-Point Governance Check indicated that there was sufficient room for development in system governance, as indicated by the mapping of the different governance aspects (e.g., coordination). While the instrument was not offered as an absolute assessment of the state of governance for the organization (system), it provided a basis for further explorations of possibilities for system governance enhancement.

What Was Learned About CSG Entry

In sum, the *CSG Entry* effort provided an *efficient, low-risk, and value-added* set of activities to introduce CSG to a participating organization (system). The 'hands-on' effort was intended to provide a 'snapshot' of several different aspects of CSG—as examined by the enterprise system practitioners who provide governance for the organization (system). CSG provided a new and novel look into the organization (system) with a different set of lenses and frame of reference, from which different thinking, decision, action, and interpretation development possibilities accrued.

Table 4 Summary results of responses to the 14-point governance check instrument (11 participants—1 less effective, 5 more effective)

Checkpoint	Explanation	1	2	3	4	5	Range	Mean
1. We have a detailed mapping of our environment and governance system that shows how we function to produce value	This mapping is like an owner's manual for your system—it shows the precise detail, like a set of blueprints that maps the environment and shows how system governance (integration, coordination, communications, control) is achieved to produce products and services	**1**	**5**	**4**	**0**	**1**	**5**	**2.5**
2. We actively perform scanning of our environment to identify events, entities, trends, or patterns that impact present system performance and future system development	Environmental scanning is essential to continuously monitor and interpret what goes on external to the system. It identifies, processes, and responds to external events, trends, or factors that can impact system performance and influence future development directions	**1**	**5**	**4**	**1**	**0**	**4**	**2.5**
3. We are well equipped to keep up with external turbulence and speed of change that exist in our environment	If the environment is changing faster than our ability to respond, we will continually feel behind and struggle to keep up. Matching this rate of environmental change is essential to avoid crisis situations and identify implications for future development	**3**	**5**	**1**	**1**	**1**	**5**	**2.3**
4. Our system design is effective in balancing accountability with resources necessary to achieve expected levels of performance	Consistency between resources provided and accountability for expected contributions is a source of stability. Continual shifts in resources, or expectations, can create an imbalance and increase the level of uncertainty and stress in the system	**1**	**7**	**2**	**1**	**0**	**4**	**2.3**

(continued)

Table 4 (continued)

Checkpoint	Explanation	1	2	3	4	5	Range	Mean
5. We routinely communicate the right information, at the right time, and at the right place to support consistent decision and action	Information is the lubricant for effective system decisions. The exponential rise in information makes the purposeful design for communications critical. Effectiveness in communications ensures availability, accuracy, and accessibility in the flow of information to support decision and action	**1**	**5**	**3**	**2**	**0**	**4**	**2.5**
6. We share and maintain our identity such that our uniqueness is clear and we have a common reference point to support consistent decision, action, and interpretation	Sharing a strong sense of system identity (e.g., vision, mission, values, strategic orientation) supports and maintains consistency in decisions, actions, and interpretations	**1**	**2**	**5**	**3**	**0**	**4**	**2.9**
7. Our strategic system performance measures are balanced, monitored, and effectively utilized for system improvement	System measures should be limited in number and balanced between the present and future. They should also monitor performance across a spectrum of technical, social, and policy considerations. System measures should also guide actionable improvement	**1**	**4**	**5**	**1**	**0**	**4**	**2.5**
8. We effectively detect, correct, and learn from our system errors, making system adjustments to preclude recurrence	Systems should not only focus on detecting and correcting compliance errors, but also on addressing deeper underlying system structural issues. Effective systems identify deep structural issues and initiate responsive actions before they become crises	**1**	**5**	**3**	**2**	**0**	**4**	**2.5**

(continued)

While the initial effort for CSG Entry met the original intent, there were several areas that were rethought and adjusted for future CSG Entry. For conciseness, we have provided these lessons in Table 5 for each phase of CSG Entry.

Table 4 (continued)

Checkpoint	Explanation	1	2	3	4	5	Range	Mean
9. Our system design provides the greatest possible degree of flexibility for making local decisions and taking action in response to their circumstances	Over constraining/regulating a system wastes resources and steals initiative from those entities closest to and in the best position to directly address the source of errors plaguing a system. Close proximity to a problem reduces the error propagation and time between source and response	**1**	**4**	**5**	**1**	**0**	**4**	**2.5**
10. We actively pursue rigorous 'self-study' of our system design and execution in pursuit of purposeful system development	Active system 'self-study' is essential to higher-level system development and requires a commitment of time, energy, and resources to follow a rigorous development plan. Potential for system development is dampened without continual examination and questioning	**2**	**4**	**4**	**1**	**0**	**4**	**2.4**
11. There is an appropriate balance between short and long-term focus for our system that continues to evolve with our changing circumstances	Overemphasis on the short term can sacrifice long-term prospects. Overemphasis on the long term can diminish short-term performance. Balance is necessary, can shift over time, and should be a source of continual examination for system development	**1**	**2**	**5**	**3**	**0**	**4**	**2.9**
12. We provide effective coordination between entities in our system such that unnecessary variability is eliminated	Effective interaction between system entities is necessary to reduce conflicts and issues stemming from poorly designed or executed coordination. Coordination helps to ensure that scarce resources are not wasted due to ineffective interrelationships between system entities	**1**	**7**	**3**	**0**	**0**	**3**	**2.2**

(continued)

Table 4 (continued)

Checkpoint	Explanation	1	2	3	4	5	Range	Mean
13. Our system design and execution effectively eliminate operation in 'crisis' or 'reactive' modes	When crisis operation is too frequent, it can indicate potential issues in system design, execution, or both. Constantly being in reaction mode creates unsustainable stresses in the system and individuals who must compensate for this mode of operation	**1**	**5**	**5**	**0**	**0**	**3**	**2.4**
14. We effectively design and account for the wide range of influences on our system (technical, human, social, organizational, managerial, political, policy, cultural, and stakeholder)	System design and execution are dependent on both 'hard' (technical) and 'soft' (sociotechnical) influences. Limited or complete absence of consideration of the spectrum of hard and soft influences risks achievement and maintenance of higher-level system performance	**1**	**3**	**5**	**2**	**0**	**4**	**2.7**

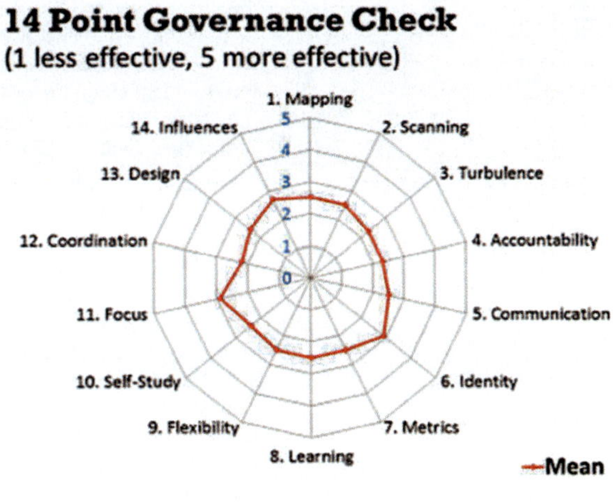

Fig. 9 Mapping of the 14-point governance check (1 less effective, 5 more effective)

The CSG Entry proved to be an effective introduction to CSG and identified several areas for refinement of CSG Entry. In addition, as a result of the initial application, the implications of systemic intervention for CSG have been revised.

Table 5 Lessons and adjustment to CSG Entry

CSG Entry Phase	Lessons and implications
Phase 1: Invitation	• Initial briefing of sponsors should include more detailed explanation as to what might accrue from the effort. In particular, value offered from the entry effort • Potential utility expectations and paths forward that might be anticipated for the effort need to be emphasized • Relationship establishment to past, present, and future system development activities, concerns, and priorities is an essential conversation to position CSG Entry, distinguish CSG from other approaches, and identity potential further CSG development in relationship to the organization
Phase 2: Overview Briefing	• Greater detail as to the 'fit' of CSG to the 'ongoing' governance activities and initiatives and distinctions of CSG development • Additional details, with simplified explanations, concerning CSG in the overview briefing were identified as an area for future focus • Greater clarity on the utility expectations and what might lie beyond the CSG Entry instruments application as value-adding potential
Phase 3: Instrument Application	• The 14-Point Governance Check, while efficiently executed, was not effectively linked to performance of the nine governance functions—this resulted in a revamping of the governance check (a 45-question web-based instrument directly aligned to CSG functions) for future applications • Establishing the range of variance within the group of participants for Systems Thinking Capacity and Environment Complexity Demand was identified as an important delineation in addition to the aggregate mean scores (was included in the final outbriefing and report)
Phase 4: Outbrief Results	• The technical results were provided concurrent with the outbriefing, with the intent not to have the group 'misconstrue' the results without guidance. In hindsight, prior distribution of the technical results might have sharpened the focus of the discussion • Some rudimentary preparation materials (e.g., short papers, video) to provide greater context for the exploration would have been beneficial • The depth of exploration necessary to properly explore the results for implications lends itself to more of an extended workshop endeavor, rather than a limited technical outbriefing • Closure to the CSG Entry effort would benefit from the examination of potential for further development, based on results and their implications as well as fit to current developmental priorities. As conducted, potential paths forward were not effectively presented or explored

Initial Systemic Intervention Beyond CSG Entry

The System Map

The objective is to map the components or subsystems of the enterprise governance system to each of the nine metasystem functions that they are, or should be, performing. The results of the mapping will probably bear little resemblance to the traditional organizational chart. Articulating the enterprise governance system architecture in terms of CSG gives the enterprise system practitioners a new perspective on their system governance and their enterprise system in general. A graphical representation utilizing images and language commonly used within the enterprise system produces enhanced understanding.

The initial system map will, in all probability, be revised over the course of the intervention as the enterprise system practitioners learn about their system governance and what changes are required to enhance its performance. It may be advantageous to generate several different representations of various types emphasizing different perspectives. However, the representations must relate to the CSG reference model to ensure continuity as the intervention progresses.

Like other forms of enterprise records and drawings, the final versions of the various representations are useful to the enterprise system practitioners beyond the intervention as they adapt the enterprise governance system to changes in the enterprise's context and environment.

Metasystem Pathologies Assessment (M-Path) method

Another investigation that is part of the initial systemic intervention suite is an assessment of the various pathologies (weaknesses) that exist within the enterprise governance system. What follows is an introduction to the M-Path method as developed in a previous chapter. It includes a method with repeatable procedures to support identification of pathologies in a system enterprise. This method extends previous research related to problem formulation [5–8]. For this discussion, a brief recap of CSG concepts is included to ensure understanding. Also, a specific set of pathologies is provided for illustrative purposes. The set of pathologies is drawn from the earlier work of [12] and supplemented by recent research into the emerging field of Complex System Governance [2, 13, 15]. Complex System Governance (CSG) is an emerging field, representing an approach to improve performance through purposeful 'Design, execution, and evolution of the metasystem functions necessary to provide control, communication, coordination, and integration of a complex system' [13]. CSG was developed at the National Centers for System of Systems Engineering and is anchored in the fields of Systems Theory and Management Cybernetics. The CSG reference model was developed to provide a detailed account of 'an organizing construct for the interrelated [nine metasystem] functions necessary to perform CSG' [14]. Table 6 elaborates on the nine interrelated metasystem functions essential to CSG and acting to enable system viability. These functions provide a 'backdrop' against which the pathologies are derived (Katina and Keating 2016). Following the

Table 6 Metasystem functions in the CSG reference model

Metasystem function	Primary role of the function
Metasystem five (M5): Policy and identity	To provide direction, oversight, accountability, and evolution of the system. Focus includes policy, mission, vision, strategic direction, performance, and accountability for the system such that: (1) the system maintains viability, (2) identity is preserved, and (3) the system is effectively projected both internally and externally
Metasystem Five Star (M5*): System context	To monitor the system context (i.e., the circumstances, factors, conditions, or patterns that enable and constrain the system)
Metasystem Five Prime (M5'): Strategic system monitoring	To monitor measures for strategic system performance and identify variance requiring metasystem-level response. Particular emphasis is on variability that may impact future system viability. Maintains system context
Metasystem Four (M4): System development	To provide for the analysis and interpretation of the implications and potential impacts of trends, patterns, and precipitating events in the environment. Develops future scenarios, design alternatives, and future focused planning to position the system for future viability
Metasystem Four Star (M4*): Learning and transformation	To provide for identification and analysis of metasystem design errors (second order learning) and suggest design modifications and transformation planning for the system
Metasystem Four Prime (M4'): Environmental scanning	To provide the design and execution of scanning for the system environment. Focus is on patterns, trends, threats, events, and opportunities for the system.
Metasystem Three (M3): System operations	To maintain operational performance control through the implementation of policy, resource allocation, and design for accountability
Metasystem Three Star (M3*): Operational performance	To monitor measures for operational performance and identify variance in system performance requiring system-level response. Particular emphasis is on variability and performance trends that may impact system viability
Metasystem Two (M2): Information and communications	To enable system stability by designing and implementing architecture for information flow, coordination, transduction, and communications within and between the metasystem, the environment, and the systems being governed

development of the CSG formulation, subsequent research [16] has resulted in development of a three-stage methodology (i.e., initialization, readiness level assessment, and governance development) for implementation to provide structured identification, assessment, and development of CSG. This development methodology relies on effective formulation of the problem domain at the 'front end' of the effort. As part of this formulation, the identification, assessment, and strategizing with respect to pathologies are fundamental.

The focus of this 'front end' initialization stage of the CSG methodology involves establishing the present state of the governance of the complex (enterprise) system through framing and context articulation. Framing involves establishing the nature and structure of the enterprise governance system. The articulation of system context involves identification of circumstances, factors, patterns, or trends that constrain/enable the system [16]. Following [12], an expanded set of pathologies (53 in total) corresponding to the nine metasystem functions for CSG are proposed in Table 7. In effect, these pathologies provide a potential set for purposeful exploration of their existence in any system of interest.

The following conclusions are drawn regarding the set of pathologies identified in Table 7 and their essential role in problem formulation and the initialization stage of the CSG methodology. First, these pathologies are not unique to any one enterprise system. They certainly could be present or absent to some degree for any given system. Instead, this set represents aberrant conditions affecting metasystem functions of complex systems in a generalized form. Therefore, the 53 pathologies in Table 3 are circumstances, conditions, factors, or patterns that can act to limit system performance, or lessen system viability, such that the likelihood of a system achieving performance expectations is reduced. However, the particular form of manifestation of the pathologies will be specific to a particular system. Second, these pathologies do not exist in a binary fashion of 'present' or 'not present.' Rather, it is best to recognize that they may exhibit themselves in 'degrees of existence.' Third, pathologies have real consequences for performance of a given system/organization which can be measured in terms of a 'range of possible effects.' While the range of effects can vary in particular systems, there are always consequences for a given pathology. Fourth, in accordance with previous research, these pathologies should be a subject of exploration during problem formulation, since bringing change to the enterprise governance system is largely dependent on understanding the current state of the system [3, 6, 19]. It is from this perspective that present research articulates the metasystem pathologies (M-Path) method for use in the identification and assessment of the conditions (listed in Table 7) that negatively impact system performance.

Phases of the M-Path Method

The proposed method consists of five phases (identification, analysis, exploration, systemic implementation, and follow-up) as shown in Fig. 10. A detailed account of the five phases is the basis for the remainder of this paper.

Table 7 Metasystem functions and corresponding CSG metasystem pathologies

Metasystem function	Corresponding set of pathologies
Metasystem five (M5): Policy and identity	M5.1. Identity of the system is ambiguous and does not effectively generate consistent system decision, action, and interpretation
	M5.2. System vision, purpose, mission, or values remain unarticulated, or articulated but not embedded in the execution of the system
	M5.3. Balance between short-term operational focus and long-term strategic focus is unexplored
	M5.4. Strategic focus lacks sufficient clarity to direct consistent system development
	M5.5. System identity is not routinely assessed, maintained, or questioned for continuing ability to guide consistency in system decision and action
	M5.6. External system projection is not effectively performed
Metasystem Five Star (M5*): System context	M5*0.1. Incompatible metasystem context constraining system performance
	M5*0.2. Lack of articulation and representation of metasystem context
	M5*0.3. Lack of consideration of context in metasystem decisions and actions
Metasystem Five Prime (M5'): Strategic system monitoring	M5'0.1. Lack of strategic system monitoring
	M5'0.2. Inadequate processing of strategic monitoring results
	M5'0.3. Lack of strategic system performance indicators
Metasystem Four (M4): System development	M4.1. Lack of forums to foster system development and transformation
	M4.2. Inadequate interpretation and processing of results of environmental scanning—non-existent, sporadic, and limited
	M4.3. Ineffective processing and dissemination of environmental scanning results
	M4.4. Long-range strategic development is sacrificed for management of day-to-day operations—limited time devoted to strategic analysis
	M4.5. Strategic planning/thinking focuses on operational-level planning and improvement
Metasystem Four Star (M4*): Learning and transformation	M4*0.1. Limited learning achieved related to environmental shifts

(continued)

Table 7 (continued)

Metasystem function	Corresponding set of pathologies
	M4*0.2. Integrated strategic transformation not conducted, limited, or ineffective
	M4*0.3. Lack of design for system learning—informal, non-existent, or ineffective
	M4*0.4. Absence of system representative models—present and future
Metasystem Four Prime (M4'): Environmental scanning	M4'0.1. Lack of effective scanning mechanisms
	M4'0.2. Inappropriate targeting/undirected environmental scanning
	M4'0.3. Scanning frequency not appropriate for rate of environmental shifts
	M4'0.4. System lacks enough control over the variety generated by the environment
	M4'0.5. Lack of current model of system environment
Metasystem Three (M3): System operations	M3.1. Imbalance between autonomy of productive elements and integration of the whole system
	M3.2. Shifts in resources without corresponding shifts in accountability/shifts in accountability without corresponding shifts in resources
	M3.3. Mismatch between resource and productivity expectations
	M3.4. Lack of clarity for responsibility, expectations, and accountability for performance
	M3.5. Operational planning frequently preempted by emergent crises
	M3.6. Inappropriate balance between short-term operational versus long-term strategic focus
	M3.7. Lack of clarity of operational direction for productive entities (i.e., subsystems)
	M3.8. Difficulty in managing integration of system productive entities (i.e., subsystems)
	M3.9. Slow to anticipate, identify, and respond to environmental shifts
Metasystem Three Star (M3*): Operational performance	M3*0.1. Limited accessibility to data necessary to monitor performance
	M3*0.2. System-level operational performance indicators are absent, limited, or ineffective

(continued)

Table 7 (continued)

Metasystem function	Corresponding set of pathologies
	M3*0.3. Absence of monitoring for system and subsystem-level performance
	M3*0.4. Lack of analysis for performance variability or emergent deviations from expected performance levels—the meaning of deviations
	M3*0.5. Performance auditing is non-existent, limited in nature, or restricted mainly to troubleshooting emergent issues
	M3*0.6. Periodic examination of system performance largely unorganized and informal in nature
	M3*0.7. Limited system learning based on performance assessments
Metasystem Two (M2): Information and communications	M2.1. Unresolved coordination issues within the system
	M2.2. Excess redundancies in the system resulting in inconsistency and inefficient utilization of resources—including information
	M2.3. System integration issues stemming from excessive entity isolation or fragmentation
	M2.4. System conflict stemming from unilateral decisions and actions
	M2.5. Excessive level of emergent crises—associated with information transmission, communication, and coordination within the system
	M2.6. Weak or ineffective communications systems among system entities (i.e., subsystems)
	M2.7. Lack of standardized methods (i.e., procedures, tools, and techniques) for routine system-level activities
	M2.8. Overutilization of standardized methods (i.e., procedures, tools, and techniques) where they should be customized
	M2.9. Overly ad hoc system coordination versus purposeful design
	M2.10. Difficulty in accomplishing cross-system functions requiring integration or standardization
	M2.11. Introduction of uncoordinated system changes resulting in excessive oscillation

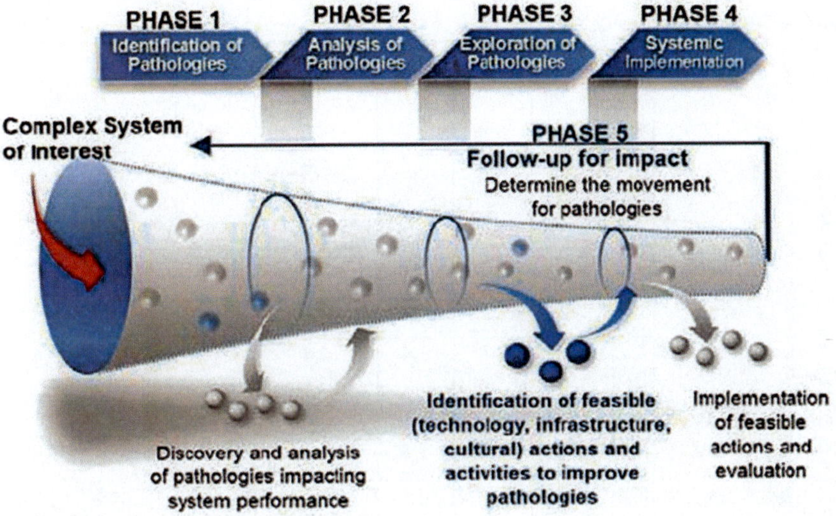

Fig. 10 Five phases of the M-Path method

Phase I: Identification

This phase involves the identification and discovery of the degree to which the 53 pathologies exist for a given situation for a system/organization. This phase produces two essential pieces of information: ***degree of existence*** and the corresponding ***impact*** of each pathology. The degree of existence is the level to which a pathology is deemed to be present—ranging from negligible to extreme and corresponding to numerical values of 1 to 7. Similarly, the measure of impact of a given pathology ranges from 1 to 7 (1 being negligible and 7 being extreme). There are a variety of tools that an analyst can use, including data mining and surveys, to ascertain information regarding the presence of pathologies in a system of interest as well as their impact on system operations. Previous research has used a web-based instrument (e.g., see Katina, 2015b). The associated pathology analysis involves an ordinal process of 'binning' pathologies based on levels of existence and potential impact. Table 8 presents a pathology matrix based on the two levels. The scale for existence is along the horizontal axis. Impact is along the vertical axis of the matrix.

The following caveats apply to Table 8.

- Each pathology must be evaluated for existence and impact on a given system. This produces a total of 53 tables (one for each potential pathology)
- The top-right most cells of the table provide higher numbers (e.g., {7,7})—these correspond to issues that are considered to be of most pressing (highest level of existence and impact)

Table 8 Pathology ordinal matrix

Range of impact associated with pathology, P_C							
Extreme	1,7	2,7	3,7	4,7	5,7	6,7	7,7
Very High	1,6	2,6	3,6	4,6	5,6	6,6	7,6
High	1,5	2,5	3,5	4,5	5,5	6,5	7,5
Moderate	1,4	2,4	3,4	4,4	5,4	6,4	7,4
Low	1,3	2,3	3,3	4,3	5,3	6,3	7,3
Very Low	1,2	2,2	3,2	4,2	5,2	6,2	7,2
Negligible	1,1	2,1	3,1	4,1	5,1	6,1	7,1
	Negligible	Very Low	Low	Moderate	High	Very High	Extreme

The degree to which pathology exists in a given system, P_E

- A decision-maker should not ignore pathologies in the top-left and bottom-right cells (e.g., {1,7} and {7,1}) since the very presence of a pathology suggests that system performance is at stake

Phase II: Analysis

The first phase only indicates presence and impact of the 53 metasystem pathologies. The second phase, analysis, involves an examination of nature and implications of the unique 'landscape' of pathologies for the system of interest. Driven by the kind of tools used in data collection of phase I, the analyst collects and synthesizes the data into meaningful information concerning pathologies. This phase provides an initial portrait, in the form of a landscape, of pathologies for the system. This landscape is unique to each system of interest and articulates the degree to which pathologies exist and affect the system.

The following caveats apply to this phase:

- Analysis in this phase includes an enumeration of metasystem pathologies using measures of existence and impact.

- Provides an indication of variability in measures of the degree of pathology existence and impact as suggested by participants. Variability is expected since each participant will not provide identical measures for the entire set of pathologies. Such variability provides insights that might be further examined in Phase III.

Phase III: Exploration

The results of phase II are made available to system participants to provide a guided investigation into the meaning of the identified pathologies as well as their implications for system development. This phase involves a two-way dialog between system participants and the analyst and involves the ***general meaning*** of pathologies and exploration of the ***meaning in context*** for the system of interest. This dialog is instrumental for articulating and/or voicing system of interest development ***implications*** in response to the discovered pathologies. It is during this phase that the existing initiatives (development activities already underway in the organization) are mapped against discovered pathologies. This mapping enables discovery of strengths and weaknesses in system development in relationship to the existing pathologies. The results of this phase include a ***prioritized enumeration*** of pathologies based on ***feasibility***—organizational ability to successfully address pathologies with a reasonable chance of success. The result is a set of ***strategies*** and corresponding ***actions*** designed to impact the identified pathologies.

Phase IV: Systemic Implementation

The purpose of this phase is to ensure that selected responsive strategies are effectively deployed. Activities in this phase are based on what is decided in the previous phase. For example, an activity such as the '*development of effective environmental scanning mechanisms*' could be identified in the previous phase due to existence of metasystem pathology M4*0.1 '*a lack of effective scanning mechanisms*' as identified in Table 7. Identifying this as an issue starts in Phase I. This issue becomes more explicit in Phase II. In Phase III, there is a follow-up to develop new initiatives to address '*a lack of effective scanning mechanisms.*' This is in conjunction with understanding ongoing initiatives, including effectiveness of the existing scanning mechanisms. Once there is agreement on the need to develop effective scanning mechanisms, a strategy to develop such mechanisms must be put in place in Phase IV. This phase is necessary to ensure that something is done in relation to a pathology. A comparative medical analogy is being prescribed medication for an illness and failing to take the medication. In such a case, an identified pathology will not 'disappear' and might even worsen if left without being addressed. In addition, this phase sets a time line for future incremental system evaluation to determine the shifting state of pathologies in response to strategies.

Phase V: Follow-up

This 'final' phase is focused on an examination of the effects of strategic actions undertaken to address pathologies. An established time line can serve as a placeholder for a re-evaluation of the system by fulfilling two primary purposes. First is to measure the effects of the strategies/actions as implemented in Phase IV, and second is the identification of new pathologies. Such efforts serve the role of continuous system development. Continuous system development is essential since an organization in question is operating within a dynamic and most likely turbulent environment. Moreover, the deployed strategies might lose effectiveness over time, new pathologies might emerge, and new technologies might shift the landscape of pathologies. Therefore, navigating through the M-Path method is truly a continuous process with each phase complementing and interrelated to previous phases.

M-Path Method implications

Applying the M-Path method to a system of interest serves to identify, analyze, explore implications, and generate a response to the systemic deficiencies (pathologies) impacting system performance. This method is consistent with [19] supposition that an analyst ought to be in a position to 'identify the problem to be studied and define its scope in such a way that he has some hope of finding an acceptable and implementable solution with the economic, political, technological, and other constraints that exist, including limitations imposed by the policy makers' span of control and the time available for decision' [19], p. 23). The value associated with the proposed M-Path method is summarized as follows:

- Identification and representation of pathologies in a given system of interest,
- Exploration of the nature of pathologies and their implications for improving system performance/viability,
- Determination of feasible actions and initiatives to impact pathologies,
- Purposefully evolving a system based on continuous assessment of development.

The proposed M-Path method echoes Dery [3] in that it does not simply offer a descriptive definition of a situation. The M-Path method does not merely describe pathologies in a situation but also helps in selection of 'certain aspects of reality as being relevant for action in order achieve certain goals' (Dery [3], p. 35). Although the developed method is a guide through problem formulation, it is also focused on generating important subsequent courses of action that are dependent on the results of execution of the M-Path method.

In summary, the M-Path method is a well-developed and rigorous procedure for exploring possible weaknesses in the enterprise governance system. The results from M-Path, presented in various ways including several visual forms currently under development, starkly illuminate weaknesses in the enterprise governance system that impede enterprise system performance.

Exploring Potential for Integration of FMECA into CSG

There is much to be gained from the development and tailoring of FMEA/FMECA to improve capabilities in the developing CSG field. For application of FMEA/FMECA to CSG, we suggest an approach outlined in Fig. 11. The essence of this approach is to move through five primary phases, including: (1) identification of existing and potential CSG failure modes, (2) exploration of contributing factors to the failure mode, (3) attribution of the consequences stemming from the failure modes, (4)

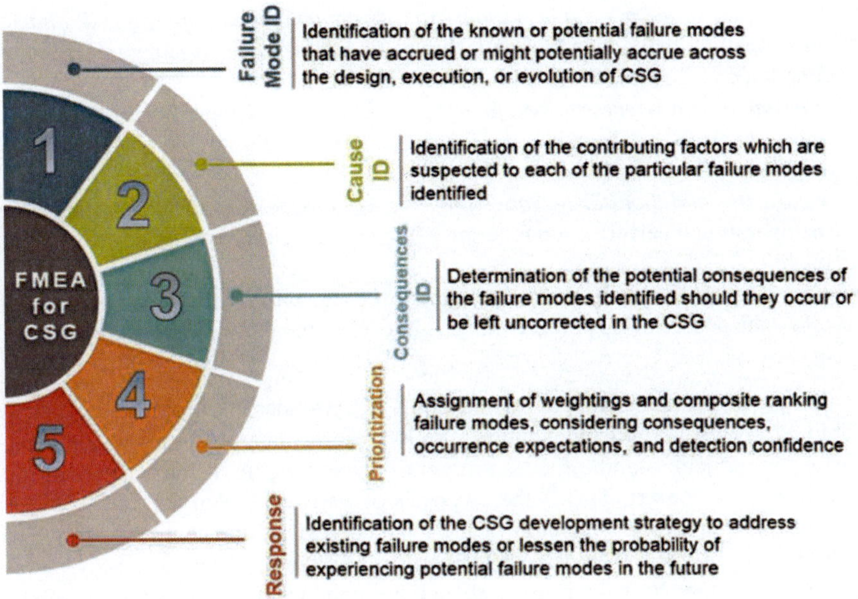

Fig. 11 Five-phased FMECA approach for CSG

prioritization of the failure modes, and (5) response for CSG modification based on results.

To provide this examination, we begin by specifying ten potential failure modes in CSG that are representative of failure across the functions we earlier defined for CSG. These failure modes include (Fig. 11):

- *Information flow does not support consistent decision and action*—this failure mode would be experienced in the Information and Communication function of CSG.
- *Lack of coordination among entities produces uncertainty and incongruence*—this potential failure mode would emanate from the Information and Communication function of CSG.
- *Stable planning and execution surrender to* ad hoc *responses*—this failure mode would be associated with the system development function.
- *Process for examination of performance variance and crises is inconsistent*—the source for this failure mode would be located in the operational performance function of CSG.
- *Future system development is sacrificed for near term operational demands*—this failure mode is more than likely associated with the system development function in CSG.
- *Resolution of issues frequently results in temporary or piecemeal relief*—this failure more is associated with the learning and transformation function of CSG.

- *Monitoring, assessment, and response to environmental shifts are sporadic and* ad hoc—this failure mode is a potential emanation from the environmental scanning function of CSG.
- *Internal system circumstances, factors, and conditions impede performance*—to understand this failure mode, the system context function of the CSG would be the likely source.
- *Measuring and monitoring long-range strategic development lack emphasis*—the operational performance function of CSG is the likely source for this failure mode.
- *Lack of clear focus creates internal inconsistency and external misunderstandings*—the source of this failure mode is likely in the policy and identify function of the CSG.

The set of ten potential failure modes in CSG provides a set to delineate applicability of FMECA for demonstration purposes. However, each system would experience the particular failure modes differently depending on the uniqueness of the system and its context. This is the emphasis of *Phase 1 Failure Mode ID. Phase 2 Cause ID* requires examination of the local context for the contributions to the failure mode in CSG. Likewise, the attribution of *Phase 3 Consequences* requires assessment of the particular impacts the failure mode would produce and should be experienced in the system. Similarly, *Phase 4 Prioritization* represents classification of the highest impact failure modes for the specific system. Finally, *Phase 5 Response* is focused on developing appropriate responses tailored to what is feasible (technologically, contextually, resources, legally, safety) given the particular CSG and the context within which it exists. The result of the FMECA application is concentrated on bringing a higher degree of rigor, critical examination, and assessment to the identification and exploration of potential/existing failures for CSG. The ultimate result would be to make design modifications based on the analysis.

The intersection of CSG and FMECA offers three important contributions to the emerging CSG field. First, FMECA is a disciplined and proven approach to identification of potential failure modes in a system. Since CSG is a systems-based articulation of governance, the rigor imposed in the 'failure modes' thinking of FMECA can prove insightful. Second, FMECA forces the analysis to assign a prioritization to the different failure modes identified. There is not an assumption that 'all' failure modes are congruent in their importance. This can assist in the allocation and targeting of scarce resources to the areas of greatest impact for CSG improvement. Third, the FMECA is ultimately about making improvements in the system, be they at concept/design, (manufacturing) execution, processes, or service provision. Therefore, even with a 'system' as unwieldy as governance, FMECA induces a disciplined consideration across a spectrum of design, execution, and production aspects of the CSG system.

However, there are several challenges that loom for the further development and modification of FMECA for use in the CSG field. Three primary challenges include (1) developing sufficient detail in the identification of governance failure modes such that subsequent analysis can be conducted, (2) the complexity of CSG is such that the interrelationship between failure modes may prove to be an important consideration, requiring an additional element of analysis for the assessment, and (3) the rigorous assignment of the prioritization is essential, while it is doubtful that for governance the Risk Prioritization Number (RPN) as utilized in traditional FMECA could be replicated. Notwithstanding limitations in incorporating FMECA into CSG, there is an opportunity to provide a rigorous approach to establishing prioritization of failure modes for FMECA application in CSG.

There is much left to develop for FMECA application to CSG. However, there are significant contributions that FMECA can provide to help advance the CSG field and to assist practitioners in providing a method to identify and address existing and potential CSG failure modes, such as the representative ten CSG failure modes introduced. Although there is still much to be developed in the application of FMECA to CSG, there is also great promise in extrapolating a proven method into a field in search of more rigorous formulation of methods, tools, and techniques. While beyond the scope of this chapter, future direction for FMECA for CSG will involve a case application to establish the application in a field setting. This emphasis will demonstrate the ability of FMECA to provide a more rigorous analysis of CSG failure across the spectrum of both 'hard' and 'soft' failure modes. This significantly extends the traditionally 'technical' failure orientation of FMECA.

Conclusion: Systemic Intervention Future Development Directions

As an emerging field, there is much that remains unknown about CSG, particularly with respect to systemic intervention to improve CSG. Much of the unknown for CSG stems from the unique demands for intervention in complex systems. CSG is somewhat unique in relationship to other systems-based approaches in three primary ways. First, CSG makes an *explicit* mapping to systems theory [22] as a grounding basis for the field. This is not to suggest that other systems-based approaches are not 'born' out of an underlying systems theory base. However, CSG is explicit in the delineation of the systems theory conceptual basis. Second, engagement in CSG is constrained by the degree of Systems Thinking Capacity of the participating group and the state of system governance that currently exists for the system in focus. Therefore, the directions and engagement will be driven by the individuals and system 'fitness' to participate across a range of CSG development activities. This range of fitness determines the nature, depth, and expectations for the level of CSG system improvement activities that might be effectively engaged. Third, CSG is not equivalent to introduction of a new program or initiative (e.g., lean six sigma, TQM, balanced scorecard, CRM, etc.) that will be engaged 'in addition to' what is already being done by the individuals/organization. Instead, all viable (continuing to exist) systems are already performing the nine CSG metasystem functions, irrespective of intervention. Whether or not these functions are purposefully explored for development, they are, and will continue to be, performed if the system continues to exist. Thus, CSG is not a temporary endeavor that exists beyond the normal scope of system activities/initiatives being engaged by the organization (system).

With respect to systemic intervention, several implications have been identified from experiences with the initial deployment of CSG Entry. These implications include the following:

1. *CSG is not the Entry Point:* As promising as CSG might be for advancing system understanding and performance, it is not the highest priority for those who might be considering engagement. Instead, the priority for enterprise system practitioners is focused on 'their problems.' Thus, first understanding their problems and then drawing the linkage to potential CSG value contributions are essential. Making this connection is critical to draw attention to the possibilities that CSG might bring related to their most vexing issues.
2. *CSG Engagement is not a Binary (all or nothing) Proposition:* Following CSG Entry and the implications that might be suggested from the results, there are many developmental paths that might be pursued. It is incorrect to have CSG postured as an all or nothing alternative. Instead, there are a spectrum of activities (training, development, modeling, etc.) and levels (practitioner, system, enterprise, problem) that might be pursued in the development path to enhance CSG.
3. *CSG is not an 'In Addition To' Endeavor:* Unlike more traditional system interventions that seek to address a new concern by introduction of a totally new initiative (e.g., lean, six sigma, TQM, CRM, etc.), CSG functions are already being performed by a system that is viable (exists). Thus, CSG is focused on understanding and potentially improving that which is already being performed by an enterprise system . Therefore, the language, thinking, and explorations of CSG are applied to existing enterprise system execution of CSG functions which are already being 'tacitly' performed.
4. *CSG Systemic Intervention Time and Risk Should Initially Fall on the Facilitator:* It is unrealistic to expect participants to fully engage a CSG initiative in terms of investment of time and acceptance of 'risk of failure.' Instead, the CSG facilitator should bear the burden of time and risk until the value of investment (time) and utility of CSG engagement combine to produce an acceptable *risk-value-cost* trade-off. In effect, CSG should be conducted in a 'safe to fail' mode.

To elaborate implications for systemic intervention, a systemic intervention framework was developed following initial applications of CSG Entry. This framework, titled the *9R framework for systemic intervention* (Fig. 11), identifies eight areas of concern that practitioners would be advised to consider as they design and execute systemic intervention initiatives for complex systems. This framework has broad implications for systemic intervention beyond CSG.

Each of the framework elements has been identified as having potential impact on systemic interventions undertaken to improve performance of complex systems. Each element provides an area that should be considered when looking to undertake an intervention into a complex system. The following discussion elaborates each of the eight elements targeted to CSG (Fig. 12).

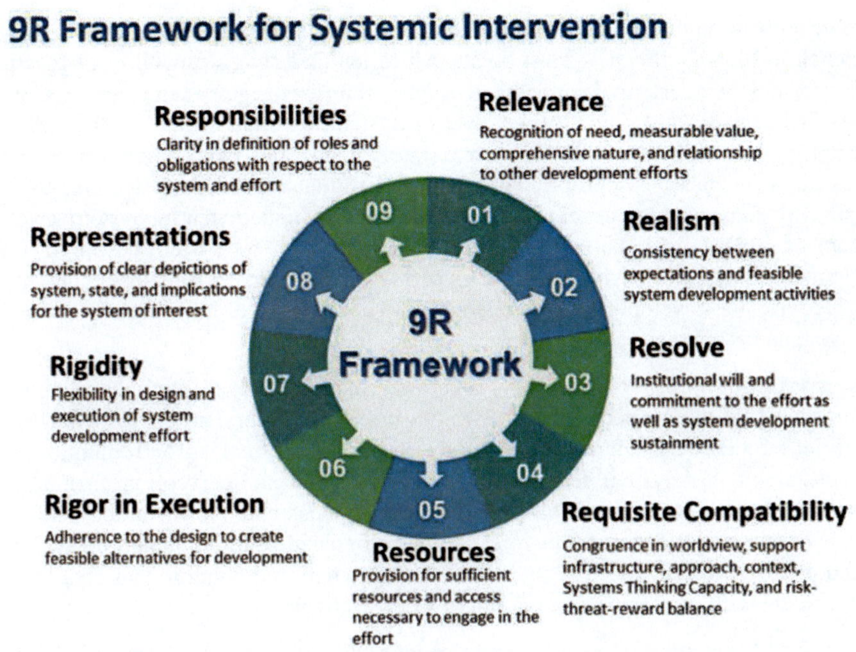

Fig. 12 9R framework for systemic intervention

Relevance: Systemic intervention is undertaken in response to a recognized need or problem situation which is unresolved and persists in a system. However, CSG is not targeted to specific problems, but rather to the 'underlying system' that must address problems. The problems might be perceived as surface manifestations stemming from deficiencies in underlying system functions. While the true value of CSG is in addressing the underlying system deficiencies, value is most recognizable as addressing the surface 'symptomatic' conditions immediately perceived by practitioners. Thus, systemic intervention must focus on: (1) translation of surface problems to the capabilities of CSG to discover the deep seated 'roots' of the problem and offer a different frame of reference for understanding potential alternative paths to resolution, (2) casting CSG in relationship to past, ongoing, and future development initiatives to better position CSG as a 'meta-initiative' that provides an integrating perspective of system development, (3) exploration of systemic intervention as *'enhancing system capabilities'* such that in the future the *'system can solve the problem(s),'* and (4) projection of CSG as system enhancement for functions that the system is already performing, without the benefit of the CSG framing of those functions, and is therefore not 'in addition to' ongoing system work, but rather facilitation of existing work.

Realism: Although CSG holds great promise to identify insights into systemic deficiencies, this identification must be subject to the underlying capability of practitioners and their system to apply those insights to fully engage systemic issues. CSG development is constrained by the level of Systems Thinking Capacity that exists within system participants and the current state of CSG. Thus, expectations for system development must be appropriately metered such that capabilities are commensurate with an appropriate level of improvement activity undertaken for system development. This defines the region of feasible engagement for system development. Knowing issues and having capabilities to address those issues must be congruent. Otherwise, the development is likely to fail and perhaps leaves the system in a worse state than before the systemic intervention was initiated.

Resolve: Commitment of resources (manpower, material, money, methods, minutes, information—M5I) is necessary for engaging in systemic intervention. However, they are not sufficient. Sufficiency is also determined by institutional will and commitment to sustainment of system development following systemic intervention. *Institution will and commitment* are not easily determined or measured. However, the willingness to increase engagement beyond simple resource allocations should be evident and escalated throughout the intervention. Thus, will and commitment should be congruent with increasing recognition of value accrued.

Requisite Compatibility: Systemic intervention for CSG is not necessarily the right approach or fit to every problematic circumstance or every system. The determination of 'fitness' for CSG appropriateness should consider compatibility with system: (1) *predominant worldview* recognized as the prevailing paradigm(s) which drive decision, actions, and interpretations related to system circumstances, (2) *support infrastructures* (e.g., procurement, human resources) that influence, and will be influenced by, system development stemming from intervention discoveries, (3) *contextual factors* (e.g., policy, power, politics, culture, management style) that influence the prospects for conducting systemic intervention and implementing modifications, (4) approach taken to conduct the systemic intervention (e.g., level of participation), and (5) *risk-threat-reward balance* that indicates willingness to engage rigorous self-examination in hopes of finding deeper sources of system development. Lacking these compatibilities, CSG is not likely to produce success. A rigorous analysis of the results following the CSG Entry effort may indicate that continuing to intervene in the enterprise system utilizing CSG will be unlikely to produce positive outcomes.

Resources: Provision for sufficient resources and access necessary to engage in the effort. This must consider the time investment of participants as well as the more mundane aspects related to sufficient levels of funding necessary to engage the desired depth of systemic intervention. Resource allocations should be consistent with expectations of value to be accrued from the effort. Additionally, shifts in resources necessary due to 'discoveries' during systemic intervention activities should be expected, scrutinized, and embraced where appropriate. Incongruence between resource allocation and expectations of value are likely to disappoint the best systemic intervention intentions.

Rigor in Execution: Systemic intervention should have sufficient detail and clarity such that it can be executed with precision. Detailed design related to data collection, analysis, and interpretation should be thorough and explicit such that *what must be done, how it will be done, who will do it, when it will be done, where it will occur, and why it is necessary* are clearly delineated. This does not preclude shifts in design or execution. However, the shifts in approach, execution, and interpretations should be clearly articulated, with the underlying assumptions and supporting logic made explicit and capable of withstanding scrutiny.

Responsibilities: Each systemic intervention is unique in the specific roles that will be played and responsibilities that are allocated to those roles. Responsibilities range across the spectrum of intervention design, execution, and implementation of decisions/actions stemming from systemic intervention activities. Sufficient clarity must exist such that accountability for achievement of different aspects of the systemic intervention can be clearly fixed. This is not to support a punitive dimension for systemic intervention, but rather to ensure that expectations for completion of assignments is unambiguous. Additionally, the pursuit of system changes stemming from a CSG endeavor should have clarity in responsibilities as well.

Rigidity: Systemic intervention follows a particular plan that lays out the design for execution. Although there might be emergent understanding that suggests alteration of the initial design, modifications should be purposeful rather than arbitrary or fickle. Execution of systemic intervention is always dynamic, emergent, and subject to shifts in direction. Reasonable and measured changes in systemic intervention should be expected and embraced, allowing for flexibility in design, execution, expectations, and trajectory of an effort.

Representation: Systemic intervention for CSG is not offered or pursued as yet another approach to improve systems. Instead, CSG and the systemic intervention that it pursues provide a theoretically grounded, application-driven, and practitioner-oriented approach to enhance prospects for better dealing with complex (enterprise) system development. While not presented as a panacea, CSG systemic intervention has shown promise to enhance system development and professional practice by: (1) development of a systems theory-based approach to engaging complex system development, and (2) providing a frame of reference for more rigorous examination of system performance. Future development of CSG and systemic interventions to develop CSG are poised to contribute to development of complex systems in new and novel ways.

Exercises

1. The introduction section of this chapter describes six characteristics of the evolving landscape for the systems engineering practitioner. Think of a complex system in your experience and describe the issues faced by that complex system using those six characteristics.
2. What are some ways that a facilitator of an intervention can identify the level of systemic thinking within an enterprise?
3. For each of the four phases of CSG Entry, please identify reasons why each phase may not be successful and strategies that might increase the probability of success.
4. What does CSG Entry contribute to enabling systemic intervention?
5. How can the 9R framework for systemic intervention be used to address the systemic deficiencies and not just readily recognizable symptoms that may appear on the surface?

References

1. Carter B (2016) Systems based architectural framework for complex systems governance, PhD Dissertation, Old Dominion University, Norfolk VA
2. Calida BY (2013) System governance analysis of complex systems. Doctoral Dissertation. Old Dominion University
3. Drery D (1984) Problem definition in policy analysis. University Press of Kansas, Lawrence, KS
4. Jaradat RM, Keating CB (2014) Fragility of oil as a critical infrastructure problem. Int J Crit Infrastruct Prot 7(2):86–99
5. Katina PF, Keating CB (2014) Metasystem pathologies: towards a systems based construct for complex system deficiencies. In: Long S, Ng EH, Downing C (eds) Proceedings of the American society for engineering management 2014 international annual conference. Virginia Beach, VA
6. Katina PF (2015) Emerging systems theory based pathologies for governance of complex systems. Int J Syst Syst Eng 6(½):144–159
7. Katina PF (2015b) Systems theory based construct for identifying metasystem pathologies for complex systems governance. PhD Dissertation, Old Dominion University, Norfolk, VA
8. Katina PF (2016) Systems theory as a foundation for discovery of pathologies for complex system problem formulation. In: Masys AJ (ed) Applications of systems thinking and soft operations research in managing complexity. Springer International Publishing, Geneva, Switzerland, pp 227–267
9. Katina PF (2018) Metasystem pathologies (M-Path) method: phases and procedures. J Manage Dev
10. Keating CB, Katina PF (2011) Systems of systems engineering: prospects and challenges for the emerging field. Int J Syst Syst Eng 2(2–3):234–256
11. Keating CB (2014) Governance implications for meeting challenges in the system of systems engineering field. In: Paper presented at the IEEE system of systems engineering conference. Adelaide, SA
12. Keating CB, Katina PF (2012) Prevalence of pathologies in system of systems. Int J Syst Syst Eng 3(¾):243–267

13. Keating CB, Katina PF, Bradley JM (2014) Complex systems governance: concepts, challenges & emerging research. Int J Syst Syst Eng 5(3):263–288
14. Keating CB, Bradley JM (2015) Complex system governance reference model. Int J Syst Syst Eng 6(1):33–52
15. Keating CB, Katina PF, Bradley JM (2015) Challenges for developing complex systems governance. In: Proceedings of the 2015 industrial and systems engineering conference. IIE, Knoxville, TN
16. Keating CB, Katina PF (2016) Complex systems governance development: a first generation methodology. Int J Syst Syst Eng 7(1/2/3):43–47
17. Keating CB, Katina PF (2017) Systemic intervention for complex system governance development. In: Proceedings of the American society for engineering management, October 18–21, 2017. Huntsville, Alabama, pp 85–95
18. Keating CB, Katina PF, Pyne JC, Jaradat RM (2017) Systemic intervention for complex systems governance development. In: Ng EH, Schotts E (eds) sProceedings of the American society for engineering management 2017 international conference
19. Quade ES (1980) Pitfalls of formulation and modeling. In: MaJone G, Quade DS (eds) Pitfalls of analysis, vol 8. Wiley Interscience, Chichester, England, pp. 22–43
20. Midgley G (2000) Systemic intervention. In: Systemic intervention. Springer US, pp 113–133
21. Midgley G (2003) Science as systemic intervention: some implications of systems thinking and complexity for the philosophy of science. Systemic Pract Action Res 16(2)
22. Whitney K, Bradley JM, Baugh DE, Chesterman CW Jr (2015) Systems theory as a foundation for governance of complex systems. Int J Syst Syst Eng 6(1):15–32

Future Challenges for Complex System Governance Research and Practice

Charles B. Keating, Polinpapilinho F. Katina, Charles W. Chesterman Jr., and James C. Pyne

Abstract This chapter provides an overview of the current state of the CSG field, including achievements, future challenges, and developmental directions. CSG is introduced as an evolution and distinction from System of Systems Engineering. Following the introduction, four primary areas of exploration are examined. First, a summary of the current state of the CSG Field is conducted. The current state of the field is critiqued with respect to what has been accomplished as well as present shortcomings. Second, advancement challenges across the spectrum of theoretical/conceptual, application, and supporting methods/tools/techniques are examined. Third, the future development of the CSG Field as '*the system-science based engineering of technologies for application to improve design, execution, and development of complex systems*' is explored. Fourth, major points of consideration for advancing practice of CSG is examined. The chapter concludes with a set of exercises to examine critical issues in the design, execution, and evolution of systems using CSG.

Keywords CSG field · Development challenges · Future directions

C. B. Keating (✉) · C. W. Chesterman Jr. · J. C. Pyne
Department of Engineering Management and Systems Engineering, Old Dominion University, Norfolk, VA, USA
e-mail: ckeating@odu.edu

C. W. Chesterman Jr.
e-mail: cwjstellami@aol.com

J. C. Pyne
e-mail: jpyne@odu.edu

P. F. Katina
Department of Informatics and Engineering Systems, University of South Carolina Upstate, Spartanburg, SC, USA
e-mail: pkatina@uscupstate.edu

C. B. Keating et al. (eds.), *Complex System Governance*, Topics in Safety, Risk, Reliability and Quality 40, https://doi.org/10.1007/978-3-030-93852-9_17

1 Introduction

Complex system governance (CSG) is an emerging field that traces its formal introduction to 2014 in an article titled *Complex system governance: concept, challenges, and emerging research* [55]. However, the seeds for CSG were sown well before the 2014 formal introduction. The formation of the National Centers for System of Systems Engineering (NCSoSE) at Old Dominion University in 2003 focused on understanding the issues related to integration of multiple complex systems. The launching of the Center coincided with the first article titled *System of Systems Engineering* [65]. In this article, System of Systems Engineering (SoSE) was defined as "The design, deployment, operation, and transformation of higher-level metasystems that must function as an integrated complex system to produce desirable results." [65, p. 41]. This early work, coupled with the evolution of the work at the Center, set in motion seven important evolutionary distinctions of CSG from SoSE (Fig. 1). First, the concept of 'metasystem' was introduced from the Management Cybernetics field [5–7]. The metasystem is a set of functions and corresponding communications channels that act to provide for control and communication in a system. The metasystem became a central construct for CSG. The metasystem, and the management cybernetics from which it emanated, followed through to the current instantiation of the CSG Reference Model and emerging CSG field.

A second pivotal developmental theme for CSG stemmed from the conclusion that the systems of interest for System of Systems Engineering (SoSE) were too complex to take a *'technology first, technology only'* approach to the development of systems

Fig. 1 Distinguishing CSG from SoSE

of systems. Instead, the work of the Center was grounded in the underlying '*holism*' suggested by systems theory, a perspective that was not embraced by the dominant SoSE community at the time and not fully understood. Grounding in systems theory and management cybernetics invoked taking a systems view and the inclusion of the range of technological, human, social, organizational, managerial, policy, and political dimensions into account for the SoSE problem space. The consideration of the holistic range of dimensions for SoSE moved beyond the *'technology first, technology only'* (e.g., technical interoperability) approaches that had dominated the early development of the SoSE field. This is not intended to disparage in any way the early work in the SoSE field focused on technology integration. This was necessary early in the instantiation of SoSE and still remains essential today. However, the exclusive view of the SoSE problem space as fundamentally technology-oriented has diminished the projection of the field to more complex 'holistic' problem spaces. Also, engaging a holistic-based paradigm required that a systemic world-view, grounded in the underlying systems theory (the paradigm and doctrine that are based in systems science) would be necessary to advance SoSE. Unfortunately, the prevailing perspectives of SoSE at the time viewed SoSE as an extension of Systems Engineering, thus requiring a reductionist mindset. This mindset required linear thinking, tightly bounded problems, objective definition, repeatable application of proven methods/tools, and seeking optimal solutions. Unfortunately, the reductionist mindset continues to show limitations when cast against the emerging holistic problem domains that are the hallmark of modern systems (e.g., health care, cybersecurity, Internet of Things, etc.). Thus, CSG emerged as a response to take a more holistic approach to the complex problem domain characteristic of the state of complex systems (of systems) and their constituent problems.

A third primary distinction sought by CSG was in the appreciation of 'context'. Context is taken as the set of circumstances, conditions, factors, trends, or patterns within which a system of interest is embedded. The separation of a system from its context for the convenience of analysis is a false separation. For example, such contextual factors as resources, power, politics, support infrastructure, and leadership style can play a substantial role in determining system performance. Absent an emphasis on context is considered incomplete framing for a complex system. The result is to create the conditions for committing a Type III Error [64], or solving the wrong problem in the most efficient way possible. Holistic framing was deemed essential to taking both system and context into consideration, as well as the interaction effects for CSG. Additionally, the wider inclusion of context introduced the appreciation of 'soft' (human, social, organizational, managerial, political, and policy) aspects of complex system development. This was in addition to the traditional 'hard' (technical/technology) aspects of development. The inclusion of 'soft' dimensions in the 'analytical SoSE space' was in contrast to prevailing mindsets in SoSE that were focused almost exclusively on the 'hard' aspect of complex SoSE. Thus, CSG was in search of an alternate paradigm, grounded in the conceptual/theoretical foundations of systems theory (the axioms and propositions that explain and provide understanding of complex system structure, behavior, and performance) and management cybernetics (the science of effective system structural organization).

A fourth distinction in CSG was the incorporation of system governance as the third conceptual underpinning to provide completeness in CSG. The system governance field helped to: (1) add an important dimension to the communication and control perspectives provided by management cybernetics, (2) engage more readily in the higher-level perspectives of establishing direction, oversight, and accountability, including 'soft' dimensions, to supplement management cybernetics, and (3) projected the 'long view' and 'fuzziness' that characterize the governance field. The intersection of management cybernetics, systems theory, and system governance provided the conceptual/theoretical foundations upon which the emerging CSG paradigm could draw. This allowed for a departure from more restrictive instantiations of SoSE.

A fifth distinction sought in separation of CSG from SoSE was found in the qualification of both participants and entities to appropriately engage the approach. Application of SoSE, as well as other systems-based methodologies (for examples see [27]) had no qualification of preparedness of individuals or the system having necessary prerequisites to effectively apply the approach. The remedy for this shortcoming in SoSE formulations was found in the CSG mandates to: (1) establish the degree of systemic thinking capacity, held by individuals and the system of interest, available to deploy CSG from a compatible systemic mindset essential to success, and (2) understanding the current state of CSG for the system to determine the types of feasible strategies/actions/initiatives that might be pursued with confidence in the probability of being successful. Thus, CSG was born of a need to separate from SoSE formulations that did not qualify either individuals or the system capacity to effectively engage the approach.

A sixth distinction of the CSG separation from SoSE had to do with the *expectations* for engaging in an endeavor. SoSE applications were primarily driven as problem-focused approaches with solution-driven expectations. At a tacit level, this mindset requires the narrow bounding of the 'problem' as opposed to the 'holistic system of systems' as the focus. The result is the engagement of SoSE as problem-centric, versus SoS-centric, in search of solutions to well-bounded problems or decision support. In contrast, CSG is targeted to system development, not whole system solutions to narrowly prescribed problems. However unsatisfying this might be to traditional SoSE perspectives, CSG is targeted to accomplishment of several different potential opportunities for system development, including: (1) development of individuals and the entity to engage in higher levels of systemic thinking, not only about the system in focus, but also with collateral extensions to other systems, issues, and contexts, (2) identification of system support infrastructure as a source that can be both enabling and disabling and targeted for development from a CSG effort, (3) the development of the system of interest across design, execution, and developmental improvement areas discovered as feasible to address, (4) identification of aspects of the 'larger system/organization/enterprise' that are in need of adjustment to more properly support the system of interest, (5) the assessment and accounting of contextual aspects for the system of interest that have a positive/negative impact on the performance of the system of interest and can be addressed within the scope of the CSG endeavor, and (6) determination of environmental considerations that constrain/enable the system of interest in ways that can be influential in directing modifications to system design, execution, or development. These expectations represented a major departure from SoSE.

A seventh distinction of CSG is the emphasis on whole system development, where learning takes precedence over 'solution' finding. In one sense, CSG provides a guided 'self-study' of a system of interest to facilitate learning—and corresponding responsive and feasible action—about the system. The exploration identifies 'deep system' issues that are the underlying source of problems in the system. Therefore, CSG was pushed to generate knowledge of the architecture of the CSG functions, the deficiencies (pathologies) in those functions, and the feasibility of addressing the disfunctions. This push of CSG was against the backdrop of establishment of the state of the system and capacity for 'thinking in systems' that is fit to the task. These discoveries are important products and artifacts in CSG endeavors.

Given the need to separate CSG from the evolving SoSE field, at a high level, CSG was targeted to focus on improving the theory and practice of more effectively 'taming' modern complex systems and their problems. CSG draws upon and exists at the intersection of three primary fields, including systems theory, management cybernetics, and system governance. *Systems theory* provides a strong intellectual foundation focused on effective integration and coordination of disparate elements into a coherent whole. This coherent whole must 'obey' the axioms and corresponding propositions of systems theory that govern behavior of systems or suffer the consequences related to deviations. *Management cybernetics* brings an emphasis on communication and control essential to provide for the continuing existence (viability) of a system as it deals with the inevitable internal flux and environmental turbulence endemic to modern complex systems. Consistent with management cybernetics, CSG appreciates and responds to the constant change in the context and environment for a governed system or system of systems. Thus, 'cybernetic steering' emphasizes control necessary to regulate and maintain system stability. This inherently acknowledges the need to monitor the potential impact of near and long-term fluctuations on continuing system viability. Finally, *governance* provides an emphasis on direction, oversight, and accountability for the execution and development of a system. While each of the three fields underpinning CSG have made substantial contributions to the state of human affairs, they have not been brought together in meaningful ways that takes advantage of their intersection to produce a novel alternative to complex system development.

This book does not represent the end state for the development, propagation, and application of CSG. On the contrary, it should be considered a waypoint. A temporary stop to take perspective on what has been accomplished, what is being accomplished, and what lies on the immediate and distant horizon for further development of the CSG field. This book has provided three primary contributions:

1. *Comprehensive Collection of the State of Knowledge for CSG*—There has been a growing body of work produced for CSG. However, the work has been somewhat

fragmented and dispersed in getting to this state. This book is a consolidation of the current state of knowledge in CSG. Therefore, it attempts to bring the body of CSG knowledge together into a coherent framework.

2. *Identification of Gaps in the Knowledge for CSG*—This work serves to collect and organize the state of knowledge for CSG. As a byproduct of this organization, the gaps in knowledge have come to the forefront. This is not a criticism of the work that has been done in CSG. Instead, it identifies CSG field developmental targets that can concentrate efforts.
3. *Definition of the Challenges for the Field and Setting Developmental Priorities*—CSG is no longer held in a limited set of works. Instead, it has amassed a growing set of research, articles, and developmental works. The timing is right for this work to critically survey the CSG knowledge stage, determine the significant challenges, and chart a coherent path forward to the next waypoint.

The purpose of this chapter is to provide an overview of the current state of the CSG field and future directions. This discussion will include achievements, future challenges, and developmental directions to advance the field. The chapter is organized (Fig. 2) to focus on four primary areas. First, a summary of the current state of the CSG Field is conducted. This current state of the field is critiqued with respect to what has been accomplished and present shortcomings. Second, advancement challenges across the spectrum of theoretical/conceptual, methodological, methods,

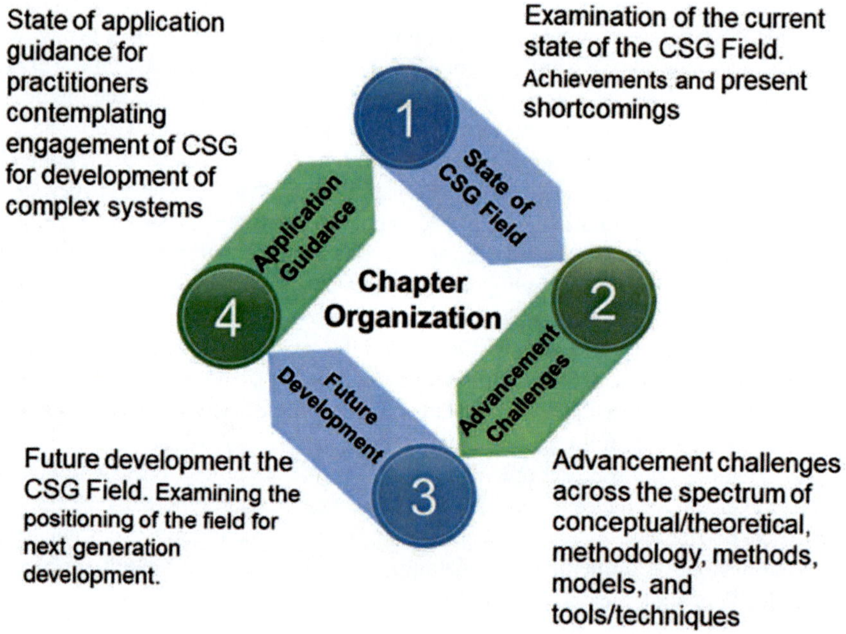

Fig. 2 Chapter organization

tools/techniques, and applications are suggested based on the current state of research and application. Third, the future development of the CSG Field is examined. The positioning of CSG as *'the system-science based engineering of technologies for application to improve design, execution, and development of complex systems'* is explored. This development culminates with a practice and practitioner-based set of guidance to continue advancement of the CSG Field. The guidance attempts to provide an integrated trajectory of the science (theoretical, conceptual, philosophical), engineering (technologies, artifacts, methods), and application (practice tools, techniques, processes) development directions for the field. Fourth, a current state of application guidance for interested practitioners is provided. This guidance is targeted to make the emerging CSG field more accessible to practitioners for the improvement of complex systems. Application emphasizes contributions across individual, organizational, system, infrastructure, and enterprise levels. The chapter concludes with a set of exercises to examine critical issues in the design, execution, and evolution of systems using CSG.

2 The State of the CSG Field

There is a growing body of knowledge related to CSG [47, 51, 52, 56]. CSG is described as the 'Design, execution, and evolution of the [nine] metasystem functions necessary to provide control, communication, coordination, and integration of a complex system.' [48, p. 228]. In this section, we examine the essence of CSG. This essence is found in the nine metasystem functions of CSG and the ten implementing communication channels. Second, an exploration of the essence of CSG as an approach to better deal with complex systems and their problems is conducted. This exploration suggests several points of emphasis that serve as a high-level articulation of the paradigm and central themes of CSG. The section concludes with an assessment of the current state of CSG. The work that has been completed is critically reviewed to set implications for moving the CSG field forward.

2.1 The Essence of CSG—Functions and Communication Channels

The essence of CSG is found in the performance of nine essential governance functions and ten corresponding communication channels. Of all that comprises CSG, the metasystem functions and the communication channels represent the greatest degree of stability. The nine governance functions [52] include the following:

- *Policy and Identity*—Metasystem Five (M5)—focused on overall steering and trajectory for the system. Maintains identity and balance between current and future focus.

- *System Context*—Metasystem Five Star (M5*)—focused on the specific context within which the metasystem is embedded. Context is the set of circumstances, factors, conditions, or patterns that enable or constrain execution of the system.
- *Strategic System Monitoring*—Metasystem Five Prime (M5′)—focused on oversight of the system performance indicators at a strategic level, identifying performance that exceeds or fails to meet established expectations.
- *System Development*—Metasystem Four (M4)—maintains the models of the current and future system, concentrating on the long range development of the system to ensure future viability.
- *Learning and Transformation*—Metasystem Four Star (M4*)—focused on facilitation of learning based on correction of design errors in the metasystem functions and planning for transformation of the metasystem.
- *Environmental Scanning*—Metasystem Four Prime (M4′)—designs, deploys, monitors, and communicates sensing of the environment for trends, patterns, or events with implications for both present and future system viability
- *System Operations*—Metasystem Three (M3)—focused on the day to day execution of the metasystem to ensure that the overall system maintains established performance levels.
- *Operational Performance*—Metasystem Three Star (M3*)—monitors system performance to identify and assess aberrant conditions, exceeded thresholds, or anomalies.
- *Information and Communications*—Metasystem Two (M2)—designs, establishes, and maintains the flow of information and consistent interpretation of exchanges (communication channels) necessary to execute metasystem functions.

The current diagram depicting the CSG functions and communication channels is provided in Fig. 3.

Communication channels represent the second major element of CSG. In substance, the communication channels provide for the flow and interpretation of information in the system, and between the system and the environment (Table 1).

The CSG functions, in concert with the communication channels, produce control, communication, coordination, and integration—in essence the governance responsible for system performance. *Control* establishes constraints necessary to ensure consistent performance and future trajectory. *Communications* provides for flow and processing of information necessary to support consistent decision, action, and interpretation throughout the system. *Coordination* provides for effective interaction to prevent unnecessary instabilities within and external to the system. *Integration* maintains system unity through common purpose, designed accountability, and maintenance of balance between system and constituent interests. Each system is unique in defining 'how' the functions are performed. CSG is concerned with understanding sources of underperforming systems in terms of issues in the design and execution of the nine essential system functions and communication channels. Although addressing underperforming systems is not new, the introduction of CSG offers a new and novel perspective, approach, and system development alternatives. CSG can aid practitioners who must contend with increasing internal flux and external turbulence characteristic of the modern organizational (system) landscape. This landscape

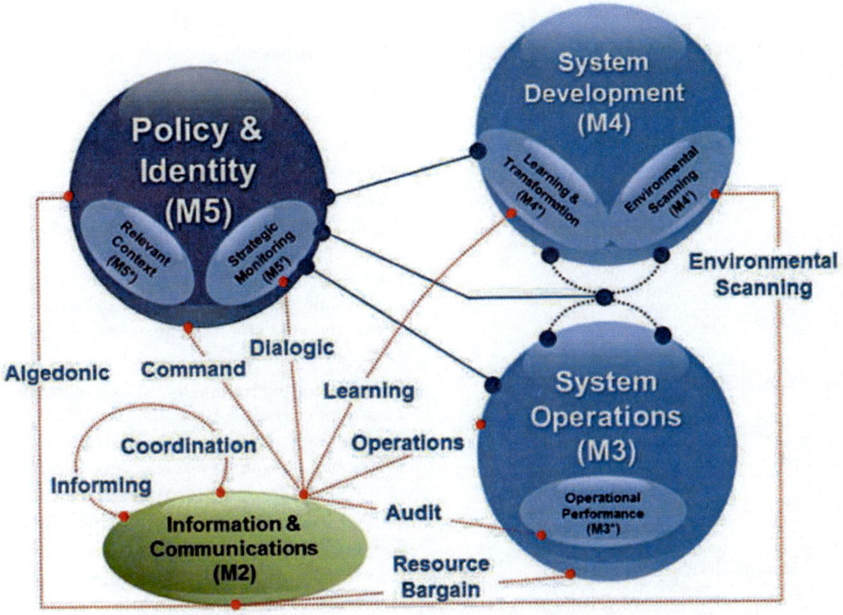

Fig. 3 CSG metasystem functions and communication channels

represents the 'new normal' for systems and their practitioners and shows no signs of subsiding in the near future.

Although the underlying theory, concepts, and execution of CSG are challenging and beyond the scope of this chapter, the essence of CSG is not difficult to grasp. The essence of CSG might be captured in the following statement and elaborated in the four points that follow:

> *Subject to fundamental systems theory propositions, all systems perform essential governance functions. System performance is determined by effectiveness in achievement of governance functions consistent with systems theory propositions. System performance can be enhanced through purposeful development of governance functions.*

There are four fundamental points that help to explain the nature and role of CSG. These include:

- *All systems are subject to the laws of systems*. Just as there are laws governing the nature of matter and energy (e.g., physics law of gravity), so too are our systems subject to laws (*propositions* which include systems theory based laws, principles, and concepts). These system laws are always there, non-negotiable, non-biased, and explain system behavior, structure, and performance.
- *All systems perform essential governance functions that determine system performance.* Nine system governance functions are performed by all systems, regardless of sector, size, or purpose. These functions define 'what' must be achieved for

Table 1 Summary of the CSG communication channels

Communications channel and responsibility	CSG metasystem role
Command (Metasystem 5)	• Provides non-negotiable direction to the metasystem and governed systems • Primarily from the metasystem 5 and disseminated throughout the system
Resource bargain/accountability (Metasystem 3)	• Determines and allocates the resources (manpower, material, money, information, support) to governed systems • Defines performance levels, responsibilities, and accountability for governed systems • Primarily an interface between Metasystem 3 to the governed systems
Operations (Metasystem 3)	• Provides for the routine interface focused on near-term operational focus • Concentrated on direction for system production (products, services, processes, information) consumed external to the system • Primarily an interface between Metasystem 3 and governed systems
Coordination (Metasystem 2)	• Provides for metasystem and governed systems balance and stability • Ensures that information concerning decisions and actions necessary to prevent disturbances are shared within the metasystem and governed systems • Primarily a channel designed and executed by metasystem 2
Audit (Metasystem 3*)	• Provides routine and sporadic feedback concerning operational performance • Investigation and reporting on problematic performance issues within the system • Primarily a Metasystem 3* channel for communicating between Metasystem 3 and governed systems concerning performance issues
Algedonic (Metasystem 5)	• Provides a 'bypass' of all channels when the integrity of the system is threatened • Compels instant alert to crisis or potentially catastrophic situations for the system • Directed to Metasystem 5 from anywhere in the metasystem or governed systems
Environmental Scanning (Metasystem 4')	• Provides design for sensing of the external environment • Identifies environmental patterns, activities, or events with system implications • Provided for access throughout the metasystem as well as governed systems

(continued)

Table 1 (continued)

Communications channel and responsibility	CSG metasystem role
Dialog (Metasystem 5')	• Provides for examination of system decisions, actions, and interpretations for consistency with system purpose and identity • Directed to Metasystem 5' from anywhere in the metasystem or governed systems
Learning (Metasystem 4*)	• Provides detection and correction of error within the metasystem as well as governed systems, focused on system design issues as opposed to execution • Directed to Metasystem 4* from anywhere in the metasystem or governed systems
Informing (Metasystem 2)	• Provides for flow and access to routine information in the metasystem or between the metasystem and governed systems • Access provided to entire metasystem and governed systems

governance of a system. Every system invokes a set of unique implementing mechanisms (means of achieving governance functions) that determine 'how' governance functions are accomplished. Mechanisms can be formal-informal, tacit-explicit, routine-sporadic, or limited-comprehensive in nature. CSG produces system performance which is a function of previously discussed communication, control, integration, and coordination.

- *Violations of systems theory propositions, in performance of governance functions, carry consequences.* Irrespective of noble intentions, ignorance, or willful disregard, violation of system theory propositions carries real consequences for system performance. In the best case, violations degrade performance. In the worst case violation can escalate to cause catastrophic consequences or even eventual system collapse.
- *System performance can be enhanced through purposeful development of governance functions and communication channels.* When system performance fails to meet expectations, identification of deficiencies in governance functions can offer novel insights into the deeper systemic sources of failure. Performance issues can be traced to governance function issues as well as violations of underlying system propositions (laws, principles, and concepts). Thus, system development can proceed in a more informed and purposeful mode.

At a high level, the paradigm for CSG can be expressed as a straightforward set of relationships and products stemming from CSG (Fig. 4). First, CSG is grounded in the underlying conceptual/theoretical underpinnings found primarily in systems theory and management cybernetics. Second, consistent with this grounding, the set of nine CSG metasystem functions and ten communication channels are performed if a system is to remain viable (continue to exist). Third, the CSG functions and communication channels are performed by the mechanisms (activities, vehicles,

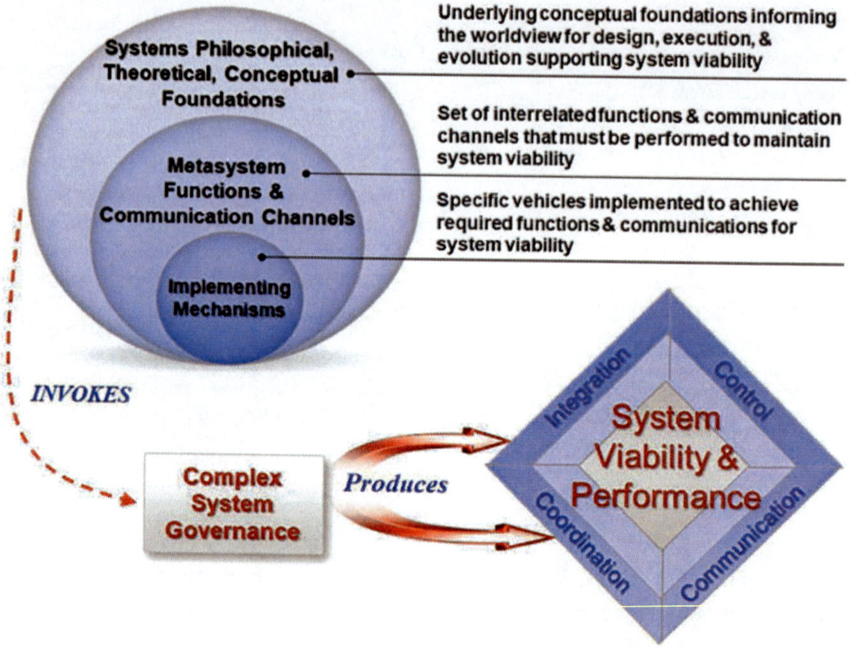

Fig. 4 The CSG paradigm

events, procedures, processes) that serve to enact them—this invokes the performance of CSG. Fourth, the performance of CSG produces system viability, and ultimately, the level of performance through communications, control, integration, and coordination. Thus, complex system performance is grounded in the degree to which CSG is effectively designed, executed, and developed.

2.2 Points of Emphasis for CSG

There are several points of emphasis for understanding the basis for CSG in design, execution, and evolution of complex systems. First, *design* accentuates the purposeful and proactive engagement in creation of the governance system. While this seems as though it should be a taken for granted proposition, we suggest that truly purposeful, holistic, and comprehensive design of governing systems represents the exceptional case rather than the norm. Although we might argue the merits of this conclusion, at this point, it suffices to say that based on the current level of system performance of our complex systems, the conclusion seems to be supported. Based on issues propagating all manner and form of our 'manmade' complex systems, the anecdotal evidence suggests that our systems are not sufficiently serving the needs or expectations intended to enhance societal wellbeing. From the CSG perspective, we can see that the integrated and purposeful design for governance is not presently being performed in many of our complex systems. The current state of CSG suggests that there is a significant opportunity to engage 'faulty' complex systems to elevate their performance by purposeful (re)design.

The second element of consideration of CSG revolves around *execution*. Irrespective of purposeful/purposeless design, execution embodies the notion that a design without effective deployment offers little more than good intention. Execution is where a design meets the harsh realities of the 'real world,' which is fraught with complexity and emergent conditions that are sure to test our most thoughtful system designs. For CSG, we suggest that execution is achieved through a multitude of entities and activities. While each of the activities undertaken in support of CSG has merit, a major emphasis of CSG is execution of the design. Lacking execution, CSG is absent an essential element for successful achievement of associated functions.

A third element of CSG, *evolution*, recognizes that systems, as well as their environments, are in constant flux and change over time. Therefore, governance must also be able to flex (evolve) in response to internal and external changes impacting the system over time. Evolution by its very nature suggests that the developmental emphasis is on long-term sustainability, irrespective of the need to operate a system in real time. In effect, governance must be capable of absorbing, processing, and responding to external turbulence and internal system flux. This can ensure the system remains viable (continues to exist). This viability is in both the short-term operational sense that delineates current system existence as well as the long-term evolutionary sense that positions the system for the future. Taking the long view of CSG development, an evolutionary perspective is essential.

CSG is an emerging field focused on helping systems and their practitioners (owners, operators, designers, performers) deal more effectively with increasingly complex systems and their problems. In a nutshell, CSG suggests that we are not inevitably 'doomed' to suffer the ill effects of poorly performing systems. CSG is not offered as a panacea promising to cure all system ills. Instead, CSG offers an alternative path forward for practitioners interested in exploration of new and novel thinking and practice for more effectively dealing with difficult complex systems and problems.

An important emphasis of CSG is that it lies at the intersection of three knowledge streams, Systems Theory (the set of laws that explain the behavior and performance of all systems), Management Cybernetics (the science of effective structuring of systems), and Governance (provision of direction, oversight, and accountability for systems). At the intersection, CSG is focused on the design, execution, and evolution of essential system functions. Proficiency in execution of these functions ultimately determines the level of system performance. The reliance on proven fields enhances the veracity of CSG as an 'intersected' field that draws on a substantial intellectual base.

2.3 *The Current State of CSG*

CSG started in earnest in 2014 [47]. Since that inception, the field has continued to grow in depth and stature. The gains in CSG have spanned the spectrum of theory, methodology, methods, models, tools, and applications. Although still in the embryonic stages, there has been significant progress. Figure 5 provides an overview of the CSG field current state of published works.

CSG has made strides across the six developmental areas necessary to advance the field. A brief accounting of what has been accomplished across each of the areas includes:

- *Conceptual/theoretical*—This represents the most advanced area of CSG development. This is to be expected as the early emphasis of CSG was directed to establishing a solid and well grounded conceptual/theoretical basis. The works in this area have stayed stable as CSG has continued development. This has allowed the other developmental levels to have a reference point that has remained relatively stable. The anchoring of this foundation in systems theory, management cybernetics, and system governance has provided this stability.
- *Methodology*—The area of methodology (the general approach that defines 'what' must be achieved to engage CSG development) has been in place for several years. It has remained relatively intact from the original development. However,

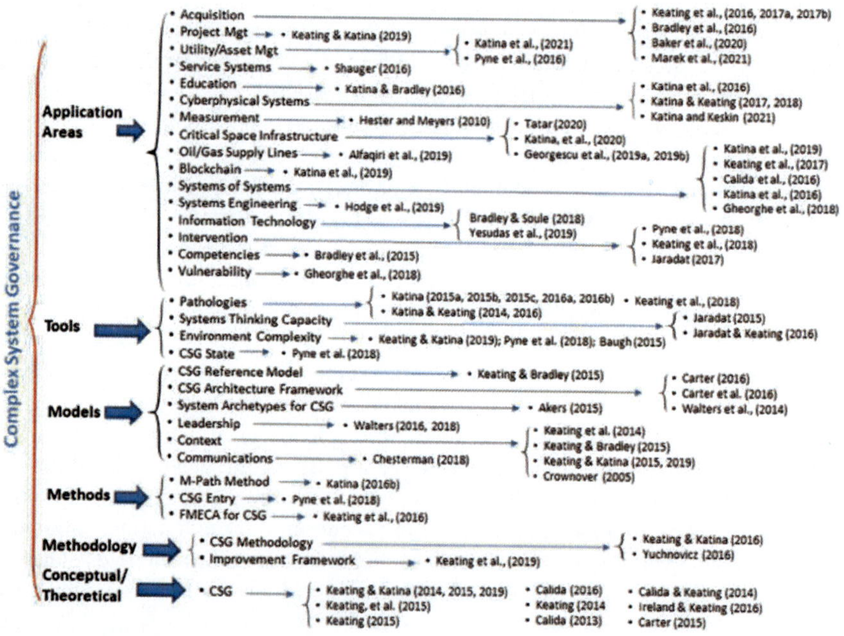

Fig. 5 The current state of the CSG field development

there are two issues with the state of the CSG development methodology. First, the methodology, although conceptually sound, has not seen full engagement for operational deployment. While there have been derivative applications of the 'front end' of the methodology, it has not gravitated to a full engagement. Second, lacking a full engagement, the validation of the CSG development methodology will be lacking.

- *Methods*—In the area of methods to support CSG initiatives there have been several developed. Chief among these are the M-Path method for the discovery and assessment of pathologies in complex systems [36], the CSG Entry method [68] to begin engagement in CSG, and the FMECA for CSG method [57] to discover and assess failure modes in complex systems. While these methods have shown promise, the number of methods developed to support the wide ranging CSG landscape is currently sparse.
- *Models*—The CSG Reference Model remains as the single most prevalent and well developed model for CSG. However, it has not translated into a sufficient set of specific tools or techniques to facilitate deployment of the model in operational settings. Other models have been developed to support specific aspects of CSG, including Communications [18], system archetypes [1], CSG Architecture Framework [17], leadership [71, 73], and Context [19, 55]. While there are models that have been generated to aid in understanding of CSG functions and communications channels, they fall short in number, scope, and operational deployment.
- *Tools*—There are multiple support tools that have been developed for different aspects of CSG. Among the tools are the ST-Cap method for the assessment of systems thinking capacity for individuals [29], Pathologies discovery [32] to identify and process CSG pathologies, Environment Complexity [4] to determine the state of the environment for a complex system, and CSG state [68] to set the current state of the CSG functions and communication channels. While there have been some developments to assist in the performance of CSG, currently there is not sufficient support for practitioners in accomplishment of CSG development.
- *Application Areas*—This is perhaps the least mature of all of the developmental areas for CSG. There have been several documented cases of limited applications of 'parts' of CSG. For example, there have been a variety of application areas (e.g., asset management [39]) identified for utility of CSG. However, actual applications of CSG have been limited, mostly targeted to the initial stages and CSG entry efforts as well as training. Unfortunately, without the demonstration of utility in operational settings, the development and propagation of CSG will be limited.

The current state of CSG is where we might expect for a field in the earliest stages of development. The conceptual foundations are well established and are reaching a point of relative stability. Additionally, there are a host of application areas identified where CSG can contribute. Also, there is the beginning formulation of methods and tools that are being developed to support operational deployment of CSG. However, the application of CSG has seen limited deployment in operational settings. Alleviating this concern will be critical for the continued development of the CSG field.

3 Advancement Challenges for CSG

While the current state of the field is impressive in productivity since 2014, there are several challenges in the continuing evolution of the CSG field. Among these challenges are: (1) early development of CSG was almost exclusively dedicated to first setting the conceptual/theoretical foundations in place—the result is a strong foundation upon which to begin expanding into other aspects, including methods, tools, and techniques, (2) the field is in a position to begin greater emphasis on getting the field directed toward development and deployment of applications for applied settings—this can provide direction for the trajectory of the field and acceleration of advancements through the coupling of theory to practice, and (3) CSG has not been 'stereotyped' to a particular domain/sector for development—while this permits flexibility in the applicability of CSG, it also slows adoption by a lack of an attentive targeted audience. These are the realities of the current state of CSG. While not insurmountable, if CSG is to continue to propagate, it must address these developmental areas.

There are a multitude of contributions that CSG can make to advance the state and practice of complex systems. These contributions can be summarized with respect to the challenges identified in Table 2.

CSG is an emerging field with great potential. Therefore, we must certainly expect challenges, additions, extensions, and insights as the field continues to evolve through research, development, and application. In examination of the challenges for advancing the CSG field, the following development areas have been identified in previous works and remain [15, 46, 47, 55].

- *Holistic field development and application*—continued development of CSG will be well served by research and practice being simultaneously developed. Research must be directed at pursuit of advances across the spectrum of Philosophical (worldviews), Theoretical (explanations concerning phenomena), Methodological (high level guiding frameworks), Axiological (values, value judgments, and beliefs), and Axiomatic (underlying principles). Enhanced practice will be the beneficiary of this holistic development of the field. It is shortsighted to only focus on either research or practice exclusively. Based on the current state of development for CSG, the emphasis on finding opportunities to apply CSG, in part or whole, in operational settings will help to suggest where the field might be enhanced.
- *Focus on Both Practice and Practitioners*—CSG is not an intellectually 'easy' body of knowledge to assimilate. While the theoretical foundations are well grounded, they do not necessarily translate easily to the world of the practitioner. The CSG field should not lose sight of the drive to improve practice and enhance

Table 2 The challenges for CSG to advance complex systems

Challenge	Challenge explanation for complex system governance development
Simultaneous emphasis on design, execution, and evolution	• **Design** is the purposeful and deliberate arrangement of the governance system to achieve desirable system performance and behavior. For complex systems, this suggests making the design explicit and enabling critique against known CSG requirements for effective design. • **Execution** is performance of the system design within the unique system context, subject to emergent conditions stemming from interactions within the system and between the system and its external environment. For complex system, execution provides a path for evaluation as to how effective the execution of the design is in producing performance/behavior desired. • **Evolution** involves the change of the governance system in response to internal and external shifts as well as revised trajectory. For complex system, evolution provides a long view and continual focus on evolving the governing system based on environmental shifts.
Articulate metasystem functions	• Metasystem functions are performed by all viable systems. They serve to provide communication, control, integration, and coordination essential to ensure continuing system performance in the wake of internal flux and environmental turbulence. For complex systems, the purposeful design of metasystem functions can provide performance that fragmented entities and mechanisms will neither be able to achieve nor maintain.
Emphasize design of communication channels	• **Communication** involves the flow, transduction, and processing of information within and external to the system, that provides for consistency in decisions, actions, interpretations, and knowledge creation made with respect to the system. For complex systems, communication is an essential element that should be developed by purposeful design and not left to fortuitous development. Additionally, communications must consider the means and activities beyond the purely technical exchange of information.

(continued)

Table 2 (continued)

Challenge	Challenge explanation for complex system governance development
Design for minimal control (regulatory capacity)	• **Control** is focused on invoking the minimal constraints necessary to ensure desirable levels of performance and maintenance of system trajectory. This is achieved by installing regulatory capacity that permits the system to maintain desired performance in the midst of internally or externally generated perturbations of the system. For complex systems, control suggests that only the constraints necessary to integrate the multiple stakeholders and systems should be invoked. Any excess constraint consumes scarce resources and unnecessarily limits constituent autonomy.
Design for integration of constituent systems	• **Integration** provides for continuous maintenance of system integrity. This requires a dynamic balance between autonomy of constituent entities and the interdependence of those entities to form a coherent whole. This interdependence produces the system identity (uniqueness) that exists beyond the identities of the individual constituents. This permits the system to produce collective behavior/performance beyond that of any of the individual constituent entities.
Design for coordination among constituents	• **Coordination** is focused on providing for interactions (relationships) between constituent entities within the system, and between the system and external entities, such that unnecessary instabilities are avoided. For CSG, coordination becomes a necessary attribute to ensure that the multiple entities, perspectives, and infrastructures are engaged to prevent unnecessary fluctuations and conflict.
Account for context in system development	• **Context** embodies the circumstances, factors, patterns, conditions, or trends within which a system is embedded. It acts to constrain or enable the system. The inclusion and accounting for context in complex systems is critical to improve system performance. All complex systems are embedded in a unique context that enables/constrains a system. Removing the 'system' elements from the context to simplify for assessment purposes creates a false separation between the 'system' and its context, as they are integral to one another.

(continued)

Table 2 (continued)

Challenge	Challenge explanation for complex system governance development
Account for environment constraining or enabling the system	• **Environment:** The aggregate of all surroundings and conditions within which a system operates. It influences, and is influenced by, a system. The environment is a source of variability and constraint for a complex system. The accounting for the environment is critical to system development. The environment is the source of input (resources) for a system and also the place where the value of the system (products, services, information) is consumed.

the capabilities of practitioners to deal more effectively with complex systems and their problems. CSG field development should include the need for methods, tools, and techniques necessary to support applications. These artifacts of CSG must also appreciate that the application of them may be by practitioners not necessarily well steeped in the theoretically underpinnings of CSG. Thus, the development pathways may be adjusted to compensate for limited practitioner knowledge. However, what cannot be lost on CSG challenges for development is the need to develop practitioners, as well as their complex systems, to effectively engage methods, tools, and techniques from a 'systems worldview'. This represents a challenge to CSG and should also be a primary development concern.

- *Emphasis on sustainable field development*—development of the CSG field should focus on long term evolutionary development. This presents a difficult challenge, given the short term views that are limited in compatibility with the 'long view' required by CSG. However, the CSG field should not be subjected to a 'faddish' development, making claims and promises that are unrealistic and not likely to be achieved. Instead, care must be taken such that the field does not create expectations that are unrealistic for the current stage of development. Unrealistic expectations at best will cause disappointment amid initial fanfare. At worst, unrealistic expectations might do harm to the reputation of CSG as an approach to improve complex systems. The result of unnecessary pressures on CSG deployment will either result in the field being minimized at best or suffering an early demise at worst.
- *Maintenance of theoretical grounding for field sustainability*—there is a propensity for the 'quick hit', large value proposition for improving practices in operational settings. While this 'instant' gratification perspective is pervasive, care must be taken to make sure that the continuing development of the CSG field is not ignored. If CSG is to maintain coherence in continued development, it will be necessary to maintain the grounding of the field in a strong conceptual/theoretical base. For CSG this involves field evolution around systems theory, governance, and management cybernetics. In addition, there should be no hesitation to pursue further elaboration of the theoretical basis of CSG as a work in progress.

CSG development, as with any emerging field, will not be without challenges and issues. However, purposeful development of the field will certainly accelerate the path of development. While the CSG field is certainly not portrayed as a panacea that can cure all of the ills of modern complex systems or produce renaissance practitioners, it offers a different systems-based approach to improve complex systems. Additionally, CSG offers practitioners an additional set of capabilities to more effectively enhance practices related to complex systems.

Many of the greatest challenges facing the CSG field development have to do with the challenges to deploy CSG, CSG methods, and CSG tools. To succinctly articulate these deployment challenges, Table 3 provides the challenge area and explanation of the challenge.

There are significant challenges for deployment of CSG. However, there are deployment challenges for any systems-based methodology. The set of deployment challenges must be factored into efforts to utilize CSG in operational settings.

Vignette–This is hard stuff and a bit threatening

This example captures the difficulties in application of CSG. In this instance, an organization (system) was interested in exploring the possibilities that CSG might hold for improvement in their operations. Through an introduction, briefings on the essence of CSG, and application of several exploratory instruments, the exploration continued. The initial 'dive' into CSG suggested several areas in the system where pathologies (systemic deficiencies) existed, and the environment was demanding more than the system could accommodate. The interest started to quickly wane with the realization that there were no quick fixes. The exploration discovered that the sources of issues stemmed from the current design of the system and the overreliance on execution to compensate for a design that had deficiencies. Then the stark realizations set in that: (1) further understanding and development of actions to address the system design issues were nontrivial, (2) significant investment of their time/energy would be necessary to operate on the system instead of continue to 'band aid' the system to maintain viability, at whatever low levels initially sparked the interest in CSG, and (3) the option to 'do nothing' was much less threatening to the 'status quo' than starting to turn over rocks, underneath which might not be pleasant findings. In short, CSG presented an interesting diversion, but the continued and escalating search for deep system improvements were beyond the capacity of the system to engage.

4 Future Development Directions

Thus far, we have examined the current state and challenges for the CSG field. In this section, we examine specific developmental directions and potential to further develop CSG. There has been significant literature that has developed the foundations of CSG as an emerging field [52]. In this section, we examine four aspects for future development of CSG. First, the three interrelated developmental areas of science, engineering, and application are examined for CSG. These three areas are examined in relationship to their joint influence on practice. Second, seven interrelated developmental thrusts are suggested for CSG. Progression of the CSG field is dependent on the joint and balanced development across the thrust areas. Third, current challenges for accelerating the development of CSG are examined. These

Table 3 Challenges for deployment of CSG

Challenges	Explanation
Sufficient level of systems thinking to engage CSG	Engagement in CSG requires significant capacity for systems thinking. Absent this requisite capacity, it is unlikely that CSG will have the anticipate results. Instead, it is likely that CSG efforts will fall short of expectations. Systems thinking must be assessed, and if short of that necessary for engaging development activities, should have methods to increase systems thinking capacity integrated into the development application
Limited patience for the long view and immersive self-study	There is limited patience for seeing results occur over a long duration. This short-term fixation works against the deployment of CSG initiatives. CSG, by design, is focused on the long-term development of systems. Lacking patience for the long view of system development is detrimental to the prospects for CSG development. CSG requires that a system be studied by those with the responsibility/accountability for governance functions and communications channels. Lacking engagement for self-study casts doubt on effectively engaging CSG
Preference for tools and applications over deep systems development	Given a propensity for superficial thinking and action in response to system development, emphasis on tools and applications are preferred. Unfortunately, this preference is not well served by CSG. Tools and applications certainly have a place in CSG. However, the deeper levels of methodology (understanding what must be done) and the grounding in systems thinking (taking a holistic/systems theory viewpoint) requires going deeper than the superficial application of tools/techniques to holistically address truly complex system issues
Overcoming the 'in addition to' syndrome	CSG is not something that is done in addition to what is already being performed by practitioners in a complex system. If the system is viable (continuing to exist), then the CSG functions and communication channels are already being performed. They may not call the functions and communications channels by their CSG nomenclature, but they are being performed. Therefore, CSG is not something that is 'in addition to' what is being done. This places CSG in a privileged position of not being totally new and novel to what is already being performed. However, the difficulty of communicating this point is challenging

(continued)

Table 3 (continued)

Challenges	Explanation
Appreciation that systemic intervention by CSG has many failure modes	There is no guarantee that a CSG systemic intervention will be successful. There are too many variabilities in the deployment of CSG to arrogantly claim that it will be successful. On the contrary, tempering expectations is essential, since the precise results from a CSG systemic intervention cannot be know or predicted in advance. Instead, the results will emerge in unpredictable ways, irrespective of the noble intentions of the intervention
Perceived threat to the status quo	CSG ultimately shifts power to resolve uncertainty from individuals to the system of interest as a whole. Additionally, identification of 'deficiencies' in the design or execution of CSG functions can 'wrongly' be assumed to indicate a failure of complex system leadership. Unfortunately, the perceived threat to the 'status quo' system operation is likely to challenge the continuing and deepening exploration into systemic deficiencies

challenges must be met if CSG is to achieve the promising potential for impact of the field. Fourth, a set of guidance considerations for practitioners contemplating engagement of CSG is provided.

4.1 CSG Development Across Science, Engineering, and Application

CSG has not been disseminated or projected to the much wider community of practitioners across multiple sectors. CSG has the potential to significantly improve capabilities for practitioners (owners, operators, performers, designers) responsible for the design, execution and development of complex systems. We suggest that the utility of CSG proceeds along three interrelated streams of development, including *science, engineering, and application, all* targeted to improvement of practice. To look at these three aspects of the development of a field as independent and mutually exclusive of one another is false and somewhat naive. The CSG field faces a major challenge to pursue parallel integrated paths of development for the science, engineering, and application of CSG. The easy, and more traditional research approach is to separate the development of underlying science from corresponding engineering technologies and eventual applications. However, there is much to be gained by permitting the triad to constrain as well as enable one another for accelerated CSG field development. The research path that emerges through the integration of science, engineering, and application may be very different than if joint development had not been considered. It is certainly arguable that the CSG field currently pursues research that engages a close correlation between science, engineering, and application domains. There is much to gained by pursuit of CSG field development that explicitly couples science, engineering, and applications by design from an integrated systems perspective (Fig. 6).

For purposes of this discussion, we take *science* broadly as the search for knowledge to develop testable theory and laws related to a field. The tenets of good science include disciplined inquiry that can withstand the scrutiny of a particular field. The results of science must be theories and laws that can be tested to determine their continued power to provide confirmation or to be refuted. For CSG, this suggests that the discovery of new tenets of science supporting CSG may be found at the intersection of CSG's foundations in systems theory, management cybernetics, and system governance. In fact, systems theory is the doctrine that instantiates system science foundations. It would be easy to dismiss development of the science thrust for CSG as nonessential or a frivolous waste of scarce resources. However, engineering of technologies and their supporting applications, without grounding in the underlying science, misses an important stable base. While engineered technologies and applications can change rapidly, the underlying theoretical/scientific basis for a field provides long-term stability. The importance of this stable science-based foundation for the emerging CSG field cannot be overstated. This is particularly the case given

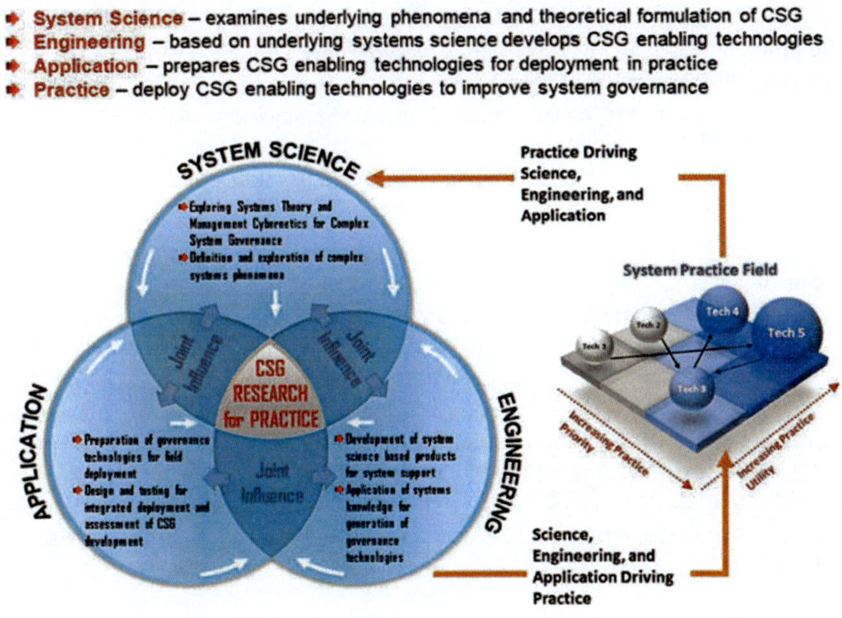

Fig. 6 CSG field development integrating science, engineering, and application for practice

the increasingly turbulent conditions faced by professionals and programs that seek to address complex systems and their problems across a multitude of sectors.

Engineering engages science to develop innovations that solve problems and increase the capabilities of practitioners to function more effectively. Thus, engineering becomes a bridge between science and application. This occurs through the development of science-based technologies for application support. Finally, applications involve putting science-based engineering technologies into action to support human purposes. Ultimately, the applications deployed by practitioners provide utility for science-based technologies. We believe that CSG research must be engaged and integrated across each of the three levels (science, engineering, applications) if it is to provide sustainable improvement of practices for the CSG field. Additionally, there must be eventual deployment in operational settings. The interrelated advancement across these three developmental thrusts for CSG will: (1) accelerate development of each of the other thrusts, (2) provide a grounding to better inform each of the thrust areas such that different development directions and insights might be possible, and (3) draw the worlds of science, engineering, application, and practice closer together to provide a more balanced development of the CSG field. Ultimately, the future development of CSG must rely on the *system-science based development of engineering technologies for application to improve practice.*

4.2 The Seven Developmental Thrusts for CSG

The future development of CSG must achieve balance across seven developmental thrusts. Following earlier work by Keating [62] related to field development (Fig. 7). These seven levels are interrelated and provide guidance to maintain a balance, ensuring a holistic treatment of the field.

The seven developmental thrusts, and associated questions that should provide a focus for CSG field research and development include:

- ***Philosophy***—research directed at developing a theoretically consistent articulation of the paradigm(s) for Complex System Governance. The emerging system of values and beliefs providing grounding for theoretical development is the primary contribution of this area.

  Questions for consideration: *(a) What are the epistemic foundations for CSG?, (b) What are the ontological predispositions for CSG?, (c) What are the existing and emerging paradigms that can serve to inform CSG?, and (d) How can philosophical disposition be identified, represented, and evolved for practitioners and entities engaging in CSG?, and (e) What are the implications for philosophy application concerning design, execution, and evolution of complex systems and CSG?*

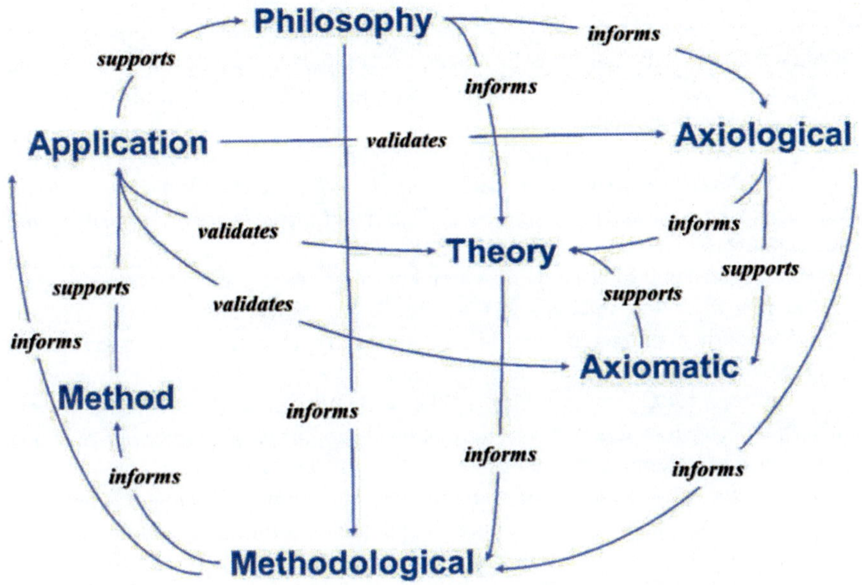

Fig. 7 Interrelated development thrusts for balanced development of the CSG field

- ***Theoretical***—research focused on explaining phenomena related to complex system governance and development of explanatory models and testable conceptual frameworks. The range of theoretical developments advances understanding of the field.

  Questions for consideration: *(a) What explanatory frameworks or models can be created to support CSG?, (b) What informing theoretical models are instructive for CSG?, (c) What are the phenomena in question with respect to CSG design, execution, and evolution?, and (d) How can prediction of CSG outcomes be supported and tested?*
- ***Axiological***—research that establishes the underlying value, value judgment frameworks, and belief propositions that are fundamental to understanding the variety of perspectives for Complex System Governance.

  Questions for consideration: *(a) What are the values informing different variants of CSG?, (b) How can axiological dispositions be measured and modeled for CSG and possibly changed?, and (c) What effect do values/value judgments have on design, execution, and evolution of CSG?*
- ***Methodological***—research undertaken to develop the theoretically informed frameworks that provide high level guidance for design, analysis, deployment, execution, and evolution of complex governance systems.

  Questions for Consideration: *(a) What frameworks can be constructed, or derived, to guide CSG design, analysis, diagnostics, and transformation?, (b) How can methodologies for CSG be tested and validated?, and (c) What technologies can be developed to support and execute CSG methodologies?*

- ***Axiomatic***—investigation into the emerging principles, concepts, and laws that define the field and constitute the "taken for granted" knowledge upon which the field rests. This also includes integration of knowledge from other informing and related fields/disciplines.

  Questions for consideration: *(a) What are the emerging areas where principles must be developed to support CSG? and (b) What existing principles can be incorporated or adapted to inform CSG, beyond those currently informing the CSG field?*
- ***Method***—research focused on development of the specific models, technologies, standards, processes, and tools for Complex System Governance. This is, in effect, the development of the toolsets and capabilities to enable practitioners to perform in the Complex System Governance domain.

  Questions for Consideration: *(a) What technologies, tools, processes, procedures, tools, or techniques can be developed to support performance of CSG? and (b) How can appropriate methods be selected for performing CSG?*
- ***Application***—advancement of the practice of Complex System Governance through the deployment of science based methodologies, technologies, and methods.

  Questions for Consideration: *(a) What standards/guidance can be developed to enhance the practice of CSG?, (b) What are best practices for CSG across different systems?, and (c) How can CSG be effectively deployed and measured in an operational setting?*

CSG is not a panacea for improving the prospects for more effective development of complex systems. However, CSG does offer a strong systems science grounded, engineering-focused, and application-oriented approach. This approach targets improving practices related to enhanced governance of complex systems in a more rigorous and purposeful manner. The development of CSG is dependent upon the degree to which there is a balance among and between the different field development levels suggested above.

4.3 Current Challenges and Research Directions to Advance the CSG Field

Based on the current state and trajectory of the CSG field, there are six challenges for research and development to accelerate advancement of the field. While these areas are not suggested as the 'complete' set of challenges and directions, they are representative of areas that need attention if CSG is to progress toward achievement of its full potential.

1. *Vigilant pursuit of practice improvement as the primary driver*—ultimately, all that is done in pursuit of the CSG field has the baseline purpose of improving

complex system performance. This must be achieved through advances in practice and enabling capabilities for practitioners. The farther away from this fundamental understanding CSG research and development gets, the less likely the CSG field will fulfill its potential.

2. *Emphasis on development of tools, methods, techniques, processes, and technologies to enable practice*—to improve practices related to CSG, practitioners must be armed with appropriate artifacts that will enable success in CSG applications. Absent these artifacts, it is unlikely that CSG will be capable of efficient/effective deployment in operational settings.
3. *Focus on making CSG approachable for application*—CSG is a difficult topic to grasp let alone master. In some sense it becomes unapproachable by practitioners who might benefit from CSG application but lack the time for protracted study of the subject. Thus, a major developmental thrust for the CSG field must revolve around making the field accessible to 'everyday' practitioners. If CSG remains unapproachable to all but a few, the objective of preparing practitioners to deploy CSG will be forfeit.
4. *CSG deployment must be capable of engaging a range of applications*—CSG must avoid being cast as a binary (all or nothing) application. Instead, CSG must be identified for application across a spectrum of potential deployment opportunities. In this sense, smaller scope applications can be included in the CSG application opportunities.
5. *Finding tenable balance in the Risk-Reward-Value tradeoff for CSG deployment*—There can be considerable perceived risk inherent in engagement of a CSG effort. Care must be taken to allay unsubstantiated fears of excessive risk from engaging CSG development. Also, the value accrued and reward for engaging CSG should be made explicit and palatable. Thus, the burden (perceived risk) of engagement in CSG can be lessened.
6. *Making CSG measurable such that improvement can be captured and monitored over time*—CSG can be resource intensive. The current state and transition possibilities enabled by CSG should be made explicit. Additionally, there should be a corresponding set of 'metrics' that can serve to measure and monitor progression in system performance being accrued from CSG development efforts.

Although this set is not presented as absolute or complete, it does suggest a necessary set of considerations related to how CSG can be better designed for engagement in complex systems.

4.4 Practitioner Guidance for Deployment of CSG

Application of CSG is a difficult endeavor to say the least. However, to gain a better grasp of preparing for successful engagement in CSG, there are several areas for consideration. These areas, although not a complete set, help prepare practitioners to embrace the prospects for successful CSG endeavors. This set of guidance for application of CSG includes:

1. *CSG development must involve the individuals who own the system, are accountable for system performance, and responsible to ensure that the system continues to develop such that viability is maintained.* CSG development pursuit without engagement of these individuals is unlikely to achieve anticipated results. There is no shortcut for system practitioners—CSG responsibility cannot be relegated or delegated.
2. *Individual capacity, organizational competence, and infrastructure compatibility to engage in systemic thinking/action will determine the degree to which system governance can enhance system performance.* Without a commensurate effort to understand the impacts, and necessity to include their development, these three areas (individual capacity, organizational competence, and infrastructure compatibility) can severely limit CSG developmental achievements.
3. *The focus on development for CSG efforts can include practitioner, support infrastructure, system, organization, or context.* It is shortsighted to only view CSG development as targeted to the system of interest. There are many potential benefactors and beneficiaries for a CSG effort. The more expansive and holistic the view of CSG development is the more likely deeper developmental impacts can be achieved.
4. *Feasible actions to improve the governance system are a function of the degree of engagement, resources, will, and the existing state of 'governance' for the system of interest.* Realization of 'full potential' for CSG development requires alignment of all of these elements. Outcome-expectation desires that are incongruent with investments of time, energy, commitment, and resources are likely to produce disappointing results.
5. *Greater understanding of a system of interest targeted for development accrues through the process of model construction.* Modeling efforts can provide insights into the structural relationships, context, and systemic deficiencies that exist for a system of interest. These insights can accrue regardless of whether or not specific actions to address issues are initiated. The models can be constructed without system modification and can range in degree of depth and sophistication. Therefore, alternative decisions, actions, and interpretations can be selectively engaged based on consideration of insights and understanding generated through system of interest modeling efforts.
6. *CSG application provides insights for alternative decisions.* CSG provides the 'big picture' view of the governance landscape. This includes identification of highest leverage strategic impact areas and their interrelationship to the larger CSG performance gaps. Thus, decisions for resource allocation can be better targeted. This allows steering away from activities that are simply 'intriguing' without demonstrating the highest substantial benefit to the larger

'systemic' governance concerns (e.g., pathologies). In light of CSG development priorities, low contribution efforts can be eliminated, or resources shifted appropriately.

7. *The design for comprehensive governance development is fallible and must be continually adjusted.* It is naïve to engage in CSG development assuming that precise outcomes can be known in advance. Instead, care must be taken to understand that the design for CSG development cannot be static. CSG development must adjust in response to changes in the system itself, the external environment, and the context within which CSG is embedded. The rate of change for CSG development design must minimally keep pace with the rate of change in the system, external environment, and context.
8. *The nature of CSG development is evolutionary rather than revolutionary.* Therefore, the implementation of CSG development requires 'the long view' and patience. CSG resists the 'quick fix' mentality prevalent in many development initiatives. Expectations for CSG development must be appreciative of the current state of governance effectiveness, which did not recently appear, but rather evolved over time. This CSG state will dictate what level of system improvement might be feasibly engaged over the near and long term.
9. *In essence, CSG development is a protracted 'self-study' of the system of interest, enacted through a new set of lenses, corresponding language, methods, and tools.* New thinking requires new language, which can produce alternative decision, action, and interpretation in route to pursuit of different outcomes (system performance levels). The willingness to engage in protracted self-study is essential for realization of the benefits of CSG development. There is no shortcut to the reflective self-study required to fully realize the potential of CSG to improve performance in complex systems and address their problems.
10. *Engaging governance development is not a trivial endeavor.* It is hard work, requiring significant investment of resources, patience to take the 'long view', and sacrifice of instant gratification for sustainable longer term performance improvement. Superficial CSG efforts are not likely to produce desirable or sustainable results, and in fact may make matters worse.

The essence of the emerging CSG field is focused on improving the ability of practitioners to more effectively deal with complex systems and their problems. CSG has the potential to significantly improve capabilities for practitioners (owners, operators, performers, designers) of complex systems. The guidance provided above offers a set of considerations for practitioners contemplating engagement of CSG.

5 Summary

This chapter has provided an overview of the genesis, current state, and future directions for the emerging CSG field. CSG was presented as a necessary evolution of the SoSE field. The CSG evolution was cast as a departure to complement traditional forms of SoSE, which were focused primarily on development of technology integration for large complex systems. The necessity to include the range of human, social, organizational, managerial, political, and policy aspects of complex systems is a major emphasis for CSG. In essence, this emphasis entails the inclusion of the 'soft' aspects of complex systems in addition to the 'hard' aspects. What was taken forward from the SoSE traditional formulation were the importance of technology in wider CSG applications and the rigorous formulation, albeit more holistic in orientation, driven from the engineering perspective.

The evolution of CSG emerged as the 'Design, execution, and evolution of the [nine] metasystem functions necessary to provide control, communication, coordination, and integration of a complex system.' [48, p. 228], which was a clear delineation of CSG from SoSE as well as other systems-based approaches for complex systems. The inclusion of *systems theory* (axioms and propositions that explain the behavior, structure, and performance of complex systems and which all complex systems are subject to), *management cybernetics* (communication and control as the science of effective structural organization), and *governance* (high-level steering of a system through direction, oversight, and accountability) were introduced as the supporting theoretical and conceptual foundations for CSG.

The essence of the CSG field stemming from the performance of nine essential governance functions and corresponding communication channels was explored. These functions and communication channels were presented as essential for a system to maintain viability, are present in any viable system, and are the source for aberrant behavior or performance (pathologies) in complex systems. The essence of CSG was established as, '*Subject to fundamental systems theory propositions, all systems perform essential governance functions. System performance is determined by effectiveness in achievement of governance functions consistent with system theory propositions. System performance can be enhanced through purposeful development of governance functions.*' This articulation capsules CSG. The CSG paradigm was introduced as the overarching depiction of CSG, linking the central aspects of CSG to production of system performance.

The current state of CSG was introduced. The CSG field was recognized, albeit emerging as a new and novel field, as having made significant and balanced progress since its formal inception in 2014. Advances across the conceptual/theoretical, methodology, methods, models, tools, and application areas were examined. While the state of CSG has been evolving rapidly, there is still much to be done as the maturation of the field continues. The advancement challenges for CSG were examined to focus complementary efforts to the contributions made thus far in development. Several development areas were suggested, including: (1) the need for holistic field development and continued emphasis on application and practices in operational settings, (2) the need to focus on the practice of CSG and emphasizing the need to enable CSG to be 'approachable' for practitioners and balancing the Risk-Reward-Value tradeoff, (3) continued emphasis on balanced and sustainable field development, (4) maintaining and evolving the theoretical grounding of CSG to ensure sustainability of the field, and (5) making CSG measurable to demonstrate system improvement. These challenges were extended by addressing the specific challenges related to deployment of CSG for operational settings. The particular 'resistance' areas for which CSG deployment must be evolved were also identified.

Future development directions for CSG were established. These development directions were captured as the balancing across science, engineering, and application to influence higher states of practice for CSG. Ultimately, the thrust of development for CSG is the improvement of practice for complex systems and enabling practitioners to more effectively engage complex systems and their problems. Seven development thrusts were suggested to mature the CSG field, including philosophy, axiological, theoretical, axiomatic, axiomatic, method, and application levels. The interrelated and balanced development in these seven areas were suggested. Development was also examined with the presentation of current challenges and research directions to advance the CSG field. The chapter concluded with a set of practitioner guidance to suggest the limitations and considerations that should be considered before engagement in CSG endeavors.

The emerging field of CSG is not presented as the 'magic elixir' or solution to all that is problematic in complex systems. CSG is not a trivial treatment that can be prescriptively applied to deficient complex systems and perform some miracles that heal the system deficiencies. Instead, CSG is an emerging field that offers a theoretically grounded, nontrivial application approach, action orientation, improvement focused, and holistic treatment for complex systems and their problems. Ultimately, CSG is about enhancing practice and enabling practitioners to engage complex systems and their problems more effectively.

Exercises

1. Discuss the evolution of CSG as a departure from SoSE formulations and why this departure was necessary.
2. Identify the three most significant contributions that CSG can make to improve complex system performance.
3. Identify the impediments to deployment of CSG in operational settings. Discuss what might be done to enhance the approachability of CSG for deployment to enhance practice.
4. Identify the three primary challenges for advancement of the CSG field and implications for research directions that should be engaged to address these challenges.

References

1. Akers W, Keating CB, Gheorghe A, Sousa-Poza A (2015) The nature and behaviour of complex system archetypes. Int J Syst Syst Eng 6(4):302–326
2. Alfaqiri A, Hossain NUI, Jaradat R, Abutabenjeh S, Keating CB, Khasawneh MT, Pinto CA (2019) A systemic approach for disruption risk assessment in oil and gas supply chains. Int J Crit Infrastruct 15(3):230–259

3. Baker W, Castelle K, Bradley J, Marino L (2020) Design for sustainment: governance engineering in major acquisition programs. In: Proceedings of the seventeenth annual acquisition research symposium. Monterey, California, pp 1-18
4. Baugh D (2015) Environmental scanning implications in the governance of complex systems. Int J Syst Syst Eng 6(1–2):127–143
5. Beer S (1979) The heart of enterprise. Wiley, New York
6. Beer S (1985) Diagnosing the system for organizations. Wiley, Hoboken, NJ
7. Beer S (1981) Brain of the firm: the managerial cybernetics of organization. Wiley, New York
8. Bradley JM, Joiner KF, Efatmaneshnik M, Keating CB (2017) Evaluating Australia's most complex system-of-systems, the future submarine: a case for using new complex systems governance. In: INCOSE international symposium, vol 27, no 1, pp 187–199
9. Bradley JM, Soule RT (2018) Information technology governance through the complex system governance lens. In: INCOSE international symposium, vol 28, no 1, pp 1237–1249
10. Bradley JM, Unal R, Pinto CA, Cavin ES (2015) Competencies for governance of complex systems of systems. Int J Syst Syst Eng 6(1–2):71–89
11. Bradley J, Katina P, Keating C (2016) Complex system governance for acquisition. Leading Change, LLC Augusta United States
12. Calida BY (2013) System governance analysis of complex systems. Doctoral Dissertation. Old Dominion University
13. Calida BY (2016) Complex system governance: moving diverse theory to practice. Int J Syst Syst Eng 7(1–3):22–42
14. Calida BY, Jaradat RM, Abutabenjeh S, Keating CB (2016) Governance in systems of systems: a systems-based model. Int J Syst Syst Eng 7(4):235–257
15. Calida BY, Keating CB (2014) System governance: emergence of practical perspectives across the disciplines. In: Gheorghe AV, Masera M, Katina PF (eds) Infranomics: sustainability, engineering design, and governance. Springer, New York, pp 269–296
16. Carter B (2015) A metasystem perspective and implications for governance. Int J Syst Syst Eng 6(1–2):90–100
17. Carter B, Moorthy S, Walters D (2016) Enterprise architecture view of complex system governance. Int J Syst Syst Eng 7(1–3):95–108
18. Chesterman CW Jr (2018) Contextual framework of communications functions supporting complex system governance. Old Dominion University
19. Crownover WM (2005) Complex system contextual framework (CSCF): a grounded-theory construction for the articulation of system context in addressing complex systems problems. Old Dominion University
20. Georgescu A, Gheorghe AV, Piso MI, Katina PF (2019) CSI—a complex system governance approach. In: Critical space infrastructures. Springer, Cham, pp 281–320
21. Georgescu A, Gheorghe AV, Piso MI, Katina PF (2019) CSG: towards CSI research. In: Critical space infrastructures. Springer, Cham, pp 321–343
22. Gheorghe AV, Vamanu DV, Katina PF, Pulfer R (2018) Governance vulnerability facets. In: Critical infrastructures, key resources, key assets. Springer, Cham, pp 39–79
23. Gheorghe AV, Vamanu DV, Katina PF, Pulfer R (2018) System of systems governance. In: Critical infrastructures, key resources, key assets. Springer, Cham, pp 93–130
24. Hester PT, Meyers TJ (2010) The role of measurement in complex system governance
25. Hodge RJ, Craig S, Bradley JM, Keating CB (2019) Systems engineering and complex systems governance–lessons for better integration. In: INCOSE international symposium, vol 29, no 1, pp 421–433
26. Ireland V, Keating C (2016) Introduction to special issue on complex system governance. Int J Syst Syst Eng 7(1/2), 3
27. Jackson MC (2019) Critical systems thinking and the management of complexity. Wiley
28. Jaradat RM (2015) Complex system governance requires systems thinking-how to find systems thinkers. Int J Syst Syst Eng 6(1–2):53–70
29. Jaradat RM, Keating CB (2016) Systems thinking capacity: implications and challenges for complex system governance development. Int J Syst Syst Eng 7(1–3):75–94

30. Jaradat RM (2017) Systemic intervention for complex system governance development. In: Proceedings of the international annual conference of the American society for engineering management. American Society for Engineering Management (ASEM), pp 1–10
31. Katina P, Keating C (2016) Metasystem pathologies: towards discovering of impediments to system performance. In: Proceedings of the 2016 industrial and systems engineering research conference, pp 1–6
32. Katina PF (2015) Systems theory-based construct for identifying metasystem pathologies for complex system governance. Old Dominion University
33. Katina PF (2015) Emerging systems theory–based pathologies for governance of complex systems. Int J Syst Syst Eng 6(1–2):144–159
34. Katina PF (2016) Systems theory as a foundation for discovery of pathologies for complex system problem formulation. In: Applications of systems thinking and soft operations research in managing complexity. Springer, Cham, pp 227–267
35. Katina PF, Keating CB, Sisti JA, Gheorghe AV (2019) Blockchain governance. Int J Crit Infrastruct 15(2):121–135
36. Katina PF (2016) Metasystem pathologies (M-Path) method: phases and procedures. J Manag Dev
37. Katina PF, Keating CB (2014) Metasystem pathologies: towards a systems-based construct for complex system deficiencies. In: Proceedings of the international annual conference of the American society for engineering management. American Society for Engineering Management (ASEM), p 1
38. Katina PF, Keating CB (2018) Cyber-physical systems governance: a framework for (meta) cybersecurity design. In: Security by design. Springer, Cham, pp 137–169
39. Katina PF, Keskin OF (2021) Complex system governance as a foundation for enhancing the cybersecurity of cyber-physical systems. Int J Cyber Warfare Terror (IJCWT) 11(3):1–14
40. Katina PF, Calida BY, Bobo JA (2020) Complex system governance: advancing prospects for critical space infrastructure applications. Space Infrastruct Risk Resil Gov 57:29
41. Katina PF, Keating CB, Bradley JM (2016) The role of 'Metasystem' in engineering a system of systems. In: Proceedings of the 2016 industrial and systems engineering research conference, pp 1–6
42. Katina PF, Keating CB, Gheorghe AV (2016) Cyber-physical systems: complex system governance as an integrating construct. In: Proceedings of the 2016 industrial and systems engineering research conference. Anaheim, CA: IISE
43. Katina PF, Keating CB, Bobo JA, Toland TS (2019) A governance perspective for system-of-systems. Systems 7(4), 54
44. Katina P, Bradley JM (2016) Towards a systems theory-based curriculum for complex systems governance. In: 2016 ASEE annual conference and exposition
45. Katina PF (2015) Systems theory-based construct for identifying metasystem pathologies for complex system governance. Doctoral dissertation, Doctoral Thesis, Old Dominion University, Norfolk, VA, USA
46. Keating CB, Katina PF (2011) System of systems engineering: Prospects and challenges for the emerging field. Int J Syst Syst Eng 2(2/3):234–256
47. Keating CB (2014) Governance implications for meeting challenges in the system of systems engineering field. Paper presented at the IEEE system of systems engineering conference, Adelaide, SA
48. Keating CB (2015) Complex system governance: theory to practice challenges for system of systems engineering. In: 2015 10th system of systems engineering conference (SoSE). IEEE, pp 226–231
49. Keating CB, Bradley JM (2015) Complex system governance reference model. Int J Syst Syst Eng 6(1):33–52
50. Keating CB, Katina PF (2015) Foundational perspectives for the emerging complex system governance field. Int J Syst Syst Eng 6(1/2):1
51. Keating CB, Katina PF (2016) Complex system governance development: a first generation methodology. Int J Syst Syst Eng 7(1–3):43–74

52. Keating CB, Katina PF (2019) Complex system governance: concept, utility, and challenges. Syst Res Behav Sci 36(5):687–705
53. Keating CB, Katina PF (2019) Enterprise governance toolset. In: Evolving toolbox for complex project management. Auerbach Publications, pp 153–181
54. Keating CB, Bradley JM, Katina PF, Jaradat REM (2017) A systems theoretic-based framework to discover pathologies in acquisition system governance
55. Keating CB, Katina PF, Bradley JM (2014) Complex system governance: concept, challenges, and emerging research. Int J Syst Syst Eng 5(3):263–288
56. Keating CB, Katina PF, Bradley JM (2015) Challenges for developing complex system governance. In: Proceedings of the IIE annual conference. Institute of Industrial and Systems Engineers (IISE), p 2943
57. Keating CB, Katina PF, Bradley JM (2016) Complex system governance: failure mode effects and criticality analysis application. In: Presented at the proceedings of the 2016 industrial and systems engineering research conference, ISERC, Anaheim, CA
58. Keating CB, Katina PF, Bradley JM, Jaradat RE, Hodge R (2018) Systemic intervention for complex system governance. In: INCOSE international symposium, vol 28, no 1, pp 1534–1548
59. Keating CB, Katina PF, Gheorghe AV, Jaradat R (2017) Complex system governance: advancing prospects for system of systems engineering applications
60. Keating CB, Katina PF, Jaradat RE, Bradley JM, Gheorghe AV (2017) Acquisition system development: a complex system governance perspective. In: INCOSE international symposium, vol 27, no 1, pp 811–825
61. Keating CB, Katina PF, Joiner KF, Bradley JM, Jaradat RM (2018) A method for identification, representation, and assessment of complex system pathologies in acquisition programs
62. Keating CB (2005) Research foundations for system of systems engineering. In: 2005 IEEE international conference on systems, man and cybernetics, vol 3, pp 2720–2725. https://doi.org/10.1109/ICSMC.2005.1571561
63. Keating CB, Bradley JM (2015) Complex system governance reference model. Int J Syst Syst Eng 6(1/2):33–52
64. Mitroff I (1997) Smart thinking for crazy times: the art of solving the right problems. Berrett-Koehler Publishers, Inc., 450 Sansome Street, Suite 1200, San Francisco, CA 94111-3320
65. Keating C, Rogers R, Unal R, Dryer D, Sousa-Poza A, Safford R, Peterson W, Rabadi G (2003) System of systems engineering. Eng Manag J 15(3):36–45
66. Marek C, Bradley J, McGowan J, Cui Q (2021) A governance model and safety management system framework for industrial fire safety during naval ship maintenance availabilities. In: Proceedings of the eighteenth annual acquisition research symposium. Monterey, California, pp 6–25
67. Pyne JC, Keating CB, Katina PF (2016) Enhancing utility manager's capability for dealing with complex issues. Proc Water Environ Fed 2016(8):4207–4232
68. Pyne JC, Keating CB, Katina PF, Bradley JM (2018) Systemic intervention methods supporting complex system governance initiatives. Int J Syst Syst Eng 8(3):285–309
69. Shauger JJ (2016) Introduction to service system governance. Int J Syst Syst Eng 7(1–3):189–206
70. Tatar U (2020) Critical space infrastructure: contributions of complex system governance. Space Infrastruct Risk Resil Gov 57:60
71. Walters D (2016) Leadership issues in governance of complex systems. Int J Syst Syst Eng 7(1–3):130–142
72. Walters D, Moorthy S, Carter B (2014) System of systems engineering and enterprise architecture: implications for governance of complex systems. Int J Syst Syst Eng 5(3):248–262
73. Walters D (2018) Leadership and foresight in complex system governance. Int J Syst Syst Eng 8(3):249–267

74. Yesudas R, Efatmaneshnik M, Joiner K (2019) Identifying focal points in IT project governance using a synthetic and systems thinking approach. In: International conference on complex systems design and management. Springer, Cham, pp 79–92
75. Yuchnovicz DE (2016) Understanding system structural tensions to support complex system governance development methodology. Int J Syst Syst Eng 7(1–3):109–129

Index

C. B. Keating et al. (eds.), *Complex System Governance*, Topics in Safety, Risk, Reliability and Quality 40, https://doi.org/10.1007/978-3-030-93852-9

T

U

V